DISCRETE MATHEMATICS AND STRUCTURES

DISCRETE MATHEMATICS AND STRUCTURES

SATINDER BAL GUPTA

UNIVERSITY SCIENCE PRESS
(An Imprint of Laxmi Publications Pvt. Ltd.)
An ISO 9001:2015 Company
BENGALURU • CHENNAI • GUWAHATI • HYDERABAD
KOCHI • KOLKATA • LUCKNOW • RANCHI
NEW DELHI

Printed and bound in India
First 2002; *Second* 2004; *Third* 2006; *Fourth* 2008, *Edition* : 2025
ISBN : 978-81-31803-52-3

Branches

©	Bengaluru	080-26 75 69 30
©	Chennai	044-24 34 47 26
©	Guwahati	0361-254 36 69
©	Hyderabad	040-27 55 53 83
©	Kochi	0484-405 13 03
©	Kolkata	033-40 04 77 79
©	Lucknow	0522-430 36 13
©	Ranchi	0651-224 24 64

PUBLISHED IN INDIA BY

UNIVERSITY SCIENCE PRESS
(An Imprint of Laxmi Publications Pvt. Ltd.)

An ISO 9001:2015 Company
113, GOLDEN HOUSE, GURUDWARA ROAD, DARYAGANJ,
NEW DELHI - 110002, INDIA

Corporate Office : B-33, Sector-5, Noida,
Gautambuddha Nagar (U.P.) - 201 301, INDIA
Telephone : +91-120-44 46 900
www.laxmipublications.com
info@laxmipublications.com

C—R/025/10
Printed at : Ajit Printing Press, Delhi

PREFACE

It gives me a great pleasure in bringing out the Fourth Edition of the book "Discrete Mathematics and Structures". The necessity for Discrete Structure in Computer Science arises due to selection of certain applications from various areas of the field. As the subject 'Discrete Mathematics or Discrete Structures' is taught in most Engineering Institutions, the students face a lot of problems in this subject as no book covers the whole syllabi and also due to lack of solved problems in the various books. This book is based on the experience gained in teaching a course on the subject. My intention is to cover all topics in the book with a number of solved problems and solutions of the questions from the previous question papers of different universities and other competitive examinations. I am sure that this book will help a student to grip each topic without any difficulty.

What is New in the Fourth Edition

I have been very pleased by the warm reception given to the first edition of the book. Many changes have been made in this edition and my goal continues to be that of maximizing the clarity of presentation.

* A new chapter has been added on Posets, Lattices and Boolean Algebra.
* A new section on Pigeonhole principle in Counting Techniques.
* Some new sections on Functions.
* Some new sections on Relations.
* Some new sections on Graphs and Groups.
* Additional material on Logic and Trees.
* Additional examples from latest question papers of various universities.

I would seek the indulence of my readers and intimating me about may errors and shortcomings noticed by then in the book.

—AUTHOR

ACKNOWLEDGMENTS

It is difficult to acknowledge adequately the help & encouragement, I have received in completion and final publication of this work which has been exercising my mind since long. First, I must express my gratitude to the *Professors & Lecturers* of the *Sant Longowal Institute of Engg. & Technology, Longowal (Punjab)* where I learnt the elements of Discrete Mathematics and Structures during the course of my various assignments on the subject.

I also owe a debt of gratitude to the Esteemed *Management of Vaish College of Engg., Rohtak* who has been a source of inspiration to me in attempting this work. I should also express my grateful thanks to my associates with whom I had occasions to discuss the work of may book at length which are in large numbers. I must express my indebtness to the students who have helped me in fixing the pattern of emphasis & in correction of proofs, to mention the least are *Manish Goel, Ashish Goel & Manish Kansal.*

Last, but not the least, I express my appreciation to my wife, *Moniks Gupta* for her assistance and constant encouragement without which the venture would have failed.

—AUTHOR

CONTENTS

To My
Little Daughter
Arshita

1

Sets

DEFINITION

A set is defined as a collection of distinct objects of same type or class of objects.

The objects of a set are called elements or members of the set. Objects can be numbers, alphabets, names etc.

e.g., $A = \{1, 2, 3, 4, 5\}$

A is a set of numbers containing elements 1, 2, 3, 4 and 5.

* A set is usually denoted by capital letters A, B, C, D, P, Q, R, S, T etc.

* Elements of the set are defined by p, q, r, t or p_1, q_1, r_1, t_1 etc.

SET FORMATION

The set can be formed by two ways :

(*i*) Tabular form of a set

(*ii*) Builder form of a set.

(*i*) **Tabular Form of a Set.** If a set is defined by actually listing its members, *e.g.*, if set P contains elements a, b, c, d then it is expressed as $P = \{a, b, c, d\}$.

This is called tabular form of a set.

(*ii*) **Builder Form of a Set.** If a set is defined by the properties which its elements must satisfy *e.g.*,

$P = \{x : x \in N, x \text{ is a multiple of } 3\}$.

$R = \{x : x > 1 \text{ and } x < 10 \text{ and } x \text{ is an odd integer}\}$.

$T = \{x : x \text{ is even less than } 9\}$.

STANDARD NOTATIONS

$x \in A$	x belongs to A or x is an element of set A.
$x \notin A$	x does not belong to set A.
ϕ	Empty set.
U	Universal set.
N	The set of all natural numbers.
I	The set of all integers.
I_0	The set of all non-zero integers.

I_+	The set of all + ve integers.
C, C_0	The sets of all complex, non-zero complex numbers respectively.
Q, Q_0, Q_+	The sets of rational, non-zero rational, + ve rational numbers respectively.
R, R_0, R_+	The sets of real, non-zero real, +ve real number respectively.

FINITE SET

If a set consists of specific number of different elements then that set is called finite set. *e.g.*,

$$P = \{x : x \in N, 3 < x < 9\}.$$
$$Q = \{2, 4, 6, 8\}.$$
$$R = \{\text{months of year}\}.$$

INFINITE SET

If a set consists of infinite number of different elements or if the counting of different elements of the set does not come to an end, the set is called infinite set. *e.g.*,

$$I = \{\text{The set of all integers}\}$$
$$E = \{x : x \in N, x \text{ is a multiple of } 2\}.$$

EQUALITY OF SETS

Two sets A and B are said to be equal and written as A = B if both have the same elements. Therefore, every element which belongs to A is also an element of the set B and every element which belongs to the set B is also an element of the set A.

$$A = B \quad \Leftrightarrow \quad (x \in A \quad \Leftrightarrow \quad x \in B).$$

If there is some element in a set A that does not belong to set B or vice-versa then $A \neq B$ *i.e.*, A is not equal to B.

* A set does not change if one or more of its elements are repeated.

* A set does not change if we change the order in which its elements are tabulated. *e.g.*,

(*a*) $A = \{x : x < 10 \text{ and } x \text{ is even}\}$

$B = \{2, 4, 6, 8\}$

$C = \{x : x > 1 \text{ and } x < 10 \text{ and } x \text{ is even}\}$

All the three sets are equal.

(*b*) $P = \{r, s, t\}$

$Q = \{r, r, s, t\}$

$R = \{s, r, t\}$

$S = \{r, s, t, t\}$

All the four sets P, Q, R, S are equal.

(*c*) $A = \{a, b, c\}$

$B = \{b, a, c\}$

$C = \{b, a\}$

The sets A and B are equal but C is not equal to either A or B.

DISJOINT SETS

Two sets A and B are said to be disjoint if no element of A is in B and no element of B is in A. *e.g.*,

$$R = \{a, b, c\}, S = \{k, p, m\}$$

R and S are disjoint sets.

FAMILY OF SETS

If a set A contains elements which are itself sets then it is called family of sets or a set of sets. *e.g.*,

$$A = \{\{1, 2\}, \{3, 4\}, \phi\}$$

A is set of sets.

SUBSET OF A SET

If every element of a set A is also an element of a set B then A is called subset of B and is written as $A \subseteq B$. B is called superset of A.

$$A \subseteq B = \{x : x \in A \Rightarrow x \in B\}$$

(*a*) **Proper Subset.** If A is subset of B and $A \neq B$ then A is said to be proper subset of B. If A is a proper subset of B then B is not subset of A *i.e.*, there is atleast one element in B which is not in A. *e.g.*,

(*i*) Let $A = \{2, 3, 4\}$

$B = \{2, 3, 4, 5\}$

A is a proper subset of B.

(*ii*) The null set ϕ is a proper subset of every set.

(*b*) **Improper Subset.** If A is subset of B and A = B, then A is said to be an improper subset of B. *e.g.*,

(*i*) $A = \{2, 3, 4\}$, $B = \{2, 3, 4\}$

A is improper subset of B.

(*ii*) Every set is improper subset of itself.

NULL SET OR EMPTY SET

The set that contains no element is called the null set or the empty set and is denoted by ϕ. *e.g.*,

(*a*) $P = \{x : x^2 = 4, x \text{ is odd}\}$.

(*b*) $Q = \{x : x^2 = 9, x \text{ is even}\}$.

(*c*) $R = \{x : x^2 = 9, 2x = 4\}$.

The set $\phi = \{0\}$ is not a null set because 0 is the element of the set.

The set $A = \{\phi\}$ is not a null set because set ϕ is the element of the set.

POWER SET

The power set of any given set A is the set of all subsets of A and is denoted by P(A). If A has n elements then P(A) has 2^n elements.

If $A = \{r, s\}$ then its subsets are ϕ, $\{r\}$, $\{s\}$, $\{r, s\}$

$P(A) = \{\phi, \{r\}, \{s\}, \{r, s\}\}$ which has $2^2 = 4$ elements.

UNIVERSAL SET

If all the sets under investigation are subsets of a fixed set U, then the set U is called universal set. *e.g.*, in plane geometry, the universal set consists of all the points in the plane.

COMPARABILITY

Two sets A and B are comparable if one of them is subset of other, *i.e.*, $A \subseteq B$ or $B \subseteq A$. If $A \subseteq B$ and $B \subseteq A$, then $A = B$. Set ϕ is comparable to every set. Every set is comparable to the universal set U.

Two sets A and B are said to be incomparable if $A \not\subseteq B$ and also $B \not\subseteq A$ *i.e.*, there is atleast one element in A not in B and vice-versa.

Theorem I. *Prove that, the null set ϕ is a subset of every set.*

Proof. Suppose that ϕ is not a subset of A *i.e.*, $\phi \not\subset A$.

Then there exists an element $x \in \phi$ such that $x \notin A$.

But ϕ is the null set and hence for every x, $x \notin \phi$ because null set does not contain any element.

From above, $x \in \phi$ and $x \notin \phi$ which is contradiction. Hence, $\phi \not\subset A$ is wrong supposition.

Therefore, ϕ is a subset of A. Hence proved.

Theorem II. *Prove that, every set is a subset of itself i.e., $A \subseteq A$.*

Proof. Let $x \in A \Rightarrow x \in A \quad \therefore \quad A \subseteq A$

As every element belonging to set A is also an element of set A. Therefore, A is subset of A or itself. Hence proved.

Theorem III. *Prove that, if $A \subseteq B$ and $B \subseteq C$, then $A \subseteq C$.*

Proof. Let x be any element of the set A.

Since, $A \subseteq B \quad \therefore \quad x \in A \Rightarrow x \in B$

$B \subseteq C \quad \therefore \quad x \in B \Rightarrow x \in C \qquad \therefore \quad x \in A \Rightarrow x \in C$

$\therefore$ $A \subseteq C$. Hence proved.

Theorem IV. *Prove that, if $A \subseteq B$, $B \subseteq C$, $C \subseteq A$, then $A = C$.*

Proof. If $A \subseteq B$ and $B \subseteq C$, then $A \subseteq C$

$C \subseteq A$ (Given)

$\therefore$ $A \subseteq C$ and $C \subseteq A$

$\therefore$ $A = C$. Hence proved.

Theorem V. *Prove that, if $A \subset \phi$, then $A = \phi$.*

Proof. $\phi \subset A$ (As ϕ is subset of every set).

But $A \subset \phi$ (Given)

$\therefore$ $A = \phi$. Hence proved.

Example 1. *How many subsets can be formed from a set of n elements ? How many of these will be proper and how many improper ?*

Sol. There are ${}^nC_1$ subsets each consisting of one of the elements of the given set.

There will be ${}^nC_2$ subsets each consisting of any two of the n elements of the given set.

Also, there will be ${}^nC_3$ subsets each consisting of any three of the n elements of given set.

At last, there will be ${}^nC_n$ subset consisting of all the n elements of the given set.

Also, there will be one set ϕ.

Hence, the total number of subsets will be

$$({}^nC_1 + {}^nC_2 + {}^nC_3 + \ldots\ldots + {}^nC_n) + 1 \qquad \text{(For the null set)}$$

or $${}^nC_0 + {}^nC_1 + {}^nC_2 + \ldots\ldots + {}^nC_n = 2^n \qquad \because \; {}^nC_0 = 1.$$

Therefore, 2^n subsets formed from n elements of a set. Out of 2^n subsets $2^n - 1$ subsets will be proper and 1 (one) subset improper *i.e.*, the set itself.

OPERATIONS ON SETS

The basic set operations are :

1. Union of Sets. Union of the sets A and B is defined to be the set of all those elements which belong to A or B or both and is denoted by $A \cup B$.

$$A \cup B = \{x : x \in A \quad \text{or} \quad x \in B\}$$

e.g., Let $$A = \{1, 2, 3\},\ B = \{3, 4, 5, 6\}$$

$$A \cup B = \{1, 2, 3, 4, 5, 6\}.$$

2. Intersection of Sets. Intersection of two sets A and B is the set of all those elements which belong to both A and B and is denoted by $A \cap B$.

$$A \cap B = \{x : x \in A \text{ and } x \in B\}$$

e.g., Let $$A = \{a, b, c, d\},\ B = \{a, b, l, m\}$$

$$A \cap B = \{a, b\}.$$

3. Difference of Sets. The difference of two sets A and B is a set of all those elements which belong to A but do not belong to B and is denoted by $A - B$.

$$A - B = \{x : x \in A \text{ and } x \notin B\}$$

e.g., Let $$A = \{a, b, c, d\},\ B = \{d, l, m, n\}$$

$$A - B = \{a, b, c\}.$$

4. Complement of a Set w.r.t. a Universal Set. The complement of a set A is a set of all those elements of the universal set which do not belong to A and is denoted by A^c.

$$A^c = U - A = \{x : x \in U \text{ and } x \notin A\} = \{x : x \notin A\}$$

e.g., Let U is the set of all natural numbers.

$$A = \{1, 2, 3\}$$

$$A^c = \{\text{all natural numbers except } 1, 2, 3\}.$$

5. Symmetric Difference of Sets. The symmetric difference of two sets A and B is the set containing all the elements that are in A or in B but not in both and is denoted by $A \oplus B$ *i.e.*,

$$A \oplus B = (A \cup B) - (A \cap B)$$

e.g., (*i*) Let $A = \{a, b, c, d\}$

$B = \{a, b, l, m\}$

$A \oplus B = \{c, d, l, m\}$

(*ii*) Let $A = \{1, 2, 3, 4\}$

$A \oplus \phi = \{1, 2, 3, 4\}$

(*iii*) Let $A = \{l, m, n, m\}$

$A \oplus A = \phi.$

ALGEBRA OF SETS

A set is obtained from a set formula by replacing the variables by definite sets. When the set variables appearing in two set formulas are replaced by any sets and both the set formulas are equal as sets, then we call that both the set formula are equal. The equality of set formulas are called set identities. Some of the identities describe certain properties of the operations involved. These properties describe an algebra called algebra of sets.

Table I

(*i*) **Idempotent Laws**	(*vi*) **Identity Laws**
(*a*) $A \cup A = A$	(*a*) $A \cup \phi = A$
(*b*) $A \cap A = A$	(*b*) $A \cap U = A$
(*ii*) **Associative Laws**	(*c*) $A \cup U = U$
(*a*) $(A \cup B) \cup C = A \cup (B \cup C)$	(*d*) $A \cap \phi = \phi$
(*b*) $(A \cap B) \cap C = A \cap (B \cap C)$	(*vii*) **Complement Laws**
(*iii*) **Commutative Laws**	(*a*) $A \cup A^c = U$
(*a*) $A \cup B = B \cup A$	(*b*) $A \cap A^c = \phi$
(*b*) $A \cap B = B \cap A$	(*c*) $U^c = \phi$
(*iv*) **Distributive Laws**	(*d*) $\phi^c = U$
(*a*) $A \cup (B \cap C) = (A \cup B) \cap (A \cup C)$	(*viii*) **Involution Law**
(*b*) $A \cap (B \cup C) = (A \cap B) \cup (A \cap C)$	(*a*) $(A^c)^c = A$
(*v*) **De Morgan's Laws**	
(*a*) $(A \cup B)^c = A^c \cap B^c$	
(*b*) $(A \cap B)^c = A^c \cup B^c$	

The Table I shows the laws of algebra of sets.

Example 2. *Prove (a) $A \cup A = A$ (b) $A \cap A = A$.*

Sol. (*a*) To prove $A \cup A = A$

Since, $B \subset A \cup B$, therefore $A \subset A \cup A$

Let $x \in A \cup A \Rightarrow x \in A$ or $x \in A \Rightarrow x \in A$

$\therefore \quad A \cup A \subset A$

As $A \cup A \subset A$ and $A \subset A \cup A \Rightarrow A = A \cup A$. Hence proved.

(*b*) To prove $A \cap A = A$

Since, $A \cap B \subset B$, therefore $A \cap A \subset A$

Let $x \in A \Rightarrow x \in A$ and $x \in A$

$\Rightarrow \quad x \in A \cap A \qquad \therefore \quad A \subset A \cap A$

As $A \cap A \subset A$ and $A \subset A \cap A \Rightarrow A = A \cap A$. Hence proved.

Example 3. *Prove (a)* $(A \cup B) \cup C = A \cup (B \cup C)$

(b) $(A \cap B) \cap C = A \cap (B \cap C)$.

Sol. (*a*) To prove $(A \cup B) \cup C = A \cup (B \cup C)$

Let some $x \in (A \cup B) \cup C$

$\Rightarrow \quad (x \in A$ or $x \in B)$ or $x \in C$

$\Rightarrow \quad x \in A$ or $x \in B$ or $x \in C$

$\Rightarrow \quad x \in A$ or $(x \in B$ or $x \in C)$

$\Rightarrow \quad x \in A$ or $x \in B \cup C$

$\Rightarrow \quad x \in A \cup (B \cup C)$.

Similarly, if some $x \in A \cup (B \cup C)$, then $x \in (A \cup B) \cup C$.

Thus, any $x \in A \cup (B \cup C) \Leftrightarrow x \in (A \cup B) \cup C$. Hence proved.

(*b*) To prove $(A \cap B) \cap C = A \cap (B \cap C)$

Let some $x \in A \cap (B \cap C) \Rightarrow x \in A$ and $x \in B \cap C$

$\Rightarrow \quad x \in A$ and $(x \in B$ and $x \in C) \Rightarrow x \in A$ and $x \in B$ and $x \in C$

$\Rightarrow \quad (x \in A$ and $x \in B)$ and $x \in C \Rightarrow x \in A \cap B$ and $x \in C$

$\Rightarrow \quad x \in (A \cap B) \cap C$.

Similarly, if some $x \in A \cap (B \cap C)$, then $x \in (A \cap B) \cap C$

Thus, any $x \in (A \cap B) \cap C \Leftrightarrow x \in A \cap (B \cap C)$. Hence proved.

Example 4. *Prove (a)* $A \cup B = B \cup A$ *(b)* $A \cap B = B \cap A$.

Sol. (*a*) To prove

$$A \cup B = B \cup A$$

$$A \cup B = \{x : x \in A \text{ or } x \in B\}$$

$$= \{x : x \in B \text{ or } x \in A\} \quad (\because \text{Order is not preserved in case of sets})$$

$$= B \cup A. \text{ Hence proved.}$$

(*b*) To prove $\quad A \cap B = B \cap A$

$$A \cap B = \{x : x \in A \text{ and } x \in B\}$$

$$= \{x : x \in B \text{ and } x \in A\}$$

($\because$ Order is not preserved in case of sets)

$$= B \cap A. \text{ Hence proved.}$$

Example 5. *Prove (a) Intersection of sets is distributive w.r.t. union of sets i.e.,*

$$A \cap (B \cup C) = (A \cap B) \cup (A \cap C)$$

(b) Union of sets is distributive w.r.t. intersection of sets i.e.,

$$A \cup (B \cap C) = (A \cup B) \cap (A \cup C).$$

Sol. (*a*) To prove $A \cap (B \cup C) = (A \cap B) \cup (A \cap C)$

Let $\quad x \in A \cap (B \cup C) \Rightarrow x \in A$ and $x \in B \cup C$

$\Rightarrow \quad (x \in A$ and $x \in A)$ and $(x \in B$ or $x \in C)$

$\Rightarrow \quad (x \in A$ and $x \in B)$ or $(x \in A$ and $x \in C)$

$\Rightarrow \quad x \in A \cap B$ or $x \in A \cap C$

$\Rightarrow \quad x \in (A \cap B) \cup (A \cap C)$

Therefore, $A \cap (B \cup C) \subset (A \cap B) \cup (A \cap C)$ ...(*i*)

Again, let $y \in (A \cap B) \cup (A \cap C) \quad \Rightarrow \quad y \in A \cap B$ or $y \in A \cap C$

$\Rightarrow \quad (y \in A \text{ and } y \in B) \text{ or } (y \in A \text{ and } y \in C)$

$\Rightarrow \quad (y \in A \text{ or } y \in A) \text{ and } (y \in B \text{ or } y \in C)$

$\Rightarrow \quad y \in A \text{ and } y \in B \cup C$

$\Rightarrow \quad y \in A \cap (B \cup C)$

Therefore, $(A \cap B) \cup (A \cap C) \subset A \cap (B \cup C)$ ...(*ii*)

Combining (*i*) and (*ii*), we get $A \cap (B \cup C) = (A \cap B) \cup (A \cap C)$. Hence proved.

(*b*) To prove $A \cup (B \cap C) = (A \cup B) \cap (A \cup C)$

Let $x \in A \cup (B \cap C) \quad \Rightarrow \quad x \in A \text{ or } x \in B \cap C$

$\Rightarrow \quad (x \in A \text{ or } x \in A) \text{ or } (x \in B \text{ and } x \in C)$

$\Rightarrow \quad (x \in A \text{ or } x \in B) \text{ and } (x \in A \text{ or } x \in C)$

$\Rightarrow \quad x \in A \cup B \text{ and } x \in A \cup C$

$\Rightarrow \quad x \in (A \cup B) \cap (A \cup C)$

Therefore, $A \cup (B \cap C) \subset (A \cup B) \cap (A \cup C)$ ...(*i*)

Again, let $y \in (A \cup B) \cap (A \cup C) \quad \Rightarrow \quad y \in A \cup B$ and $y \in A \cup C$

$\Rightarrow \quad (y \in A \text{ or } y \in B) \text{ and } (y \in A \text{ or } y \in C)$

$\Rightarrow \quad (y \in A \text{ and } y \in A) \text{ or } (y \in B \text{ and } y \in C)$

$\Rightarrow \quad y \in A \text{ or } y \in B \cap C$

$\Rightarrow \quad y \in A \cup (B \cap C)$

Therefore, $(A \cup B) \cap (A \cup C) \subset A \cup (B \cap C)$ ...(*ii*)

Combining (*i*) and (*ii*), we get $A \cup (B \cap C) = (A \cup B) \cap (A \cup C)$. Hence proved.

Example 6. *Prove De Morgan's Laws*

(*a*) $(A \cup B)^c = A^c \cap B^c$ (*b*) $(A \cap B)^c = A^c \cup B^c$.

Sol. (*a*) To prove $(A \cup B)^c = A^c \cap B^c$

Let $x \in (A \cup B)^c \quad \Rightarrow \quad x \notin A \cup B$ $(\because a \in A \Leftrightarrow a \notin A^c)$

$\Rightarrow \quad x \notin A \text{ and } x \notin B$

$\Rightarrow \quad x \in A^c \text{ and } x \in B^c$

$\Rightarrow \quad x \in A^c \cap B^c$

Therefore, $(A \cup B)^c \subset A^c \cap B^c$ ...(*i*)

Again, let $x \in A^c \cap B^c \quad \Rightarrow \quad x \in A^c \text{ and } x \in B^c$

$\Rightarrow \quad x \notin A \text{ and } x \notin B$

$\Rightarrow \quad x \notin A \cup B$

$\Rightarrow \quad x \in (A \cup B)^c$

Therefore, $A^c \cap B^c \subset (A \cup B)^c$ ...(*ii*)

Combining (*i*) and (*ii*), we get $A^c \cap B^c = (A \cup B)^c$. Hence proved.

(*b*) To prove $(A \cap B)^c = A^c \cup B^c$

Let $x \in (A \cap B)^c \quad \Rightarrow \quad x \notin A \cap B$ $(\because a \in A \Leftrightarrow a \notin A^c)$

$\Rightarrow \quad x \notin A \text{ or } x \notin B$

$\Rightarrow \quad x \in A^c \text{ or } x \in B^c$

$\Rightarrow \quad x \in A^c \cup B^c$

$\therefore \quad (A \cap B)^c \subset A^c \cup B^c$ ...(*i*)

Again, let $x \in A^c \cup B^c \Rightarrow x \in A^c$ or $x \in B^c$

$\Rightarrow x \notin A$ or $x \notin B$

$\Rightarrow x \notin A \cap B$

$\Rightarrow x \in (A \cap B)^c$

$\therefore \quad A^c \cup B^c \subset (A \cap B)^c$ ....(*ii*)

Combining (*i*) and (*ii*), we get $(A \cap B)^c = A^c \cup B^c$. Hence proved.

Example 7. *Prove*

(*a*) $A \cup \phi = A$ (*b*) $A \cap \phi = \phi$ (*c*) $A \cup U = U$ (*d*) $A \cap U = A$.

Sol. (*a*) To prove $A \cup \phi = A$

Let $x \in A \cup \phi \Rightarrow x \in A$ or $x \in \phi$

$\Rightarrow x \in A$ ($\because x \notin \phi$, as ϕ is the null set)

Therefore, $x \in A \cup \phi \Rightarrow x \in A$

Hence $A \cup \phi \subset A$.

We know that $A \subset A \cup B$ for any set B.

But for $B = \phi$, we have $A \subset A \cup \phi$

From above, $A \subset A \cup \phi$, $A \cup \phi \subset A \Rightarrow A = A \cup \phi$. Hence proved.

(*b*) To prove $A \cap \phi = \phi$

If $x \in A$, then $x \notin \phi$ ($\because \phi$ is null set)

Therefore, $x \in A$, $x \notin \phi \Rightarrow A \cap \phi = \phi$. Hence proved.

(*c*) To prove $A \cup U = U$

Every set is a subset of universal set

$\therefore \quad A \cup U \subseteq U$

Also, $U \subseteq A \cup U$

Therefore, $A \cup U = U$. Hence proved.

(*d*) To prove $A \cap U = A$

We know $A \cap U \subset A$ ...(*i*)

So we have to show that $A \subset A \cap U$

Let $x \in A \Rightarrow x \in A$ and $x \in U$ ($\because A \subset U$ so $x \in A \Rightarrow x \in U$)

$\therefore \quad x \in A \Rightarrow x \in A \cap U$

$\therefore \quad A \subset A \cap U$ ...(*ii*)

From (*i*) and (*ii*), we get $A \cap U = A$. Hence proved.

Example 8. *Prove* (*a*) $A \cup A^c = U$ (*b*) $A \cap A^c = \phi$

(*c*) $U^c = \phi$ (*d*) $\phi^c = U$.

Sol. (*a*) To prove $A \cup A^c = U$

Every set is a subset of U

$\therefore \quad A \cup A^c \subset U$ ...(*i*)

We have to show that $U \subseteq A \cup A^c$

Let $x \in U \Rightarrow x \in A$ or $x \notin A$

$\Rightarrow x \in A$ or $x \in A^c \Rightarrow x \in A \cup A^c$

$\therefore$ $U \subseteq A \cup A^c$ ...(*ii*)

From (*i*) and (*ii*), we get $A \cup A^c = U$. Hence proved.

(*b*) To prove $A \cap A^c = \phi$

As ϕ is a subset of every set

$\therefore$ $\phi \subseteq A \cap A^c$ ...(*i*)

We have to show that $A \cap A^c \subseteq \phi$

Let $x \in A \cap A^c$ $\Rightarrow$ $x \in A$ and $x \in A^c$

$\Rightarrow$ $x \in A$ and $x \notin A$ (A contradictory statement)

$\Rightarrow$ $x \in \phi$

$\therefore$ $A \cap A^c \subset \phi$ ...(*ii*)

From (*i*) and (*ii*), we get $A \cap A^c = \phi$. Hence proved.

(*c*) To prove $U^c = \phi$

Let $x \in U^c \Leftrightarrow x \notin U \Leftrightarrow x \in \phi$

$\therefore$ $U^c = \phi$. Hence proved. (As U is a universal set)

(*d*) To prove $\phi^c = U$

Let $x \in \phi^c \Leftrightarrow x \notin \phi \Leftrightarrow x \in U$ (As ϕ is an empty set)

$\therefore$ $\phi^c = U$. Hence proved.

Example 9. *Prove* $(A^c)^c = A$.

Sol. Let $x \in (A^c)^c \Leftrightarrow x \notin A^c \Leftrightarrow x \in A$

$\therefore$ $(A^c)^c = A$. Hence proved.

CARDINALITY OF A SET

The total numbers of unique elements in the set is called the cardinality of the set. The cardinality of the countably infinite set is countably infinite *e.g.*,

1. Let $P = \{k, l, m, n\}$

The cardinality of the set P is 4.

2. Let A is the set of all non-negative even integers *i.e.*,

$$A = \{0, 2, 4, 6, 8, 10, \ldots\ldots\}$$

As A is countably infinite set hence the cardinality.

VENN DIAGRAMS

The Venn diagram represents sets as regions in a plane. These are used to understand the relationships between sets. Results from a Venn diagram usually are not considered to constitute a proof, but the results in the Venn diagram can be converted into a proof.

Venn Diagrams and Set Operations. Venn diagrams are used to represent set operations. In Venn diagrams, shadings are used to indicate which regions represent which sets. The results are obtained by determining which regions in the diagram are shaded in which fashion.

To draw Venn diagrams following points are to be taken into consideration :

(*i*) All the sets are usually represented by circles.

(*ii*) The universal set is represented by a rectangle.

(*iii*) The complement of the set is represented by that portion of the universal set which is not in the set.

Example 10. *Draw Venn diagram to represent the sets A, A^c and U.*

Sol. The shaded portion represents the set A. (Fig. 1).

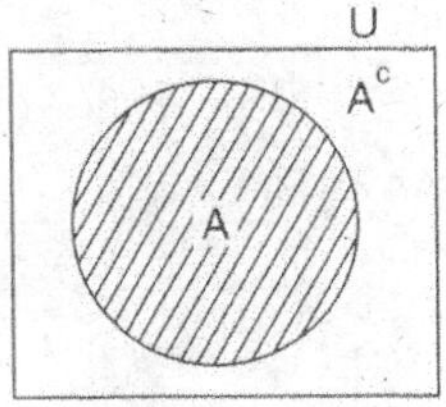

Fig. 1

The unshaded portion represents the set A^c w.r.t. the universal set represented by the rectangle.

Example 11. *Draw the Venn diagrams for the following :*

(*i*) $A \cup B$ (*ii*) $A \cap B$.

Sol. (*i*) The Venn diagram for $A \cup B$ is shown in Fig. 2.

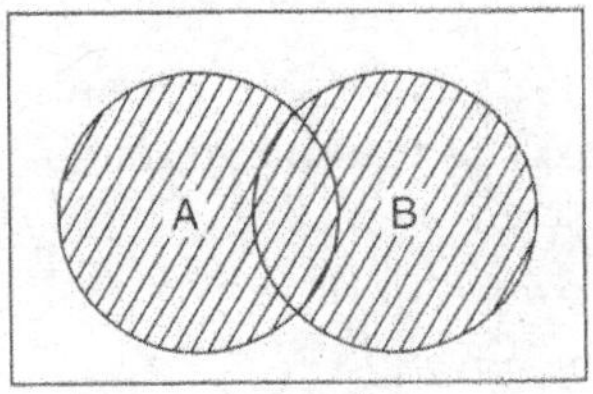

A ∪ B

Fig. 2

(*ii*) The Venn diagram for $A \cap B$ is shown in Fig. 3.

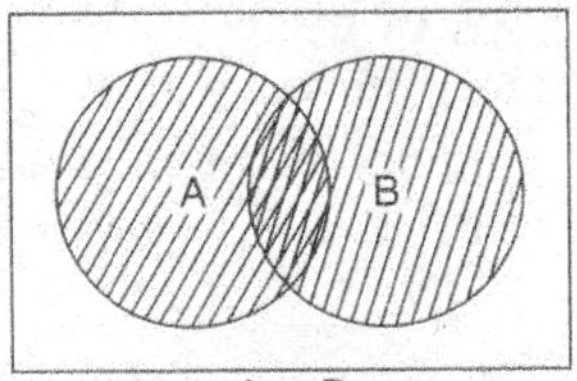

A ∩ B

Fig. 3

The region shaded in both directions represents $A \cap B$.

Example 12. *Draw the Venn diagram for $A^c \cap B^c \cap C^c$.*

Sol. The Venn diagram for $A^c \cap B^c \cap C^c$ is shown in Fig. 4.

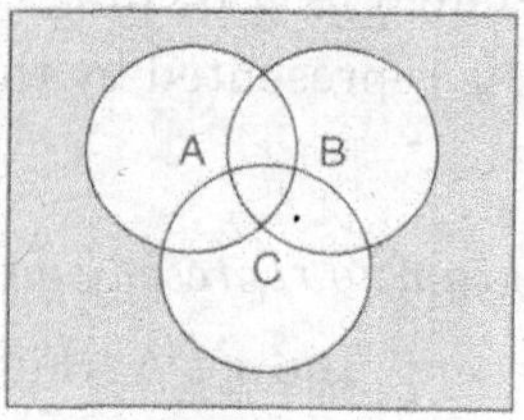

$A^c \cap B^c \cap C^c$

Fig. 4

The shaded portion represents $A^c \cap B^c \cap C^c$.

MULTISETS

Definition. A multiset is an unordered collection of elements, in which the multiplicity of an element may be one or more than one or zero. The multiplicity of an element is the number of times the element repeated in the multiset. In other words, we can say that an element can appear any number of times in a set *e.g.*,

$$A = \{l, l, m, m, n, n, n, n\}$$
$$B = \{a, a, a, a, a, c\}.$$

OPERATIONS ON MULTISETS

1. Union of Multisets. The union of two multisets A and B, is a multiset such that the multiplicity of an element is equal to the maximum of the multiplicity of an element in A and B and is denoted by $A \cup B$, *e.g.*,

Let $A = \{l, l, m, m, n, n, n, n\}$

$B = \{l, m, m, m, n\}$, then $A \cup B = \{l, l, m, m, m, n, n, n, n\}$.

2. Intersection of Multisets. The intersection of two multisets A and B, is a multiset such that the multiplicity of an element is equal to the minimum of the multiplicity of the element in A and B, and is denoted by $A \cap B$, *e.g.*,

Let $A = \{l, l, m, n, p, q, q, r\}$

$B = \{l, m, m, p, q, r, r, r, r\}$, then $A \cap B = \{l, m, p, q, r\}$.

3. Difference of Multisets. The difference of two multisets A and B, is a multiset such that the multiplicity of an element is equal to the multiplicity of the element in A minus the multiplicity of the element in B if the difference is +ve, and is equal to 0 if the difference is 0 or negative *e.g.*,

Let $A = \{l, m, m, m, n, n, n, p, p, p\}$

$B = \{l, m, m, m, m, n, r, r, r\}$, then $A - B = \{n, n, p, p, p\}$.

4. Sum of Multisets. The sum of two multisets A and B, is a multiset such that the multiplicity of an element is equal to the sum of the multiplicities of the element in A and B, *e.g.*,

Let $A = \{l, m, n, p, r\}$

$B = \{l, l, m, n, n, n, p, r, r\}$

then $A + B = \{l, l, l, m, m, n, n, n, n, p, p, r, r, r\}$

5. Cardinality of a Multiset. The cardinality of a multiset is the number of distinct elements in a multiset without considering the multiplicity of an element *e.g.*,

Let $A = \{l, l, m, m, n, n, n, p, p, p, p, q, q, q\}$

The cardinality of the multiset A is 5.

ORDERED PAIRS

An ordered pair consists of two elements such that one of them is designated as first member and other as second member. If p is the first element and q is the second element, then the ordered pair is written as (p, q).

e.g., Let $A = \{1, 2\}$, then possible ordered pairs of $A \times A$

$$A \times A = \{(1, 1), (1, 2), (2, 1), (2, 2)\}$$

CARTESIAN PRODUCT OF TWO SETS

The cartesian product of two sets P and Q in that order is the set of all ordered pairs whose first member belongs to the set P and second member belongs to set Q and is denoted by $P \times Q$ *i.e.,*

$$P \times Q = \{(x, y) : x \in P, y \in Q\}.$$

Example 13. *Let $P = \{a, b, c\}$ and $Q = \{k, l, m, n\}$. Determine the cartesian product of P and Q.*

Sol. The cartesian product of P and Q is

$$P \times Q = \begin{Bmatrix} (a, k), (a, l), (a, m), (a, n) \\ (b, k), (b, l), (b, m), (b, n) \\ (c, k), (c, l), (c, m), (c, n) \end{Bmatrix}.$$

Example 14. *Let $R = \{1, 2, 3\}$ and $S = \{4, 5, 6\}$. Determine the cartesian product.*

Sol. The cartesian product of R and S is

$$R \times S = \begin{Bmatrix} (1, 4), (1, 5), (1, 6) \\ (2, 4), (2, 5), (2, 6) \\ (3, 4), (3, 5), (3, 6) \end{Bmatrix}.$$

Example 15. *Prove that $(A \times B) \cap (P \times Q) = (A \cap P) \times (B \cap Q)$.*

Sol. Let $(x, y) \in (A \times B) \cap (P \times Q)$

$\Rightarrow$ $(x, y) \in (A \times B)$ and $(x, y) \in (P \times Q)$

$\Rightarrow$ $(x \in A$ and $y \in B)$ and $(x \in P$ and $y \in Q)$

$\Rightarrow$ $(x \in A$ and $x \in P)$ and $(y \in B$ and $y \in Q)$

$\Rightarrow$ $x \in (A \cap P)$ and $y \in (B \cap Q)$

$\Rightarrow \quad (x, y) \in (A \cap P) \times (B \cap Q)$

Therefore, $(A \times B) \cap (P \times Q) \subset (A \cap P) \times (B \cap Q)$ ...(*i*)

Now, conversely let $(x, y) \in (A \cap P) \times (B \cap Q)$

$\Rightarrow \quad x \in (A \cap P)$ and $y \in (B \cap Q)$

$\Rightarrow \quad (x \in A$ and $y \in P)$ and $(y \in B$ and $y \in Q)$

$\Rightarrow \quad (x \in A$ and $y \in B)$ and $(x \in P$ and $y \in Q)$

$\Rightarrow \quad (x, y) \in (A \times B)$ and $(x, y) \in (P \times Q)$

$\Rightarrow \quad (x, y) \in (A \times B) \cap (P \times Q)$

Therefore, $(A \cap P) \times (B \cap Q) \subset (A \times B) \cap (P \times Q)$ ...(*ii*)

From (*i*) and (*ii*), we have

$(A \times B) \cap (P \times Q) = (A \cap P) \times (B \cap Q)$. Hence proved.

Example 16. *Prove that $A \times (B \cap C) = (A \times B) \cap (A \times C)$.*

Sol. Let $(x, y) \in A \times (B \cap C) \quad \Rightarrow \quad x \in A$ and $y \in B \cap C$

$\Rightarrow \quad x \in A$ and $(y \in B$ and $y \in C)$

$\Rightarrow \quad x \in A$ and $y \in B$ and $x \in A$ and $y \in C$

$\Rightarrow \quad (x, y) \in A \times B$ and $(x, y) \in A \times C$

$\Rightarrow \quad (x, y) \in (A \times B) \cap (A \times C)$

Therefore, $A \times (B \cap C) \subset (A \times B) \cap (A \times C)$ ...(*i*)

Now, conversely let $(x, y) \in (A \times B) \cap (A \times C)$

$\Rightarrow \quad (x, y) \in A \times B$ and $(x, y) \in A \times C$

$\Rightarrow \quad (x \in A$ and $y \in B)$ and $(x \in A$ and $y \in C)$

$\Rightarrow \quad x \in A$ and $y \in B$ and $y \in C$

$\Rightarrow \quad x \in A$ and $y \in B \cap C$

$\Rightarrow \quad (x, y) \in A \times (B \cap C)$

Therefore, $(A \times B) \cap (A \times C) \subset A \times (B \cap C)$ ...(*ii*)

From (*i*) and (*ii*), we have $A \times (B \cap C) = (A \times B) \cap (A \times C)$. Hence proved.

Example 17. *Prove that $A \times (B \cup C) = (A \times B) \cup (A \times C)$.*

Sol. Let $(x, y) \in A \times (B \cup C) \quad \Rightarrow \quad x \in A$ and $y \in B \cup C$

$\Rightarrow \quad x \in A$ and $(y \in B$ or $y \in C) \quad \Rightarrow \quad (x \in A$ and $y \in B)$ or $(x \in A$ and $y \in C)$

$\Rightarrow \quad (x, y) \in A \times B$ or $(x, y) \in (A \times C) \quad \Rightarrow \quad (x, y) \in (A \times B) \cup (A \times C)$

Therefore, $A \times (B \cup C) \subset (A \times B) \cup (A \times C)$ ...(*i*)

Now, conversely let $(x, y) \in (A \times B) \cup (A \times C)$

$\Rightarrow \quad (x, y) \in (A \times B)$ or $(x, y) \in (A \times C) \quad \Rightarrow \quad (x \in A$ and $y \in B)$ or $(x \in A$ and $y \in C)$

$\Rightarrow \quad x \in A$ and $y \in B$ or $y \in C \quad \Rightarrow \quad x \in A$ and $y \in (B \cup C)$

$\Rightarrow \quad (x, y) \in A \times (B \cup C)$

Therefore, $(A \times B) \cup (A \times C) \subset A \times (B \cup C)$ ...(*ii*)

From (*i*) and (*ii*), we have $A \times (B \cup C) = (A \times B) \cup (A \times C)$. Hence proved.

Example 18. *Prove that if $A \subset B$, then $A \times C \subset B \times C$.*

Sol. Let $(x, y) \in A \times C \quad \Rightarrow \quad x \in A$ and $y \in C$

$\Rightarrow \quad x \in B$ and $y \in C$ $\quad (\because \ A \subset B)$

$\Rightarrow$ $(x, y) \in B \times C$

Therefore, $A \times C \subset B \times C$. Hence proved.

Example 19. *If S and T have n elements in common. Show that $S \times T$ and $T \times S$ have n^2 elements in common.*

Sol. Suppose, a set R, consisting of n common elements of S and T.

Therefore, $R \subset S$ and $R \subset T$

Let $(x, y) \in (R \times R)$ $\Leftrightarrow x \in R$ and $y \in R$

$\Leftrightarrow (x \in R$ and $y \in R)$ and $(x \in R$ and $y \in R)$

$\Leftrightarrow (x \in S$ and $y \in T)$ and $(x \in T$ and $y \in S)$ $(\because R \subset S ; R \subset T)$

$\Leftrightarrow (x, y) \in (S \times T)$ and $(x, y) \in (T \times S)$

$\Leftrightarrow (x, y) \in (S \times T) \cap (T \times S)$

Therefore, $(R \times R) = (S \times T) \cap (T \times S)$.

The right hand side contains ordered pairs common to both $S \times T$ and $T \times S$.

The left hand side $R \times R$ has n^2 elements $(\because$ R has n elements)

Since, the two sets are equal, both have the same number of elements.

Hence, $S \times T$ and $T \times S$ have n^2 common elements.

SOLVED PROBLEMS

Problem 1. *Write the following sets in builder form :*

(*a*) $A = \{1, 3, 5, 7, 9, ...\}$ (*b*) $B = \{1, 8, 27, 64, ...\}$

(*c*) $R = \{a, e, i, o, u\}$ (*d*) $S = \{2, 3, 5, 7, 11, 13, 17, 19, ...\}$

(*e*) *T = {JAWAHAR LAL NEHRU, INDIRA GANDHI, ..., ATAL BEHARI VAJPAI}.*

Sol. (*a*) $\{x : x$ is an odd +ve integer$\}$.

(*b*) $\{x : x$ is a +ve integer, x is a perfect cube$\}$.

(*c*) $\{x : x$ is a letter of alphabet and x is a vowel$\}$.

(*d*) $\{x : x$ is a +ve integer and x is a prime number$\}$.

(*e*) $\{x : x$ was prime minister of India$\}$.

Problem 2. *Write the following sets in builder from :*

(*a*) $A = \{2, 4, 6, 8, 10, 12, 14\}$ (*b*) $K = \{3, 6, 9, 12, 15, 18, ...\}$

(*c*) *L = {PUNJAB, HARYANA, DELHI, UP, ..., BIHAR}.*

Sol. (*a*) $\{x : x$ is a +ve integer divisible by 2 and less than 15$\}$.

(*b*) $\{x : x$ is +ve integer and is multiple of 3$\}$.

(*c*) $\{x : x$ is the stage of India$\}$.

Problem 3. *Write the following sets in tabular form :*

(*a*) $A = \{x : x^2 = 9\}$ (*b*) $B = \{x : x$ *is a multiple of 3 and* $0 < x < 20\}$

(*c*) $C = \{x : x$ *is a +ve even integer*$\}$ (*d*) $D = \{x : x$ *is a multiple of 5*$\}$.

Sol. (*a*) $A = \{3\}$ (*b*) $B = \{3, 6, 9, 12, 15, 18\}$

(*c*) $C = \{2, 4, 6, 8, 10, ...\}$ (*d*) $D = \{5, 10, 15, 20, 25, ...\}$

Problem 4. *Write the following sets in tabular form :*
(*a*) $A = \{x : x \text{ is a +ve integer and a perfect square}\}$
(*b*) $B = \{x : x \text{ is a divisor of } 24\}$
(*c*) $C = \{x : x \text{ is a multiple of 3 and } 5\}$
(*d*) $D = \{x : x \text{ is a multiple of 3 or } 5\}$
(*e*) $T = \{x : x \text{ is a letter in the alphabet}\}$.

Sol. (*a*) A = {1, 4, 9, 16, 25, 36, ...} (*b*) B = {1, 2, 3, 4, 6, 8, 12, 24}
(*c*) C = {15, 30, 45, 60, ...} (*d*) D = {3, 5, 6, 10, 15, ...}
(*e*) T = {*a*, *b*, *c*, *d*, *e*, ...}.

Problem 5. *Determine finite or infinite sets among the sets in Example 4.*

Sol. (*a*) Infinite. There are infinite integers that are perfect square.
(*b*) Finite. Divisor of 24 are finite.
(*c*) Infinite. There are infinite numbers that are multiple of 3 and 5.
(*d*) Infinite. There are infinite numbers that are multiple of 3 or 5.
(*e*) Finite. There are 26 letters in the alphabet.

Problem 6. *Determine finite or infinite sets among the following :*
(*i*) R = {*month in a year*} (*ii*) S = {*lines through the origin*}
(*iii*) T = {*positive odd integer*} (*iv*) K = {*cities in India*}
(*v*) L = {*members of the Parliament*} (*vi*) M = {*positive integer between 1 and – 1*}.

Sol. (*i*) Finite. Twelve months in a year.
(*ii*) Infinite. There are infinite number of lines passing through origin.
(*iii*) Infinite. There are infinite number of +ve odd integers.
(*iv*) Finite. There are finite number of cities in India.
(*v*) Finite. The members of the parliament are finite.
(*vi*) Finite. The set is empty. There is no +ve integer between 1 and – 1.

Problem 7. *Which of the following sets are equal :*
(*i*) $A = \{a, b, c, d\}$ (*ii*) $B = \{a, a, b, b, c, d, d, d\}$
(*iii*) $K = \{d, a, c, b\}$ (*iv*) $L = \{x : x \text{ is the letter of alphabet before } e\}$
(*v*) M = {*First four letters of the alphabet*}.

Sol. All the sets are equal. In set B multiple number of same elements does not affect the set. In set K, the order does not change the set.

Problem 8. *Which of the following sets are equal :*
(*a*) $R = \{x : x^2 = 7\}$ (*b*) $S = \{x : x + 4 = 4\}$
(*c*) $T = \{x : x^2 + 2 = 8\}$.

Sol. The sets R and T are equal as they are empty sets. The set S is not, because element a belongs to set S.

Problem 9. *Determine the power set P(A) of the set A = {1, 2, 3}.*

Sol. P(A) = {{1}, {2}, {3}, {1, 2}, {1, 3}, {2, 3}, {1, 2, 3}, ϕ}.

Problem 10. *Determine the power sets of the following sets :*

(*a*) $\{a\}$ (*b*) $\{\{a\}\}$ (*c*) $\{\phi, \{\phi\}\}$.

Sol. (*a*) $\{\{a\}, \phi\}$ (*b*) $\{\{\{a\}\}, \phi\}$ (*c*) $\{\{\phi\}, \{\{\phi\}\}, \{\phi, \{\phi\}\}, \phi\}$.

Problem 11. *Let* $A = \{\phi, \{\phi\}\}$. *Determine whether the following statements are true or false.*

(*a*) $\phi \in P(A)$ (*b*) $\phi \subseteq P(A)$ (*c*) $\{\phi\} \subseteq P(A)$

(*d*) $\{\phi\} \subseteq A$ (*e*) $\{\phi\} \in P(A)$ (*f*) $\{\phi\} \in A$

(*g*) $\{\{\phi\}\} \subseteq P(A)$ (*h*) $\{\{\phi\}\} \subseteq A$ (*i*) $\{\{\phi\}\} \in P(A)$

(*j*) $\{\{\phi\}\} \in A$.

Sol. The power $P(A) = \{\{\phi\}, \{\{\phi\}\}, \{\phi, \{\phi\}\}, \phi\}$

(*a*) True (*b*) True (*c*) True (*d*) True

(*e*) True (*f*) True (*g*) True (*h*) True

(*i*) True (*j*) False.

Problem 12. *Let* $A = \{a, \{a\}\}$. *Determine whether each of the following is true or false :*

(*a*) $\phi \in P(A)$ (*b*) $\phi \subseteq P(A)$ (*c*) $\{a\} \in P(A)$ (*d*) $\{a\} \subseteq P(A)$

(*e*) $\{\{a\}\} \in P(A)$ (*f*) $\{\{a\}\} \subseteq P(A)$ (*g*) $\{a, \{a\}\} \in P(A)$ (*h*) $\{a, \{a\}\} \subseteq P(A)$

(*i*) $\{\{\{a\}\}\} \in P(A)$ (*j*) $\{\{\{a\}\}\} \subseteq P(A)$.

Sol. The power set $P(A) = \{\{a\}, \{\{a\}\}, \{a, \{a\}\}, \phi\}$

(*a*) True (*b*) True (*c*) True (*d*) False

(*e*) True (*f*) True (*g*) True (*h*) False

(*i*) False (*j*) True.

Problem 13. *Let* $A = \{\phi\}$. *Let* $B = P(P(A))$.

(*a*) *Is* $\phi \in B$? $\phi \subseteq B$? (*b*) *Is* $\{\phi\} \in B$? $\{\phi\} \subseteq B$?

(*c*) *Is* $\{\{\phi\}\} \in B$? $\{\{\phi\}\} \subseteq B$?

Sol. The power set of A *i.e.*, $P(A) = \{\{\phi\}, \phi\}$

$$B = P(P(A)) = \{\{\phi\}, \{\{\phi\}\}, \{\{\phi\}, \phi\}, \phi\}$$

(*a*) Yes, $\phi \in$ B and $\phi \subseteq$ B as ϕ is a subset of every set.

(*b*) Yes, $\{\phi\} \in$ B. Also, $\{\phi\} \subseteq$ B as ϕ is an element in B.

(*c*) Yes, $\{\{\phi\}\} \in$ B and also $\{\{\phi\}\} \subseteq$ B since $\{\phi\}$ is an element of set B.

Problem 14. *Suppose*

$U = \{1, 2, 3, 4, 5, 6, 7, 8, 9\}$

$A = \{1, 4, 9\}$

$P = \{x : x \in U \text{ and } x \text{ is a perfect square}\}$

$R = \{1, 2, 3, 5, 7, 9\}$

$D = \{2, 3, 5, 7\}$

$N = \{x : x \in U \text{ and } x \text{ is a prime number}\}$

ϕ

(*a*) *Determine which sets are subsets of others.*

(*b*) *Determine which sets are proper subsets of others.*

(*c*) *Determine pair of sets which are disjoint.*

(*d*) *Determine pair of sets which are comparable.*

(*e*) *Determine pair of sets which are incomparable.*

Sol. (*a*) All the sets are subsets of U since the elements of every set belong to U. Set ϕ is subset of all other sets. Set A is subset of P. Set D is subset of R and N.

(*b*) All the sets are proper subsets of U since they are not equal to U. Set D is proper subset of R. Set N is proper subset of R. Set ϕ is proper subset of all the other sets.

(*c*) The pairs A and D, A and N, P and D, and P and N are disjoint sets.

(*d*) All the sets are comparable with the set U.

Set D is comparable with R as $D \subseteq R$

Set N is comparable with R as $N \subseteq R$

All sets comparable with ϕ as $\phi \subseteq$ Every set.

(*e*) The pairs A and R, A and D, A and N, P and R, P and D, P and N, are incomparable.

Problem 15. *Determine the cardinalities of the sets :*

(*a*) $P = \{n^7 : n \text{ is a positive integer}\}$ (*b*) $Q = \{n^{109} : n \text{ is a positive integer}\}$

(*c*) $P \cup Q$ (*d*) $P \cap Q$.

Sol. (*a*) The cardinality of the set P is infinite as the number of +ve integers are infinite.

(*b*) The cardinality of the set Q is infinite as the number of positive integers are infinite.

(*c*) Since sets P and Q are both infinite, their union is also infinite and hence cardinality.

(*d*) The cardinality of $P \cap Q$ is one because for $n = 1$

$$1^7 = 1^{109}$$

But for $n = 2$; $2^7 \neq 2^{109}$ and so on.

Hence, $P \cap Q$ contains only one element.

Problem 16. *Let N is the set of all natural numbers. Let P denotes all finite subsets of N. What is the cardinality of set P ? Give reason.*

Sol. The number of subsets of any set is given by 2^n, where n is number of elements in the set. As the number of subsets of set N is 2^n, hence there are 2^n number of subsets in set P. Therefore, the cardinality of set S is 2^n.

Problem 17. *Let* $U = \{1, 2, 3, 4, 5, 6, 7, 8, 9\}$ *and suppose that*

$$A = \{1, 2, 3, 4, 5\}, B = \{4, 5, 6, 7\}, C = \{5, 6, 7, 8, 9\}, D = \{1, 3, 5, 7, 9\}.$$

Find the following subsets of U :

(*a*) $A \cup B$ (*b*) $A \cap B$ (*c*) A'

(*d*) $C - D$ (*e*) $(C \cup D)'$.

Sol. (*a*) $A \cup B = \{1, 2, 3, 4, 5, 6, 7\}$ (*b*) $A \cap B = \{4, 5\}$

(*c*) $A' = \{6, 7, 8, 9\}$ (*d*) $C - D = \{6, 8\}$

(*e*) $(C \cup D)' = \{2, 4\}$.

Problem 18. *Determine the following sets :*

(*a*) $\phi \cup \{\phi\}$ (*b*) $\phi \cap \{\phi\}$ (*c*) $\{\phi\} \cup \{a, \phi, \{\phi\}\}$ (*d*) $\{\phi\} \cap \{a, \phi, \{\phi\}\}$

(*e*) $\phi \oplus \{a, \phi, \{\phi\}\}$ (*f*) $\{\phi\} \oplus \{a, \phi, \{\phi\}\}$.

Sol. (*a*) $\{\phi\}$ (*b*) ϕ

(*c*) $\{a, \phi, \{\phi\}\}$ (*d*) $\{\phi\}$

(*e*) $\{a, \phi, \{\phi\}\}$ (*f*) $\{a, \{\phi\}\}$.

Problem 19. *Determine whether each of the following statements is true or false. Explain your answer.*

(*a*) $A \cup P(A) = P(A)$ (*b*) $A \cap P(A) = A$

(*c*) $\{A\} \cup P(A) = P(A)$ (*d*) $\{A\} \cap P(A) = A$

(*e*) $A - P(A) = A$ (*f*) $P(A) - \{A\}$.

Sol. (*a*) $A \cup P(A) = P(A)$.

False. P(A) contains all subsets of A but does not contain all elements of A. Therefore, $A \cup P(A) \neq P(A)$.

(*b*) $A \cap P(A) = A$.

False. Since power set of A contains all subsets of A but no elements of A, hence there is no element common to both the sets.

(*c*) $\{A\} \cup P(A) = P(A)$

True. P(A) contains all subsets of A and {A} is also an element of P(A).

(*d*) $\{A\} \cap P(A) = A$.

False. {A} is common to both {A} and P(A) because P(A) contains all subsets of A but their intersection is not A but {A}.

(*e*) $A - P(A) = A$ **False.**

(*f*) $P(A) - \{A\} = P(A)$.

False. {A} is also an element of P(A). Therefore, $P(A) - \{A\} \neq P(A)$.

Problem 20. *Let* $A = \{\phi, a\}$. *Construct the following sets :*

(*a*) $A - \phi$ (*b*) $\{\phi\} - A$ (*c*) $A \cup P\{A\}$ (*d*) $A \cap P\{A\}$.

Sol. (*a*) $\{\phi, a\}$ (*b*) ϕ

(*c*) $\{\phi, a\} \cup \{\{\phi\}, \{a\}, \{\phi, a\}, \phi\} = \{\phi, a, \{\phi\}, \{a\}, \{\phi, a\}\}$

(*d*) $\{\phi, a\} \cap \{\{\phi\}, \{a\}, \{\phi, a\}, \phi\} = \{\phi\}$.

Problem 21. *Let A, B, C be sets. Under what conditions is each of the following true ?*

(*a*) $(A - B) \cup (A - C) = \phi$ (*b*) $(A - B) \cup (A - C) = \phi$

(*c*) $(A - B) \cap (A - C) = \phi$ (*d*) $(A - B) \oplus (A - C) = \phi$.

Sol. (*a*) This is true if B and C are null sets or A and B are disjoint sets and A and C are disjoint sets.

(*b*) This is true if A = B = C or A, B and C are null sets.

(*c*) This is true if A = B = C or all three sets are null sets.

(*d*) This is true if $A \subseteq B$ and $A \subseteq C$.

It is also true if $A \subseteq B$ and $A \not\subseteq C$ but B = C.

Problem 22. *What can you say about P and Q if*

(*a*) $P \cap Q = P$ (*b*) $P \cup Q = P$ (*c*) $P \oplus Q = P$ (*d*) $P \cap Q = P \cap Q$.

Sol. (*a*) This tells that $P \subseteq Q$. (*b*) This tells that P = Q or $Q \subseteq P$.

(*c*) This tells that Q is a null set. (*d*) This tells that P = Q.

Problem 23. *Let A, B, C be arbitrary sets :*

(*a*) *Show that* $(A - B) - C = A - (B \cup C)$.

(*b*) *Show that* $(A - B) - C = (A - C) - B$.

(*c*) *Show that* $(A - B) - C = (A - C) - (B - C)$.

Sol. (*a*) $(A - B) - C = A - (B \cup C)$

Let $x \in (A - B) - C \Rightarrow x \in (A - B), x \notin C$

$\Rightarrow x \in A, x \notin B, x \notin C$

$\Rightarrow x \in A, x \notin B$ or C

$\Rightarrow x \in A, x \notin B \cup C$

$\Rightarrow x \in A - (B \cup C)$.

$\therefore \quad A - (B \cup C) \subseteq (A - B) - C$ ...(*i*)

Conversely, let $x \in A - (B \cup C)$

$\Rightarrow x \in A, x \notin B \cup C$

$\Rightarrow x \in A, x \notin B, x \notin C$

$\Rightarrow x \in A - B, x \notin C$

$\Rightarrow x \in (A - B) - C$

$\therefore \quad (A - B) - C \subseteq A - (B \cup C)$ ...(*ii*)

From (*i*) and (*ii*), we get $(A - B) - C = A - (B \cup C)$.

(*b*) $(A - B) - C = (A - C) - B$

Let $x \in (A - B) - C$

$\Rightarrow x \in A - B, x \notin C \quad \Rightarrow x \in A, x \notin B, x \notin C$

$\Rightarrow x \in A - C, x \notin B \quad \Rightarrow x \in (A - C) - B$

$\therefore \quad (A - C) - B \subseteq (A - B) - C$ ...(*i*)

Conversely, let $x \in (A - C) - B$

$\Rightarrow x \in (A - C), x \notin B \quad \Rightarrow x \in A, x \notin C, x \notin B$

$\Rightarrow x \in A, x \notin B, x \notin C \quad \Rightarrow x \in (A - B), x \notin C$

$\Rightarrow x \in (A - B) - C$

$\therefore \quad (A - B) - C \subseteq (A - C) - B$ ...(*ii*)

From (*i*) and (*ii*), we get $(A - B) - C = (A - C) - B$.

(*c*) $(A - B) - C = (A - C) - (B - C)$

Let $x \in (A - B) - C \quad \Rightarrow x \in (A - B), x \notin C$

$\Rightarrow x \in A, x \notin B, x \notin C \quad \Rightarrow x \in A, x \notin C, x \notin B, x \notin C$

$\Rightarrow x \in A - C, x \notin B - C \quad \Rightarrow x \in (A - C) - (B - C)$

$\therefore \quad (A - C) - (B - C) \subseteq (A - B) - C$ ...(*i*)

Conversely, let $x \in (A - C) - (B - C)$

$\Rightarrow x \in (A - C), x \notin (B - C) \quad \Rightarrow x \in A, x \notin C, x \notin B, x \notin C$

$\Rightarrow x \in A, x \notin B, x \notin C \quad \Rightarrow x \in A - B, x \notin C$

$\Rightarrow x \in (A - B) - C$

$\therefore \quad (A - B) - C \subseteq (A - C) - (B - C)$ ...(*ii*)

From (*i*) and (*ii*), we get $(A - B) - C = (A - C) - (B - C)$.

Problem 24. *Given that $P \cup Q = P \cup R$, is it necessary that $Q = R$? Justify your answer.*

Sol. This is not necessary.

$P \cup Q \quad \Rightarrow$ all elements of P or Q are in this set.

$P \cup R \quad \Rightarrow$ all elements of P or R are in this set.

If $Q \subseteq P$ and also $R \subseteq P$. Then $P \cup Q = P \cup R = P$.

Therefore, it is not necessary that $Q = R$.

e.g., Let $P = \{1, 2, 3, 4\}$, $Q = \{1, 2\}$, $R = \{3, 4\}$, then $P \cup Q = P \cup R = P$

But here $Q = R$.

Problem 25. *Given that $P \cap Q = P \cap R$, is it necessary that $Q = R$? Justify your answer.*

Sol. This is not necessary

$P \cap Q$ contains elements common to both P and Q.

$P \cap R$ contains elements common to both P and R.

But it is not necessary that $Q = R$ because set R can have elements of set Q which are elements of set P as well, but it can also have elements other than those in set Q.

e.g., Let $P = \{1, 2, 5, 6\}$, $Q = \{1, 2, 8\}$, $R = \{1, 2, 9\}$

Then $P \cap Q = P \cap R = \{1, 2\}$; but $Q \neq R$.

Problem 26. *Given that $P \oplus Q = P \oplus R$, is it necessary that $Q = R$? Justify your answer.*

Sol. It is necessary that $Q = R$.

Because $P \oplus Q = (P \cup Q) - (P \cap Q)$ *i.e.,* this set contains elements which are in sets P or Q but does not contain elements common to both sets.

Similarly, $P \oplus R = (P \cup R) - (P \cap R)$ *i.e.,* this set contains elements which are in sets P or R but does not contain elements common to both sets.

As $P \oplus Q = P \oplus R$

$\therefore \quad (P \cup Q) - (P \cap Q) = (P \cup R) - (P \cap R)$

$\Rightarrow \quad P \cup Q = P \cup R$ and $P \cap Q = P \cap R$

$\Rightarrow \quad Q = R$

e.g., Let $P = \{1, 2, 3, 4\}$, $Q = \{1, 5\}$, $R = \{1, 5\}$

$P \oplus Q = (P \cup Q) - (P \cap Q) = \{1, 2, 3, 4, 5\} - \{1\} = \{2, 3, 4, 5\}$

$P \oplus R = (P \cup R) - (P \cap R) = \{1, 2, 3, 4, 5\} - \{1\} = \{2, 3, 4, 5\}$

$\Rightarrow \quad Q = R$.

Problem 27. *Prove that $(A - B) \cap B = \phi$.*

Sol. $(A - B) \cap B = \phi$

If $x \in B$, then $x \notin A - B$ due to definition of $A - B$

$x \in B, x \notin A - B \Rightarrow (A - B) \cap B = \phi$. Hence proved.

Problem 28. *Prove that $A \cup B = \phi \Rightarrow A = \phi, B = \phi$.*

Sol. Let $A \cup B = \phi$

Now ϕ is a subset of every set $\Rightarrow \phi \subset A, \phi \subset B$

Since, $A \subset A \cup B, B \subset A \cup B$

Hence, $A \subset \phi, B \subset \phi$

So $A \subset \phi, \phi \subset A \Rightarrow A = \phi$

$B \subset \phi, \phi \subset B \Rightarrow B = \phi$. Hence proved.

Problem 29. *Prove that $A - B \subset B'$.*

Sol. Let $x \in A - B \Rightarrow x \in A$ and $x \notin B$

$\Rightarrow x \in A$ and $x \in B' \Rightarrow x \in A \cap B'$ as $A \cap B' \subset B'$

$\Rightarrow x \in B'$

Therefore, any $x \in A - B \Rightarrow x \in B'$

So $A - B \subset B'$. Hence proved.

Problem 30. *Given that*

$$(A \cap C) \subseteq (B \cap C)$$

$$(A \cap \overline{C}) \subseteq (B \cap \overline{C})$$

Show that $A \subseteq B$.

Sol. For $(A \cap C) \subseteq (B \cap C)$, all elements of set A which are also elements of set C are contained in a set which contains elements common to both B and C. It implies $A \subseteq B$ for elements which are common in A and C, and B and C.

For $(A \cap \overline{C}) \subseteq (B \cap \overline{C})$, elements common to sets A and $\overline{C}$ are contained in a set containing elements common to both B and C. It implies $A \subseteq B$ for elements common in A and $\overline{C}$ and also in B and $\overline{C}$.

Hence, from both conditions it is shown that $A \subseteq B$.

2

Principle of Inclusion and Exclusion

As we know the cardinality of the set P is the number of unique elements in set P. It is denoted as $|P|$ and read as cardinality of set P.

FIRST PRINCIPLE

If P and Q are disjoint sets, then

$$|P \cup Q| = |P| + |Q|.$$

Theorem I. *Let P and Q be any two non-disjoint sets. Then*

$$|P \cup Q| = |P| + |Q| - |P \cap Q|.$$

Proof. Draw Venn diagram for the above as shown in Fig. 1.

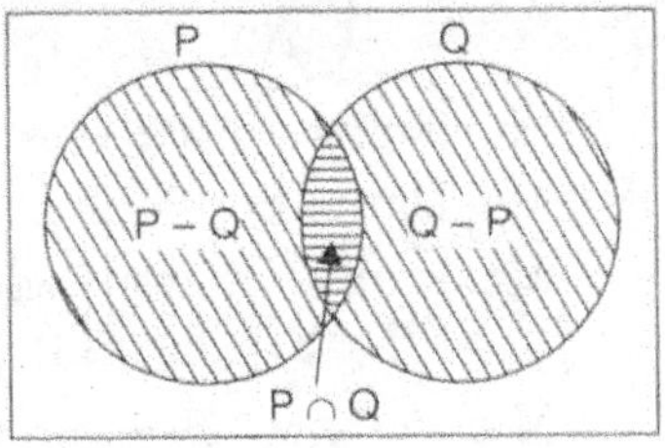

Fig. 1

From figure, we see that $P \cup Q$ can be seen to be the union of three disjoint sets $P - Q$, $Q - P$ and $P \cap Q$.

So $|P \cup Q| = |P - Q| + |Q - P| + |P \cap Q|$ ...(*i*)

Also, $|P| = |P - Q| + |P \cap Q|$ ...(*ii*)

and $|Q| = |Q - P| + |P \cap Q|$ ...(*iii*)

Combining (*ii*) and (*iii*)

$$|P| + |Q| = |P - Q| + |Q - P| + 2|P \cap Q|$$

$$|P| + |Q| = |P \cup Q| + |P \cap Q|$$

(As $|P \cup Q| = |P - Q| + |Q - P| + |P \cap Q|$)

$$|P \cup Q| = |P| + |Q| - |P \cap Q|$$

Hence proved.

Theorem II. *Let P, Q and R are three finite sets. Then*

$$|P \cup Q \cup R| = |P| + |Q| + |R| - |P \cap Q| - |P \cap R| - |Q \cap R| + |P \cap Q \cap R|.$$

Proof. Using theorem I, we have

$$|P \cup (Q \cup R)| = |P| + |Q \cup R| - |P \cap (Q \cup R)|$$

$$= |P| + |Q| + |R| - |Q \cap R| - |P \cap (Q \cup R)| \quad ...(i)$$

As $P \cap (Q \cup R) = (P \cap Q) \cup (P \cap R)$

So $|P \cap (Q \cup R)| = |P \cap Q| + |P \cap R| - |(P \cap Q) \cap (P \cap R)|$

$$= |P \cap Q| + |P \cap R| - |P \cap Q \cap R| \quad ...(ii)$$

Putting (*ii*) in (*i*), we get

$$|P \cup Q \cup R| = |P| + |Q| + |R| - |P \cap Q| - |P \cap R| - |Q \cap R| + |P \cap Q \cap R|$$

Hence proved.

INCLUSION-EXCLUSION PRINCIPLE IN GENERAL

Let $P_1, P_2, \ldots\ldots, P_n$ are finite sets. Then $|P_1 \cup P_2 \cup \ldots\ldots P_n| = \sum_{1 \le i \le n} |P_i| - \sum_{1 \le i < j \le n} |P_i \cap P_j| + \sum_{1 \le i < j < k \le n} |P_i \cap P_j \cap P_k| \ldots\ldots + (-1)^{n-1} |P_1 \cap P_2 \cap \ldots\ldots P_n|$

SOLVED PROBLEMS

Problem 1. *In a survey of 200 musicians, it was found that 40 wore gloves on the left hand and 39 wore gloves on the right hand. If 160 wore no gloves at all, how many wore a glove on only the right hand ? Only the left hand ? On both hands ?*

Sol. Total number of musicians wore gloves on left, right or both hands *i.e.*,

$$|L \cup R| = 200 - 160 = 40$$

Musicians wore gloves on left hand $|L| = 40$

Musicians wore gloves on right hand $|R| = 39$

Musicians who wore gloves on both hands

$$|L \cap R| = |L| + |R| - |L \cup R| = 40 + 39 - 40 = 39$$

Musicians who wore gloves only on right hand

$$= 40 - 39 = 1$$

Musicians who wore gloves only on left hand

$$= 39 - 39 = 0.$$

Problem 2. *Out of 1200 students at a college*

582 took Economics
627 took English
543 took Mathematics
217 took both Economics and English
307 took both Economics and Mathematics
250 took both Mathematics and English
222 took all three courses.

How many took none of the three ?

Sol. Suppose $|A| = 582$ $\quad |B| = 627$

$|C| = 543$ $\quad |A \cap B| = 217$

$|A \cap C| = 307$ $\quad |B \cap C| = 250$

$|A \cap B \cap C| = 222$

The total number of students who took any of three subjects

$$|A \cup B \cup C| = 582 + 627 + 543 - 217 - 307 - 250 + 222 = 1200$$

Students who took none of three

= (total students in the college) – (total students who took any of three subjects)

= 1200 – 1200 = 0.

Problem 3. *40 computer programmers interviewed for a job. 25 knew JAVA, 28 knew ORACLE, and 7 knew neither language. How many knew both languages ?*

Sol. Now $| J | = 25$

$| O | = 28$

$| J \cup O | = 40 - 7 = 33$

Computer programmers who knew both languages are

$| J \cap O | = 25 + 28 - 33 = 20.$

Problem 4. *Among 100 students, 32 study Mathematics, 20 study Physics, 45 study Biology, 15 study Mathematics and Biology, 7 study Mathematics and Physics, 10 study Physics and Biology and 30 do not study any of three subjects.*

(*a*) *Find the number of students studying all three subjects.*

(*b*) *Find the number of students studying exactly one of the three subjects.*

Sol. $| M | = 32$ $\quad | P | = 20$ $\quad | B | = 45$

$| M \cap B | = 15$ $\quad | M \cap P | = 7$ $\quad | P \cap B | = 10$

$| M \cup P \cup B | = 100 - 30 = 70.$

(*a*) Number of students studying all three subjects

$| M \cap P \cap B | = 70 - 32 - 20 - 45 + 15 + 7 + 10 = 5.$

(*b*) 5 study all three subjects.

15 – 5 = 10 study Mathematics and Biology but not all the three.

7 – 5 = 2 study Mathematics and Biology but not all the three.

10 – 5 = 5 study Physics and Biology but not all the three.

32 – 10 – 2 – 5 = 15 study only Mathematics.

20 – 2 – 5 – 5 = 8 study only Physics.

45 – 10 – 5 – 5 = 25 study only Biology.

Number of students studying exactly one of three subjects

= 15 + 8 + 25 = 48.

Problem 5. *A survey of 550 television watchers produced the following information :*

285 watch football games

195 watch hockey games

115 watch baseball games

45 watch football and baseball games

70 watch football and hockey games

50 watch hockey and baseball games

100 do not watch any of the three games.

(*a*) *How many people in the survey watch all three games ?*

(*b*) *How many people watch exactly one of the three games ?*

Sol. $| F | = 285$; $| H | = 195$; $| B | = 115$

$| F \cap B | = 45$; $| F \cap H | = 70$; $| H \cap B | = 50$

$$| F \cup H \cup B | = 550 - 100 = 450$$

(*a*) The number of people watch all three games

$$| F \cap H \cap B | = 450 - 285 - 195 - 115 + 45 + 70 + 50 = 20.$$

(*b*) 20 watch all three games.

45 – 20 = 25 watch football and baseball but not all three.

70 – 20 = 50 watch football and hockey but not all three.

50 – 20 = 30 watch hockey and baseball but not all three.

285 – 25 – 50 – 20 = 190 watch only football.

195 – 50 – 30 – 20 = 95 watch only hockey.

115 – 25 – 30 – 20 = 40 watch only baseball.

Number of people exactly watch one of the three games

$$= 190 + 95 + 40 = 325.$$

Problem 6. *In a survey of 300 students,*

64 had taken a Mathematics course
94 had taken a English course
58 had taken a Computer course
28 had taken both a Mathematics and a Computer course
26 had taken both a English and a Mathematics course
22 had taken both a English and a Computer course
14 had taken all three courses.

(*a*) *How many students were surveyed who had taken non of the three courses ?*

(*b*) *How many had taken only a Computer course ?*

Sol.

$$| M | = 64 \; ; \; | E | = 94 \; ; \; | C | = 58$$
$$| M \cap C | = 28 \; ; \; | M \cap E | = 26 \; ; \; | E \cap C | = 22$$
$$| M \cap E \cap C | = 14$$

(*a*)

$$| M \cup E \cup C | = | M | + | E | + | C | - | M \cap C | - | M \cap E | - | E \cap C | + | M \cap E \cap C |$$
$$= 64 + 94 + 58 - 28 - 26 - 22 + 14 = 154$$

Students who had taken none of the courses

$$= 300 - 154 = 146.$$

(*b*) 14 had taken all three courses.

28 – 14 = 14 had taken both a Mathematics and a Computer but not all three

22 – 14 = 8 had taken both a English and a Computer courses but not all three

58 – 14 – 8 – 14 = 22 had taken only Computer course.

Problem 7. *A survey was conducted among 1000 people. Of these 595 like Metro Channel, 595 like Star Movies and 550 like Zee TV ; 395 of them like Metro Channel and Star Movies, 350 of them like Metro Channel and Zee TV and 400 of them like Star Movies and Zee TV ; 250 of them like Metro Channel, Star Movies and Zee TV.*

(*a*) *How many of them who do not like Metro Channel, do not like Star Movies and do not like Zee TV ?*

(*b*) *How many of them who like Metro Channel, do not like Star Movies and do not like Zee TV ?*

Sol. $|M| = 595$; $|S| = 595$; $|Z| = 550$

$|M \cap S| = 395$; $|M \cap Z| = 350$; $|S \cap Z| = 400$

$|M \cap S \cap Z| = 250$

(*a*) Number of people like at least one of the channels

$$|M \cup S \cup Z| = 595 + 595 + 550 - 395 - 350 - 400 + 250 = 845$$

The number of people who do not like Metro Channel, do not like Star Movies and do not like Zee TV

= total number of people surveyed – number of people like at least one of the channels

= 1000 – 845 = 155.

(*b*) 250 people like all the three channels

395 – 250 = 145 like Metro and Star Movies but not all the three

350 – 250 = 100 like Metro and Zee TV but not all the three

595 – 145 – 100 – 250 = 100 people like Metro Channel but do not like Star Movies and do not like Zee TV.

Problem 8. *Among the first 500 positive integers :*

(*a*) *Determine the integers which are not divisible by 2, nor by 3, nor by 5.*

(*b*) *Determine the integers which are exactly divisible by one of them.*

Sol. Let A is the number of integers divisible by 2

B is the number of integers divisible by 3

C is the number of integers divisible by 5.

$$|A| = \left[\frac{500}{2}\right] = 250 ; \qquad |B| = \left[\frac{500}{3}\right] = 166 ; \qquad |C| = \left[\frac{500}{5}\right] = 100$$

$$|A \cap B| = \left[\frac{500}{2 \times 3}\right] = 83 ; \qquad |A \cap C| = \left[\frac{500}{2 \times 5}\right] = 50$$

$$|B \cap C| = \left[\frac{500}{3 \times 5}\right] = 33 ; \quad |A \cap B \cap C| = \left[\frac{500}{3 \times 3 \times 5}\right] = 16.$$

(*a*) $|A \cup B \cup C| = 250 + 166 + 50 - 83 - 100 - 33 + 16 = 366$

The integers not divisible by 2, 3 and 5 = 500 – 366 = 134.

(*b*) The integers divisible by all the three = 16

83 – 16 = 67 integers are divisible by 2 and 3 but not all the three

50 – 16 = 34 integers are divisible by 2 and 5 but not by all the three

33 – 16 = 17 integers are divisible by 3 and 5 but not by all the three

250 – 67 – 34 – 16 = 133 integers are only divisible by 2

166 – 67 – 17 – 16 = 66 integers are only divisible by 3

100 – 34 – 17 – 16 = 33 integers are only divisible by 5

Total number of integers only divisible by 2, 3 and 5

= 133 + 33 + 66 = 232.

Problem 9. *Among the first 1000 positive integers :*

(*a*) *Determine the integers which are not divisible by 5, nor by 7, nor by 9.*

(*b*) *Determine the integers divisible by 5, but not by 7, not by 9.*

Sol. Let A is the number of integers divisible by 5

B is the number of integers divisible by 7

C is the number of integers divisible by 9.

So $$|A| = \left|\frac{1000}{5}\right| = 200; \qquad |B| = \left\lfloor\frac{1000}{7}\right\rfloor = 142$$

$$|C| = \left\lfloor\frac{1000}{9}\right\rfloor = 111; \qquad |A \cap B| = \left\lfloor\frac{1000}{5 \times 7}\right\rfloor = 28$$

$$|A \cap C| = \left\lfloor\frac{1000}{5 \times 9}\right\rfloor = 22; \qquad |B \cap C| = \left\lfloor\frac{1000}{7 \times 9}\right\rfloor = 15$$

$$|A \cap B \cap C| = \left\lfloor\frac{1000}{5 \times 7 \times 9}\right\rfloor = 3.$$

(*a*) The number of integers divisible by 5, 7 and 9

$$|A \cup B \cup C| = 200 + 142 + 111 - 28 - 22 - 15 + 3$$
$$= 391.$$

The number of integers not divisible by 5, nor by 7, nor by 9

= Total number of integers – integers divisible by 5, 7 and 9

= 1000 – 391 = 609.

(*b*) The integers divisible by all the three integers = 3

28 – 3 = 25 integers divisible by 5 and 7 but not by all the three

22 – 3 = 19 integers divisible by 5 and 9 but not by all the three

∴ 200 – 25 – 19 – 3 = 153 integers divisible by 5 but not by 7, not by 9.

3

Mathematical Induction

PRINCIPLE

The process to establish the validity of a general result involving natural numbers is the principle of mathematical induction.

WORKING RULE

Let n_0 be a fixed integer. Suppose P(n) is a statement involving the natural number n and we wish to prove that P(n) is true for all $n \geq n_0$.

1. Basis of Induction. P(n_0) is true *i.e.*, P(n) is true for $n = n_0$.

2. Induction Step. Assume that the P(k) is true for $n = k$. $(k \geq n_0)$

Then P(k + 1) must also be true.

Then P(n) is true for all $n \geq n_0$.

PEANO'S AXIOMS

Peano, a mathematician defined the natural number in the following way. According to him, the member of set N satisfying the following properties are called natural number.

1. $1 \in$ N.

2. For all $n \in$ N, there exists a unique $n' \in$ N, such that

(*a*) $m' = n' \quad \Leftrightarrow \quad m = n$

(*b*) There exists no element p in N such that $p' = 1$.

3. If S $\subset$ N and

(*a*) $1 \in$ S

(*b*) $x \in$ S $\Rightarrow$ $x' \in$ S then S = N.

x' is called the successor of x and x is known as predecessor of x'.

These properties are called Peano's Axioms.

The number x' and $x + 1$.

Therefore $2 = 1'$, $3 = 2' = (1')'$, $4 = 3' = (2')' = ((1')')'$ and so on.

SOLVED PROBLEMS

Prove the following by Mathematical Induction :

Problem 1. $1 + 3 + 5 + ... + 2n - 1 = n^2$.

Sol. Basis of Induction. Let us assume that

$$P(n) = 1 + 3 + 5 + ... + (2n - 1) = n^2.$$

For $n = 1$, $\quad P(1) = 1 = 1^2 = 1$

It is true for $n = 1$. ...(*i*)

Induction Step. For $n = r$,

$$P(r) = 1 + 3 + 5 + ... + 2r - 1 = r^2 \text{ is true} \quad ...(ii)$$

Adding $2r + 1$ in both sides

$$P(r + 1) = 1 + 3 + 5 + ... + 2r - 1 + 2r + 1$$
$$= r^2 + (2r + 1) = r^2 + 2r + 1 = (r + 1)^2 \quad ...(iii)$$

As $P(r)$ is true. Hence $P(r + 1)$ is also true.

From (*i*), (*ii*) and (*iii*), we conclude that

$$1 + 3 + 5 + ... + 2n - 1 = n^2$$

is true for $n = 1, 2, 3, 4, 5, ...$ Hence proved.

Problem 2. $1^2 + 2^2 + 3^2 + ... + n^2 = \dfrac{n(n + 1)(2n + 1)}{6}$.

Sol. Basis of Induction. For $n = 1$,

$$P(1) = 1^2 = \frac{1(1 + 1)(2(1) + 1)}{6} = \frac{6}{6} = 1$$

It is true for $n = 1$. ...(*i*)

Induction Step. For $n = r$,

$$P(r) = 1^2 + 2^2 + 3^2 + ... + r^2 = \frac{r(r + 1)(2r + 1)}{6} \text{ is true} \quad ...(ii)$$

Adding $(r + 1)^2$ on both sides, we get

$$P(r + 1) = 1^2 + 2^2 + 3^2 + ... + r^2 + (r + 1)^2 = \frac{r(r + 1)(2r + 1)}{6} + (r + 1)^2$$
$$= \frac{r(r + 1)(2r + 1) + 6(r + 1)^2}{6} = (r + 1)\left[\frac{r(2r + 1) + 6(r + 1)}{6}\right]$$
$$= \frac{(r + 1)}{6}[r(2r + 1) + 6(r + 1)] = \frac{(r + 1)}{6}[2r^2 + 7r + 6]$$
$$= \frac{(r + 1)(r + 2)(2r + 3)}{6} \quad ...(iii)$$

As $P(r)$ is true, hence $P(r + 1)$ is true.

From (*i*), (*ii*) and (*iii*), we conclude that

$$1^2 + 2^2 + 3^2 + ... + n^2 = \frac{n(n + 1)(2n + 1)}{6}$$

is true for $n = 1, 2, 3, 4, 5, ...$ Hence proved.

Problem 3. $1^2 - 2^2 + 3^2 + ... + (-1)^{n+1} n^2 = \frac{(-1)^{n+1} n(n+1)}{2}$.

Sol. Let $\quad P(n) = 1^2 - 2^2 + 3^2 + ... + (-1)^{n+1} n^2 = \frac{(-1)^{n+1} n(n+1)}{2}$.

Basic Step. For $n = 1$,

$$P(1) = (-1)^2 . 1^2 = 1 = \frac{(-1)^{1+1} . 1 . (1+1)}{2} = 1$$

It is true for $n = 1$. ...(*i*)

Induction Step. For $n = r$,

$$P(r) = 1^2 - 2^2 + 3^2 + ... + (-1)^{r+1} r^2 = \frac{(-1)^{r+1} r(r+1)}{2} \text{ is true} \quad ...(ii)$$

Adding $(-1)^{r+2} . (r+1)^2$ in both sides, we get

$$P(r+1) = 1^2 - 2^2 + 3^2 + ... + (-1)^{r+1} . r^2 + (-1)^{r+2} (r+1)^2$$

$$= \frac{(-1)^{r+1} r(r+1)}{2} + (-1)^{r+2} (r+1)^2$$

$$= \frac{(-1)^{r+1} r(r+1) + 2\{(-1)^{r+2} (r+1)^2\}}{2}$$

$$= \frac{(-1)^{r+2} (r+1)\left\{(-1)^{-1} r + 2(r+1)\right\}}{2}$$

$$= \frac{(-1)^{r+2} (r+1)(r+2)}{2} \quad ...(iii)$$

As $P(r)$ is true, hence $P(r+1)$ is also true. From (*i*), (*ii*) and (*iii*), we conclude that

$$1^2 - 2^2 + 3^2 + ... + (-1)^{n+1} n^2 = \frac{(-1)^{n+1} n(n+1)}{2}$$

is true for $n = 1, 2, 3, 4, 5, ...$. Hence proved.

Problem 4. $1^3 + 2^3 + 3^3 + ... + n^3 = \left[\frac{n(n+1)}{2}\right]^2$.

Sol. Let $\quad P(n) = 1^3 + 2^3 + 3^3 + ... + n^3 = \left[\frac{n(n+1)}{2}\right]^2$.

Basic Step. For $n = 1$,

$$P(1) = 1^3 = \left[\frac{1(1+1)}{2}\right]^2 = (1)^2 = 1$$

Hence $P(n)$ is true for $n = 1$. ...(*i*)

Induction Step. For $n = r$,

$$P(r) = 1^3 + 2^3 + 3^3 + ... + r^3 = \left[\frac{r(r+1)}{2}\right]^2 \text{ is true} \quad ...(ii)$$

Adding $(r+1)^3$ on both sides

$$P(r+1) = 1^3 + 2^3 + 3^3 + ... + r^3 + (r+1)^3 = \left[\frac{r(r+1)}{2}\right]^2 + (r+1)^3$$

$$= \frac{r^2(r+1)^2 + 4(r+1)(r+1)^2}{4} = \frac{(r+1)^2\{r^2 + 4(r+1)\}}{4}$$

$$= \frac{(r+1)^2[r^2 + 4r + 4]}{4} = \frac{(r+1)^2(r+2)^2}{4}$$

$$= \left[\frac{(r+1)(r+2)}{2}\right]^2 \qquad ...(iii)$$

As P(r) is true, hence P(r + 1) is also true.

From (*i*), (*ii*) and (*iii*), we conclude that

$$1^3 + 2^3 + 3^3 + ... + n^3 = \left[\frac{n(n+1)}{2}\right]^2$$

is true for n = 1, 2, 3, 4, 5, ... Hence proved.

Problem 5. $1 \,.\, 2 + 2 \,.\, 3 + 3 \,.\, 4 + ... + n(n+1) = \dfrac{n(n+1)(n+2)}{3}$.

Sol. Let $\quad P(n) = 1 \,.\, 2 + 2 \,.\, 3 + 3 \,.\, 4 + ... + n(n+1) = \dfrac{n(n+1)(n+2)}{3}$.

Basis of induction. For $n = 1$,

$$P(1) = 1 \,.\, 2 = \frac{1(1+1)(1+2)}{3} = 1 \,.\, 2$$

It is true for $n = 1$. ...(*i*)

Induction Step. For $n = r$,

$$P(r) = 1 \,.\, 2 + 2 \,.\, 3 + 3 \,.\, 4 + ... + r(r+1) = \frac{r(r+1)(r+2)}{3} \text{ is true} \qquad ...(ii)$$

Adding $(r + 1)(r + 2)$ on both sides,

$$P(r+1) = 1 \,.\, 2 + 2 \,.\, 3 + 3 \,.\, 4 + ... + r(r+1) + (r+1)(r+2)$$

$$= \frac{r(r+1)(r+2)}{3} + (r+1)(r+2)$$

$$= \frac{r(r+1)(r+2) + 3(r+1)(r+2)}{3} = \frac{(r+1)(r+2)(r+3)}{3} \qquad ...(iii)$$

As P(r) is true, hence P(r + 1) is also true.

From (*i*), (*ii*) and (*iii*), we conclude that

$$1 \,.\, 2 + 2 \,.\, 3 + 3 \,.\, 4 + ... + n(n+1) = \frac{n(n+1)(n+2)}{3}$$

is true for n = 1, 2, 3, 4, ... Hence proved.

Problem 6. $\dfrac{1}{1 \,.\, 3} + \dfrac{1}{3 \,.\, 5} + \dfrac{1}{5 \,.\, 7} + ... + \dfrac{1}{(2n-1)(2n+1)} = \dfrac{n}{2n+1}$.

Sol. Basic Step. For $n = 1$,

$$\frac{1}{(2-1)(2+1)} = \frac{1}{3} = \frac{1}{2(1)+1} = \frac{1}{3}$$

Hence P(n) is true for $n = 1$. ...(*i*)

Induction Step. For $n = r$,

$$P(r) = \frac{1}{1.3} + \frac{1}{3.5} + \frac{1}{5.7} + \ldots + \frac{1}{(2r-1)(2r+1)} = \frac{r}{2r+1} \text{ is true} \quad \ldots(ii)$$

Adding $\frac{1}{(2r+1)(2r+3)}$ on both sides

$$P(r+1) = \frac{1}{1.3} + \frac{1}{3.5} + \frac{1}{5.7} + \ldots + \frac{1}{(2r-1)(2r+1)} + \frac{1}{(2r+1)(2r+3)}$$

$$= \frac{r}{2r+1} + \frac{1}{(2r+1)(2r+3)} = \frac{r(2r+3)+1}{(2r+1)(2r+3)}$$

$$= \frac{2r^2+3r+1}{(2r+1)(2r+3)} = \frac{(2r+1)(r+1)}{(2r+1)(2r+3)} = \frac{(r+1)}{(2r+3)} \quad \ldots(iii)$$

As P(r) is true, hence P(r + 1) is also true.

From (i), (ii) and (iii), we conclude that

$$\frac{1}{1.3} + \frac{1}{3.5} + \frac{1}{5.7} + \ldots + \frac{1}{(2n-1)(2n+1)} = \frac{n}{2n+1}$$

is true for n = 1, 2, 3, 4, ... Hence proved.

Problem 7. *1(1 !) + 2(2 !) + ... + n(n !) = (n + 1) ! – 1.*

Sol. Let P(n) = 1(1 !) + 2(2 !) + ... + n(n !) = (n + 1) ! – 1.

Basis of Induction. For n = 1,

$$P(1) = 1(1\,!) = (1+1)\,! - 1 \Rightarrow 2\,! - 1 \Rightarrow 1 = 1$$

It is true for n = 1. ...(i)

Induction Step. For $n = r$,

$$P(r) = 1(1\,!) + 2(2\,!) + \ldots + r(r\,!) = (r+1)\,! - 1 \text{ is true.} \quad \ldots(ii)$$

Adding (r + 1) [(r + 1) !] on both sides,

$$P(r+1) = 1(1\,!) + 2(2\,!) + \ldots + r(r\,!) + (r+1)(r+1)\,!$$
$$= (r+1)\,! - 1 + (r+1)(r+1)\,!$$
$$= (r+1)\,!\,(r+1+1) - 1 = (r+1)\,!\,(r+2) - 1$$
$$= (r+2)\,.\,(r+1)\,! - 1 = (r+2)\,! - 1 \quad \ldots(iii)$$

As P(r) is true, hence P(r + 1) is also true.

From (i), (ii) and (iii), we conclude that

$$1(1\,!) + 2(2\,!) + \ldots + n(n\,!) = (n+1)\,! - 1$$

is true for n = 1, 2, 3, 4, 5, ... Hence proved.

Problem 8. *Prove 1 + 3 + 5 + ... + (2n – 1) = n^2 by in induction ($n \geq 1$).*

Sol. Consider P(n) = 1 + 3 + 5 + ... + (2n – 1) = n^2

Basis of Induction. For n = 1,

$$P(1) = 2 \times 1 - 1 = 1 = 1^2 = 1$$

It is true for n = 1. ...(i)

Induction Step. For $n = r$,

$$P(r) = 1 + 3 + 5 + \ldots + (2r-1) = r^2 \text{ is true} \quad \ldots(ii)$$

Adding $(2r + 1)$ to both sides,

$$P(r + 1) = 1 + 3 + 5 + ... + (2r - 1) + (2r + 1)$$
$$= r^2 + (2r + 1) = (r + 1)^2 \quad ...(iii)$$

As P(r) is true, hence P(r + 1) is also true.

From (i), (ii) and (iii), we conclude that

$$1 + 3 + 5 + ... + (2n - 1) = n^2$$

is true for n = 1, 2, 3, ... Hence proved.

Problem 9. *Show that* $1 + 2 + 2^2 + 2^3 + ... + 2^n = 2^{n+1} - 1$ *by induction (for* $n \geq 0$*).*

Sol. Consider $P(n) = 1 + 2 + 2^2 + 2^3 + ... + 2^n = 2^{n+1} - 1$.

Basis of Induction. For $n = 0$,

$$P(0) = 1 = 2^1 - 1 = 1$$

It is true for $n = 0$. ...(i)

Induction Step. For $n = r$,

$$P(r) = 1 + 2 + 2^2 + 2^3 + ... + 2^r = 2^{r+1} - 1 \text{ is true.} \quad ...(ii)$$

Adding 2^{r+1} to both sides,

$$P(r + 1) = 1 + 2 + 2^2 + 2^3 + ... + 2^r + 2^{r+1} = 2^{r+1} - 1 + 2^{r+1}$$
$$= 2(2^{r+1}) - 1 = 2^{r+2} - 1 \quad ...(iii)$$

As P(r) is true, hence P(r + 1) is also true.

From (i), (ii) and (iii), we conclude that

$$1 + 2 + 2^2 + ... + 2^n = 2^{n+1} - 1.$$

is true for n = 1, 2, 3, ...

Problem 10. *Prove by induction that for* $n \geq 0$ *and* $a \neq 1$; $1 + a + a^2 + ... + a^n = \dfrac{1 - a^{n+1}}{1 - a}$.

Sol. Basis of Induction. For $n = 1$,

$$1 + a^1 = \frac{1 - a^{1+1}}{1 - a} = \frac{1 - a^2}{1 - a} = 1 + a$$

It is true for $n = 1$. ...(i)

Induction Step. For $n = r$,

$$P(r) = 1 + a + a^2 + ... + a^r = \frac{1 - a^{r+1}}{1 - a} \text{ is true} \quad ...(ii)$$

Adding a^{r+1} to both sides,

$$P(r + 1) = 1 + a + a^2 + ... + a^r + a^{r+1} = \frac{1 - a^{r+2}}{1 - a}$$
$$= \frac{1 - a^{r+1}}{1 - a} + a^{r+1} = \frac{1 - a^{r+1} + a^{r+1} - a^{r+2}}{1 - a} = \frac{1 - a^{r+2}}{1 - a} \quad ...(iii)$$

As P(r) is true, hence P(r + 1) is also true.

From (i), (ii) and (iii), we conclude that

$$1 + a + a^2 + ... + a^n = \frac{1 - a^{n+1}}{1 - a} \text{ is true for } n \geq 0. \text{ Hence proved.}$$

Problem 11. *Prove*

$$1.2.3+2.3.4+3.4.5+...+n(n+1)(n+2)=\frac{1}{4}n(n+1)(n+2)(n+3).$$

Sol. Basis of Induction. For $n = 1$,

$$\text{L.H.S.} = 1(1+1)(1+2) = 1.2.3 = 6$$

$$\text{R.H.S.} = \frac{1}{4}.1(1+1)(1+2)(1+3) = \frac{1}{4}.1.2.3.4 = 6$$

It is true for $n = 1$.

Induction Step. For $n = r$,

$$1.2.3+2.3.4+3.4.5+...+r(r+1)(r+2)=\frac{1}{4}r(r+1)(r+2)(r+3) \text{ is true.}$$

For $n = r + 1$,

$$\text{L.H.S.} = 1.2.3+2.3.4+...+r(r+1)(r+2)+(r+1)(r+2)(r+3)$$

$$= \frac{1}{4}r(r+1)(r+2)(r+3)+(r+1)(r+2)(r+3)$$

$$= \frac{1}{4}(r+1)(r+2)(r+3)(r+4)$$

$$= \text{R.H.S., for } n = r+1.$$

Hence, $1.2.3+2.3.4+3.4.5+...+n(n+1)(n+2)=\frac{1}{4}n(n+1)(n+2)(n+3)$. Hence proved.

Problem 12. *Show that for any integer n*

$11^{n+2} + 12^{2n+1}$ is divisible by 133.

Sol. Let $P(n) = 11^{n+2} + 12^{2n+1}$

Basis of Induction. For $n = 1$,

$$P(1) = 11^3 + 12^3 = 3059 = 133 \times 23$$

So, 133 divides P(1). ...(*i*)

Induction Step. For $n = r$,

$$P(r) = 11^{r+2} + 12^{2r+1} = 133 \times s \quad ...(ii)$$

Now, for $n = r + 1$,

$$P(r+1) = 11^{r+2+1} + 12^{2(r)+3} = 11[133s - 12^{2r+1}] + 144 . 12^{2r+1}$$

$$= 11 \times 133s + 12^{2r+1} . 133 = 133[11s + 12^{2r+1}] = 133 \times t \quad ...(iii)$$

As (*i*), (*ii*) and (*iii*) all are true, hence P(n) is divisible by 133.

Problem 13. *Prove by induction that the sum of the cubes of three consecutive integers is divisible by 9.*

Sol. Let $P(n) = n^3 + (n+1)^3 + (n+2)^3$

P(n) is divisible by 9

$$P(1) = 1 + 8 + 27 = 36$$

which is divisible by 9. ...(*i*)

For $n = r$,

$$P(r) = r^3 + (r+1)^3 + (r+2)^3 = 9.q \quad ...(ii)$$

For $n = r + 1$, $P(r + 1) = (r + 1)^3 + (r + 2)^3 + (r + 3)^3$

$$= r^3 + (r + 1)^3 + (r + 2)^3 + [9r^2 + 27r + 27]$$

$$= 9q + 9(r^2 + 3r + 3) = 9[q + r^2 + 3r + 3]$$

$$= 9 \,.\, q \qquad \text{...}(iii)$$

From (*i*), (*ii*) and (*iii*), we have the required result by induction. Hence proved.

Problem 14. *Prove* $\frac{1}{n+1} + \frac{1}{n+2} + \ldots + \frac{1}{2n} > \frac{13}{24}$, *for* $n \geq 2$.

Sol. For $n = 2$,

$$\text{L.H.S.} = \frac{1}{2+1} + \frac{1}{2+2} = \frac{7}{12} > \frac{13}{24} = \text{R.H.S.} \qquad \text{...}(i)$$

It is true for $n = 2$.

Now, for $n = r$, where $r > 2$

$$\frac{1}{r+1} + \frac{1}{r+2} + \ldots + \frac{1}{2r} > \frac{13}{24} \qquad \text{...}(ii)$$

For $n = r + 1$,

$$\text{L.H.S.} = \frac{1}{r+2} + \frac{1}{r+3} + \ldots + \frac{1}{2r} + \frac{1}{2r+1} + \frac{1}{2r+2} > \frac{13}{24} + \frac{1}{2r+1} + \frac{1}{2r+2} - \frac{1}{r+1}$$

[Using (*ii*)]

$$= \frac{13}{24} + \frac{2r+2+2r+1-2(2r+1)}{(2r+1)(2r+2)} = \frac{13}{24} + \frac{1}{(2r+1)(2r+2)} > \frac{13}{24}$$

As $n = r$ is true. Thus, the result is also true for $n = r + 1$. Hence, we have the result for $n \geq 2$ by induction. Hence proved.

Problem 15. *Prove* $\frac{1}{1\times 2} + \frac{1}{2\times 3} + \frac{1}{3\times 4} + \ldots + \frac{1}{n(n+1)} = \frac{n}{n+1}$.

Sol. Basis of Induction. For $n = 1$,

$$\frac{1}{1(1+1)} = \frac{1}{2} = \frac{1}{1+1} = \frac{1}{2}$$

It is true for $n = 1$.

Induction Step. For $n = r$,

$$P(r) = \frac{1}{1\times 2} + \frac{1}{2\times 3} + \frac{1}{3\times 4} + \ldots + \frac{1}{r(r+1)} = \frac{r}{r+1} \text{ is true.}$$

For $n = r + 1$,

$$\text{L.H.S.} = \frac{1}{1\times 2} + \frac{1}{2\times 3} + \frac{1}{3\times 4} + \ldots + \frac{1}{(r+1)(r+1+1)}$$

$$= \frac{1}{1\times 2} + \frac{1}{2\times 3} + \ldots + \frac{1}{r(r+1)} + \frac{1}{(r+1)(r+2)}$$

$$= \frac{r}{r+1} + \frac{1}{(r+1)(r+2)} = \frac{1}{r+1}\left(r + \frac{1}{r+2}\right)$$

$$= \frac{r^2+2r+1}{(r+1)(r+2)} = \frac{r+1}{r+2}$$

$$\text{R.H.S.} = \frac{r+1}{r+1+1} = \frac{r+1}{r+2}$$

L.H.S. = R.H.S.

Therefore, $\frac{1}{1\times 2} + \frac{1}{2\times 3} + \frac{1}{3\times 4} + \ldots + \frac{1}{n(n+1)} = \frac{n}{n+1}$ is true for $n = r + 1$.

Hence proved.

Problem 16. *Show that $n^3 + 2n$ is divisible by 3 for all $n \geq 1$ by induction.*

Sol. Basis of Induction. For $n = 1$,

$P(1) = 1^3 + 2 \times 1 = 3$. It is divisible by 3.

Induction Step. For $n = r$,

$$P(r) = r^3 + 2r = 3r$$

For $n = r + 1$,

$$\begin{aligned} P(r+1) &= (r+1)^3 + 2(r+1) = r^3 + 1 + 3r(r+1) + 2r + 2 \\ &= r^3 + 1 + 3r^2 + 3r + 2r + 2 \\ &= r^3 + 3 + 5r + 3r^2 = r^3 + 2r + 3 + 3r + 3r^2 \qquad (\because\ r^3 + 2r = 3r) \\ &= 3r + 3 + 3r + 3r^2 = 3r^2 + 6r + 3 \\ &= 3(r^2 + 2r + 1) = 3(r+1)^2. \end{aligned}$$

It is divisible by 3.

Hence, we have the required result by induction.

Problem 17. *Show that $2^n \times 2^n - 1$ is divisible by 3 for all $n \geq 1$ by induction.*

Sol. Basis of Induction. For $n = 1$,

$2^1 \times 2^1 - 1 = 3$ divisible by 3.

It is true for $n = 1$.

Induction Step. For $n = r$,

$$2^r\, 2^r - 1 = 3r \quad i.e., \quad 2^{r+1} - 1 = 3r$$

For $n = r + 1$,

$$\begin{aligned} 2^{r+1} \,.\, 2^{r+1} - 1 &= 2^{r+1}(3r + 1) - 1 \\ &= 3 \,.\, 2^{r+2} + 2^{r+1} - 1 = 3 \,.\, 2^{r+2} + 3r \\ &= 3(2^{r+2} + r). \end{aligned}$$

It is divisible by 3.

Hence, we have the required result by induction.

Problem 18. *Show that $1^2 + 3^2 + 5^2 + \ldots + (2n-1)^2 = \frac{n(2n-1)(2n+1)}{3}$.*

Sol. Basis of Induction. For $n = 1$,

$$(2 \times 1 - 1)^2 = 1 = \frac{1(2\times 1 - 1)(2 \times 1 + 1)}{3} = 1$$

It is true of $n = 1$.

Induction Step. For $n = r$,

$$1^2 + 3^2 + 5^2 + \ldots + (2r-1)^2 = \frac{r(2r-1)(2r+1)}{3} \text{ is true.}$$

For $n = r + 1$,

$$P(r+1) = 1^2 + 3^2 + 5^2 + ... + (2r-1)^2 + [2(r+1)-1]^2 = \frac{(r+1)(2r+1)(2r+3)}{3}$$

$$\text{L.H.S.} = \frac{r(2r-1)(2r+1)}{3} + (2r+1)^2 = \frac{r(2r-1)(2r+1) + 3(2r+1)^2}{3}$$

$$= \frac{(2r+1)[2r^2 - r + 6r + 3]}{3} = \frac{(2r+1)[2r^2 + 5r + 3]}{3}$$

$$= \frac{(2r+1)[2r^2 + 3r + 2r + 3]}{3} = \frac{(2r+1)(2r+3)(r+1)}{3}$$

L.H.S. = R.H.S.

So, $1^2 + 3^2 + 5^2 + ... + (2n-1)^2 = \dfrac{n(2n-1)(2n+1)}{3}$.

Problem 19. *Show that* $\dfrac{1^2}{1.3} + \dfrac{2^2}{3.5} + ... + \dfrac{n^2}{(2n-1)(2n+1)} = \dfrac{n(n+1)}{2(2n+1)}$.

Sol. Basis of Induction. For $n = 1$,

$$\frac{1}{1.3} = \frac{1.2}{2.3} = \frac{1}{3}$$

It is true for $n = 1$.

Induction Step. For $n = r$,

$$\frac{1^2}{1.3} + \frac{2^2}{3.5} + ... + \frac{r^2}{(2r-1)(2r+1)} = \frac{r(r+1)}{2(2r+1)}$$

Now, for $n = r + 1$,

$$\frac{1^2}{1.3} + \frac{2^2}{3.5} + ... + \frac{r^2}{(2r-1)(2r+1)} + \frac{(r+1)^2}{(2r+1)(2r+3)} = \frac{(r+1)(r+2)}{2(2r+3)}$$

$$\text{L.H.S.} = \frac{r(r+1)}{2(2r+1)} + \frac{(r+1)^2}{(2r+1)(2r+3)} = \frac{r+1}{2r+1}\left[\frac{r}{2} + \frac{(r+1)}{(2r+3)}\right]$$

$$= \frac{r+1}{2r+1}\left[\frac{2r^2 + 3r + 2r + 2}{2(2r+3)}\right]$$

$$= \frac{(r+1)(2r^2 + 5r + 2)}{2(2r+1)(2r+3)} = \frac{(r+1)(r+2)}{2(2r+3)}$$

L.H.S. = R.H.S.

So, $\dfrac{1^2}{1.3} + \dfrac{2^2}{3.5} + ... + \dfrac{n^2}{(2n-1)(2n+1)} = \dfrac{n(n+1)}{2(2n+1)}$.

Problem 20. *Prove by mathematical induction*

$$1 + 2 + 3 + 4 + ... + n = \frac{n(n+1)}{2}.$$

Sol. Let $P(n) = 1 + 2 + 3 + ... + n = \dfrac{n(n+1)}{2}$

Basis of Induction. For $n = 1$,

$$P(1) = 1 = \frac{1(1+1)}{2} = 1$$

It is true for $n = 1$. ...(*i*)

Induction Step. For $n = r$,

$$P(r) = 1 + 2 + 3 + ... + r = \frac{r(r+1)}{2} \quad ...(ii)$$

Now, for $n = r + 1$,

$$P(r+1) = 1 + 2 + 3 + 4 + ... + r + (r+1) = \frac{(r+1)(r+2)}{2}$$

Consider, $\text{L.H.S.} = 1 + 2 + 3 + ... + r + (r+1)$

$$= \frac{r(r+1)}{2} + (r+1) = \frac{r(r+1) + 2(r+1)}{2}$$

$$= \frac{r+1}{2}[r+2] = \frac{(r+1)(r+2)}{2} = \text{R.H.S.} \quad ...(iii)$$

As $P(r)$ is true. Hence, $P(r + 1)$ is also true.

From (*i*), (*ii*) and (*iii*), we conclude that

$$1 + 2 + 3 + 4 + ... + n = \frac{n(n+1)}{2}. \text{ Hence proved.}$$

Problem 21. *Prove that $n(n + 1)(n + 2)$ is a multiple of 6.*

Sol. $P(n) = n(n + 1)(n + 2)$

Basis of Induction. For $n = 1$,

$$P(1) = 1 . (1 + 1)(1 + 2) = 1 . 2 . 3 = 6$$

which is true.

Induction Step. For $n = r$,

$$P(r) = r(r + 1)(r + 2) \text{ be a multiple of 6}$$

For $n = r + 1$,

$$P(r + 1) = (r + 1)(r + 2)(r + 3) \text{ is also a multiple of 6.}$$

Now, $P(r + 1) = (r + 1)(r + 2)(r + 3) = (r + 1)(r + 2)\, r + 3(r + 1)(r + 2)$

$$= r(r + 1)(r + 2) + 3(2r')$$

where $(r + 1)(r + 2) = \text{even} = 2r' = r(r + 1)(r + 2) + 6r'$

which is a multiple of 6.

Thus, $P(r + 1)$ is true as $P(r)$ is true.

$\therefore$ $P(n)$ is true for all natural numbers n

i.e., $n(n + 1)(n + 2)$ is a multiple of 6. Hence proved.

Problem 22. *Prove that $\forall n \in N$*

$$\frac{1}{5}n^5 + \frac{1}{3}n^3 + \frac{7}{15}n \text{ is a natural number.}$$

Sol. Let $P(n) = \frac{1}{5}n^5 + \frac{1}{3}n^3 + \frac{7}{15}n$ is a natural number.

Basis of Induction. For $n = 1$,

$$\frac{1}{5} + \frac{1}{3} + \frac{7}{15} = 1, \text{ which is a natural number.}$$

It is true for $n = 1$.

Induction Step. For $n = r$,

$$\left(\frac{1}{5}r^5 + \frac{1}{3}r^3 + \frac{7}{15}r\right) \text{ is a natural number} \quad ...(i)$$

For $n = r + 1$,

$$\frac{1}{5}(r+1)^5 + \frac{1}{3}(r+1)^3 + \frac{7}{15}(r+1)$$

$$= \frac{1}{5}(r^5 + 5r^4 + 10r^3 + 10r^2 + 5r + 1) + \frac{1}{3}(r^3 + 3r^2 + 3r + 1) + \frac{7}{15}(r+1)$$

$$= \left(\frac{1}{5}r^5 + \frac{1}{3}r^3 + \frac{7}{15}r\right) + (r^4 + 2r^3 + 3r^2 + 2r) + 1$$

= (a natural number) + (a natural number) + 1 [Using (*i*)]

= a natural number

As (P(r)) is true, hence P(r + 1) is also true.

Hence, by induction, P(n) is true $\forall n \in$ N. Hence proved.

Problem 23. *Prove that $n(n + 1)(2n + 1)$ is divisible by 6.*

Sol. Let P(n) = $n(n + 1)(2n + 1)$ is divisible by 6.

Basis of Induction. For $n = 1$,

P(1) = 1(1 + 1)(2 + 1) = 1 . 2 . 3 = 6 which is true for $n = 1$ ...(*i*)

Induction Step. For $n = r$,

P(r) = $r(r + 1)(2r + 1)$ be divisible by 6 ...(*ii*)

For $n = r + 1$,

$$P(r + 1) = (r + 1)(r + 2)(2r + 3)$$

$$\text{L.H.S.} = (r + 1)(r + 2)(2r + 3) = 2r^3 + 9r^2 + 13r + 6$$

$$= (2r^3 + 3r^2 + r) + (6r^2 + 12r + 6)$$

$$= r(r + 1)(2r + 1) + 6(r^2 + 2r + 1)$$

which is divisible by 6. [From (*ii*)]

As (P(r)) is true, hence P(r + 1) is also true.

∴ P(n) is true for all natural numbers n

i.e., $n(n + 1)(2n + 1)$ is divisible by 6. Hence proved.

4

Relations

BINARY RELATION

Let P and Q be two non-empty sets. A binary relation R is defined to be a subset of P × Q from a set P to Q. If $(a, b) \in R$ and $R \subseteq P \times Q$ then a is related to b by R *i.e.*, aRb. If sets P and Q are equal, then we say $R \subseteq P \times P$ is a relation on P *e.g.*,

(*i*) Let $A = \{a, b, c\}$

$B = \{r, s, t\}$

Then $R = \{(a, r), (b, r), (b, t), (c, s)\}$

is a relation from A to B.

(*ii*) Let $A = \{1, 2, 3\}$ and $B = A$

$R = \{(1, 1), (2, 2), (3, 3)\}$

is a relation (equal) on A.

Example 1. *If a set A has n elements, how many relations are there from A to A.*

Sol. If a set A has n elements, A × A has n^2 elements. So, there are 2^{n^2} relations from A to A.

Example 2. *If A has m elements and B has n elements. How many relations are there from A to B and vice-versa ?*

Sol. There are $m \times n$ elements, hence there are $2^{m \times n}$ relations from A to B.

Example 3. *If a set A = {1, 2}. Determine all relations from A to A.*

Sol. There are $2^2 = 4$ elements *i.e.*, {(1, 2), (2, 1), (1, 1), (2, 2)} in A × A. So, there are $2^4 = 16$ relations from A to A. *i.e.*,

{(1, 2), (2, 1), (1, 1), (2, 2)}, {(1, 2), (2, 1)}, {(1, 2), (1, 1)}, {(1, 2), (2, 2)}, {(2, 1), (1, 1)}, {(2, 1), (2, 2)}, {(1, 1), (2, 2)}, {(1, 2), (2, 1), (1, 1)}, {(1, 2), (1, 1), (2, 2)}, {(2, 1), (1, 1), (2, 2)}, {(1, 2), (2, 1), (2, 2)}, {(1, 2), (2, 1), (1, 1), (2, 2)} and ϕ.

DOMAIN OF RELATION

The domain of relation R is the set of elements in P which are related to some element in Q or it is the set of all first entries of the ordered pairs in R. It is denoted by DOM (R).

RANGE OF RELATION

The range of a relation R is the set of elements in Q which are related to some element in P or it is the set of all second entries of the ordered pairs in R. It is denoted by RAN (R).

For example : Let A = {1, 2, 3, 4}, B = {a, b, c, d} and R = {(1, a), (1, b), (1, c), (2, b), (2, c), (2, d)}.

Then $\quad$ DOM (R) = {1, 2}

RAN (R) = {a, b, c, d}

COMPLEMENT OF A RELATION

Consider a relation R from a set A to B. The complement of relation R denoted by $\overline{R}$ is a relation from A to B such that

$$\overline{R} = \{(a, b) : (a, b) \notin R\}.$$

Example 4. *Consider the relation R from X to Y*

$$X = \{1, 2, 3\} ; Y = \{8, 9\} \text{ and } R = \{(1, 8), (2, 8), (1, 9), (3, 9)\}.$$

Find the complement of relation R.

Sol. First we find the universal relation X × Y *i.e.,*

$$X \times Y = \{(1, 8), (2, 8), (3, 8), (1, 9), (2, 9), (3, 9)\}$$

Now, we find the complement relation $\overline{R}$ w.r.t. X × Y

$$\overline{R} = \{(3, 8), (2, 9)\}.$$

Example 5. *Let A = {7, 8, 9} and B = {k, l, m, n} and R is relation from A to B.*

$$R = \{(7, k), (8, k), (8, l), (8, m), (9, m), (9, n)\}.$$

Determine its complement.

Sol. The relation A × B from the set A to B is

$$A \times B = \{(7, k), (7, l), (7, m), (7, n), (8, k), (8, l), (8, m), (8, n), (9, k), (9, l), (9, m), (9, n)\}$$

The complement of relation R w.r.t. A × B is

$$\overline{R} = \{(7, l), (7, m), (7, n), (8, n), (9, k), (9, l)\}.$$

INVERSE OF A RELATION

Consider a relation R from a set A to B. The inverse of a relation R, denoted by R^{-1}, is a relation from B to A such that $bR^{-1} a$ iff aRb *i.e.,*

$$R^{-1} = \{(b, a) : (a, b) \in R\}.$$

Example 6. *Consider the relation '≤' on the set A = {2, 3, 4, 5}. Determine its inverse.*

Sol. The relation '≤' is defined by $(a,b) \in R$ if $a \le b$, $a, b \in A$. Then

$$R = \{(2, 2), (2, 3), (2, 4), (2, 5), (3, 3), (3, 4), (3, 5), (4, 4), (4, 5), (5, 5)\}$$

and

$$R^{-1} = \{(2, 2), (3, 2), (4, 2), (5, 2), (3, 3), (4, 3), (5, 3), (4, 4), (5, 4), (5, 5)\}.$$

Example 7. *Consider the following relation R on the set of +ve integers. Find its inverse*

$$R = \{(1, 1), (1, 2), (1, 3), (2, 1), (3, 1) (3, 2), (2, 3)\}.$$

Sol. The inverse of relation R is

$$R^{-1} = \{(1, 1), (2, 1), (3, 1), (1, 2), (1, 3), (2, 3), (3, 2)\}.$$

REPRESENTATION OF RELATIONS

Relations can be represented in many ways. Some of which are as follows :

1. Relation as a Matrix. Let $P = \{a_1, a_2, \ldots, a_m\}$ and $Q = \{b_1, b_2, \ldots, b_n\}$ are finite sets, containing m and n number of elements respectively. R is a relation from P to Q. The relation R can be represented by $m \times n$ matrix $M = [M_{ij}]$, defined as

$$M_{ij} = \begin{cases} 0 & \text{if } (a_i, b_j) \notin R \\ 1 & \text{if } (a_i, b_j) \in R \end{cases}$$

e.g., Let $P = \{1, 2, 3, 4\}$, $Q = \{a, b, c, d\}$

and $R = \{(1, a), (1, b), (1, c), (2, b), (2, c), (2, d)\}$.

The matrix of the relation R is shown in Fig. 1.

$$M_R = \begin{array}{c} \\ 1 \\ 2 \\ 3 \\ 4 \end{array} \begin{array}{c} \begin{array}{cccc} a & b & c & d \end{array} \\ \begin{bmatrix} 1 & 1 & 1 & 0 \\ 0 & 1 & 1 & 1 \\ 0 & 0 & 0 & 0 \\ 0 & 0 & 0 & 0 \end{bmatrix} \end{array}$$

Fig. 1.

2. Relation as a Directed Graph. If P is a finite set and R is a relation on P, R can be represented as a directed graph as follows :

Draw a small circle for each element of P and label the circle with corresponding element of P. These circles are called vertices. Draw a directed line from vertex a_i to vertex a_j if a_i R a_j. These directed lines are called edges. The obtained picture is the directed graph.

For example : Let $P = \{1, 2, 3, 4\}$

and $R = \{(1, 1), (1, 2), (2, 1), (2, 2), (3, 2), (3, 4), (4, 3), (4, 4)\}$

which is a relation on P.

Then the directed graph of relation R is shown in Fig. 2.

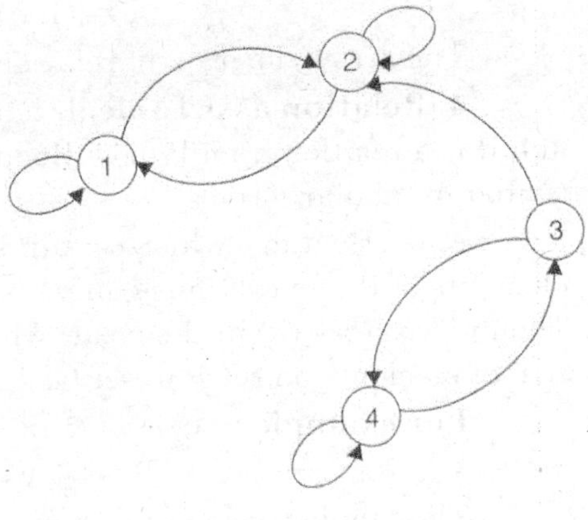

Fig. 2.

From Fig. 2, we see that the edges in the directed graph of R corresponds to the pairs in R, and the vertices correspond to the elements in the set P.

Example 8. *Find the relations determined by the following figures.*

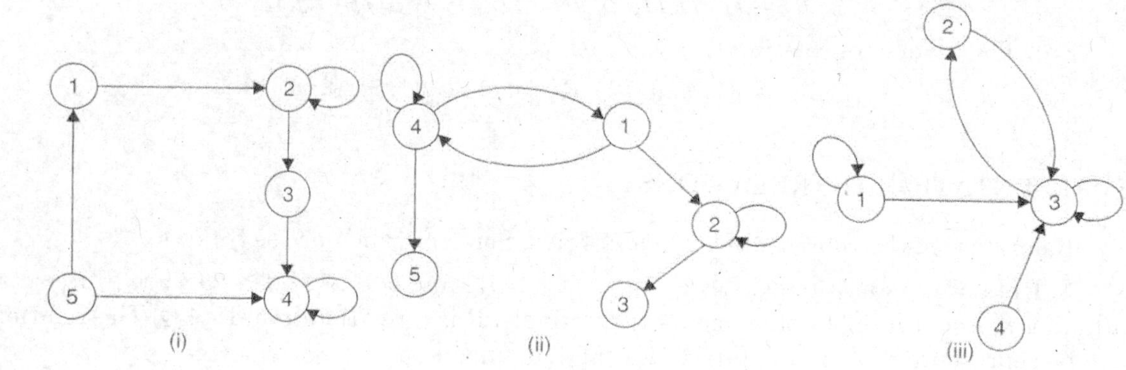

Fig. 3.

Sol. The relations determined by above figures are as follows :

(*i*) R = {(1, 2), (2, 2), (2, 3), (3, 4), (4, 4), (5, 4), (5, 1)}

(*ii*) R = {(1, 2), (1, 4), (2, 2), (2, 3), (4, 1), (4, 4), (4, 5)}

(*iii*) R = {(1, 1), (1, 3), (2, 3), (3, 2), (3, 3), (4, 3)}.

3. Relation as an Arrow Diagram. If P and Q are finite sets and R is a relation from P to Q. Relation R can be represented as an arrow diagram as follows.

Draw two ellipses for the sets P and Q. Write down the elements of P and elements of Q columnwise in three ellipses. Then draw an arrow from first ellipse to second ellipse if a is related to b and $a \in$ P and $b \in$ Q.

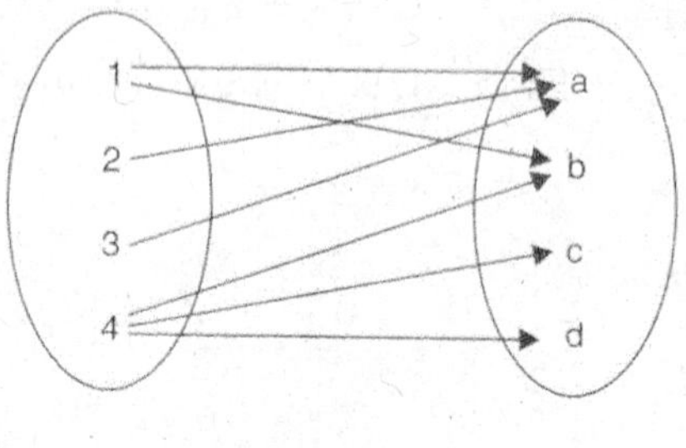

Fig. 4.

For example : Let P = {1, 2, 3, 4}, Q = {a, b, c, d}

and R = {(1, a), (2, a), (3, a), (1, b), (4, b), (4, c), (4, d)}

The arrow diagram of relation R is shown in Fig. 4.

4. Relation as a Table. If P and Q are finite sets and R is a relation from P to Q. Relation R can be represented in tabular form.

Make the table which contains rows equivalent to elements of P and columns equivalent to elements of Q. Then place cross (X) in the boxes which represents relation of elements on set P to set Q.

	x	y	z	k
1	×	×		
2			×	
3			×	
4				×

Fig. 5.

For example : Let P = {1, 2, 3, 4}, Q = {x, y, z, k}

and R = {(1, x), (1, y), (2, z), (3, z), (4, k)}.

The tabular form of relation is as shown in Fig. 5.

COMPOSITION OF RELATIONS

Consider a relation R_1 from A to B and R_2 be a relation from B to C. The composition of relation R_1 and R_2 denoted by R_1 o R_2, is the relation from A to C and is defined by R_1 o R_2 = {(a, c) : $(a, b) \in R_1$ and $(b, c) \in R_2$ for some $b \in$ B}.

Example 9. *Let P and Q be the relations on set A = {1, 2, 3, 4} defined by*

P = {(1, 2), (2, 2), (2, 3), (2, 4), (3, 2), (4, 2), (4, 3)}

and *Q = {(2, 2), (2, 3), (3, 2), (3, 3), (3, 4), (4, 1), (4, 2)}*

Find (i) P o P (ii) P o Q (iii) P o P o Q.

Sol. (*i*) P o P = {(1, 2), (1, 3), (1, 4), (2, 3), (2, 4), (2, 2), (3, 2), (3, 3), (3, 4), (4, 2), (4, 3), (4, 4)}

(*ii*) P o Q = {(1, 2), (1, 3), (2, 2), (2, 3), (2, 4), (2, 1), (3, 2), (3, 3), (4, 2), (4, 3), (4, 4)}

(*iii*) P o P = {(1, 2), (1, 3), (1, 4), (2, 2), (2, 3), (2, 4), (3, 3), (4, 2), (3, 2), (3, 4), (4, 3), (4, 4)}

P o P o Q = {(1, 2), (1, 3), (1, 4), (1, 1), (2, 2), (2, 3), (2, 4), (2, 1), (3, 1), (3, 2), (3, 3), (3, 4), (4, 1), (4, 2), (4, 3)}.

Example 10. *Let X = {4, 5, 6}, Y = {a,b, c} and Z = {l, m, n}. Consider the relation R_1 from X to Y and R_2 from Y to Z.*

R_1 = {(4, a), (4, b), (5, c), (6, a), (6, c)}

R_2 = {(a, l), (a, n), (b, l), (b, m), (c, l), (c, m), (c, n)}

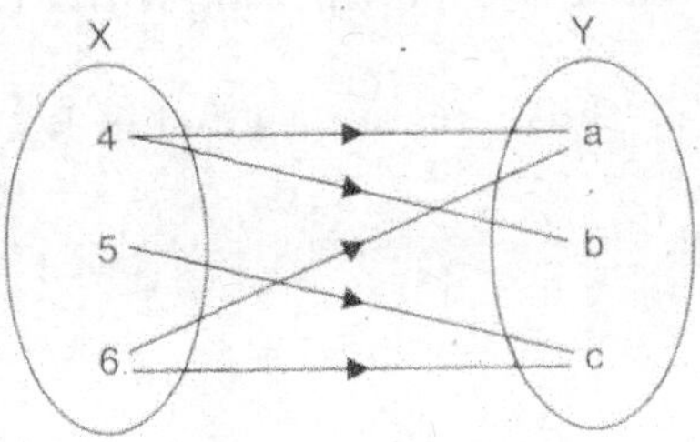

Fig. 6.

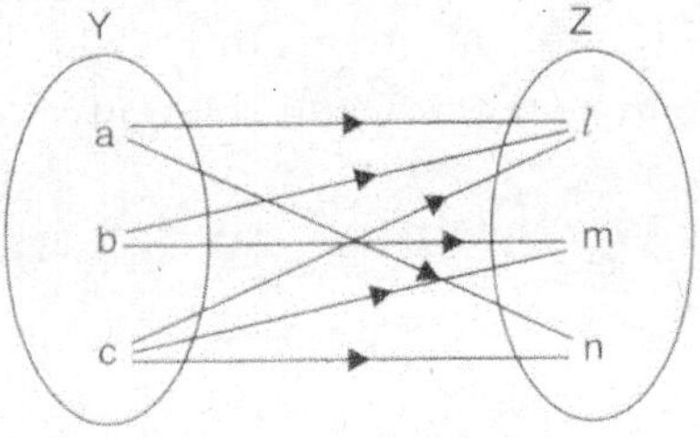

Fig. 7.

Find the composition relation (i) R_1 o R_2 (ii) R_1 o R_1^{-1}.

Sol. (*i*) The composition relation R_1 o R_2 is as shown in Fig. 8.

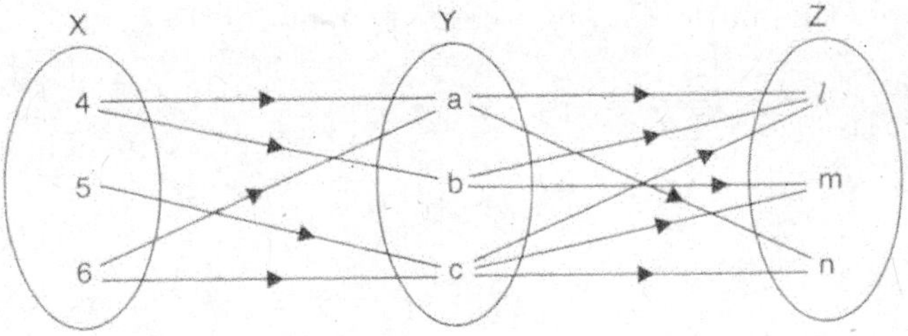

Fig. 8. R_1 o R_2.

R_1 o R_2 = {(4, *l*), (4, *n*), (4, *m*), (5, *l*), (5, *m*), (5, *n*), (6, *l*), (6, *m*), (6, *n*)}

(*ii*) The composition relation R_1 o R^{-1} is shown in Fig. 9.

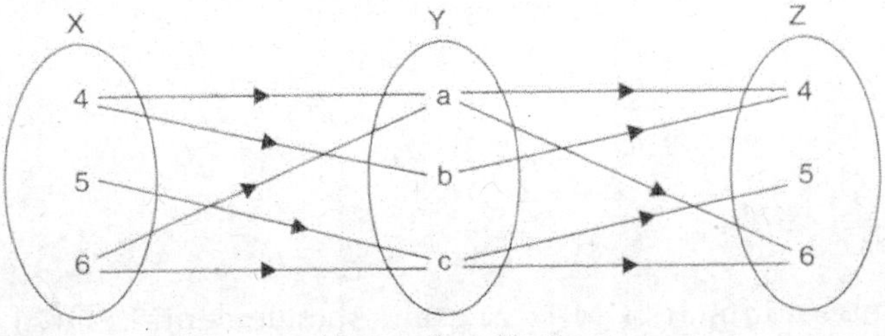

Fig. 9. R_1 o R_1^{-1}.

R_1 o R_1^{-1} = {(4, 4), (5, 5), (5, 6), (6, 4), (6, 5), (4, 6), (6, 6)}.

Example 11. *Let* $P = \{2, 3, 4, 5\}$. *Consider the relation R and S on P defined by*

$$R = \{(2, 2), (2, 3), (2, 4), (2, 5), (3, 4), (3, 5), (4, 5), (5, 3)\}$$

and

$$S = \{(2, 3), (2, 5), (3, 4), (3, 5), (4, 2), (4, 3), (4, 5), (5, 2), (5, 5)\}.$$

Find the matrices of the above relations.

Use matrices to find the following compositions of the relations R and S.

(*i*) *R o S* (*ii*) *R o R* (*iii*) *S o R.*

Sol. The matrices of the relations R and S are as shown in Fig. 10.

$$M_R = \begin{array}{c} \\ 2 \\ 3 \\ 4 \\ 5 \end{array}\begin{array}{c} \begin{array}{cccc} 2 & 3 & 4 & 5 \end{array} \\ \begin{bmatrix} 1 & 1 & 1 & 1 \\ 0 & 0 & 1 & 1 \\ 0 & 0 & 0 & 1 \\ 0 & 1 & 0 & 0 \end{bmatrix} \end{array} \quad \text{and} \quad M_S = \begin{array}{c} \\ 2 \\ 3 \\ 4 \\ 5 \end{array}\begin{array}{c} \begin{array}{cccc} 2 & 3 & 4 & 5 \end{array} \\ \begin{bmatrix} 0 & 1 & 0 & 1 \\ 0 & 0 & 1 & 1 \\ 1 & 1 & 0 & 1 \\ 1 & 0 & 0 & 1 \end{bmatrix} \end{array}$$

Fig. 10.

(*i*) To obtain the composition of relations R and S. First multiply M_R with M_S to obtain the matrix $M_R \times M_S$ as shown in Fig. 11.

The non-zero entries in the matrix $M_R \times M_S$ tells the elements related in R o S. So,

$$M_R \times M_S = \begin{array}{c} \\ 2 \\ 3 \\ 4 \\ 5 \end{array}\begin{array}{c} \begin{array}{cccc} 2 & 3 & 4 & 5 \end{array} \\ \begin{bmatrix} 2 & 2 & 1 & 4 \\ 2 & 1 & 0 & 2 \\ 1 & 0 & 0 & 1 \\ 0 & 0 & 1 & 1 \end{bmatrix} \end{array}$$

Fig. 11.

Hence, the composition R o S of the relation R and S is

$$R \text{ o } S = \{(2, 2), (2, 3), (2, 4), (3, 2), (3, 3), (4, 2), (4, 5), (5, 2), (5, 3), (5, 4), (5, 5)\}.$$

(*ii*) First, multiply the matrix M_R by itself, as shown in Fig. 12.

$$M_R \times M_S = \begin{array}{c} \\ 2 \\ 3 \\ 4 \\ 5 \end{array}\begin{array}{c} \begin{array}{cccc} 2 & 3 & 4 & 5 \end{array} \\ \begin{bmatrix} 1 & 2 & 2 & 3 \\ 0 & 1 & 0 & 1 \\ 0 & 1 & 0 & 0 \\ 0 & 0 & 1 & 1 \end{bmatrix} \end{array}$$

Fig. 12.

Hence, the composition R o R of the relation R is

$$R \text{ o } R = \{(2, 2), (3, 2), (3, 3), (3, 4), (4, 2), (4, 5), (5, 2), (5, 3), (5, 5)\}.$$

(*iii*) Multiply the matrix M_S with M_R to obtain the matrix $M_S \times M_R$ as shown in Fig. 13.

$$M_S \times M_R = \begin{array}{c} \\ 2 \\ 3 \\ 4 \\ 5 \end{array}\begin{array}{c} \begin{array}{cccc} 2 & 3 & 4 & 5 \end{array} \\ \begin{bmatrix} 0 & 1 & 1 & 1 \\ 0 & 1 & 0 & 1 \\ 1 & 2 & 2 & 2 \\ 1 & 1 & 1 & 1 \end{bmatrix} \end{array}$$

Fig. 13.

The non-zero entries in matrix $M_S \times M_R$ tells the elements related in S o R.

Hence, the composition S o R of the relation S and R is

$$S \text{ o } R = \{(2, 4), (2, 5), (3, 3), (3, 4), (3, 5), (4, 2), (4, 4), (4, 5), (5, 2), (5, 3), (5, 4), (5, 5)\}.$$

PATH IN RELATIONS

Consider a relation R on set A. A path of length n in R from a to b, is a finite sequence $a, Y_1, Y_2, Y_3, \ldots\ldots, Y_{n-1}, b$ which begins with a and ends with b such that

$$aRY_1, Y_1RY_2, Y_2RY_3, \ldots\ldots, Y_{n-1}Rb$$

The path which has length n must have $n + 1$ elements of A. The elements may be distinct or same .

A path that begins and ends at the same vertex is called a **cycle.**

As we have seen earlier that a relation can be uniquely represented by a directed graph. So, the path in a directed graph is a sequence of edges in the indicated direction. Hence, the path length is the number of edges in the path.

Example. *Consider the graph shown in Fig. 13(i).*

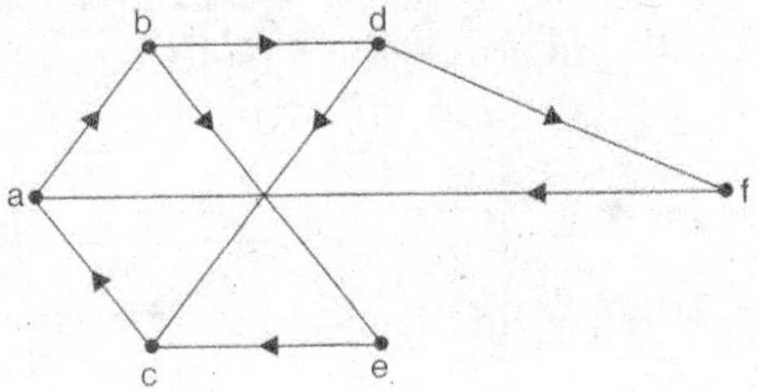

Fig. 13(*i*).

The graph has a path $P_1 = a, b, d, f$ of length 3.
The graph has a path $P_2 = a, b, e, c, d, f$ of length 5.
The graph has a cycle $P_3 = a, b, d, f, a$ of length 4.
The graph has a cycle $P_4 = a, b, e, c, d, f, a$ of length 6.

Example 12. *Consider the directed graph as shown in Fig. 13 (ii).*

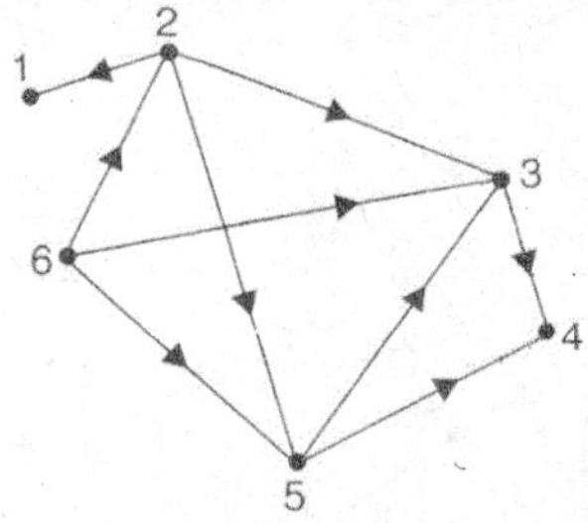

Fig. 13(*ii*).

Determine all the paths of length 1 and 2.

Sol. Paths that have length 1 are as follows :

$P_1 = 1, 2$ $P_2 = 2, 3$ $P_3 = 3, 4$ $P_4 = 5, 4$
$P_5 = 6, 5$ $P_6 = 6, 3$ $P_7 = 2, 5$ $P_8 = 5, 3$ $P_9 = 6, 2.$

Paths that have length 2 are as follows :

$P_{10} = 1, 2, 3$ $P_{11} = 6, 2, 3$ $P_{12} = 6, 3, 4$ $P_{13} = 6, 5, 3$
$P_{14} = 6, 5, 4$ $P_{15} = 2, 5, 4$ $P_{16} = 2, 5, 3$ $P_{17} = 2, 3, 4$ $P_{18} = 6, 2, 5.$

Paths in a relation R can be used to define following relation on the set A.

R^n. The relation R^n on the set A is defined as a path of length n from a to b in R and is denoted by R^n b.

R^∞. The relation R^∞ on the set A is defined as some path from a to b in R. The length of the path will depend on a and b. It is also called the connectivity relation for R.

$R^n(x)$. It is defined as a set consisting of all the vertices that can be reached from x by means of a path in R of length n.

$R^\infty(x)$. It is defined as a set consisting of all the vertices that can be reached from x by some path in R.

R^*. Consider a set A having n elements and let R is a relation on A. Then relation R^* of R is defined as a relation such that aR^*b if $a = b$ or $a\ R^\infty b$. It is also called the reachability relation. The meaning of this is that b is reachable from a if either $b = a$ or there is some path from a to b.

Example 13. *Consider the set A = {a, b, c, d, e}. Let R be the relation whose diagraph is shown in Fig. 13(iii). Determine the diagraph of the relation*

(i) R^2 *(ii)* R^3 *on A.*

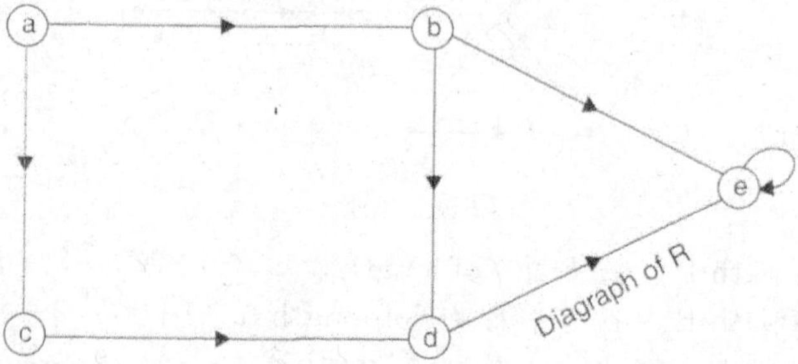

Fig. 13(*iii*).

Sol. (*i*) To obtain the diagraph of the relation R^2, we will find all lines that connects two vertices in R^2, when there is a path of length two connecting those vertices in R. The diagraph of R^2 is shown in Fig. 13(*iv*).

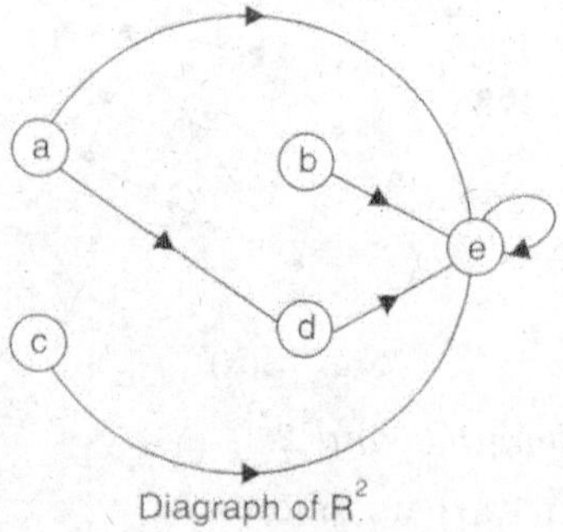

Fig. 13(*iv*).

(*ii*) To obtain the diagraph of the relation R^3, we will find all paths of length 3 in R from any vertex to any other vertex and then join those vertices in R^3 by a direct line. The diagraph of R^3 is shown in Fig. 13(*v*).

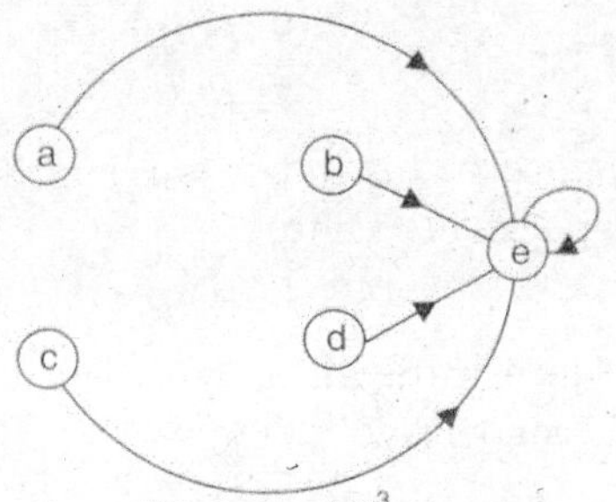

Diagraph of R^3

Fig. 13(*v*).

Example 14. *Consider the set A = {4, 5, 6, 7} and the relation R on A is given by*

$$R = \{(4, 5), (5, 6), (5, 7), (6, 6), (6, 7), (7, 6), (7, 7)\}.$$

Determine (*a*) R^3 (*b*) R^∞.

Sol. The diagraph of R is shown in Fig. 13(*vi*).

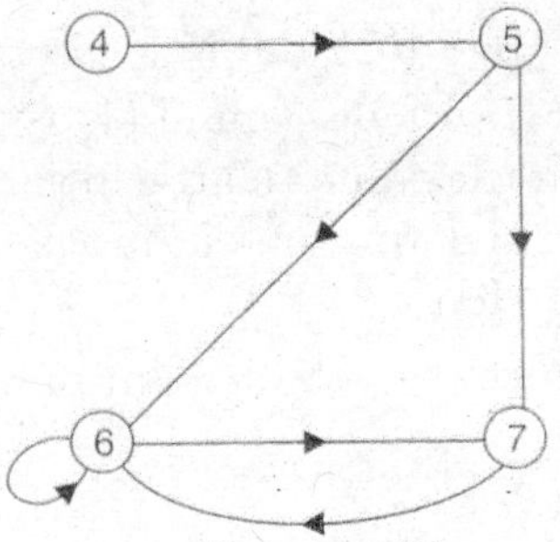

Diagraph of R

Fig. 13(*vi*)

(*a*) Now find all the paths of length 3 from any vertex to any other vertex in R and write these vertices by pair vertices in R^3. Thus, the R^3 is given by

$$R^3 = \{(4, 6), (4, 7), (5, 6), (5, 7), (6, 6), (6, 7), (6, 7), (7, 7)\}.$$

(*b*) To determine R^∞, find all the ordered pairs of vertices for which there is a path of any length from any vertex to any other vertex of the set A. Thus, R^∞ is given by

$$R^\infty = \{(4, 5), (4, 6), (4, 7), (5, 6), (5, 7), (6, 6), (6, 7), (7, 6), (7, 7)\}.$$

Example 15. *Consider the diagraph of the relation R on set A = {a, b, c, d, e, f, g} as shown in Fig. 13(vii).*

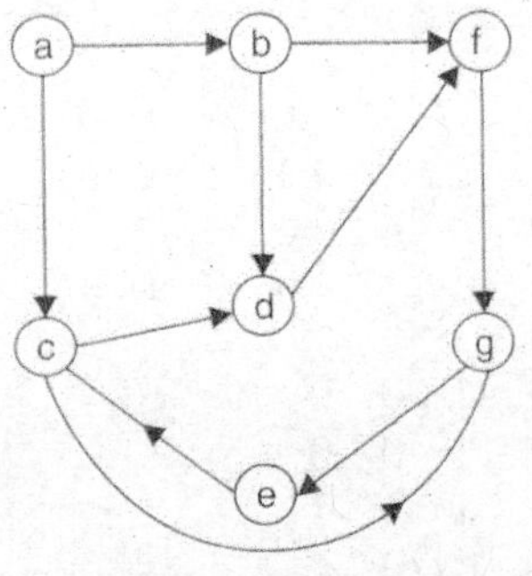

Fig. 13(*vii*).

Determine (*a*) $R^2(d)$ (*b*) $R^2(g)$

(*c*) $R^{\infty}(e)$ (*d*) $R^{\infty}(b)$.

Sol. To find $R^2(d)$ and $R^2(g)$ determine all the vertices which can be reached from d and g by means of a path of length 2 respectively. Thus,

(*a*) $R^2(d) = \{g\}$ (*b*) $R^2(g) = \{c\}$

Now to find $R^{\infty}(e)$ and $R^{\infty}(b)$, determine all the vertices which can be reached from e and b by means of a path of some length respectively. Thus,

(*c*) $R^{\infty}(e) = \{c, d, f, g\}$ (*d*) $R^{\infty}(b) = \{d, f, g, e, c\}$.

Computation of $M_R{}^2$ using M_R. Consider a relation R on the set $A = \{a_1, a_2, ..., a_n\}$.

Also let M_R be the matrix representation of R, we can compute the matrix $M_R{}^2$ of R^2 from M_R.

Theory. *Show that if R is a relation on the set $A = \{a_1, a_2, ..., a_n\}$ and M_R is the corresponding matrix of R, Then $M_{R^2} = M_R \otimes M_R$ {The symbol $\otimes$ means not an ordinary matrix product}.*

Proof. Let us consider that $M_R = [b_{ij}]$ and $M_R{}^2 = [C_{ij}]$

Now from definition, we have i, jth element of $M_R \otimes M_R$ is equal to 1 iff row i of first M_R and column j of second M_R have 1 in the same relative position (say k). Thus, $b_{ik} = 1$ and $b_{kj} = 1$ for some k, where $1 \le k \le n$. From the definition of matrix and using the above constraints we have $a_i R a_k$ and $a_k R a_j$. Thus, $a_i R^2 a_j$. Hence $c_{ij} = 1$.

In the above, we have shown that i, jth element of $M_R \otimes M_R$ is 1 iff $c_{ij} = 1$. Thus, $M_R \otimes M_R = M_{R^2}$.

Example 16. *Consider the set $A = \{1, 2, 3, 4\}$ and the relation*

$R = \{(1, 2), (2, 1), (2, 2), (3, 2), (3, 3), (3, 4), (4, 4)\}$ on set A. Determine M_{R^2} from R.

Sol. The matrix representation of R is given by

$$M_R = \begin{array}{c} \\ 1 \\ 2 \\ 3 \\ 4 \end{array} \begin{array}{c} \begin{array}{cccc} 1 & 2 & 3 & 4 \end{array} \\ \begin{bmatrix} 0 & 1 & 0 & 0 \\ 1 & 1 & 0 & 0 \\ 0 & 1 & 1 & 1 \\ 0 & 0 & 0 & 1 \end{bmatrix} \end{array}$$

Now, compute M_{R^2} using M_R, which is given as follows :

$$M_{R^2} = M_R \otimes M_R = \begin{bmatrix} 0 & 1 & 0 & 0 \\ 1 & 1 & 0 & 0 \\ 0 & 1 & 1 & 1 \\ 0 & 0 & 0 & 1 \end{bmatrix} \otimes \begin{bmatrix} 0 & 1 & 0 & 0 \\ 1 & 1 & 0 & 0 \\ 0 & 1 & 1 & 1 \\ 0 & 0 & 0 & 1 \end{bmatrix}$$

$$M_{R^2} = \begin{bmatrix} 1 & 1 & 0 & 0 \\ 1 & 1 & 0 & 0 \\ 1 & 1 & 1 & 1 \\ 0 & 0 & 0 & 1 \end{bmatrix}$$

Thus, the result M_{R^2} obtained by using the matrix is same as if we obtain it by using diagraph of R.

Computation of the Matrix of R^n. Consider a relation R on a finite set A. Also let $n \geq 2$. The matrix M_{R^n} of R is given by

$$M_{R^n} = M_R \otimes M_R \otimes M_R......\otimes M_R \text{ (} n \text{ times)}$$

Computation of the Matrix of R^∞. Consider a relation R on a finite set A. A vertex is $aR^\infty b$ iff both the vertices are connected by a path in R of length n. Thus, $aR^\infty b$ iff aRb or aR^2b or aR^3b or......Hence,

$$M_{R^\infty} = M_R \vee M_R{}^2 \vee M_R{}^3 \vee ......$$

Note 1. If R and S are two relations on A, then $R \cup S$ is a relation such that $a(R \cup S)b$ iff aRb or aSb.

2. The relation R^∞ can be defined also as $R^\infty = R \cup R^2 \cup R^3 \cup R^4...... = \bigcup_{n=1}^{\infty} R^n$.

3. The matrix of $R \cup S$ denoted by $\{M_{R \cup S}$ can be defined as $M_{R \cup S} = M_R \vee M_S$.

Computation of the Matrix of R*. As discussed earlier, we know R* is a reachability relation, which mean b is reachable from a if either $b = a$ or there is a path from a to b. Thus, M_{R^*} is defined as :

$$M_{R^*} = I_n \vee M_R{}^\infty$$

[Here I_n is the identity matrix of $n \times n$ and n is the number of elements in A]

or putting the value of M_{R^∞}, we have

$$M_{R^*} = I_n \vee M_R \vee M_R{}^2 \vee M_R{}^3 \vee .... .$$

Example 17. *Consider the diagraph of the relation R on set A = {1, 2, 3, 4, 5} as shown in Fig. 13(viii).*

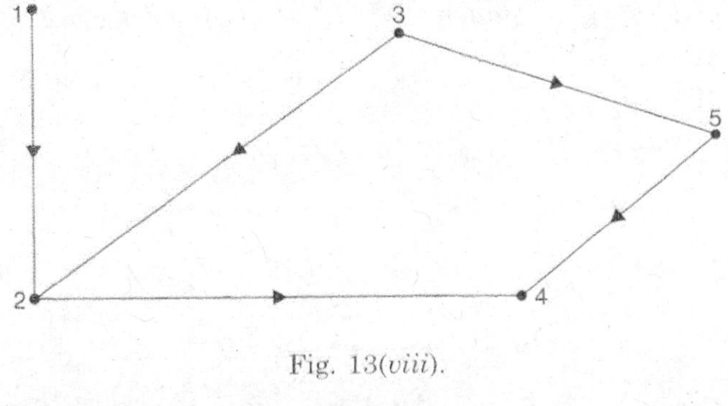

Fig. 13(*viii*).

*Determine the following (a) R** (b) M_{R^*}.

Sol. (*a*) The relation R* of R on set A is given by

R* = {(1, 1), (1, 2), (1, 4), (2, 2), (2, 4), (3, 2), (3, 3), (3, 4), (3, 5), (4, 4), (5, 4), (5, 5)}

The matrix M_{R^*} is obtained by finding the value of M_{R^∞} and then ORing with I_n (Identity matrix), which is as follows :

$$M_{R^\infty} = \begin{array}{c} \\ 1 \\ 2 \\ 3 \\ 4 \\ 5 \end{array}\begin{array}{c} \begin{array}{ccccc} 1 & 2 & 3 & 4 & 5 \end{array} \\ \begin{bmatrix} 0 & 1 & 0 & 1 & 0 \\ 0 & 0 & 0 & 1 & 0 \\ 0 & 1 & 0 & 1 & 1 \\ 0 & 0 & 0 & 0 & 0 \\ 0 & 0 & 0 & 1 & 0 \end{bmatrix} \end{array}$$

Now,
$$M_{R^*} = \begin{bmatrix} 1 & 0 & 0 & 0 & 0 \\ 0 & 1 & 0 & 0 & 0 \\ 0 & 0 & 1 & 0 & 0 \\ 0 & 0 & 0 & 1 & 0 \\ 0 & 0 & 0 & 0 & 1 \end{bmatrix} \vee \begin{bmatrix} 0 & 1 & 0 & 1 & 0 \\ 0 & 0 & 0 & 1 & 0 \\ 0 & 1 & 0 & 1 & 1 \\ 0 & 0 & 0 & 0 & 0 \\ 0 & 0 & 0 & 1 & 0 \end{bmatrix}$$

$$= \begin{bmatrix} 1 & 1 & 0 & 1 & 0 \\ 0 & 1 & 0 & 1 & 0 \\ 0 & 1 & 1 & 1 & 1 \\ 0 & 0 & 0 & 1 & 0 \\ 0 & 0 & 0 & 1 & 1 \end{bmatrix}$$

which is same as we obtained from R*.

COMPOSITION OF PATHS

Consider a relation R on a set A. Let $p_1 : x, a_1, a_2, a_3, \ldots, a_{n-1}$, Y be a path in the relation R whose length is n from x to y. Also let $p_2 : y, b_1, b_2, b_3, b_4, \ldots, b_{m-1}, z$ be a path in the relation R whose length is m from y to z.

Then, the composition of p_1 and p_2 denoted $p_2{}^0 p_1$ is a path $x, a_1, a_2, a_3, \ldots, v, b_1, b_2, b_3, \ldots, b_{m-1}, z$ of length $n + m$, which is a path from x to z.

Example 18. *Consider the diagraph of the relation R on the set* $A = \{a, b, c, d, e, f, g\}$ *as shown in Fig. 13 (ix). Also let* $p_1 : a, d, b, c$ *and* $p_2 : c, e, f, g$.

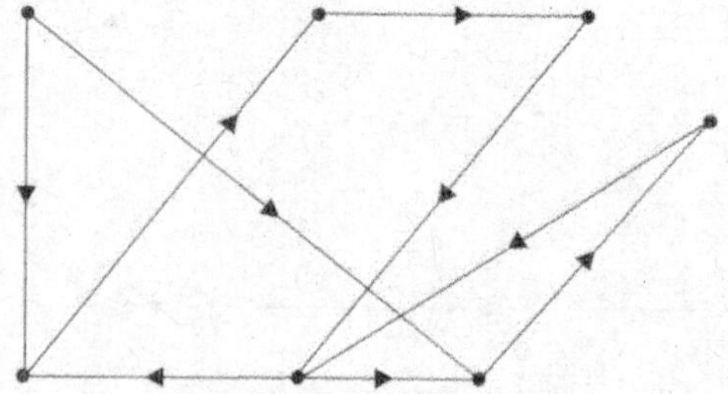

Fig. 13(*ix*)

Determine the composition on $p_2 \circ p_1$.

Sol. The composition of p_1 and p_2 is a path from a to g of length 6 given by

$$p_2 \circ p_1 : a, d, b, c, e, f, g.$$

COMPUTER REPRESENTATION OF RELATIONS

There are two most common methods for starting data items. These are :

1. Array representation
2. Linked list representation.

1. Array Representation

It refers to storing consecutive homogeneous data items in sequential storage locations in the memory.

Example. *The Fig. 13(x) shows the 10 integers stored in memory in consecutive memory locations.*

I_0	I_1	I_2	I_3	I_4	I_5	I_6	I_7	I_8	I_9

Fig. 13(*x*).

If A is the name of the array then the first, second, third, ..., data items can be stored and accessed in locations A[0], A[1], A[2], A[3],...., and so on.

Advantages. (*i*) We can access any data items I_i by simply supplying its index *i*. So direct access to any item.

(*ii*) Sorting and searching problems are very efficient.

Disadvantages. (*i*) Insertions are difficult as we have to move a large number of data items to new positions to accommodate the new data item.

(*ii*) Deletions are also difficult as we have to move a large number of data items to take the position vacated by the deleted item.

(*iii*) Once an array is declared and data is stored we cannot change the size of the array.

2. Linked List Representation

Here the data is represented by a linear list of nodes, each one is connected by a pointer or link with another node.

Every element consists of two parts called info and linked.

node

info	link

The info part contains the data item and link part contains the address of the next node. So, the nodes are in linear sequence but the data items that they represent may not be in same sequence.

Example. *The Fig. 13(xi) shows a linked list of 10 data items :*

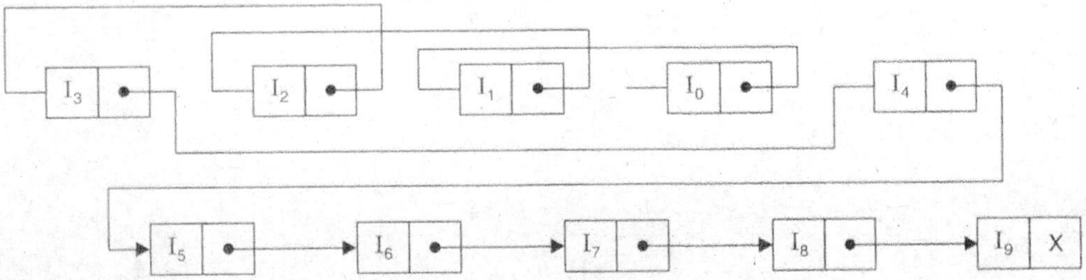

Fig. 13(*xi*)

The starting node is I_0 and the last node contain X in the link part, which tells that the linked list ends here.

Advantages. The insertions and deletions are very easy. No movement of records.

Disadvantages. (*i*) Sorting is very difficult.

(*ii*) No direct access to any element of the list.

REPRESENTATION OF RELATION IN COMPUTER

There are two possible ways to store a relation :

1. Two dimensional array representation.
2. Linked list representation.

1. Two Dimensional Array Representation

Consider a set S and relation R on the set S. The relation R can be represented by an $n \times n$ matrix M, where n is the number of elements in S. The matrix M is a boolean matrix containing only 0 and 1.

Example. *Let S = {1, 2, 3} and R = {(1, 1), (2, 2), (3, 3), (1, 2), (2, 1)}.*

Then the relation R on S can be represented by an $n \times n$ array.

$$M_R = \begin{matrix} & \begin{matrix} 1 & 2 & 3 \end{matrix} \\ \begin{matrix} 1 \\ 2 \\ 3 \end{matrix} & \begin{bmatrix} 1 & 1 & 0 \\ 1 & 1 & 0 \\ 0 & 0 & 1 \end{bmatrix} \end{matrix}$$

The value in matrix is 1 if the corresponding pair is in R otherwise zero.

2. Linked List

We can create a linked list that contains all the edges of the diagraph of a relation. The data can be represented by three arrays : TAIL, HEAD and NEXT. The Tail and Head array contains the beginning vertex and end vertex, for all rows, respectively. The array NEXT is an array of pointers from each edge to the next edge, which make the edge data into a linked list.

Example : *Consider the relation represented by diagraph as shown in Fig. 13(xii).*

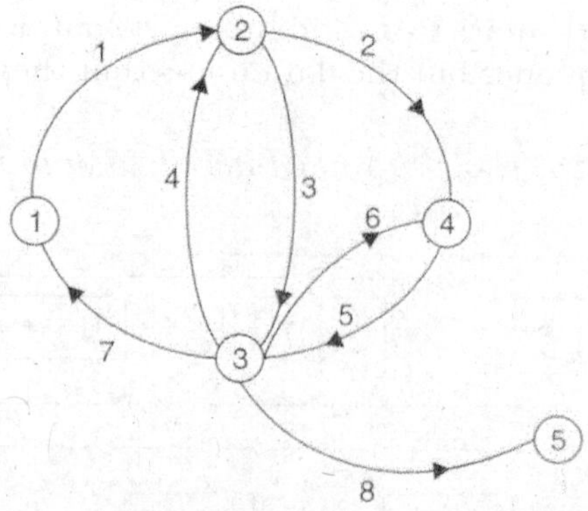

Fig. 13(*xii*).

The diagraph shown in Fig. 13(xii) is stored in linked list form as shown in Fig. 13 (xiii). The logical ordering coincides with the numbering of edges.

START

1

Index Tail	HEAD	NEXT
1	2	6
3	2	4
3	4	5
4	3	3
3	1	8
2	4	7
2	3	2
3	5	0

Fig. 13(*xiii*).

In the Fig. 13(*xiii*), the START contains the index 1 *i.e.*, the address of the first edge which is started at index 1. The NEXT (1) contains 6, so we locate the next edge in position 6. The NEXT [6] contains 7, so we locate the next edge at index 7. The NULL pointer.is represented by 0, which indicates no further data.

Disadvantages. If we want to investigate the edges that begin or end with a particular vertex, it is not possible in this type of storage scheme.

Modified linked list. The modified linked list representation consists of an additional array VERT having one position for each vertex in the diagraph. Here, VERT[J] is the index for each vertex *j*, in TAIL and HEAD. The VERT contains pointers to the edges.

For every vertex *j*, the pointers are arranged in the NEXT in such a way that they link together all edges leaving *j*, starting with the edge pointed to by VERT[*j*]. The last of these edges point to zero in each case. Thus, several linked lists of edges (one for each vertex) contained in the data array TAIL and HEAD.

Example. *Consider the relation represented by diagraph as shown in Fig. 13(xiv).*

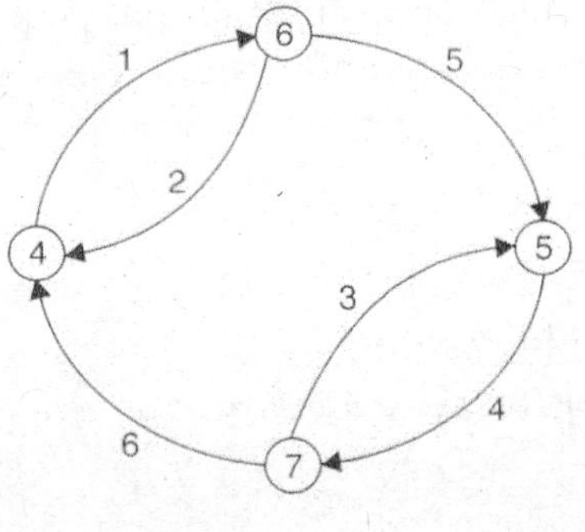

Fig. 13(*xiv*).

The diagraph shown in Fig.13(*xiv*) is stored in modified linked list representation as shown in Fig.13(*xv*).

	VERT
1	5
2	1
3	3
4	2

	TAIL	HEAD	NEXT
1	5	7	4
2	7	5	0
3	6	4	0
4	5	6	0
5	4	6	6
6	4	7	0
7			
8			

Fig. 13(*xv*).

In the modified linked list representation, VERT[1] contains 5, thus the first edge leaving vertex 1 must be stored in the 5th data position *i.e.*, the edge (4, 6). Again NEXT[5] contains 6, so the next edge leaving vertex 1 is (4, 7) in the 6th data position. Again NEXT[6] contains 0. Thus, we have come to the end of all those edges that begin at vertex 1.

Proceeding similarly we can trace through the edges coming from each vertex.

Example 19. *Consider a set A = {k, l, m, n} and let R be a relation on A defined by the matrix* M_R *as shown :*

$$M_R = \begin{array}{c} \\ k \\ l \\ m \\ n \end{array}\begin{array}{c} \begin{array}{cccc} k & l & m & n \end{array} \\ \begin{bmatrix} 1 & 0 & 0 & 1 \\ 0 & 1 & 1 & 0 \\ 0 & 0 & 0 & 1 \\ 1 & 0 & 0 & 0 \end{bmatrix} \end{array}$$

Construct a modified linked list representation for the relation R.

Sol. The modified linked list representation has four arrays *i.e.*, VERT, TAIL, HEAD and NEXT. We have four vertices *k*, *l*, *m* and *n* and six edges. The four arrays are as shown :

VERT = [3, 6, 4, 5]

TAIL = [*k*, *l*, *k*, *m*, *n*, *l*]

HEAD = [*n*, *m*, *k*, *n*, *n*, *l*]

NEXT = [0, 0, 1, 0, 0, 2]

The indexes defined in VERT are arbitrary. You can choose any index ordering.

Example 20. *Consider the set A = {a, b, c, d} and let R be the relation whose diagraph is shown in Fig. 13(xvi).*

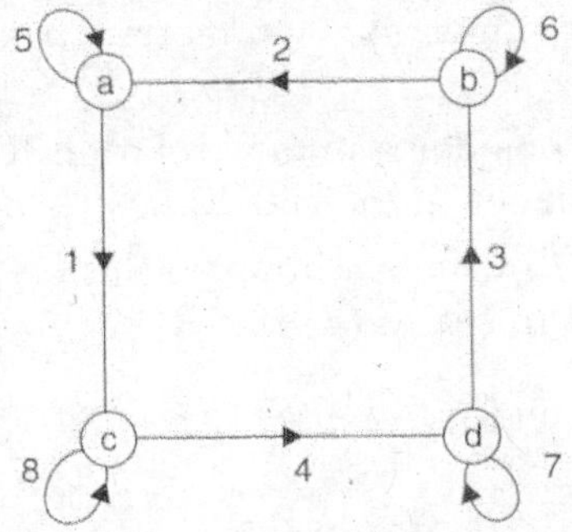

Fig. 13(*xvi*)

Construct a modified linked list representation for the relation R.

Sol. The modified linked list representation has four arrays *i.e.*, VERT, TAIL, HEAD and NEXT, which are as shown below :

VERT = [6, 8, 3, 5]
TAIL = [*a*, *b*, *c*, *d*, *d*, *a*, *c*, *b*]
HEAD = [*a*, *b*, *d*, *d*, *b*, *c*, *c*, *a*]
NEXT = [0, 0, 7, 0, 4, 1, 0, 2]

Example 21. *Consider the arrays as shown :*

VERT = [5, 4, 3, 7, 6]
TAIL = [a, a, c, b, a, e, d]
HEAD = [c, d, e, e, b, d, e]
NEXT = [2, 0, 0, 0, 1, 0, 0]

These arrays describe a relation R on the set A = {a, b, c, d, e}. Determine the diagraph of R and the corresponding matrix M_R.

Sol. The diagraph described by the arrays is shown in Fig. 13(*xvii*).

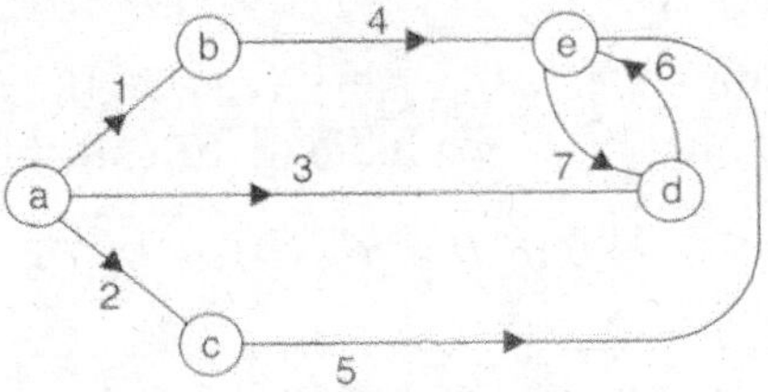

Fig. 13(*xvii*).

The corresponding matrix M_R of the relation R is as shown in Fig. 13(*xviii*).

$$M_R = \begin{array}{c} \\ a \\ b \\ c \\ d \\ e \end{array} \begin{array}{c} \begin{array}{ccccc} a & b & c & d & e \end{array} \\ \begin{bmatrix} 0 & 1 & 1 & 1 & 0 \\ 0 & 0 & 0 & 0 & 1 \\ 0 & 0 & 0 & 0 & 1 \\ 0 & 0 & 0 & 0 & 1 \\ 0 & 0 & 0 & 1 & 0 \end{bmatrix} \end{array}$$

Fig. 13(*xviii*).

PROPERTIES OF RELATIONS

Property of the relation tells the nature or type of the relation. There are many properties of relations *i.e.*, reflexive, symmetric, asymmetric, transitive, antisymmetric etc. which are as follows :

1. Reflexive Relation. Consider a binary relation R on a set A. Relation R is called a reflexive relation if, for every a in set A, $(a, a) \in R$, *i.e.*, $(a, a) \in R, \forall a \in A$.

2. Irreflexive Relation. Consider a binary relation R on a set A. Relation R is called irreflexive relation if, for every a in set A, $(a, a) \notin R$, *i.e.*, $(a, a) \notin R, \forall a \in A$.

Example 22. *Let A = {1, 2} and R = {(1, 1), (1, 2), (2, 2)}. Is the relation reflexive or irreflexive ?*

Sol. The relation is reflexive as for every $a \in A$. $(a, a) \in R$, *i.e.*, $(1, 1), (2, 2) \in R$.

The relation is not irreflexive.

Example 23. *Let A = {1, 2, 3, 4} and R = {(1, 2), (2, 1), (3, 1), (4, 1)}. Is the relation reflexive or irreflexive ?*

Sol. The relation R is not reflexive as for every $a \in A$ $(a, a) \notin R$.

The relation is irreflexive as for every $a \in A$, $(a, a) \notin R$.

Example 24. *Let A = {1, 2, 3} and R = {(1, 2), (2, 2), (3, 1), (1, 3)}. Is the relation reflexive or irreflexive ?*

Sol. The relation R is not reflexive as for every $a \in a$, $(a, a) \notin R$, *i.e.*, $(1, 1)$ and $(3, 3) \notin R$.

The relation is not irreflexive as $(a, a) \notin R$, for some $a \in A$, *i.e.*, $(2, 2) \in R$.

3. Symmetric Relation. Consider a binary relation R on a set A. Relation R is called a symmetric relation if for every $(a, b) \in R$ implies that (b, a) also belong to R.

4. Asymmetric Relation. Consider a binary relation R on a set A. Relation R is called an asymmetric relation if for every $(a, b) \in R$ implies that (b, a) does not belongs to R.

Example 25. *Let A = {1, 2, 3} and R = {(1, 2), (2, 1), (3, 2), (2, 3)}. Is the relation symmetric or asymmetric ?*

Sol. The relation is symmetric as for every $(a, b) \in R$, we have $(b, a) \in R$.

The relation is not asymmetric as we have $(b, a) \in R$, for some $(a, b) \in R$.

Example 26. *Let A = {4, 5, 6} and R = {(4, 3), (5, 6), (6, 4)}. Is the relation symmetric or asymmetric ?*

Sol. The relation is not symmetric as for every $(a, b) \in R$, we have $(b, a) \notin R$ *i.e.*, $(4, 3) \in R$ but $(3, 4) \notin R$.

The relation is asymmetric as for every $(a, b) \in R$, we have $(b, a) \notin R$.

Example 27. *Let A = {7, 8, 9} and R = {(7, 8), (8, 7), (7, 9), (8, 9)}. Is the relation symmetric or asymmetric ?*

Sol. The relation is not symmetric as for every $(a, b) \in R$, we have $(b, a) \notin R$ *i.e.*, $(7, 9) \in R$ but $(9, 7) \notin R$.

The relation is not asymmetric as for some $(a, b) \in R$, we have $(b, a) \in R$ *i.e.*, $(7, 8) \in R$ and $(8, 7) \in R$.

5. Transitive Relations. Consider a binary relation R on a set A. Relation R is called transitive relation if whenever both (a, b) and (b, c) belong to R, implies that (a, c) also belongs to R *i.e.*, $(a, b), (b, c) \in R \Rightarrow (a, c) \in R$.

Example 28. *Let* $A = \{1, 2, 3\}$ *and* $R = \{(1, 2), (2, 1), (1, 1), (2, 2)\}$. *Is the relation transitive.*

Sol. The relation R is transitive as for every $(a, b), (b, c)$ belong to R, we have $(a, c) \in R$ *i.e.*, $(1, 2), (2, 1) \in R \Rightarrow (1, 1) \in R$.

Example 29. *Let* $A = \{3, 4, 5\}$ *and* $R = \{(3, 4), (4, 3), (5, 4), (5, 3)\}$. *Is the relation R transitive ?*

Sol. The relation R is not transitive as for every $(a, b), (b, c) \in R$. We have $(a, c) \notin R$ *i.e.*, $(3, 4), (4, 3) \in R$ but $(3, 3) \notin R$.

6. Antisymmetric Relations. Consider a binary relation R on a set A. Relation R is called antisymmetric relation if $(a, b) \in R$ implies that $(b, a) \notin R$ unless $a = b$. In other words, we can say if (a, b) and (b, a) belong to R, then $a = b$.

Example 30. *Let* $A = \{4, 5, 6\}$ *and* $R = \{(4, 4), (4, 5), (5, 4), (5, 6), (4, 6)\}$. *Is the relation R antisymmetric ?*

Sol. The relation R is not antisymmetric as $4 \neq 5$ but $(4, 5)$ and $(5, 4)$ both belong to R.

Example 31. *Let* $A = \{1, 2, 3\}$ *and* $R = \{(1, 1), (2, 2)\}$. *Is the relation R antisymmetric ?*

Sol. The relation R is antisymmetric as $a = b$ when (a, b) and (b, a) both belong to R.

Example 32. *Let* $A = \{4, 5, 6, 7\}$. *Determine whether the following relation are reflexive, symmetric, transitive or antisymmetric.*

(*i*) $R_1 = \{(4, 4), (5, 5), (6, 6), (7, 7)\}$ (*ii*) $R_2 = \{(4, 4), (5, 5)\}$

(*iii*) $R_3 = \phi$ (*iv*) $R_4 = \{(4, 5), (5, 4), (7, 6), (6, 7)\}$.

Sol. (*i*) **Reflexive.** The relation is reflexive as for every $a \in A$, $(a, a) \in R_1$, *i.e.*, $(4, 4), (5, 5), (6, 6), (7, 7) \in R_1$.

Symmetric. The relation is symmetric as there is no such pair $(a, b) \in R_1$, such that (b, a) also belongs to R_1.

Transitive. The relation is transitive as there is no such pairs *i.e.*, $(a, b) \in R_1$ and $(b, c) \in R_1$ for which we have $(a, c) \in R_1$.

Antisymmetric. The relation is antisymmetric as if (a, b) and (b, a) belong to R_1, we have $a = b$.

(*ii*) **Reflexive.** The relation is not reflexive as for every a, $(a, a) \notin R_2$, *i.e.*, $(b, b) \notin R_2$.

Symmetric. The relation is symmetric as there is no such pair $(a, b) \in R_2$, such that (b, a) also belong to R_2.

Transitive. The relation is transitive as there is no such pair $(a, b) \in R_2$ and $(b, c) \in R_2$, such that (a, c) also belong to R_2.

Antisymmetric. The relation is antisymmetric because whenever $(a, b) \in R_2$ and $(b, a) \in R_2$, we have $a = b$.

(*iii*) **Reflexive.** The relation is not reflexive as for every $a \in A$, $(a, a) \notin R_3$.

Symmetric. The relation is symmetric as there is no such element $(a, b) \in R_3$ for which we have $(b, a) \in R_3$.

Transitive. The relation is transitive as there is no such elements $(a, b) \in R_3$ and $(b, c) \in R_3$ for which we have $(a, c) \in R_3$.

Antisymmetric. The relation is antisymmetric as there is no such elements $(a, b) \in R_3$ and $(b, a) \in R_3$, such that $a = b$.

(iv) **Reflexive.** The relation is not reflexive as for every $a \in A$, $(a, a) \notin R_4$.

Symmetric. The relation is symmetric as for each $(a, b) \in R_4$, we have $(b, a) \in R_4$.

Transitive. The relation is not transitive as for elements $(a, b) \in R_4$ and $(b, c) \in R_4$, we have $(a, c) \notin R_4$ *i.e.*, $(4, 5) \in R_4$ and $(5, 4) \in R_4$ but $(4, 4) \notin R_4$.

Antisymmetric. The relation is not antisymmetric as for pairs $(a, b) \in R_4$ and $(b, a) \in R_4$, we have $a \neq b$ *i.e.*, $(7, 6) \in R_4$ and $(6, 7) \in R_4$ but $6 \neq 7$.

Example 33. *Determine whether the relation R on set A = {a, b, c, d} whose directed graph is given in Fig. 14 is (i) reflexive (ii) symmetric (iii) transitive (iv) antisymmetric.*

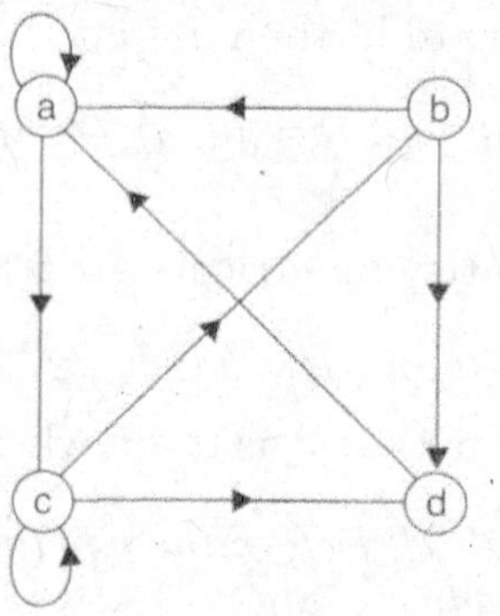

Fig. 14.

Sol. *(i)* **Reflexive.** The relation is not reflexive as each vertex does not contain an edge on itself *i.e.*, a self-loop.

(ii) **Symmetric.** The relation is not symmetric because when there is an edge $(a, b) \in R$, we have no edge $(b, a) \in R$ *e.g.*, $(a, c) \in R$ but $(c, a) \notin R$.

(iii) **Transitive.** The relation is not transitive because when there is an edge $(a, b) \in R$ and $(b, c) \in R$, we have no edge $(a, c) \in R$ *e.g.*, $(a, c) \in R$ and $(c, d) \in R$ but $(a, d) \notin R$.

(iv) **Antisymmetric.** The relation is antisymmetric because we have no such edges $(a, b) \in R$ and $(b, a) \in R$ for which $a \neq b$.

CLOSURE PROPERTIES OF RELATIONS

Consider a relation R on some set A. Suppose the relation R does not possess the desired property. We will add as few new pairs as possible to relation R to get a relation that does have the desired property.

The smallest relation S on A that contains R and the desired property is called the closure of relation R with the desired property.

1. Reflexive Closure. Consider a relation R on a set A. Relation R_F is called reflexive closure of R if R_F is the smallest relation containing R, having the reflexive property *i.e.*, $R_F = R \cup \Delta$ where Δ is a diagonal relation.

Example 34. *Consider the relation R on A = {7, 8, 9} defined by R = {(7, 8), (7, 9), (8, 8), (9, 7)}. Find the reflexive closure of R.*

Sol. The smallest relation containing R having the reflexive property is

$$R \cup \Delta = \{(7, 7), (7, 8), (7, 9), (8, 8), (9, 7), (9, 9)\}.$$

Example 35. *Let A = {k, l, m}. Let R is a relation on A defined by*

$$R = \{(k, k), (k, l), (l, m), (m, k)\}$$ *. Find the reflexive closure of R.*

Sol. $R \cup \Delta$ is the smallest relation having reflexive property. Hence,

$$R_F = R \cup \Delta = \{(k, k), (k, l), (l, l), (l, m), (m, m), (m, k)\}.$$

2. Symmetric Closure. Consider a relation R on set A. Relation R_s is called the symmetric closure of R if R_S is the smallest relation containing R, having the symmetric property. The relation $R_S = R \cup R^{-1}$ is the smallest symmetric relation containing R.

Example 36. *Consider the relation R on A = {4, 5, 6, 7} defined by*

$$R = \{(4, 5), (5, 5), (5, 6), (6, 7), (7, 4), (7, 7)\}$$

Find the symmetric closure of R.

Sol. The smallest relation containing R, having the symmetric property is $R \cup R^{-1}$ *i.e.*, $R_S = R \cup R^{-1} = \{(4, 5), (5, 4), (5, 5), (5, 6), (6, 5), (6, 7), (7, 6), (7, 4), (4, 7), (7, 7)\}$.

Example 37. *Let A = { Maths, Physics, English, Chemistry}. Let R is relation 'combination of subjects' on the set A, defined by*

R = {(Maths, Physics), (Physics, English), (Chemistry, English), (Chemistry, Physics), (Physics, Chemistry)}.

Find the symmetric closure R_S *of R.*

Sol. $R \cup R^{-1}$ is the smallest relation having the symmetric property. Thus,

$R_S = R \cup R^{-1}$ = {(Maths, Physics), (Physics, Math), (Physics, English), (English, Physics), (Chemistry, English), (English, Chemistry), (Chemistry, Physics), (Physics, Chemistry)}.

3. Transitive Closure. Consider a relation R on a set A. The transitive closure R of a relation R is the smallest transitive relation containing R.

Method to find transitive closure. Draw directed graph of the relation R. Find whether there is a path from Ist vertex to any other vertex of the graph. Include the ordered pair for that vertices R* if there is path from Ist vertex to any other vertex. Similarly, we proceed to check all the paths from all vertices to all other vertices and include the corresponding ordered pairs in R*. At the end we got the transitive closure R* of relation R.

Example 38. *Let A = {a, b, c, d}. Let R = {(a, b), (a, c), (b, a), (b, c), (c, d), (d, a)}.*

Find the transitive closure of R.

Sol. The directed graph of relation R is shown in Fig. 15.

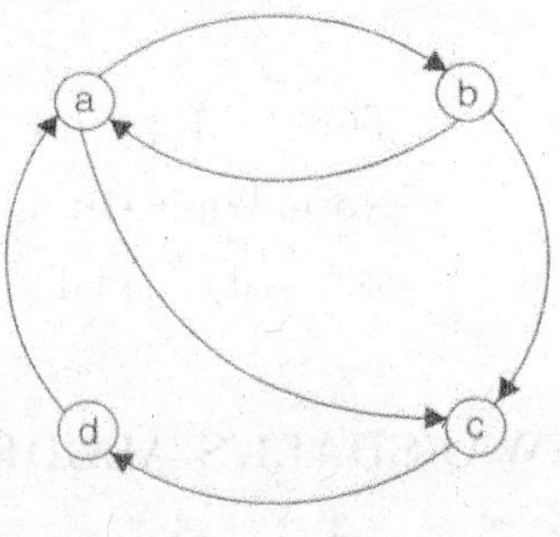

Fig. 15.

From vertex *a*, we have paths to vertices *a*, *b*, *c* and *d*. So, the ordered pairs included in R* is (*a*, *a*), (*a*, *b*), (*a*, *c*) and (*a*, *d*). Similarly, from vertex *b*, we have paths to vertices *a*, *b*, *c* and *d*. So, the ordered pairs included in R* is (*b*, *b*), (*b*, *c*), (*b*, *d*), (*b*, *a*). From vertex *c*, we have paths to vertices *c*, *b*, *d* and *a* from vertex *d*, we have paths to vertices *d*, *a*, *b* and *c*. So, the vertices included in R* is (*c*, *c*), (*c*, *d*), (*c*, *b*), (*c*, *a*) and (*d*, *d*), (*d*, *a*), (*d*, *b*), (*d*, *c*). Hence, transitive closure R* is

R* = {(a, a), (b, b), (c, c), (d, d), (a, b), (a, c), (a, d), (b, c), (b, a), (b, d), (c, a), (c, b), (c, d), (d, a), (d, b), (d, c)}.

Example 39. *Let A = {4, 6, 8, 10} and R = {(4, 4), (4, 10), (6, 6), (6, 8), (8, 10)} is a relation on set A. Determine transitive closure of R.*

Sol. The matrix of relation R is shown in Fig. 16.

$$M_R = \begin{array}{c} \\ 4 \\ 6 \\ 8 \\ 10 \end{array} \begin{array}{c} \begin{array}{cccc} 4 & 6 & 8 & 10 \end{array} \\ \begin{bmatrix} 1 & 0 & 0 & 1 \\ 0 & 1 & 1 & 0 \\ 0 & 0 & 0 & 1 \\ 0 & 0 & 0 & 0 \end{bmatrix} \end{array}$$

Fig. 16.

Now, find the powers of M_R as in Figs. 17, 18 and 19.

$$M_{R^2} = \begin{array}{c} \\ 4 \\ 6 \\ 8 \\ 10 \end{array} \begin{array}{c} \begin{array}{cccc} 4 & 6 & 8 & 10 \end{array} \\ \begin{bmatrix} 1 & 0 & 0 & 1 \\ 0 & 1 & 1 & 1 \\ 0 & 0 & 0 & 0 \\ 0 & 0 & 0 & 0 \end{bmatrix} \end{array}$$

Fig. 17.

$$M_{R^3} = \begin{array}{c} \\ 4 \\ 6 \\ 8 \\ 10 \end{array} \begin{array}{c} \begin{array}{cccc} 4 & 6 & 8 & 10 \end{array} \\ \begin{bmatrix} 1 & 0 & 0 & 1 \\ 0 & 1 & 1 & 1 \\ 0 & 0 & 0 & 0 \\ 0 & 0 & 0 & 0 \end{bmatrix} \end{array}$$

Fig. 18.

$$M_{R^4} = \begin{array}{c} \\ 4 \\ 6 \\ 8 \\ 10 \end{array} \begin{array}{c} \begin{array}{cccc} 4 & 6 & 8 & 10 \end{array} \\ \begin{bmatrix} 1 & 0 & 0 & 1 \\ 0 & 1 & 1 & 1 \\ 0 & 0 & 0 & 0 \\ 0 & 0 & 0 & 0 \end{bmatrix} \end{array}$$

Fig. 19.

Hence, the transitive closure of M_R is M_{R^*} as shown in Fig. 20. (where M_{R^*} is the ORing of powers of M_R).

$$M_{R^*} = M_R \vee M_{R^2} \vee M_{R^3} \vee M_{R^4}; \qquad M_{R^*} = \begin{array}{c} \\ 4 \\ 6 \\ 8 \\ 10 \end{array} \begin{array}{c} \begin{array}{cccc} 4 & 6 & 8 & 10 \end{array} \\ \begin{bmatrix} 1 & 0 & 0 & 1 \\ 0 & 1 & 1 & 1 \\ 0 & 0 & 0 & 1 \\ 0 & 0 & 0 & 0 \end{bmatrix} \end{array}$$

Fig. 20.

Thus, R* = {(4, 4), (4, 10), (6, 8), (6, 6), (6, 10), (8, 10)}.

Note. While ORing the powers of the matrix R, we can eliminate M_{R^n} because it is equal to M_{R^2} if n is even and is equal to M_{R^3} if n is odd.

WARSHALL'S ALGORITHM

1. Let n be the number of elements in a given set A *i.e.*, $n \to |A|$
2. Let $w_0, w_1, w_2, \ldots\ldots, w_n$ be Warshall sets.

3. To find transitive closure of relation R on A, maximum of n Warshall sets of which w_n is the last need to be computed.

The procedure to compute w_k from w_{k-1} is as follows :

(*i*) Copy 1 to all entries in w_k from w_{k-1}, where there is 1 in w_{k-1}.

(*ii*) Find the row numbers $p_1, p_2, p_3, \ldots\ldots$ for which there is 1 in column k in w_{k-1} and the column numbers $q_1, q_2, q_3, \ldots\ldots$ for which there is 1 in row k in w_{k-1}.

(*iii*) Mark entries in w_k as 1 for (p_i, q_i) of w_k if there are not already 1.

4. Stop when w_n is obtained and it is the needed result (transitive closure).

Example 40. *Consider the example 39, where A = {4, 6, 8, 10} and R = {(4, 4), (4, 10), (6, 6), (6, 8), (8, 10)} is a relation on set A. Determine the transitive closure of R using Warshall's algorithm.*

Sol. The diagraph of relation R and its corresponding matrix M_R is shown in Fig. 21.

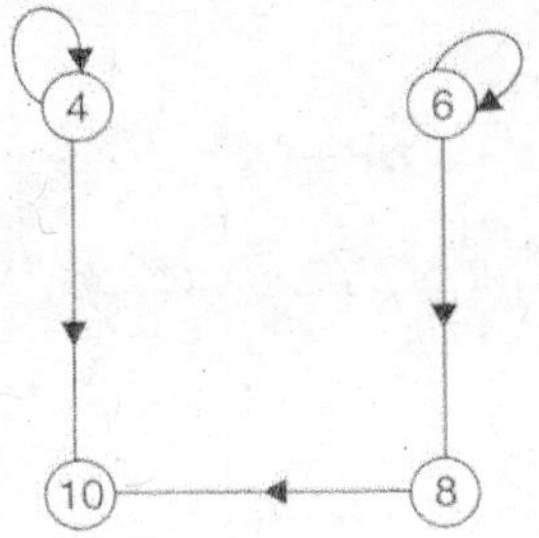

$$M_R = \begin{array}{c} \\ 4 \\ 6 \\ 8 \\ 10 \end{array} \begin{array}{c} \begin{array}{cccc} 4 & 6 & 8 & 10 \end{array} \\ \begin{bmatrix} 1 & 0 & 0 & 1 \\ 0 & 1 & 1 & 0 \\ 0 & 0 & 0 & 1 \\ 0 & 0 & 0 & 0 \end{bmatrix} \end{array}$$

Fig. 21.

The value of $n = |A| = 4$. Thus, we have to find the Warshall's sets w_0, w_1, w_2, w_3 and w_4.

The first set w_0 is same as M_R, which is shown below :

$$w_0 = M_R = \begin{bmatrix} 1 & 0 & 0 & 1 \\ 0 & 1 & 1 & 0 \\ 0 & 0 & 0 & 1 \\ 0 & 0 & 0 & 0 \end{bmatrix}$$

Now to find w_1 from w_0 we have row number 1 for column 1 in w_0 and column number 1 and 4 for row 1 in w_0. Thus, new entries in w_1 are (4, 4) and (4, 10) which are already one. Thus, w_1 is same as w_0, which is as follows :

$$w_1 = \begin{bmatrix} 1 & 0 & 0 & 1 \\ 0 & 1 & 1 & 0 \\ 0 & 0 & 0 & 1 \\ 0 & 0 & 0 & 0 \end{bmatrix}$$

To find w_2 from w_1, we have row number 2 for column 2 in w_1 and column number 2 and 3 for row 2 in w_1. Thus, new entries in w_2 are (6, 6) and (6, 8), which are already one. Thus, w_2 is same as w_1, which is as follows :

$$w_2 = \begin{bmatrix} 1 & 0 & 0 & 1 \\ 0 & 1 & 1 & 0 \\ 0 & 0 & 0 & 1 \\ 0 & 0 & 0 & 0 \end{bmatrix}$$

Similarly, w_3 is obtained from w_2. Here, we have row number 2 in column 3 and column number 4 in row 3. Thus, the new entries in w_3 are (6, 10). So, w_3 is as follows :

$$w_3 = \begin{bmatrix} 1 & 0 & 0 & 1 \\ 0 & 1 & 1 & 1 \\ 0 & 0 & 0 & 1 \\ 0 & 0 & 0 & 0 \end{bmatrix}$$

Similarly, w_4 is obtained from w_3. But there are no new entries of 1's in w_4. Hence, $M_{R^\infty} = w_4 = w_3$, which is the transitive closure of R, and is same as in example 39.

EQUIVALENCE RELATIONS

A relation R on a set A is called an equivalence relation if it satisfies following three properties :

1. Relation R is reflexive *i.e.*, $aRa \ \forall \ a \in A$.
2. Relation R is symmetric *i.e.*, $aRb \Rightarrow bRa$.
3. Relation R is transitive *i.e.*, aRb and $bRc \Rightarrow aRc$.

Example 41. *Let A = {1, 2, 3, 4} and R = {(1, 1), (1, 3), (2, 2), (2, 4), (3, 1), (3, 3), (4, 2), (4, 4)}.*

Show that R is an equivalence relation.

Sol. Reflexive. Relation R is reflexive as (1, 1), (2, 2), (3, 3), and (4, 4) ∈ R.

Transitive. Relation R is transitive because whenever (a, b) and (b, c) belong to R, (a, c) also belong to R *e.g.*, (3, 1) ∈ R and (1, 3) ∈ R ⇒ (3, 3) ∈ R.

Symmetric. Relation R is symmetric because whenever $(a, b) \in R$, (b, a) also belong to R *e.g.*, (2, 4) ∈ R ⇒ (4, 2) ∈ R.

So, as R is reflexive, symmetric and transitive. Hence, R is an equivalence relation.

Example 42. *Let S be the set of all points in a plane. Let R be a relation such that for any two points a and b ; (a, b) ∈ R if b is within two centimeter from A. Show that R is an equivalence relation.*

Sol. Reflexive. We have $(a, a) \in R$, for every $a \in S$ [∵ a lies within 2 cm from a]

Therefore, R is reflexive.

Symmetric. Assume that $(a, b) \in R$

⇒ b lies within 2 cm from a ⇒ a lies within 2 cm from b.

⇒ $(b, a) \in R$.

Thus, $(a, b) \in R \Rightarrow (b, a) \in R$.

Therefore, R is symmetric.

Transitive. Assume that (a, b) and $(b, c) \in R$.

Therefore, b lies within 2 cm from a.

c lies within 2 cm from b.

It does not implies c lies within 2 cm from a. Therefore, $(a, b) \in R$, $(b, c) \in R$ then (a, c) may not belong to R.

Hence, relation R is not always transitive. Therefore, relation is not an equivalence relation.

PARTIAL ORDER RELATION

A relation R on a set A is called a partial order relation if it satisfies the following three properties :

1. Relation R is reflexive *i.e.*, $a\text{R}a \ \forall \ a \in \text{A}$.
2. Relation R is antisymmetric $a\text{R}b$ and $b\text{R}a \ \Rightarrow \ a = b$.
3. Relation R is transitive $a\text{R}b$ and $b\text{R}c \ \Rightarrow \ a\text{R}c$.

Example 43. *Show whether the relation* $(x, y) \in \text{R}$, if $x \geq y$ *defined on the set of + ve integers is a partial order relation.*

Sol. Consider the set A = {1, 2, 3, 4} containing four + ve integers. Find the relation for this set such as R = {(2, 1), (3, 1), (3, 2), (4, 1), (4, 2), (4, 3), (1, 1), (2, 2), (3, 3), (4, 4)}.

Reflexive. The relation is reflexive as for every $a \in \text{A}$ $(a, a) \in \text{R}$ *i.e.*, (1, 1), (2, 2), (3, 3) and (4, 4) ∈ R.

Antisymmetric. The relation is antisymmetric as whenever (a, b) and $(b, a) \in \text{R}$, we have $a = b$.

Transitive. The relation is transitive as whenever (a, b) and $(b, c) \in \text{R}$, we have $(a, c) \in \text{R}$ *e.g.*, (4, 2) ∈ R and (2, 1) ∈ R, implies (4, 1) ∈ R.

As the relation is reflexive, antisymmetric and transitive. Hence, it is a partial order relation.

Example 44. *Show that the relation 'Divides' defined on N is a partial order relation.*

Sol. Reflexive. We have a divides a, $\forall \ a \in \text{N}$. Therefore, relation 'Divides' is reflexive.

Antisymmetric. Let $a, b \in \text{N}$, such that a divides b. It implies b divides a iff $a = b$.

So, relation is antisymmetric.

Transitive. Let $a, b, c \in \text{N}$, such that a divides b and b divides c.

Then a divides c. Hence, relation is transitive. Thus, the relation being reflexive, antisymmetric and transitive, the relation 'divides' is a partial order relation.

PARTIAL ORDER SET (POSET)

The set A together with a partially order relation R on the set A and is denoted by (A, R) is called a partially ordered set or POSET.

TOTAL ORDER RELATION

Consider the relation R on the set A. If it is also the case that for all $a, b \in \text{A}$, we have either $(a, b) \in \text{R}$ or $(b, a) \in \text{R}$ or $a = b$, then the relation R is called total order relation on set A.

Example 45. *Show that the relation '<' (less than) defined on N, the set of + ve integers is neither an equivalence relation nor partially ordered relation but is a total order relation.*

Sol. Reflexive. Let $a \in \text{N}$, then $a < a$

$\Rightarrow$ '<' is not reflexive.

As, the relation '<' (less than) is not reflexive, it is neither an equivalence relation nor the partial order relation.

But, as $\forall a, b \in N$, we have either $a < b$ or $b < a$ or $a = b$. So, the relation is a total ordered relation.

PARTITION

A partition $(A_1, A_2, ..., A_i)$ of a non-empty set A is defined as a collection of non-empty subsets of A, such that

(*a*) Every element of A belongs to one of A_i *i.e.*, the union of A_i is equal to set A.

(*b*) If A_i and A_j are distinct, then $A_i \cap A_j = \phi$ *i.e.*, the partitions divides the elements of the set A into disjoint subsets.

The subsets in a partition are called blocks or cells.

Example 46. *Let A = {7, 8, 9). Determine all the partitions of the set A.*

Sol. We have three elements in set A. So, in a partition there can be 1, 2 or 3 elements at the maximum. Hence, the partitions are as follows :

(*i*) [{7}, {8}, {9}] (*ii*) [{7}, {8, 9}]
(*iii*) [{8}, {7, 9}] (*iv*) [{9}, {7, 8}] (*v*) [{7, 8, 9}].

Example 47. *Let A = {1, 2, 3, 4, 5, 6, 7, 8}. Determine whether the following is a partition of A or not*

(*i*) P_1 = [{1, 2, 3, 4}, {1, 3, 5, 6, 7, 8}]

(*ii*) P_2 = [{1, 3, 5, 7}, {2, 4}, {6, 8}]

(*iii*) P_3 = [{1, 2, 3, 4}, {6, 8}]

(*iv*) P_4 = [{1, 3, 5, 6, 7, 8}, {2, 4}].

Sol. (*i*) Not a partition because $1 \in A$ belongs to both the cells.

(*ii*) P_2 is a partition.

(*iii*) P_3 is not a partition because $7 \in A$ does not belong to any cell of P_3.

(*iv*) P_4 is a partition.

EQUIVALENCE CLASS

Consider, an equivalence relation R on a set A. The equivalence class of an element $a \in A$, is the set of elements of A to which element a is related. It is denoted by $[a]$.

Example 48. *Let R be an equivalence relation on the set A = {4, 5, 6, 7} defined by*

$$R = \{(4, 4), (5, 5), (6, 6), (7, 7), (4, 6), (6, 4)\}.$$

Determine its equivalence classes.

Sol. The equivalence classes are as follows :

$$[4] = [6] = \{4, 6\}$$
$$[5] = \{5\}$$
$$[7] = \{7\}.$$

Example 49. *Let R be an equivalence relation on set A = {1, 2, 3, 4, 5} defined by*

$$R = \{(1, 1), (2, 2), (3, 3), (4, 4), (5, 5), (1, 4), (4, 1), (2, 4), (4, 2), (1, 2), (2, 1)\}$$

Determine equivalence classes of R.

Sol. The equivalence classes of R is

$$[1] = [2] = [4] = \{1, 2, 4\}$$
$$[3] = \{3\}$$
$$[5] = \{5\}.$$

Example 50. *Let R be an equivalence relation on the set A = {6, 7, 8, 9, 10} defined by*

R = {(6, 6), (7, 7), (8, 8), (9, 9), (10, 10), (6, 7), (7, 6), (8, 9), (9, 8), (9, 10), (10, 9), (8, 10), (10, 8)}.

Find the equivalence classes of R and hence find the partition of A corresponding to R.

Sol. The equivalence class of R are as follows. Element 6 is related with 6 and 7, therefore

$$[6] = [7] = \{6, 7\}.$$

Element 8 is related with 8, 9 and 10, therefore, $[8] = [9] = [10] = \{8, 9, 10\}$.

Hence, the partition corresponding to R is

$$P = \{(6, 7), (8, 9, 10)\}.$$

CIRCULAR RELATION

Consider a binary relation R on a set A. Relation R is called circular if $(a, b) \in R$ and $(b, c) \in R$ implies $(c, A) \in R$.

Example 51. *Show that relation R is reflexive and circular if and only if it is an equivalence relation.*

Sol. Consider, the relation is reflexive and circular.

Therefore, $(a, a) \in R$ for every $a \in A$ (∵ R is reflexive)

Also, let $(a, b) \in R$ and $(b, c) \in R$ implies $(c, a) \in R$ (∵ R is circular)

Now, as $(c, a) \in R$ and $(a, a) \in R$ implies $(a, c) \in R$ (∵ R is circular)

⇒ R is transitive.

Let $(a, b) \in R$ implies $(a, a) \in R$ and $(b, b) \in R$ (∵ R is reflexive)

It implies $(b, a) \in R$ (∵ R is circular)

Therefore, R is symmetric.

Therefore, relation R being reflexive, transitive and symmetric, hence it is an equivalence relation.

Example 52. *Consider R is an equivalence relation. Show that R is reflexive and circular.*

Sol. Reflexive. As, the relation R is an equivalence relation. So, reflexivity is the property of an equivalence relation. Hence, R is reflexive.

Circular. Let $(a, b) \in R$ and $(b, c) \in R$

⇒ $(a, c) \in R$ (∵ R is transitive)

⇒ $(c, a) \in R$ (∵ R is symmetric)

Thus, R is circular.

SOLVED PROBLEMS

Problem 1. *Let R be a relation on set A = {k, l, m, n} defined by*

R = {(k, l), (m, l), (n, l), (l, l), (k, k), (m, k), (l, k), (n, k)}.

Find domain and range of relation R.

Sol. The domain of relation R is

$$\text{Dom (R)} = \{k, l, m, n\} = \text{A}$$

The range of relation R is

$$\text{RAN (R)} = \{k, l\}.$$

Problem 2. *Let R be a relation on set A = {1, 2, 3, 4} defined by*

$$R = \{(1, 1), (2, 2), (3, 3), (4, 4), (4, 3), (4, 2), (4, 1), (3, 2), (3, 1)\}.$$

Find the matrix and directed graph of relation R.

Sol. The matrix of relation R is a 4 × 4 matrix as shown in Fig. 22.

$$M_R = \begin{array}{c} \\ 1 \\ 2 \\ 3 \\ 4 \end{array}\begin{array}{c} \begin{array}{cccc} 1 & 2 & 3 & 4 \end{array} \\ \begin{bmatrix} 1 & 0 & 0 & 0 \\ 0 & 1 & 0 & 0 \\ 1 & 1 & 1 & 0 \\ 1 & 1 & 1 & 1 \end{bmatrix} \end{array}$$

Fig. 22.

The directed graph of relation R is as shown in Fig. 23.

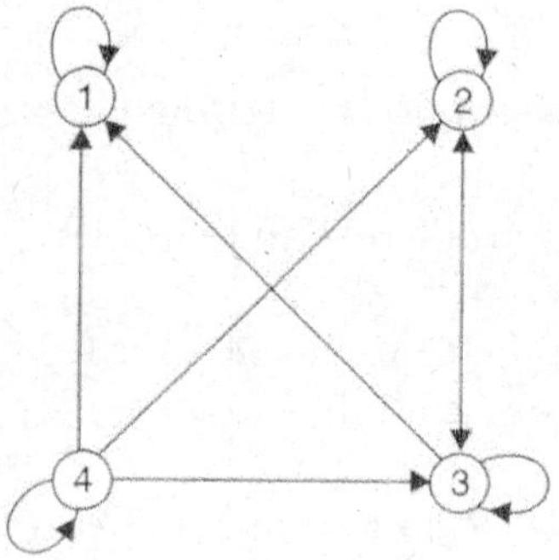

Fig. 23.

Problem 3. *Let R be a relation on set A = {1, 2, 3, 4, 5}, whose matrix is shown in Fig. 24. Find the directed graph of relation R.*

$$M_R = \begin{bmatrix} 1 & 1 & 1 & 1 & 1 \\ 0 & 0 & 0 & 0 & 0 \\ 1 & 1 & 0 & 0 & 0 \\ 0 & 0 & 0 & 0 & 0 \\ 1 & 1 & 1 & 1 & 1 \end{bmatrix}$$

Fig. 24.

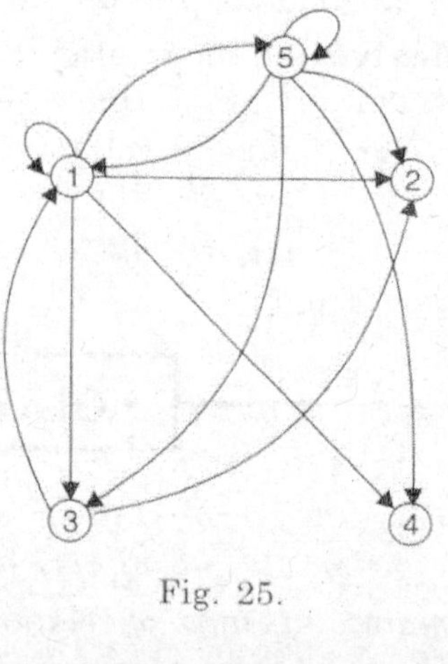

Fig. 25.

Sol. The directed graph of relation R whose matrix M_R is given is as shown in Fig. 25.

Problem 4. *Let R be a relation on set A whose directed graph is as shown in Fig. 26. Determine its matrix.*

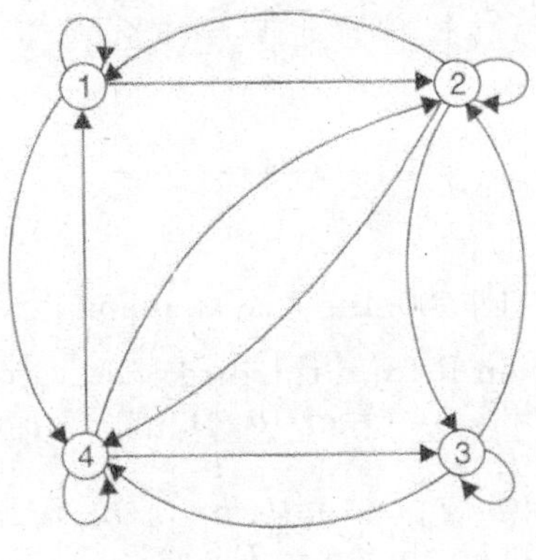

Fig. 26.

Sol. The matrix M_R of the relation R whose directed graph is shown above is as shown in Fig. 27.

$$M_R = \begin{array}{c} \\ 1 \\ 2 \\ 3 \\ 4 \end{array} \begin{array}{c} \begin{array}{cccc} 1 & 2 & 3 & 4 \end{array} \\ \begin{bmatrix} 1 & 1 & 0 & 1 \\ 1 & 1 & 1 & 1 \\ 0 & 1 & 1 & 1 \\ 1 & 1 & 1 & 1 \end{bmatrix} \end{array}$$

Fig. 27.

Problem 5. *Let R be a relation on set A = {x, y, z} defined by*

$$R = \{(x, x), (y, y), (z, z), (x, z), (x, y), (y, z)\}.$$

Write the relation as a table and also find its arrow diagram.

Sol. Relation R as a table shown in Fig. 28.

	x	*y*	*z*
x	×	×	×
y		×	×
z			×

Fig. 28.

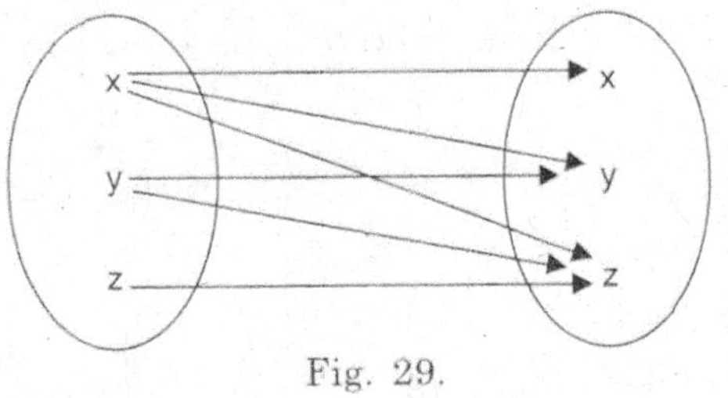

Fig. 29.

The arrow diagram of relation R is shown in Fig. 29.

Problem 6. *Consider a relation R whose directed graph is shown in Fig. 30. Determine its inverse R^{-1} and complement $\overline{R}$.*

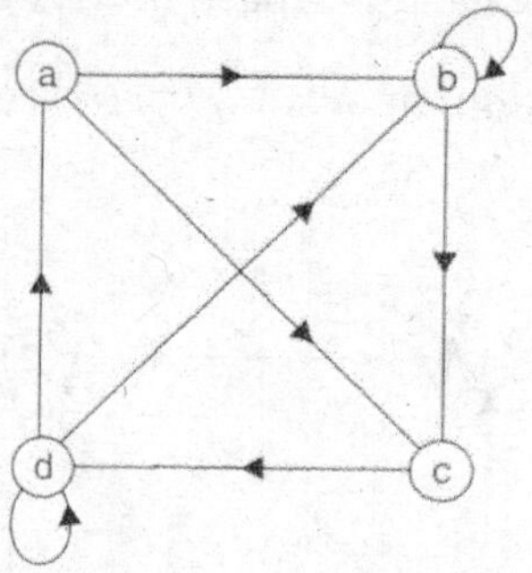

Fig. 30. Directed Graph of R.

Sol. The elements of relation R from the directed graph is

$$R = \{(a, b), (a, c), (b, b), (b, c), (c, d), (d, d), (d, a), (d, b)\}$$

(*i*) The inverse of relation R is

$$R^{-1} = \{(b, a), (c, a), (b, b), (c, b), (d, c), (d, d), (a, d), (b, d)\}.$$

The directed graph of R^{-1} is as shown in Fig. 31.

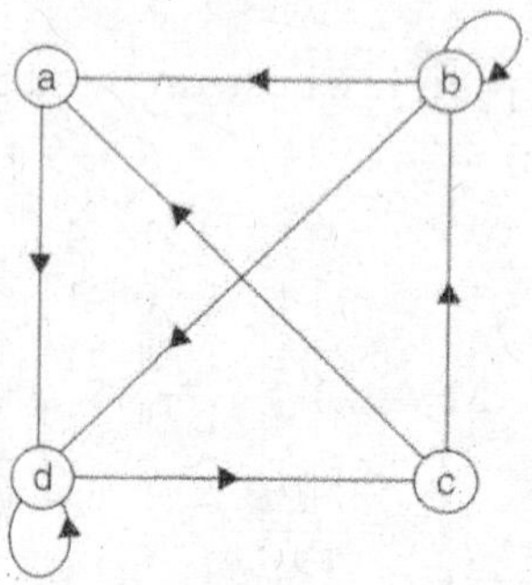

Fig. 31. Directed Graph of R^{-1}.

(*ii*) The universal relation R × R on set {*a*, *b*, *c*, *d*} is

$R \times R = \{(a, a), (a, b), (a, c), (a, d), (b, b), (b, c), (b, d), (b, a), (c, c), (c, d), (c, b), (c, a), (d, d), (d, a), (d, b), (d, c)\}$.

The complement of R is

$$\overline{R} = [(a, a), (a, d), (b, d), (b, a), (c, a), (c, b), (c, c), (d, c)].$$

The directed graph of $\overline{R}$ is shown in Fig. 32.

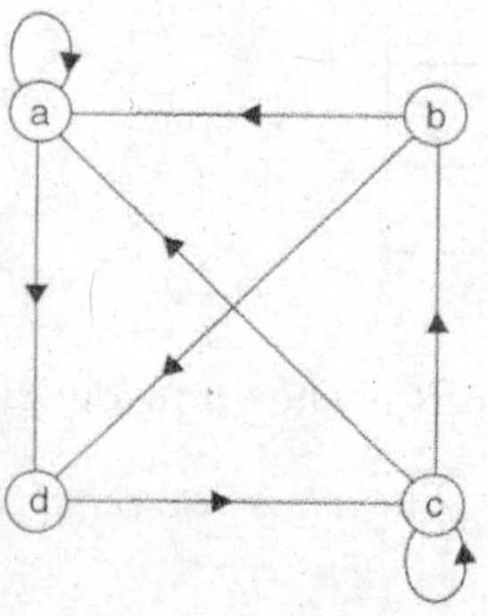

Fig. 32. Directed Graph of $\overline{R}$.

Problem 7. *Consider a relation R from a set A to B whose matrix is shown in Fig. 33. Determine is inverse R^{-1} and complement $\overline{R}$.*

$$M_R = \begin{bmatrix} 1 & 0 & 0 & 1 \\ 1 & 1 & 1 & 1 \\ 0 & 0 & 1 & 1 \\ 1 & 1 & 0 & 0 \end{bmatrix}$$

Fig. 33.

Sol. The elements of relation R is

$$R = \{(1, 1), (1, 4), (2, 1), (2, 2), (2, 3), (2, 4), (3, 3), (3, 4), (4, 1), (4, 2)\}$$

(*i*) The inverse of relation R is

$$R^{-1} = \{(1, 1), (4, 1), (1, 2), (2, 2), (3, 2), (4, 2), (3, 3), (4, 3), (1, 4), (2, 4)\}.$$

For writing in the matrix form, transpose the original matrix *i.e.*, (Fig. 34.).

$$M_R^{-1} = \begin{bmatrix} 1 & 1 & 0 & 1 \\ 0 & 1 & 0 & 1 \\ 0 & 1 & 1 & 0 \\ 1 & 1 & 1 & 0 \end{bmatrix}$$

Fig. 34.

(*ii*) The complement of relation R is taken by exchanging 0's with 1's and 1's with 0's. (Fig. 35).

$$M_{\overline{R}} = \begin{bmatrix} 0 & 1 & 1 & 0 \\ 0 & 0 & 0 & 0 \\ 1 & 1 & 0 & 0 \\ 0 & 0 & 1 & 1 \end{bmatrix}$$

Fig. 35.

So, $\overline{R} = \{(1, 2), (1, 3), (3, 1), (3, 2), (4, 3), (4, 4)\}$.

Problem 8. *Let R and S be two relation from A to B. Show that*

(*i*) $(R \cap S)^{-1} = R^{-1} \cap S^{-1}$ (*ii*) $(R \cup S)^{-1} = R^{-1} \cup S^{-1}$.

Sol. (*i*) Let $(a, b) \in (R \cap S)^{-1}$. So, we have $(b, a) \in (R \cap S)$. Now $(b, a) \in R$ and $(b, a) \in S$. This means $(a, b) \in R^{-1}$ and $(a, b) \in S^{-1}$.

Hence, $(a, b) \in R^{-1} \cap S^{-1}$ ....(*i*)

Conversely, let $(a, b) \in R^{-1} \cap S^{-1}$. So, we have $(a, b) \in R^{-1}$ and $(a, b) \in S^{-1}$. This means $(b, a) \in R$ and $(b, a) \in S$. So, $(b, a) \in (R \cap S)$.

Hence, $(a, b) \in (R \cap S)^{-1}$ ....(*ii*)

Form (*i*) and (*ii*), we have

$$(R \cap S)^{-1} = R^{-1} \cap S^{-1}$$

Hence proved.

(*ii*) Let $(a, b) \in (R \cup S)^{-1}$. So, we have $(b, a) \in (R \cup S)$

Now, $(b, a) \in R$ or $(b, a) \in S$. This means $(a, b) \in R^{-1}$ or $(a, b) \in S^{-1}$

Hence, $(a, b) \in R^{-1} \cup S^{-1}$ ...(*i*)

Conversely, let $(a, b) \in R^{-1} \cup S^{-1}$. So, we have $(a, b) \in R^{-1}$ or $(a, b) \in S^{-1}$. This means $(b, a) \in R$ or $(b, a) \in S$. So, $(b, a) \in R \cup S$.

Hence, $(a, b) \in (R \cup S^{-1})$ ....(*ii*)

From (*i*) and (*ii*), we have

$$(R \cup S)^{-1} = R^{-1} \cup S^{-1}.$$

Hence proved.

Problem 9. *Let* $A = \{k, l, m, n\}$

Let $R = \{(k, k), (l, l), (m, m), (k, l), (k, m), (l, m), (m, n), (n, k)\}$

Let $S = \{(n, k), (n, l), (n, m), (m, k), (m, l), (l, k), (k, k)\}$.

Find the composition (*i*) *RoR* (*ii*) *SoS.*

Sol. (*i*)The composition RoR is shown in Fig. 36.

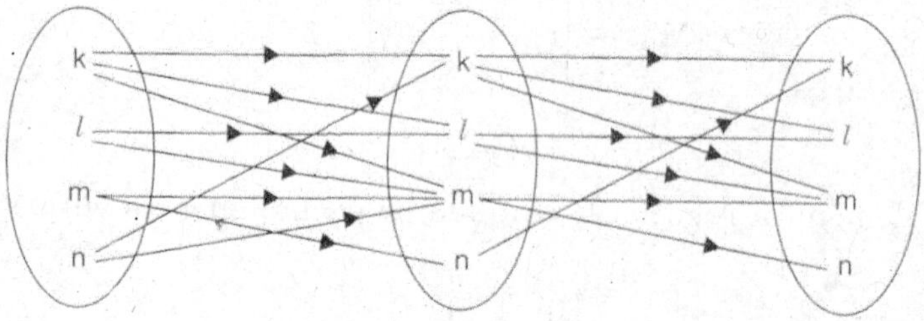

Fig. 36. RoR.

RoR = $\{(k, k), (k, l), (k, m), (k, n), (l, l), (l, m), (l, n), (m, m), (m, n), (m, k), (n, k), (n, l), (n, m)\}$

(*ii*) The composition SoS is shown in Fig. 37.

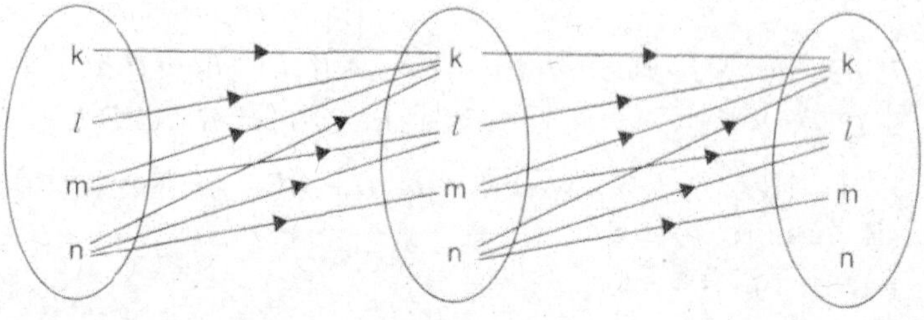

Fig. 37. SoS.

$$SoS = \{(k, k), (l, k), (m, k), (n, k), (n, l)\}$$

Problem 10. *Give an example of relations* R_1, R_2, R_3, R_4 *and* R_5 *on* $A = \{4, 5, 6, 7, 8\}$ *having property*

(*i*) R_1 *is reflexive and transitive but not symmetric.*

(*ii*) R_2 *is symmetric and antisymmetric.*

(*iii*) R_3 *is antisymmetric but not reflexive.*

(*iv*) R_4 *is neither symmetric nor antisymmetric.*

(*v*) R_5 *is neither symmetric and asymmetric nor antisymmetric.*

Sol. (*i*) $R_1 = \{(4, 4), (5, 5), (6, 6), (7, 7), (8, 8), (4, 5), (5, 6), (4, 6)\}$.

(*ii*) $R_2 = \{(4, 4), (5, 5), (6, 6), (7, 7)\}$.

(*iii*) $R_3 = \{(4, 4), (5, 5), (6, 6)\}$.

(*iv*) $R_4 = \{(4, 4), (4, 3), (3, 4), (4, 2)\}$.

(*v*) $R_5 = \{(4, 4), (4, 5), (5, 4), (4, 6)\}$.

Problem 11. *Give an example of relations R_1, R_2, R_3 and R_4 on $A = \{a, b, c, d\}$ having property*

(*i*) *R_1 is irreflexive and antisymmetric.*

(*ii*) *R_2 is asymmetric and antisymmetric.*

(*iii*) *R_3 is asymmetric but $R_3 \cup R_3^{-1}$ is symmetric.*

(*iv*) *R_4 is transitive but $R_4 \cup R_4^{-1}$ is not transitive.*

Sol. (*i*) $R_1 = \{(a, b), (b, c), (c, d), (d, a)\}$ (*ii*) $R_2 = \{(a, a), (b, b), (c, d), (b, d), (a, d)\}$

(*iii*) $R_3 = \{(a, b), (b, c), (c, d), (d, a)\}$ (*iv*) $R_4 = \{(a, c), (d, d)\}$.

Problem 12. *Give an example of a relation R on the set $A = \{4, 5, 6, 7\}$ which have no property i.e., the relation R is not reflexive, irreflexive, symmetric, asymmetric, transitive and antisymmetric.*

Sol. The relation R having no property is $R = \{(4, 4), (5, 5), (4, 5), (5, 4), (4, 7)\}$.

Problem 13. *Give an example of a relation R on the set $A = \{5, 6, 7\}$ which is reflexive, symmetric, transitive and antisymmetric i.e., a relation having all these four properties.*

Sol. The relation R having all the above four properties is

$$R = \{(5, 5), (6, 6), (7, 7)\}.$$

Problem 14. *Let A be a set of books.*

(*i*) *Let R_1 be a binary relation on A, such that (a, b) is in R_1 if book 'a' costs more and contains fewer pages than book 'b'. In general, is R_1 reflexive ? symmetric ? antisymmetric ? and transitive ?*

(*ii*) *Let R_2 is a binary relation on A, such that (a, b) is in R_2 if book 'a' costs more or contains fewer pages than book 'b'. In general is R_2 reflexive ? symmetric ? antisymmetric ? and transitive ?*

Sol. (*i*) **Reflexive.** A book does not costs more from itself and does not contain fewer pages than itself. Hence, $(a, a) \notin R_1$. Therefore, the relation is not reflexive.

Symmetric. If book a costs more and contains fewer pages than book *b*, it implies $(a, b) \in R_1$. But it is not possible that book *b* also costs more and contains fewer pages than book *a*. Hence $(b, a) \notin R_1$. Therefore, relation is not symmetric.

Antisymmetric. There is no such condition in the relation that $(a, b) \in R_1$ and $(b, a) \in R_1$ such that $a \neq b$. Hence, relation is antisymmetric.

Transitive. Let us assume that $(a, b) \in R_1$ and $(b, c) \in R_1$ *i.e.*, book a costs more and contains fewer pages than book *b* and costs more and contains fewer pages than book *c*. So, it is sure that book *a* costs more and contains fewer pages than book *c* *i.e.*, $(a, c) \in R_1$. Therefore, the relation is transitive.

(*ii*) **Reflexive.** A book *a* doesn't costs more or contains fewer pages than itself. Hence $(a, a) \notin R_2$. Therefore, the relation is not reflexive.

Symmetric. Let $(a, b) \in R_2$ *i.e.*, book a costs more or contains fewer pages than book b. But it does not imply that book b costs more or contains fewer pages than book b. Hence, $(b, a) \notin R_1$. Therefore, relation is not always symmetric.

Antisymmetric. Sometimes, there exists condition that $(a, b) \in R$ and $(b, a) \in R$ but $a \neq b$. Hence, the relation is not antisymmetric.

Transitive. Let $(a, b) \in R$ and $(b, c) \in R$ *i.e.*, book a costs more than b and b contains fewer pages than book c. So, we can have that book a costs less or does not contains fewer pages than book c. Hence, the relation is not always transitive.

Problem 15. *Determine whether the relation* $R = \{(a, b) \in R, a - b \leq 1$ *on the set* I_+ *(set of +ve integer)} is*

(*i*) *reflexive* (*ii*) *symmetric* (*iii*) *transitive*

(*iv*) *antisymmetric* (*v*) *a partial order relation* (*vi*) *an equivalence relation.*

Sol. (*i*) **Reflexive.** The relation is reflexive because $a - a \leq 1$, for every $a \in I_+$. So $(a,a) \in R$ for every $a \in I_+$.

(*ii*) **Symmetric.** The relation is symmetric because whenever $a - b \leq 1$, we have $b - a \in 1$. So, for $(a, b) \in I_+$ we also have $(b, a) \in I_+$.

(*iii*) **Transitive.** The relation is not transitive because when some $a - b \leq 1$ and $(b - c) \leq 1$ we have $a - c > 1$ *e.g.*, $(1, 2) \in R$ and $(2, 3) \in R$ but $(1, 3) \notin R$.

(*iv*) **Antisymmetric.** The relation is not antisymmetric because whenever $a - b \leq 1$ and $b - a < 1$, we have $a \neq b$ *e.g.*, $(1, 2) \in R$ and $(2, 1) \in R$ but $1 \neq 2$.

(*v*) **Partial Order Relation.** The relation is not partial order relation as it is reflexive but not transitive and antisymmetric.

(*vi*) **Equivalence Relation.** The relation is not an equivalence relation as it is reflexive and symmetric but not transitive.

Problem 16. *Determine whether the relation*

$R = \{(a, b) \in R : a + b$ *is even. On the set* I_+ *(set of +ve integers) is* (*i*) *reflexive* (*ii*) *symmetric* (*iii*) *transitive* (*iv*) *antisymmetric* (*v*) *a partial order relation* (*vi*) *an equivalence relation.*

Sol. (*i*) **Reflexive.** The relation is reflexive because when a number whether it is even or odd added to itself, it always is even. So, for every $a \in I_+$ $(a, a) \in R$.

(*ii*) **Symmetric.** The relation is symmetric because whenever $a + b$ is even, $b + a$ is also even. So, for each $(a, b) \in R$ we have $(b, a) \in R$.

(*iii*) **Transitive.** The relation is transitive because whenever $a + b$ is even $b + c$ is even, we have $a + c$ is even. So, whenever $(a, b) \in R$ and $(b, c) \in R$ we have $(a, c) \in R$.

(*iv*) **Antisymmetric.** The relation is not antisymmetric because we have elements $(a, b) \in R$ and $(b, a) \in R$ such that $a \neq b$ *e.g.*, $(2, 4) \in R$ and $(4, 2) \in R$ but $2 \neq 4$.

(*v*) **Partial Order Relation.** The relation is not a partial order relation as it is reflexive and transitive but not antisymmetric.

(*vi*) **Equivalence Relation.** The relation is an equivalence relation as it is reflexive, symmetric and transitive also.

Problem 17. *Let R be the relation on set* $A = \{a, b, c, d\}$ *defined by*

$$R = [(a, b), (b, c), (d, c), (d, a), (a, d), (d, d)]$$

Determine (*i*) *Reflexive closure of R*

(*ii*) *Symmetric closure of R*

(*iii*) *Transitive closure of R.*

Sol. (*i*) The smallest relation R having the reflexive property is

$$R \cup \Delta = \{(a, a), (b, b), (c, c), (d, d), (a, b), (b, c), (d, c), (d, a), (a, d)\}$$

(*ii*) The smallest relation R having symmetric property is

$$R \cup R^{-1} = \{(a, b), (b, a), (b, c), (c, b), (d, c), (c, d), (d, a), (a, d), (d, d)\}$$

(*iii*) To find the transitive closure of R, draw the directed graph of R as shown in Fig. 38.

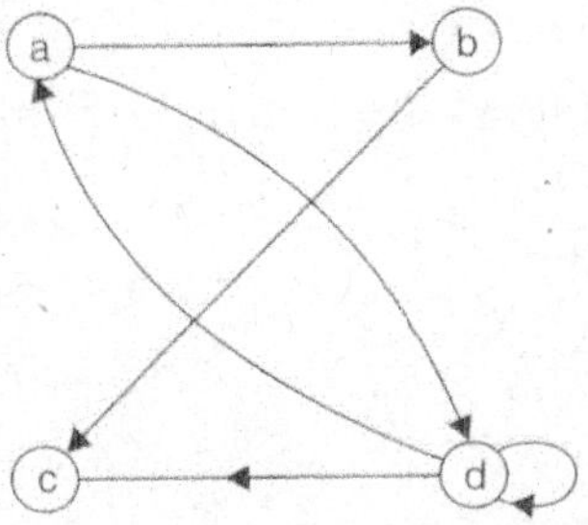

Fig. 38.

Now, from vertex a, we have paths to the vertices a, b, c and d. So, the ordered pairs included in R* are (a, a) (a, b) (a, c) and (a, d). From vertex b, we have paths to vertex c. So, ordered pairs included in R* are (b, c). From vertex c, we have no paths to any other vertex. Hence, no ordered pair is included in R*. Similarly, from vertex d, we have paths to vertices a, b, c and d. So, the ordered pairs included in R* are (d, a), (d, b), (d, c) and (d, d).

Therefore, transitive closure of R is

$$R^* = \{(a, a), (a, b), (a, c), (a, d), (b, c), (d, a), (d, b), (d, c), (d, d)\}.$$

Problem 18. *Determine whether the relation R on the set A of all triangles in the plane defined by*

$$R = \{(a, b) : \text{triangle } a \text{ is similar to triangle } b\} \text{ is an equivalence relation.}$$

Sol. Reflexive. Triangle a is similar to itself. Hence $(a, a) \in R$ for every $a \in A$. Therefore, R is reflexive.

Symmetric. If triangle a is similar to b, then it is sure that triangle b is also similar to a. Hence, whenever $(a, b) \in R$, we have (b, a) also belong to R. So, R is symmetric.

Transitive. It triangle a is similar to b and triangle b is similar to c.

We have, triangle a similar to c.

Hence, whenever $(a, b) \in R$ and $(b, c) \in R$, we have $(a, c) \in R$.

Therefore, R is transitive.

So, being reflexive, symmetric and transitive ; the relation R is an equivalence relation.

Problem 19. *Determine whether the relation* $S = \{(a, b) : a \geq b\}$ *on the set R of real numbers is an equivalence relation.*

Sol. Reflexive. We have $(a, a) \in S$, for every $a \in R$ because $a = a$ for every $a \in R$. Hence, the relation is reflexive.

Symmetric. Let $(a, b) \in S$. So, we have $a \geq b$ but for this it is not possible that $b \geq a$. Hence, $(b, a) \notin S$. Therefore, relation is not symmetric.

Transitive. Let $(a, b) \in S$ and $(b, c) \in S$. So, when we have $a \geq b$ and $b \geq c$, it is sure that $a \geq c$. Hence, $(a, c) \in S$. Therefore, the relation is transitive.

So, being the relation reflexive and transitive but not symmetric, it is not an equivalence relation.

Problem 20. *Prove that the relation R "$a - b$ is divisible by 5" $\forall\ a, b \in I_+$ (a set of + ve integers) is an equivalence relation.*

Sol. Reflexive. The relation is reflexive because $a - a$ is divisible by 5 for every $a \in I_+$.

Symmetric. Let $(a, b) \in R$.

If $a - b$ is divisible by 5 then it is sure that $b - a$ is also divisible by 5. Hence, $(b, a) \in R$. Therefore, the relation is symmetric.

Transitive. Let $(a, b) \in R$ and $(b, c) \in R$

i.e., $a - b$ is divisible by 5.

$b - c$ is divisible by 5.

It implies $a - c$ is divisible by 5.

Hence, $(a, c) \in R$.

Therefore, the relation is transitive.

So, being reflexive, symmetric and transitive, the relation "$a - b$ is divisible by 5" is an equivalence relation. Hence proved.

Problem 21. *If R and S are equivalence relations on the set A. Show that following are equivalence relations*

(*i*) $R \cap S$ (*ii*) $R \cup S$.

Sol. (*i*) **Reflexive.** The relations R and S are equivalence relations. We have $(a, a) \in R$ and $(a, a) \in S$, for every $a \in A$. So, $(a, a) \in R \cap S$ for every $a \in A$. Therefore, $R \cap S$ is reflexive.

Symmetric. Let $(a, b) \in R \cap S$

So, we have $(a, b) \in R$ also $(b, a) \in R$ ($\because$ R is symmetric)

Similarly, we have $(a, b) \in S$

Also $(b, a) \in S$ ($\because$ S is symmetric)

Thus, from above, we conclude that $(b, a) \in R \cap S$

Therefore, $R \cap S$ is symmetric.

Transitive. Let $(a, b) \in R \cap S$ and $(b, c) \in R \cap S$.

So, we have $(a, b) \in R$ and $(b, c) \in R$

Also, $(a, c) \in R$ ($\because$ R is transitive)

Similarly, we have $(a, b) \in S$ and $(b, c) \in S$

Also, $(a, c) \in S$ ($\because$ S is transitive)

Thus, from above we conclude that $(a, c) \in R \cap S$.

Therefore, $R \cap S$ is transitive.

So, being reflexive, symmetric and transitive, $R \cap S$ is an equivalence relation.

(*ii*) **Reflexive.** We have $(a, a) \in R$ and $(a, a) \in S$, for every $a \in A$. Hence, $(a, a) \in R \cup S$. Therefore, the $R \cup S$ is reflexive.

Symmetric. Let $(a, b) \in R \cup S$.

So, we have $(a, b) \in R \Rightarrow (b, a) \in R$ ($\because$ R is symmetric)

or $(a, b) \in S \Rightarrow (b, a) \in S$ ($\because$ S is symmetric)

Hence, from above we conclude that $(b, a) \in R \cup S$

Therefore, $R \cup S$ is symmetric.

Transitive. Let $(a, b) \in R \cup S$ and $(b, c) \in R \cup S$.

So, we have $(a, b) \in R$ and $(b, c) \in R \Rightarrow (a, c) \in R$ ($\because$ R is transitive)

or $(a, b) \in S$ and $(b, c) \in S \Rightarrow (a, c) \in S$ ($\because$ S is transitive)

Hence, from above we conclude that $(a, c) \in R \cup S$

Therefore, $R \cup S$ is transitive.

So, relation $R \cup S$ being reflexive, symmetric and transitive is an equivalence relation.

Problem 22. *Let* $A = \{a, b, c, d, e\}$ *and*

$$R = \{(a, b), (a, a), (b, a), (b, b), (c, c), (d, d), (d, e), (e, d), (e, e)\}$$

$$S = \{(a, a), (b, b), (c, c), (d, d), (e, e), (a, c), (c, a), (d, e), (e, d)\}$$

be equivalence relations on A. Determine the partitions corresponding to following (if it is an equivalence relation).

(*i*) R^{-1} (*ii*) $R \cup S$ (*iii*) $R \cap S$.

Sol. (*i*) The relation R^{-1} is

$$R^{-1} = \{(b, a), (a, a), (a, b), (b, b), (c, c), (d, d), (e, d), (d, e), (e, e)\}$$

The partition corresponding R^{-1} is

$$P = \{\{a, b\}, \{c\}, \{d, e\}\}.$$

(*ii*) The relation $R \cup S$ is

$$R \cup S = \{(a, a), (b, b), (c, c), (d, d), (e, e), (a, b), (b, a), (a, c), (c, a), (e, d), (d, e)\}$$

The relation $R \cup S$ is not an equivalence relation.

(*iii*) The relation $R \cap S$ is

$$R \cap S = \{(a, a), (b, b), (c, c), (d, d), (e, e), (d, e), (e, d)\}$$

The partition corresponding to $R \cap S$ is

$$P = \{\{a\}, \{b\}, \{c\}, \{d, e\}\}.$$

Problem 23. *Let R is an equivalence relation on the set* $A = \{p, q, r, s\}$ *defined by partition* $P = \{\{p, s\}, \{q, r\}\}$. *Determine the elements of equivalence relation and also find the equivalence classes of R.*

Sol. The elements of equivalence relation defined by partition P is

$$R = \{(p, p), (s, s), (p, s), (s, p), (q, q), (r, r), (q, r), (r, q)\}$$

The equivalence classes of R are

$$[p] = [s] = \{p, s\}$$

$$[q] = [r] = \{q, r\}.$$

Problem 24. *Let R is an equivalence relation on the set* $A = \{7, 8, 9, 10\}$ *defined by partition* $P = \{(7), (8), (9), (10)\}$. *Determine the elements of equivalence relation and also find the equivalence classes of R.*

Sol. The elements of equivalence relation defined by partition P is

$$R = \{(7, 7), (8, 8), (9, 9), (10, 10)\}.$$

The equivalence classes of R are

$$[7] = \{7\}, \qquad [8] = \{8\}$$

$$[9] = \{9\}, \qquad [10] = \{10\}.$$

Problem 25. *Let A = {1, 2, 3, 4} and R = {(2, 1), (2, 3), (3, 2), (4, 3)}. Find the transitive closure of R using Warshall's algo.*

Sol. The diagraph of relation R and its corresponding matrix M_R is shown in Fig. 39.

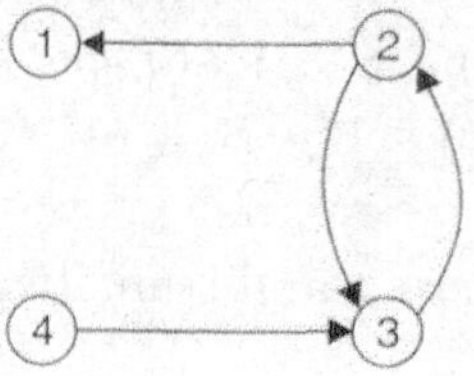

$$M_R = \begin{array}{c} \\ 1 \\ 2 \\ 3 \\ 4 \end{array} \begin{array}{c} \begin{matrix} 1 & 2 & 3 & 4 \end{matrix} \\ \begin{bmatrix} 0 & 0 & 0 & 0 \\ 1 & 0 & 1 & 0 \\ 0 & 1 & 0 & 0 \\ 0 & 0 & 1 & 0 \end{bmatrix} \end{array}$$

Fig. 39.

The value of $n = |A| = 4$. Thus, we have to find the Warshall sets w_0, w_1, w_2, w_3, w_4. The first set w_0 is same as M_R and is shown below :

$$w_0 = M_R = \begin{bmatrix} 0 & 0 & 0 & 0 \\ 1 & 0 & 1 & 0 \\ 0 & 1 & 0 & 0 \\ 0 & 0 & 1 & 0 \end{bmatrix}$$

Now to find w_1 from w_0, we have row number 2 from column 1 in w_0. But there are no column numbers for row 1 in w_0. Thus, no new entries of 1 are added in w_1. So, w_1 is same as w_0 and is shown below :

$$w_1 = \begin{bmatrix} 0 & 0 & 0 & 0 \\ 1 & 0 & 1 & 0 \\ 0 & 1 & 0 & 0 \\ 0 & 0 & 1 & 0 \end{bmatrix}$$

To find w_2 from w_1, we have row number 3 for column 2 and column numbers 1 and 3 for row 2. Thus, the new entries in w_2 are (3, 1) and (3, 3). Thus, w_2 is shown as follows :

$$w_2 = \begin{bmatrix} 0 & 0 & 0 & 0 \\ 1 & 0 & 1 & 0 \\ 1 & 1 & 1 & 0 \\ 0 & 0 & 1 & 0 \end{bmatrix}$$

Similarly, w_3 is obtained from w_2. Here we have row number 2, 3 and 4 for column 3 and column numbers 1, 2 and 3 for row 3. Thus, the possible new entries in w_3 are (2, 1), (2, 2), (2, 3), (3, 1), (3, 2), (3, 3), (4, 1), (4, 2) and (4, 3). Thus, w_3 is as follows :

$$w_3 = \begin{bmatrix} 0 & 0 & 0 & 0 \\ 1 & 1 & 1 & 0 \\ 1 & 1 & 1 & 0 \\ 1 & 1 & 1 & 0 \end{bmatrix}$$

Similarly, w_4 can be obtained from w_3. But there are no new entries of 1's in w_4 as there is row number for column 4, where there is 1 in w_3. Hence,

$$M_{R^\infty} = w_4 = w_3$$

which is the transitive closure of R.

5

Functions

DEFINITION

A function f from a set P into a set Q is a relation from P to Q such that each element of P is related to exactly one element of the set Q. It is denoted as $f : P \to Q$ and read as "f is a function from P to Q".

FUNCTIONS AS A SET

If P and Q are any two non-empty sets, then a function f from P to Q is a subset of $P \times Q$, with two important restrictions

(*i*) $\forall a \in P, (a, b) \in f$ for some $b \in Q$

(*ii*) If $(a, b) \in f$ and $(a, c) \in f$ then $b = c$.

Note 1. There may be some elements of the Q which are not related to any element of set P.

2. Every element of P must be related with at least one element of Q.

Example 1. *If a set A has n elements, how many functions are there from A to A.*

Sol. If a set A has n elements, then there are n^n functions from A to A.

Example 2. *If A has m elements and B has n elements, how many functions are there from A to B and from B to A.*

Sol. A has m elements and B has n elements. So, the total number of functions from A to B are n^m and the total number of functions from B to A are m^n.

Example 3. *If A = {2, 3, 4} and B = {5, 6}. Determine all functions from A to B.*

Sol. The total number of functions from $A \to B$ are $2^3 = 8$

(*i*) {(2, 5), (3, 5), (4, 5)}

(*ii*) {(2, 6), (3, 6), (4, 6)}

(*iii*) {(2, 5), (3, 5), (4, 6)}

(*iv*) {(2, 5), (3, 6), (4, 6)}

(*v*) {(2, 6), (3, 5), (4, 5)}

(*vi*) {(2, 6), (3, 6), (4, 5)}

(*vii*) {(2, 6), (3, 5), (4, 6)}

(*viii*) {(2, 5), (3, 6), (4, 5)}.

DOMAIN OF A FUNCTION

Let f be a function from P to Q. The set P is called domain of the function f.

CO-DOMAIN OF A FUNCTION

Let f be a function from P to Q. The set Q is called co-domain of the function f.

IMAGE OF AN ELEMENT

If the element x of P corresponds to y under function f, then y is the image of x under f and is written as

$$f(x) = y.$$

If $f(x) = y$, then we say that x is a per image of y.

If $f : X \rightarrow Y$, then each element of P has unique image in Q, whereas every element in Q need not be image of some x in P.

Example 4. *Let A = {1, 2, 3, 4} and the function f : A → A as shown in Fig. 1. Find (i) the image of each element of A (ii) the image f(A) of the function f.*

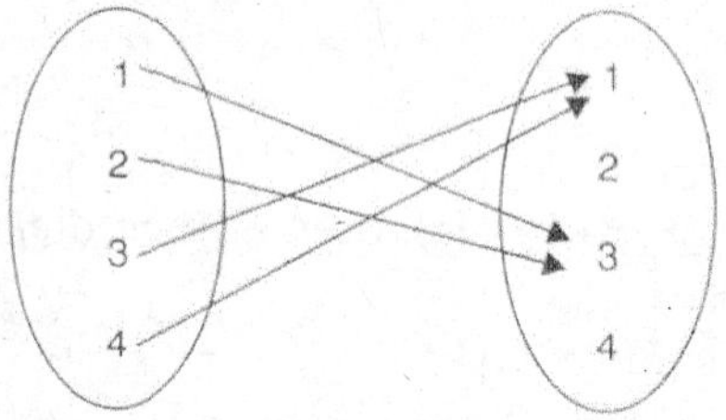

Fig. 1.

Sol. (*i*) The image of $f(1) = 3$, $f(2) = 3$, $f(3) = 1$ and $f(4) = 1$.

(*ii*) The image of $f(A)$ of f consists of all the image values.

Therefore, $f(A) = \{1, 3\}$.

RANGE OF A FUNCTION

The range of a function is the set of images of its domain. In other words, we can say it is a subset of its co-domain. It is denoted as f(domain).

If $f : P \rightarrow Q$, then $f(P) = \{f(x) : x \in P\} = \{y : y \in Q \mid \exists\, x \in P, \text{ such that } f(x) = y\}$.

Example 5. *Let P = {x, y, z, u} and Q = {a, b, c, d} and f : P → Q, such that*

$$f = \{(x, a), (y, b), (z, c), (u, c)\}.$$

Find the domain, co-domain and range of function.

Sol. Domain of function f is the set P

Co-domain of function f is the set Q and range of the function f is $\{a, b, c\}$.

Example 6. *Let A = {2, 3, 4} and B = {a, b, c} and f = {(2, a), (3, b), (4, b)}*

Find domain, co-domain and range of the function.

Sol. Domain of function is Domain $(f) = \{2, 3, 4\}$

Co-domain of function is co-domain $(f) = \{a, b, c\}$

Range of function is Range $(f) = \{a, b\}$.

REPRESENTATION OF A FUNCTION

We can represent a function by a diagram also. The two sets P and Q are represented by two circles. The function $f : P \to Q$ is represented by a collection of arrows joining the points which represent the elements of P and corresponding elements of Q

e.g., Let $P = \{a, b, c\}$ and $Q = \{x, y, z\}$ and $f : P \to Q$ such that

$$f = \{(a, x), (b, z), (c, x)\}$$

Then f can be represented diagrammatically as follows (Fig. 2).

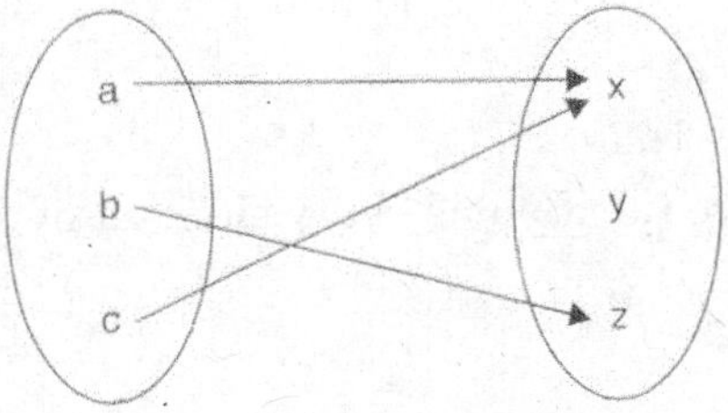

Fig. 2.

Example 7. *Let $X = \{x, y, z, k\}$ and $Y = \{1, 2, 3, 4\}$. Determine which of the following are functions. Give reasons if it is not. Find range if it is a function*

(*i*) $f = \{(x, 1), (y, 2), (z, 3), (k, 4)\}$ (*ii*) $g = \{(x, 1), (y, 1), (k, 4)\}$

(*iii*) $h = \{(x, 1), (x, 2), (x, 3), (x, 4)\}$ (*iv*) $l = \{(x, 1), (y, 1), (z, 1), (k, 1)\}$

(*v*) $d = \{(x, 1), (y, 2), (y, 3), (z, 4), (z, 4)\}$.

Sol. (*i*) It is a function. Range $(f) = \{1, 2, 3, 4\}$.

(*ii*) It is not a function because every element of X does not relate with some element of Y *i.e.*, Z is not related with any element of Y.

(*iii*) h is not a function because $h(x) = \{1, 2, 3, 4\}$ *i.e.*, element x has more than one image in set Y.

(*iv*) It is a function. Range $(l) = \{1\}$.

(*v*) d is not a function because $d(y) = \{2, 3\}$ *i.e.*, element y has more than one image in set Y.

EVERYWHERE DEFINED FUNCTION

Consider a function f from A to B. Then the function f is everywhere defined if dom $(f) = A$.

Example 8. *Let $A = \{1, 2, 3, 4\}$ and $B = \{a, b, c\}$ and $C = \{\alpha, \beta, \gamma\}$. Consider the following two functions from B to C and A to C.*

(*i*) $f = \{(a, \alpha), (b, \beta), (c, \gamma)\}$ (*ii*) $g = \{(1, \alpha), (3, \beta), (2, \gamma)\}$

Determine whether or not each function is everywhere defined.

Sol. (*i*) f is everywhere defined since dom $(f) = B$.

(*ii*) g is not everywhere defined since dom $(f) = \{1, 2, 3\}$ and $A = \{1, 2, 3, 4\}$ which is not equal.

TYPES OF FUNCTIONS

(*a*) **Injective (One-to-One) Functions.** Let $f : X \to Y$. The function f is called one-to-one or injective if different elements in X have different images in Y *i.e.*, if $f(a) = f(a') \Rightarrow a = a'$, $\forall\, a, a' \in X$.

Another way of defining injective function is that every element of domain X has a unique image in the co-domain Y and there is no element of Y which is image of more than one element of domain X.

For example : Consider, $X = \{x, y, z, k\}$ and $Y = \{1, 2, 3, 4\}$ and f is function from X to Y such that

$$f = \{(x, 1), (y, 2), (z, 3), (k, 4)\}$$

The function f is injective function as every element of domain X has a unique image in the co-domain Y (Fig. 3).

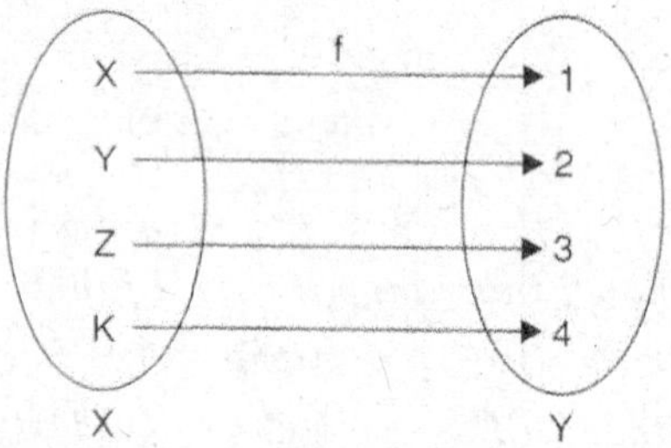

Fig. 3.

(*b*) **Surjective (Onto) Functions.** Let $f : X \to Y$. The function f is called surjective function if each element in Y, is the image of at least one element in X. In other words, in surjective functions, the range of f is equal to co-domain Y *i.e.*, $\forall\, b \in Y$, $b = f(a)$ for some $a \in X$.

For example : Consider, $X = \{1, 2, 3, 4, 5\}$, $Y = \{a, b, c, d\}$ and $f = \{(1, a), (2, a), (3, b), (4, c), (5, d)\}$.

It is a surjective function, as every element of Y is the image of some of X (Fig. 4).

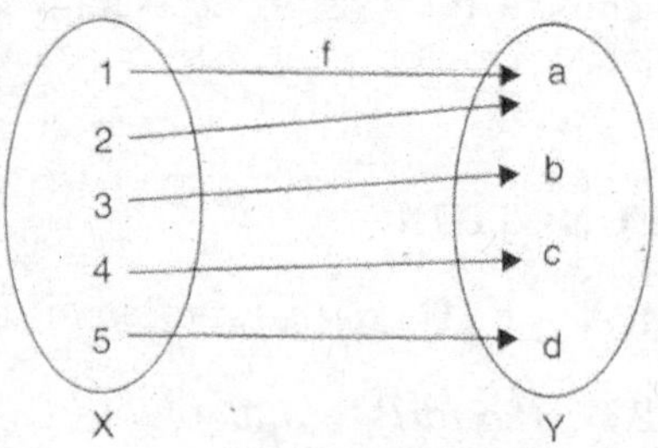

Fig. 4.

(*c*) **Bijective (One-to-One Onto) Functions.** A function which is both injective (one-to-one) and surjective (onto) is called a bijective (one-to-one-onto) function.

For example : Consider, $P = \{x, y, z\}$, $Q = \{a, b, c\}$ and $f : P \to Q$ such that

$$f = \{(x, a), (y, b), (z, c)\}$$

The f is one-to-one and also it is onto. So it is a bijective function (Fig. 5).

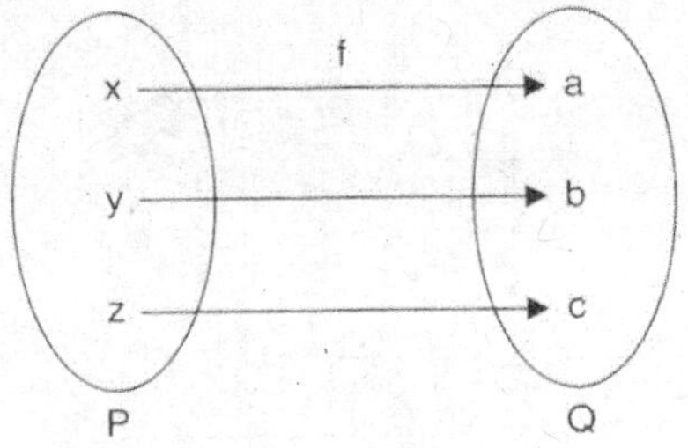

Fig. 5.

(*d*) **Into Functions.** Let $f : X \to Y$. The function f is called an into function if the range of f is not equal to the co-domain Y. Therefore, there must be an element of co-domain Y which is not the image of any element of domain X.

For example : Consider, $X = \{1, 2, 3\}$, $Y = \{k, l, m, n, p\}$ and $f : X \to Y$ such that

$$f = \{(1, k), (2, n), (3, p)\}.$$

In the function f, the range *i.e.*, $\{k, n, p\} \neq$ co-domain of Y *i.e.*, $\{k, l, m, n, p\}$

Therefore, it is an into function (Fig. 6).

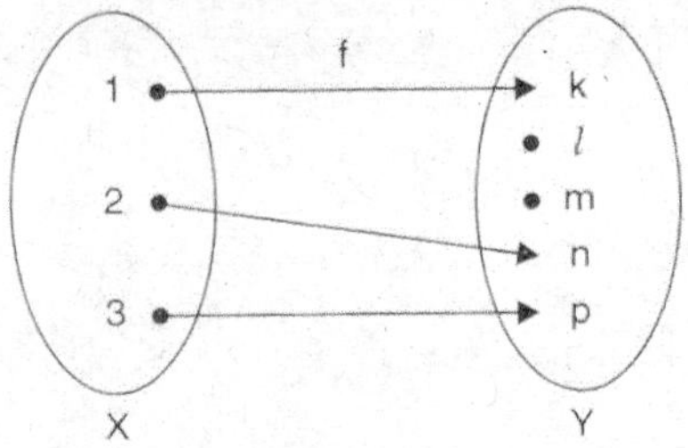

Fig. 6.

(*e*) **One-One Into Functions.** Let $f : X \to Y$. The function f is called one-one into function if different elements of X have different unique images of Y.

For example : Consider, $X = \{k, l, m\}$, $Y = \{1, 2, 3, 4\}$ and $f : X \to Y$ such that

$$f = \{(k, 1), (l, 3), (m, 4)\}$$

The function f is one-one into function (Fig. 7).

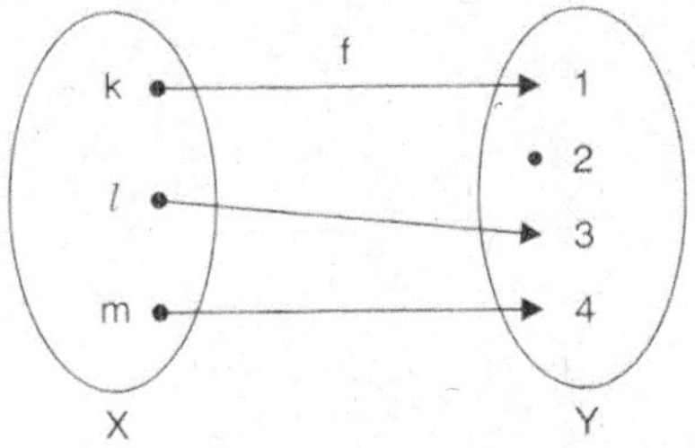

Fig. 7.

(*f*) **Many One Functions.** Let $f : X \to Y$. The function f is said to be many one function if there exists two or more than two different elements in X having the same image in Y.

For example : Consider $X = \{1, 2, 3, 4, 5\}$, $Y = \{x, y, z\}$ and $f : X \to Y$ such that

$$f = \{(1, x), (2, x), (3, x), (4, y), (5, z)\}$$

The function f is a many one function. (Fig. 8)

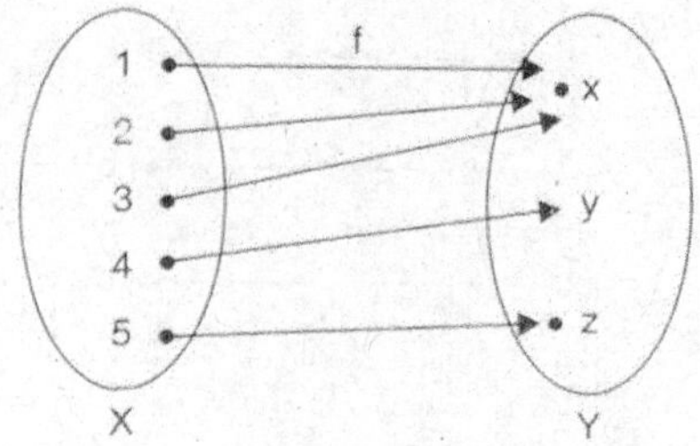

Fig. 8.

(g) **Many One Into Functions.** Let $f : X \to Y$. The function f is called many-one into function if and only if it is both many one and into function.

For example : Consider $X = \{a, b, c\}$, $Y = \{1, 2\}$ and $f : X \to Y$ such that

$$f = \{(a, 1), (b, 1), (c, 1)\}$$

As the function f is many-one and into, so it is a many-one into function (Fig. 9).

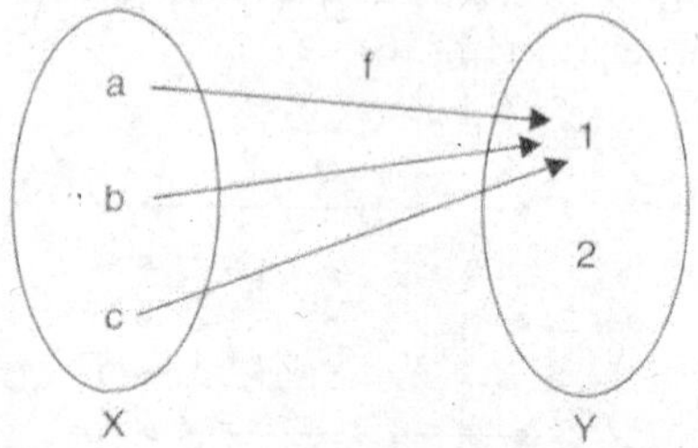

Fig. 9.

(h) **Many One Onto Functions.** Let $f : X \to Y$. The function f is called many-one-onto function if and only if it is both many one and onto.

For example : Consider, $X = \{1, 2, 3, 4\}$, $Y = \{k, l\}$ and $f : X \to Y$ such that

$$f = \{(1, k), (2, k), (3, l), (4, l)\}$$

The function f is many-one (as two elements have the same image in Y) and it is onto (as every element of Y is the image of some element X). So, it is many-one onto function (Fig. 10).

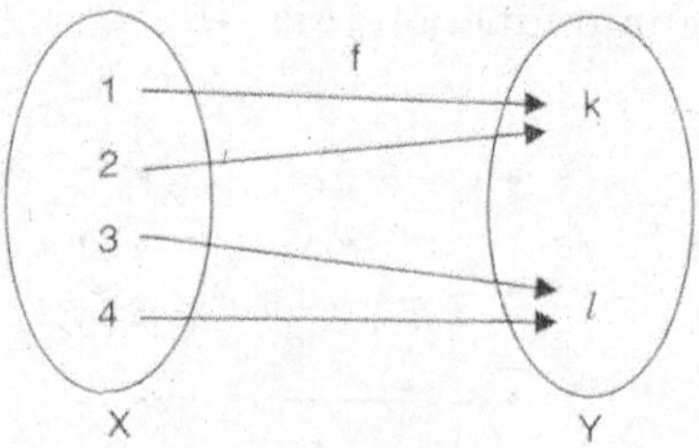

Fig. 10.

EQUAL FUNCTIONS

Consider two functions f and h from a set X to a set Y. The functions f and h are called equal functions if and only if $f(a) = h(a)$, for every $a \in X$.

The functions f and h are called unequal functions if there exist at least one element $a \in X$ such that $f(a) \neq h(a)$.

Example 9. *Let X = {1, 2, 3} and Y = {a, b, c}. Consider the function* $f : X \to Y$, $g : X \to Y$ *and* $h : X \to Y$ *such that*

$$f = \{(1, a), (2, a), (3, c)\}$$
$$g = \{(1, b), (2, a), (3, c)\}$$
$$h = \{(1, a), (2, a), (3, c)\}$$

Determine which functions are equal and which are unequal.

Sol. The functions f and h are equal functions. The functions f and g and g and h are unequal functions.

IDENTITY FUNCTIONS

Consider any set A. Let the function $f : A \to A$. The function f is called the identity function if each element of set A has image on itself *i.e.*, $f(a) = a \ \forall \ a \in A$

It is denoted by I.

For example : Consider, A = {1, 2, 3, 4, 5} and $f : A \to A$ such that

$$f = \{(1, 1), (2, 2), (3, 3), (4, 4), (5, 5)\}.$$

The function f is an identity function as each element of A is mapped onto itself. The function f is one-one and onto (Fig. 11).

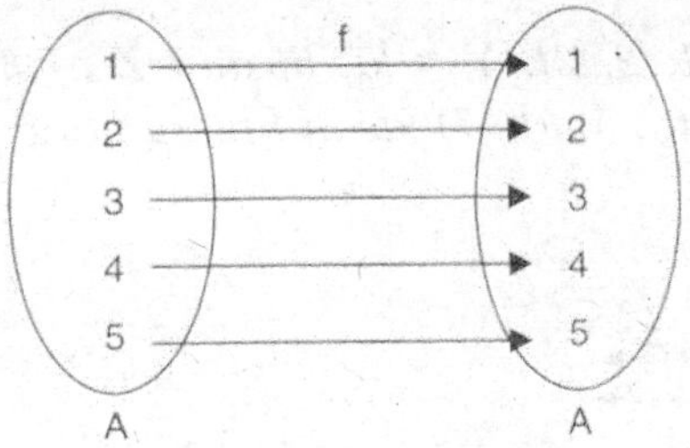

Fig. 11.

INVERTIBLE (INVERSE) FUNCTIONS

A function $f : X \to Y$ is invertible if and only if it is a bijective function.

Consider the bijective (one to one onto) function $f : X \to Y$. As f is one-to-one, therefore, each element of X corresponds to a distinct element of Y. As f is onto, there is no element of Y which is not the image of any element of X *i.e.*, range = co-domain Y.

The inverse function for f exists if f^{-1} is a function from Y to X.

For example : Consider, X = {1, 2, 3}, Y = {k, l, m} and $f : X \to Y$, such that

$$f = \{(1, k), (2, m), (3, l)\} \text{ as in Fig. 12.}$$

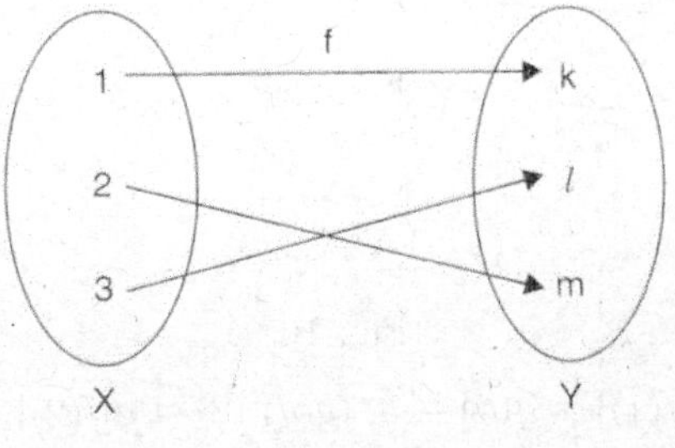

Fig. 12.

The inverse function of f is shown in Fig. 13.

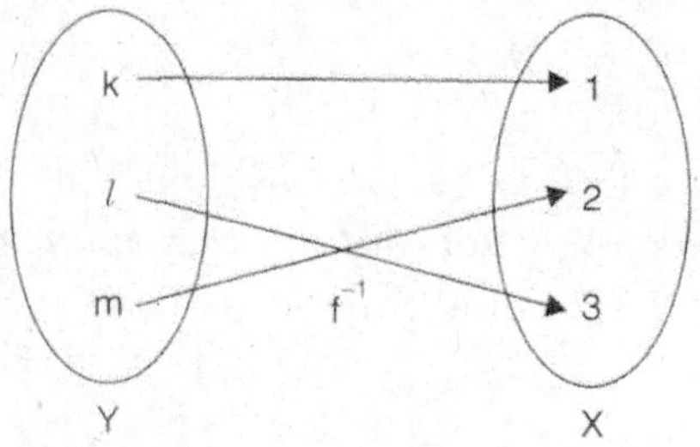

Fig. 13.

COMPOSITION OF FUNCTIONS

Consider functions, $f : A \to B$ and $g : B \to C$. The composition of f with g is a function from A into C defined by $(gof)(x) = g[f(x)]$ and is denoted by gof.

To find the composition of f and g, first find the image of x under f and then find the image of $f(x)$ under g.

Example 10. *Let* $X = \{1, 2, 3\}$, $Y = \{a, b\}$ *and* $Z = \{5, 6, 7\}$. *Consider the functions* $f = \{(1, a), (2, a), (3, b)\}$ *and* $g = \{(a, 5), (b, 7)\}$ *as in Figs. 14 and 15. Find the composition of gof.*

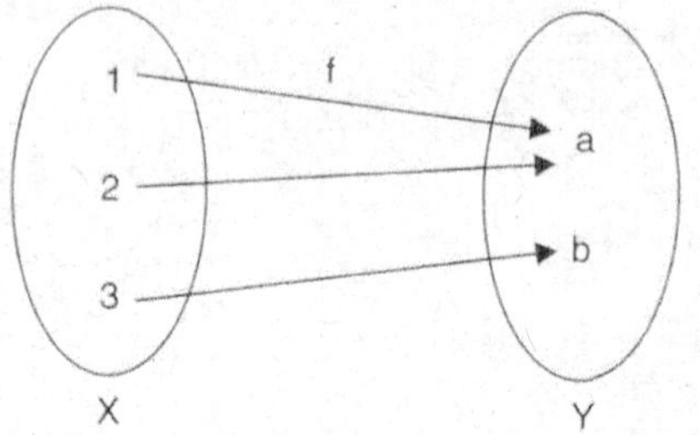

Fig. 14.

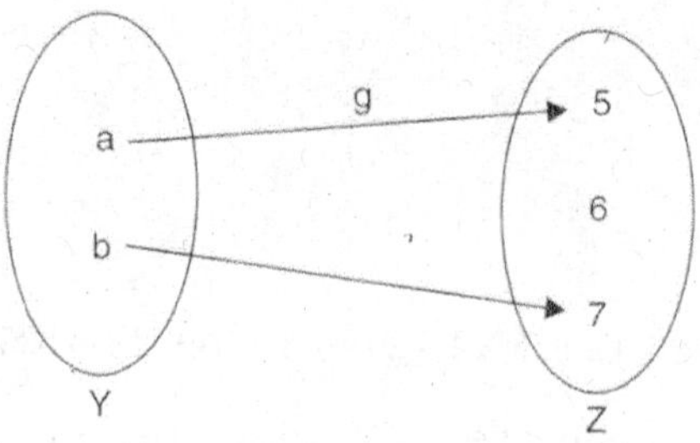

Fig. 15.

Sol. The composition function gof is shown in Fig. 16.

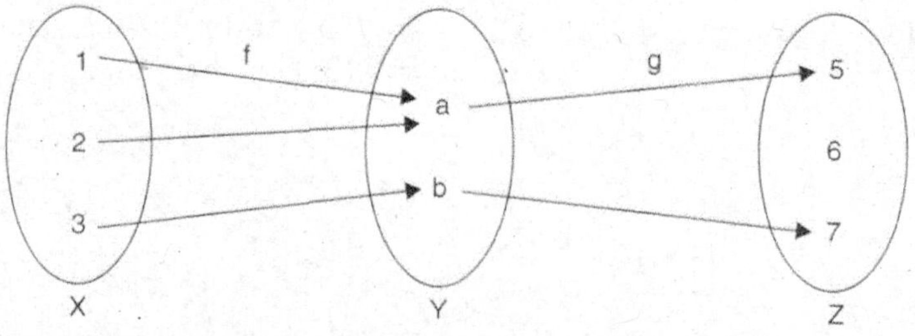

Fig. 16.

$(gof)(1) = g[f(1)] = g(a) = 5$, $(gof)(2) = g[f(2)] = g(a) = 5$

$(gof)(3) = g[f(3)] = g(b) = 7$.

Example 11. *Consider f, g and h, all functions on the integers, by $f(n) = n^2$, $g(n) = n + 1$ and $h(n) = n - 1$.*

Determine (i) hofog (ii) gofoh (iii) fogoh.

Sol. (*i*) $hofog(n) = n + 1,$

$$hofog(n + 1) = (n + 1)^2$$

$$h[(n + 1)^2] = (n + 1)^2 - 1 = n^2 + 1 + 2n - 1 = n^2 + 2n.$$

(*ii*) $gofoh(n) = n - 1,\ gof(n - 1) = (n - 1)^2$

$$g[(n - 1)^2] = (n - 1)^2 + 1 = n^2 + 1 - 2n + 1 = n^2 - 2n + 2.$$

(*iii*) $fogoh(n) = n - 1$

$$fog(n - 1) = (n - 1) + 1$$

$$f(n) = n^2.$$

Theorem I. *Prove that if $f : X \to Y$ and $g : Y \to Z$ are one to one functions, then gof is one-to-one.*

Proof. Consider two distinct elements a_1 and $a_2 \in X$.

Now $a_1 \neq a_2 \Rightarrow f(a_1) \neq f(a_2)$ ($\because$ f is one to one)

$\Rightarrow$ $g[f(a_1)] \neq g[f(a_2)]$ ($\because$ g is one to one)

$\Rightarrow$ $(gof)(a_1) \neq (gof)(a_2)$.

Therefore, the function *gof* has distinct elements $\in$ X that are mapped to distinct element of Z. Hence, *gof* is one-to-one function.

Theorem II. *Prove that if $f : X \to Y$ and $g : Y \to Z$ are into functions, then gof is onto.*

Proof. Consider an element $c \in Z$. As g is an onto function from $Y \to Z$, therefore, there exists $b \in Y$, such that $g(b) = c$.

Now, since f is onto function from X to Y, therefore, there exists $a \in X$ such that $f(a) = b$.

Thus $\forall\ c \in Z$, there exists $a \in X$ such that $C = g(b) = g[f(a)] = (gof)(a)$.

Thus, *gof* is a function of X onto Z. Hence, *gof* is an onto function.

Theorem III. *Prove that if $f : X \to Y$ and $g : Y \to Z$ be two one-to-one onto function, then gof is also one-to-one onto function.*

Proof. Assume two elements $a_1, a_2 \in A$ with $a_1 \neq a_2$.

As $f : X \to Y$ is a one-to-one onto function therefore, there exists distinct elements $b_1, b_2 \in Y$.

So, $f(a_1) = b_1,\ f(a_2) = b_2$ and $b_1 \neq b_2$ ...(1)

The function $g : Y \to Z$ is also one-to-one onto.

Therefore, there exists unique elements $c_1, c_2 \in Z$.

So $g(b_1) = c_1,\ g(b_2) = c_2$ and $c_1 \neq c_2$ ...(2)

The composition of (1) and (2) gives

$$(gof)(a_1) = g[f(a_1)] = g(b_1) = c_1$$

and $$(gof)(a_2) = g[f(a_2)] = g(b_2) = c_2$$

We know $c_1 \neq c_2$ so $(gof)(a_1) \neq (gof)(a_2)$

As $a_1 \neq a_2 \Rightarrow (gof)(a_1) = (gof)(a_2)$.

Therefore, *gof* is one-to-one function.

Now, assume an element $c \in Z$.

As $g : Y \to Z$ is a one-to-one onto function.

Therefore, there exists a distinct element $b \in Y \Rightarrow g(b) = c$

As $f : X \to Y$ is one-to-one onto function.

Therefore, there exists a distinct element $a \in X \Rightarrow f(a) = b$

Therefore, $(gof)(a) = g[f(a)] = g(b) = c$

It implies every element of c has image under *gof*. Therefore, *gof* is an onto function.

So, we conclude that *gof* is one-to-one onto function.

FUNCTIONS APPLICABLE IN COMPUTER SCIENCE

The following are the functions which are widely used in computer science.

Characteristic Function of A. Consider a subset A of the universal set $U = \{u_1, u_2, u_3, u_4, ..., u_{n-1}, u_n\}$. The characteristic function of A is a function from U to {0, 1} defined as follows :

$$f_A(u_j) = \begin{cases} 1, & \text{if } u_j \in A \\ 0, & \text{if } u_j \notin A \end{cases}$$

Note. The characteristic function of A *i.e.*, f_A is not one to one but is everywhere defined and onto.

Example 12. *Let $A = \{a, c, e\}$ and $U = \{a, b, c, d, e, f, g\}$. Compute the following function values :*

(*i*) $f_A(a)$ (*ii*) $f_A(g)$ (*iii*) $f_A(e)$ (*iv*) $f_A(t)$ (*v*) $f_A(d)$.

Sol. (*i*) $f_A(a) = 1$ (*ii*) $f_A(g) = 0$ (*iii*) $f_A(e) = 1$

(*iv*) $f_A(t)$ = undefined (*v*) $f_A(d) = 0$.

Floor Functions. The floor function for any real number x is defined as $f(x)$ is the greatest integer 1 less than or equal to x. It is denoted by $[x]$.

Example 13. *Determine the value of*

(*i*) $[3 . 5]$ (*ii*) $[-2 . 4]$ (*iii*) $[3 . 143]$.

Sol. (*i*) $[3 . 5] = 3$ (*ii*) $[-2 . 4] = -3$ (*iii*) $[3 . 143] = 3$.

Example 14. *Determine the value of*

(*i*) $[\sqrt{27}]$ (*ii*) $[-13]$ (*iii*) $[\sqrt{6}]$.

Sol. (*i*) $[\sqrt{27}] = 5$ (*ii*) $[-13] = -13$ (*iii*) $[\sqrt{6}] = 2$.

Ceiling Function. The ceiling function for any real number x is defined as $h(x)$ is the smallest integer greater than or equal to x. It is denoted by $[x]$.

Example 15. *Determine the value of*

(*i*) $[3 . 5]$ (*ii*) $[-2 . 4]$ (*iii*) $[3 . 143]$.

Sol. (*i*) $[3 . 5] = 4$ (*ii*) $[-2 . 4] = -2$ (*iii*) $[3 . 143] = 4$.

Example 16. *Determine the value of*

(*i*) $[\sqrt{27}]$ (*ii*) $[-13]$ (*iii*) $[\sqrt{6}]$.

Sol. (*i*) $[\sqrt{27}\,] = 6$ (*ii*) $[-13] = -13$ (*iii*) $[\sqrt{6}\,] = 3.$

Remainder Function. The integer remainder is obtained when some a is divided by m. It is denoted by a (MOD M). We can also define it as, a (MOD M) is the unique integer t such that $a = \mathrm{M}q + t$. Here q is the quotient $0 \le r < \mathrm{M}$.

Example 17. *Determine the value of the following :*

(*i*) *35(MOD 7)* (*ii*) *20(MOD 3)* (*iii*) *4(MOD 9)*

(*iv*) *25(MOD 9)* (*v*) *1111(MOD 11)* (*vi*) *2424(MOD 20).*

Sol. (*i*) 35(MOD 7) = 0 (*ii*) 20(MOD 3) = 2 (*iii*) 4(MOD 9) = 4

(*iv*) 25(MOD 9) = 7 (*v*) 1111(MOD 11) = 0 (*vi*) 2424(MOD 20) = 4.

Boolean Function. Consider a set Y = {true, false}. Then a function from a set X to Y is called a Boolean function.

Example. *Consider the following : P(x) : x is –ve integer and Q(y) : y is + ve integer.*

Then, we have P(– 18) is true

P(18) is false

and Q(– 16) is false

Q(16) is true.

Example 18. *Consider the following : R(x) : x is odd and S(y) : y is even.*

Also let T(x, y) : x is odd and y is even.

Determine the Boolean function T(3, 4), T(4, 3), T(4, 4).

Sol. Since T(x, y) is a Boolean function with two variables. Hence the solution of above functions is

T(3, 4) is true.

T(4, 3) is false.

T(4, 4) is false.

Exponential Functions. Consider two sets A and B. Let $\mathrm{A} = \mathrm{B} = \mathrm{I}_+$ and also let $f : \mathrm{A} \to \mathrm{B}$ be defined by $f(n) = k^n$. Here n is a +ve integer. The function f is called the base k exponential function.

Note 1. $k^t = k.k.k.......k$ (t times).

2. $k^0 = 1,\ k^{-\mathrm{M}} = \dfrac{1}{k^{\mathrm{M}}}$.

3. For rational number, a/b, the exponential function is $k^{a/b} = b\sqrt{k^a} = (b\sqrt{k}\,)^{\mathrm{M}}$.

Example 19. *Determine the value of the following :*

(*i*) 2^6 (*ii*) 10^3 (*iii*) 10^{-4}

(*iv*) $5^{1/2}$ (*v*) 3^{-5} (*vi*) $9^{-3/2}$.

Sol. (*i*) $2^6 = 2\,.\,2\,.\,2\,.\,2\,.\,2\,.\,2 = 64$ (*ii*) $10^3 = 10\,.\,10\,.\,10 = 1000$

(*iii*) $10^{-4} = \dfrac{1}{10^4} = \dfrac{1}{10\,.\,10\,.\,10\,.\,10} = \dfrac{1}{10000}$ (*iv*) $5^{1/2} = 2.23607$

(*v*) $3^{-5} = \dfrac{1}{3^5} = \dfrac{1}{3\,.\,3\,.\,3\,.\,3\,.\,3} = \dfrac{1}{243}$ (*vi*) $9^{-3/2} = 0.037$.

Logarithmic Functions. Consider two set A and B. Let A = B = R (the set of real no.'s) and also let $f_n : A \to B$ be defined for each positive integer $n > 1$ as $f_n(x) = \log_n(x)$ the logarithm to the base n of x.

Note 1. $k = \log_n x$ and $n^k = x$ are equivalent.

2. For any base n, $\log_n 1 = 0$ as $n^0 = 1$.

3. For any base n, $\log_n n = 1$ as $n^1 = n$.

Example 20. *Determine the value of the following :*

(*i*) $\log_2 16$ (*ii*) $\log_2 256$ (*iii*) $\log_{10} 1000$ (*iv*) $\log_{10} 0.0001$

(*v*) $\log_2 100$ (*vi*) $\log_2 (1/32)$ (*vii*) $\log_2 0.001$.

Sol. (*i*) $\log_2 16 = 4$ as $2^4 = 16$

(*ii*) $\log_2 256 = 8$ and $2^8 = 256$

(*iii*) $\log_{10} 1000 = 3$ as $10^3 = 1000$

(*iv*) $\log_{10} 0.0001 = -4$ as $10^{-4} = 0.0001$

(*v*) $\log_2 100 = 6$ as $2^6 = 64$ but $2^7 = 128$ which is greater

(*vi*) $\log_2 (1/32) = -5$ as $2^{-5} = \frac{1}{2^5} = \frac{1}{32}$

(*vii*) $\log_2 (0.001) = -9$ as $2^{-9} = \frac{1}{512}$ but $2^{-10} = \frac{1}{1024}$ which is greater.

Hashing Functions. Let us assume that there is a file of n records with a set of keys k_T which uniquely determine the records in file. Also, assume that the file is maintained in memory by a table of m memory locations and L is the set of memory addresses of the locations in the table, T(0 ...... $m - 1$). The set of keys k_1 and the addresses in L are integers.

A hash function h is used to compute the address of a record from the set of keys k_T. Here h maps the set of keys k_0 into the set of memory addresses called slots of a hash table T(0 ...... $m - 1$). Thus, a function $h : k_T \to L$ is called a hash function. But the major problem with such a function h is that two keys may hash to the same slot *i.e.*, two distinct keys will generate the same hash address called collision.

A good hash function is a function h which satisfies the following two principle criterias :

(*i*) The function h should be very easy and quick to complete.

(*ii*) The function h should uniformally distribute the hash addresses, so that there are a minimum number of collisions.

There are three most popular methods for creating hash functions. These are :

(*i*) Division Method (*ii*) Midsquare Method

(*iii*) Universal or Folding Method.

1. Division Method. Select a number M(usually a prime number near m) larger than the number n of keys in k_T then, hash function h is defined as

$$h(k) = k(\text{Mod M})$$

where k (MOD M) denotes the remainder when k is divided by M.

Using a prime number M reduces the number of collisions.

2. Midsquare Method. In this method, the key k is squared *i.e.*, k^2 and then the address is obtained by deleting or choping the digits from both ends of k^2. The hash function h is defined as

$$h(k) = p.$$

Here p is obtained by choping digits from both ends of k^2 and the same positions of k^2 is used for all the keys.

3. Universal Method or Folding Method. The key k is divided into a number of parts $k_1, k_2, \ldots\ldots k_t$. Here each part consists of the same number of digits as the desired address except the last one. After that these parts are added together, ignoring the last carry. Thus, the hash function h is defined as

$$h(k) = k_1 + k_2 + \ldots\ldots k_t.$$

The leading digit carries are ignored, if any. To get more precise values, the even-numbered parts *i.e.*, k_2, k_4, k_6 ......can be reversed before addition.

Example 21. *Consider an organisation which has 53 employes and each one is assigned a unique 3-digit employee number. Suppose the L consists of 100 two digit addresses (00, 01, ..., 99). Determine the hash addresses generated by different hash function for the following employee numbers*

100, 213, 710, 934.

Sol. 1. Division Method. The prime number close to 99 is 97. Thus, the hash address generated by employee number 100 is obtained by dividing 100 by 97 which gives remainder 3 which is required hash address. All other hash addresses are generated similarly

$$h(100) = 3,\ h(213) = 19,\ h(710) = 31,\ h(934) = 61.$$

2. Midsquare Method. Calculate the square of the employee address and then chop off the digits from both ends. The remaining digits is the hash address obtained. The third and fourth digits, from right are chosen to be the hash address

k :	100	213	710	934
k^2 :	10,000	45,369	50,4100	87,2356
$h(k)$:	00	53	41	23.

3. Universal Method. Divide the key k into two parts and then add to obtain the hash address by this method. Thus, the hash addresses generated by the employee numbers are as follows :

$$h(100) = 10 + 0 = 10 \qquad h(213) = 21 + 3 = 24$$
$$h(710) = 71 + 0 = 71 \qquad h(934) = 93 + 4 = 97.$$

Note. If sum of two digits is more than 99, delete the leading digit and retain only two least significant digits.

Permutation Functions. A permutation on a set A is a bijection from A into A (*i.e.*, itself). Here the set A is finite. If the set A contains n elements, then there are n ! different permutations on A.

The matrix form for describing the function on a finite set is to list the domains across the top row and the image of each element exactly below it.

Example. *Consider a set* $A = \{x_1, x_2, \ldots\ldots, x_n\}$ *is a finite set and let p is a bijection on A. Then the elements of A and the corresponding images* $p(x_1), p(x_2), \ldots\ldots, p(x_n)$ *can be represented in the matrix form as follows :*

$$\begin{pmatrix} x_1 & x_2 & \ldots\ldots & x_n \\ p(x_1) & p(x_2) & \ldots\ldots & p(x_n) \end{pmatrix}$$

The above matrix representation describes p completely because it gives the value of p for every element of set A. In general, it is written as

$$P = \begin{pmatrix} x_1 & x_2 & \ldots\ldots & x_n \\ p(x_1) & p(x_2) & \ldots\ldots & p(x_n) \end{pmatrix}$$

So, if P represents a permutation of a set A (finite set), then the sequence $p(x_1)$, $p(x_2)$, ..., $p(x_n)$ is a rearrangement of the elements of set A which corresponds to exactly one permutation of A.

Example 22. *Consider A = {a, b}. Determine all the permutations of A.*

Sol. Since, the set A has two elements, hence, it has 2 ! = 2 permutations which are as follows :

$$p_1 = \begin{pmatrix} a & b \\ a & b \end{pmatrix}$$

$$p_2 = \begin{pmatrix} a & b \\ b & a \end{pmatrix}.$$

Example 23. *Consider A = {x_1, x_2, x_3}. Determine all permutation of A.*

Sol. Since the set A has three elements, hence, it has 3 ! = 6 permutations which are as follows :

$$p_1 = \begin{pmatrix} x_1 & x_2 & x_3 \\ x_1 & x_2 & x_3 \end{pmatrix} \qquad p_1 = \begin{pmatrix} x_1 & x_2 & x_3 \\ x_1 & x_3 & x_2 \end{pmatrix}$$

$$p_2 = \begin{pmatrix} x_1 & x_2 & x_3 \\ x_2 & x_1 & x_3 \end{pmatrix} \qquad p_3 = \begin{pmatrix} x_1 & x_2 & x_3 \\ x_2 & x_3 & x_1 \end{pmatrix}$$

$$p_4 = \begin{pmatrix} x_1 & x_2 & x_3 \\ x_3 & x_1 & x_2 \end{pmatrix} \qquad p_5 = \begin{pmatrix} x_1 & x_2 & x_3 \\ x_3 & x_2 & x_1 \end{pmatrix}.$$

Inverse of a Permutation. The inverse of a permutation (which is written in matrix form) can be obtained by exchanging the two rows and rearranging the columns so that the two row is in order.

Example 24. *Consider the permutation* $p = \begin{pmatrix} a & b & c & d \\ d & c & a & b \end{pmatrix}$. *Determine inverse* p^{-1}.

Sol. The p^{-1} is obtained by exchanging the rows and then ordering the first row. So

$$p^{-1} = \begin{pmatrix} d & c & a & b \\ a & b & c & d \end{pmatrix} = \begin{pmatrix} a & b & c & d \\ c & d & b & a \end{pmatrix}.$$

Composition (Product) of two Permutations. Consider two permutations p_a and p_b of a finite set A. The composition of p_a and p_b, written $p_b \circ p_a$ is another permutation defined by

$$(p_b \circ p_a)(a) = p_b(p_a(a)).$$

It means to find the image of a under $p_b \circ p_a$, first find the image of a under p_a and then find the image of $p_a(a)$ under p_b. It is also called the product of two permutations.

Example 25. *Consider a finite set A = {5, 6, 7}. Let* $p_1 = \begin{pmatrix} 5 & 6 & 7 \\ 5 & 7 & 6 \end{pmatrix}$ *and* $p_2 = \begin{pmatrix} 5 & 6 & 7 \\ 6 & 5 & 7 \end{pmatrix}$ *be two permutation of A. Determine* $p_1 \circ p_2$ *and* $p_2 \circ p_1$.

Sol.

$$p_1 \circ p_2 = \begin{pmatrix} 5 & 6 & 7 \\ 6 & 5 & 7 \end{pmatrix} \circ \begin{pmatrix} 5 & 6 & 7 \\ 5 & 7 & 6 \end{pmatrix} = \begin{pmatrix} 5 & 6 & 7 \\ 7 & 5 & 6 \end{pmatrix}$$

$$p_2 \circ p_1 = \begin{pmatrix} 5 & 6 & 7 \\ 5 & 7 & 6 \end{pmatrix} \circ \begin{pmatrix} 5 & 6 & 7 \\ 6 & 5 & 7 \end{pmatrix} = \begin{pmatrix} 5 & 6 & 7 \\ 6 & 7 & 5 \end{pmatrix}$$

From the above result, we have seen that the result obtained from the product of two compositions is again a composition.

Cyclic Permutation. Consider a finite set $A = \{x_1, x_2, ..., x_n\}$. Also let $t_1, t_2, ..., t_k$ *be k elements of the set A and the permutation* $p : A \to A$ is defined by

$$p(t_1) = t_2$$
$$p(t_2) = t_3$$
$$\cdots\cdots$$
$$\cdots\cdots$$
$$\cdots\cdots$$
$$p(t_{r-1}) = t_r$$
$$p(t_r) = t_1$$

is called a cyclic permutation of length t.

In general, $p(a) = a$ if $a \in A$, but $a \notin t_1, t_2, \ldots\ldots, t_k$.

It is also called cycle of length t. It is generally denoted by $(t_1, t_2, t_3, \ldots\ldots t_r)$. We can arrange the elements $t_1, t_2, t_3, \ldots\ldots, t_k$ on a circle as shown in Fig. 17.

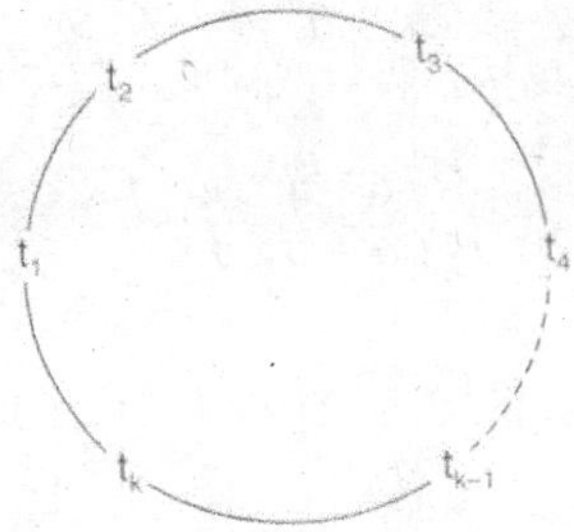

Fig. 17.

A cycle p of length k moves the elements in clockwise direction. It means t_1 sent to t_2, t_2 sent to t_3, ……, t_k sent to t_1. The remaining elements of set A are fixed in permutation p.

Example 26. *Consider the set* $A = \{1, 2, 3, 4, 5, 6, 7, 8\}$. *Determine the permutation p denoted by the cycle (1, 8, 3, 7).*

Sol. The permutation denoted by cycle (1, 8, 3, 7) is as follows :

$$p = \begin{pmatrix} 1 & 2 & 3 & 4 & 5 & 6 & 7 & 8 \\ 8 & 2 & 7 & 4 & 5 & 6 & 1 & 3 \end{pmatrix}.$$

Example 27. *Consider the permutation* $p = \begin{pmatrix} 1 & 2 & 3 & 4 & 5 & 6 & 7 \\ 7 & 4 & 6 & 1 & 5 & 3 & 2 \end{pmatrix}$ *on the set* $A = \{1, 2, 3, 4, 5, 6, 7\}$. *Determine all the cycles of permutation p. Also determine its length.*

Sol. The permutation $p = \begin{pmatrix} 1 & 2 & 3 & 4 & 5 & 6 & 7 \\ 7 & 4 & 6 & 1 & 5 & 3 & 2 \end{pmatrix}$ has three cycles as follows :

(*i*) (1, 7, 2, 4) is a cycle of length 4.

(*ii*) (3, 6) is a cycle of length 2.

(*iii*) (5) is a cycle of length 1.

Note 1. A cycle (1, 7, 2, 4) can also be written as (7, 2, 4, 1) or (2, 4, 1, 7) or (4, 1, 7, 2).

2. A cycle does not give exact number of elements in the set A. We have to explicitly write the set on which a cycle is defined.

3. A cycle of length 1 in a set A is identity permutation.

4. Since cycles are permutations, their composition (product) can be formed.

5. The composition (product) of two cycles need not be a cycle.

6. The composition (product) of two cycles is not always commutative.

Example 28. *Show that composition (product) of two cycles may not be a cycle.*

Sol. We will discuss this by taking an example. Consider a set A = {*a, b, c, d, e, f, g, h*}. Now find the product {*e, f, c, d, a*} o {*a, b, h, g*}

We know that

$$(e, f, c, d, a) = \begin{pmatrix} a & b & c & d & e & f & g & h \\ e & b & d & a & f & c & g & h \end{pmatrix}$$

and
$$(a, b, h, g) = \begin{pmatrix} a & b & c & d & e & f & g & h \\ b & h & c & d & e & f & a & g \end{pmatrix}$$

Now, the product of two cycles is given by

$$(e, f, c, d, a) \circ (a, b, h, g) = \begin{pmatrix} a & b & c & d & e & f & g & h \\ e & b & d & a & f & c & g & h \end{pmatrix} \circ \begin{pmatrix} a & b & c & d & e & f & g & h \\ b & h & c & d & e & f & a & g \end{pmatrix}$$

$$= \begin{pmatrix} a & b & c & d & e & f & g & h \\ b & h & d & a & f & c & e & g \end{pmatrix}$$

Its product is a cycle.

Now, again consider a set A = {7, 8, 9, 10, 11, 12, 13}. Now find the product (10, 13, 8, 7). (8, 10).

We know that

$$(10, 13, 8, 7) = \begin{pmatrix} 7 & 8 & 9 & 10 & 11 & 12 & 13 \\ 10 & 7 & 9 & 13 & 11 & 12 & 8 \end{pmatrix}$$

and
$$(8, 10) = \begin{pmatrix} 7 & 8 & 9 & 10 & 11 & 12 & 13 \\ 7 & 10 & 9 & 8 & 11 & 12 & 13 \end{pmatrix}$$

Now, the product of two cycles is given by

$$(10, 13, 8, 7) \circ (8, 10) = \begin{pmatrix} 7 & 8 & 9 & 10 & 11 & 12 & 13 \\ 10 & 7 & 9 & 13 & 11 & 12 & 8 \end{pmatrix} \circ \begin{pmatrix} 7 & 8 & 9 & 10 & 11 & 12 & 13 \\ 7 & 10 & 9 & 8 & 11 & 12 & 13 \end{pmatrix}$$

$$= \begin{pmatrix} 7 & 8 & 9 & 10 & 11 & 12 & 13 \\ 10 & 13 & 9 & 7 & 11 & 12 & 8 \end{pmatrix}$$

Thus, its product is not a cycle.

Example 29. *Show that composition (product) of two cycles is not commutative.*

Sol. We will show the non-commutativity by taking example. Consider the set A = {*r, s, t, u, v, w*}

Now find the product of (*r, t, v, w*) o (*w, s, t*) and (*w, s, t*) o (*r, t, v, w*).

We know that
$$(r, t, v, w) = \begin{pmatrix} r & s & t & u & v & w \\ t & s & v & u & w & r \end{pmatrix}$$

and
$$(w, s, t) = \begin{pmatrix} r & s & t & u & v & w \\ r & t & w & u & v & s \end{pmatrix}$$

Now, the product of first one is as below :

$$(r, t, v, w) \circ (w, s, t) = \begin{pmatrix} r & s & t & u & v & w \\ t & s & v & u & w & r \end{pmatrix} \circ \begin{pmatrix} r & s & t & u & v & w \\ r & t & w & u & v & s \end{pmatrix}$$

$$= \begin{pmatrix} r & s & t & u & v & w \\ t & v & r & u & w & s \end{pmatrix} \quad ...(i)$$

Also, the product of second one is as below :

$$(w, s, t) \circ (r, t, v, w) = \begin{pmatrix} r & s & t & u & v & w \\ r & t & w & u & v & s \end{pmatrix} \circ \begin{pmatrix} r & s & t & u & v & w \\ t & s & v & u & w & r \end{pmatrix}$$

$$= \begin{pmatrix} r & s & t & u & v & w \\ w & t & v & u & s & r \end{pmatrix} \quad ...(ii)$$

From (*i*) and (*ii*), we can see that

$$(r, t, v, w) \circ (w, s, t) \neq (w, s, t)\ (r, t, v, w)$$

Hence, product of two cycles is not commutative.

Disjoint Cycles. Two cycles are called disjoint cycles of a given set A if there does not exist an element $a \in A$ such that a appears in both cycles.

Example. *Consider a finite set A = {a, b, c, d}. Then the cycles (a, b) and (c, d) are disjoint cycles and (a, b, c) and (a, b, d) are non-disjoint cycles.*

Note 1. Composition of two cycles is commutative if the two cycles are disjoint *i.e.*, $p_a \circ p_b = p_b \circ p_a$.

2. A permutation of a finite set that is not the identity or a cycle can be written as a product of disjoint cycles of length ≥ 2.

Example 30. *Show that composition of two disjoint cycles is commutative.*

Sol. We will show the commutativity of two disjoint cycles by taking an example.

Consider the set A = {*a*, *b*, *c*, *d*, *e*, *f*, *g*}.

Now find the product of (*a*, *d*, *g*) o (*b*, *c*, *f*) and (*b*, *c*, *f*) o (*a*, *d*, *g*).

We know that

$$(a, d, g) = \begin{pmatrix} a & b & c & d & e & f & g \\ d & b & c & g & e & f & a \end{pmatrix} \text{ and } (b, c, f) = \begin{pmatrix} a & b & c & d & e & f & g \\ a & c & f & d & e & b & g \end{pmatrix}$$

Now, the product of first one is as follows :

$$(a, d, g) \circ (b, c, f) = \begin{pmatrix} a & b & c & d & e & f & g \\ d & b & c & g & e & f & a \end{pmatrix} \circ \begin{pmatrix} a & b & c & d & e & f & g \\ a & c & f & d & e & b & g \end{pmatrix}$$

$$= \begin{pmatrix} a & b & c & d & e & f & g \\ d & c & f & g & e & b & a \end{pmatrix} \quad ...(i)$$

Also, the product of second one is as below :

$$(b, c, f) \circ (a, d, g) = \begin{pmatrix} a & b & c & d & e & f & g \\ a & c & f & d & e & b & g \end{pmatrix} \circ \begin{pmatrix} a & b & c & d & e & f & g \\ d & b & c & g & e & f & a \end{pmatrix}$$

$$= \begin{pmatrix} a & b & c & d & e & f & g \\ d & c & f & g & e & b & a \end{pmatrix} \quad ...(ii)$$

Since (*i*) and (*ii*) are equal, hence composition of two disjoint cycles is commutative.

Example 31. *Consider the permutation* $p = \begin{pmatrix} a & b & c & d & e & f \\ b & c & a & f & d & e \end{pmatrix}$ *on the set A = {a, b, c, d, e, f}. Write p as a product of disjoint cycles.*

Sol. Begin with the first member of the set A *i.e.*, a and find the mapping $p(a) = b$, then find mapping $p(b) = c$, then find $p(c) = a$, so we have got the cycle (a, b, c). To find another cycle, choose the first element of A that does not contained in previous cycle. So, d is the first element that does not contained in the previous cycle. So, we have $p(d) = f$, and $p(f) = e$ and $p(e) = d$. We got another cycle *i.e.*, (d, e, f). Continue this process until all the elements of A appeared in any of the cycles. After obtaining all the cycles, write p as a product of disjoint cycles as below :

$$p = (d, e, f) \circ (a, b, c).$$

EVEN AND ODD PERMUTATIONS

Transposition. It is defined as a cycle of length 2 *i.e.*, a transposition is a cycle $p = (x_i, x_j)$. Here, $p(x_i) = x_j$ and $p(x_j) = x_i$.

For a given transposition $p = (x_i, x_j)$ of a finite set A, then composition $pop = I_A$ (Identity permutation of set A).

Note 1. Every cycle can be written as a product of transpositions.

2. Every permutation of a finite set with at least two elements can be written as a product of transpositions and the transpositions may or may not be disjoint.

3. If a permutation of a finite set can be written as a product of an even no. of transpositions, then it can never be written as a product of an odd number of transpositions and vice-versa.

Example 32. *Show that every cycle can be written as a product of transpositions i.e.,*

$$(x_1, x_2, x_3, \ldots\ldots, x_n) = (x_1, x_n) \circ (x_1, x_{n-1}) \circ \ldots \circ (x_1, x_3) \circ (x_1, x_2).$$

Sol. The above statement can be proved by the method of induction on n.

Basis Step. When $n = 2$, then the cycle becomes (x_1, x_2) which is in required form.

Induction Step. Let the result is true for k. Let us assume that $(x_1, x_2, x_3, \ldots\ldots, x_k, x_{k+1})$ is a cycle of length $k + 1$. Then we have

$$(x_1, x_2, \ldots\ldots, x_k, x_{k+1}) = (x_1, x_{k+1}) \circ (x_1, x_2, \ldots\ldots, x_k) \qquad \ldots(i)$$

This can be verified by computing the composition.

Now by the induction composition

$$(x_1, x_2, \ldots\ldots, x_k) = (x_1, x_k) \circ (x_1, x_{k-1}) \circ \ldots\ldots \circ (x_1, x_2) \qquad \ldots(ii)$$

Substituting eqn. (*ii*) in eqn. (*i*), we have

$$(x_1, x_2, x_3, \ldots\ldots, x_k, x_{k+1}) = (x_1, x_{k+1}) \circ (x_1, x_k) \circ \ldots\ldots \circ (x_1, x_3) \circ (x_1, x_2)$$

which is the required result.

Hence, by the principle of mathematical induction, the result holds for every cycle.

Example 33. *Write the cycle (a, b, c, d, e, f) on the set $A = \{a, b, c, d, e, f\}$ as a product of transpositions.*

Sol. The cycle $(a, b, c, d, e, f) = (a, f) \circ (a, e) \circ (a, d) \circ (a, c) \circ (a, b)$ can be written as above in product of transposition form.

Example 34. *Write the permutation* $p = \begin{pmatrix} k & l & m & n & o & p & q \\ l & k & o & p & q & n & m \end{pmatrix}$ *as a product of transpositions.*

Sol. First of all, write the permutation p as a product of disjoint cycles, which is given by

$$p = (k, l) \circ (m, o, q) \circ (n, p)$$

Now, the cycle (m, o, q) can be written as

$$(m, o, q) = (m, q) \text{ o } (m, o).$$

Now, the permutation p in product of transpositions form can be written as

$$p = (k, l) \text{ o } (m, q) \text{ o } (m, o) \text{ o } (n, p).$$

Even permutation. A permutation p of a finite set A is said to be EVEN if the permutation can be written as a product of an even no. of transpositions.

Odd Permutation. A permutation p of a finite set A is said to be odd if the permutation can be written as a product of an odd no. of transpositions.

Example 35. *Determine whether the following permutation is odd or even*

$$p = \begin{pmatrix} a & b & c & d & e & f & g & h \\ b & c & a & e & f & d & h & g \end{pmatrix}.$$

Sol. First of all, write p as a product of disjoint cycles, which is as follows :

$$p = (a, b, c) \text{ o } (d, e, f) \text{ o } (g, h)$$

Now, write all of the cycles as a product of transpositions, which is as follows :

$$(a, b, c) = (a, c) \text{ o } (a, b)$$

$$(d, e, f) = (d, f) \text{ o } (d, e) \quad \text{and} \quad (g, h) = (g, h)$$

Then, we have $p = (a, c) \text{ o } (a, b) \text{ o } (d, f) \text{ o } (d, e) \text{ o } (g, h)$. Since, p is a product of an odd no. of permutations, hence it is an odd permutation.

Note 1. The product of two odd permutations is even.

2. The product of two even permutations is even.

3. The product of an even and an odd permutation is odd.

SOLVED PROBLEMS

Problem 1. *Let* $X = \{x, y, z, k\}$ *and* $Y = \{a, b, c\}$. *Determine whether the relation S from X to Y is a function. If it is a function, give its domain and range.*

(*i*) $S = \{(x, a), (y, a), (x, b), (y, b)\}$ (*ii*) $S = \{(x, c), (y, a), (z, b)\}$

(*iii*) $S = \{(x, a), (y, b), (z, c), (k, b), (x, b)\}$ (*iv*) $S = \{(x, c), (y, a), (3, a), (k, c)\}$.

Sol. (*i*) Not a function because domain of S ≠ X *i.e.*, there is no image of Z, K in set Y.

(*ii*) Not a function because domain of S ≠ X *i.e.*, there is no image of K in set Y.

(*iii*) It is a function.

Domain (S) = $\{x, y, z, k\}$, Range (S) = $\{a, b, c\}$

(*iv*) It is a function.

Domain (S) = $\{x, y, z, k\}$, Range (S) = $\{a, c\}$.

Problem 2. *Let* $A = B = \{1, 2, 3, 4\}$. *Define functions* $f : A \to B$ *(if possible) such that*

(*i*) *f is one-to-one and onto*

(*ii*) *f is neither one-to-one nor onto*

(*iii*) *f is onto but not one-to-one*

(*iv*) *f is one-to-one but not onto.*

Sol. (*i*) The function $f = \{(1, 1), (2, 4), (3, 2), (4, 3)\}$ is one-to-one and onto (Fig. 18).

(*ii*) The function $f = \{(1, 1), (2, 1), (3, 2), (4, 3)\}$ is neither one-to-one nor onto (Fig.19).

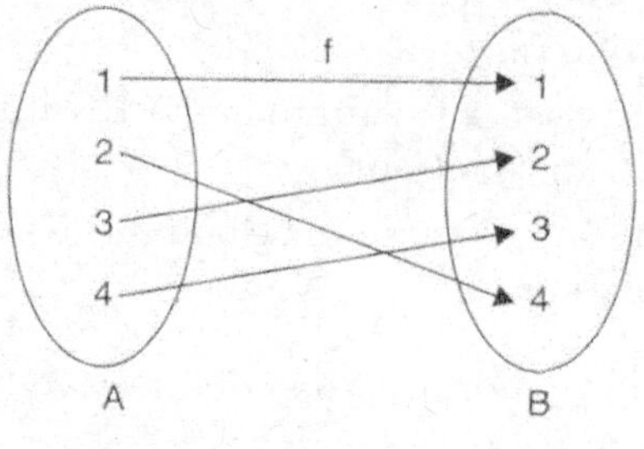

Fig. 18.

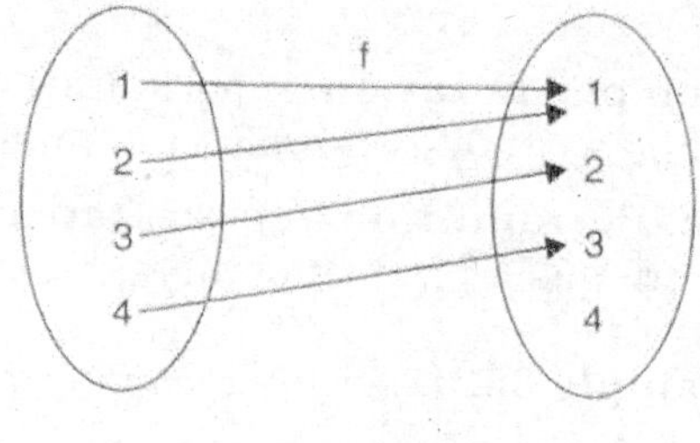

Fig. 19.

(*iii*) The function f which is onto but not one-to-one is not possible on the set A = B = {1, 2, 3, 4}.

(*iv*) The function f which is one-to-one but not onto is not possible on the set A = B = {1, 2, 3, 4}.

Problem 3. *Consider the sets X and Y and let $f : X \to Y$. Determine whether the following functions are :*

(*a*) *one-to-one function* (*b*) *onto function*

(*c*) *one-to-one onto* (*d*) *neither one-to-one nor onto.*

(*i*) $X = \{x, y, z\}$; $Y = \{1, 2, 3, 4\}$, $f = \{(x, 1), (y, 1), (z, 3)\}$

(*ii*) $X = \{a, b, c, d\} = Y$, $f = \{(a, a), (b, c), (c, d), (d, b)\}$

(*iii*) $X = \left\{\frac{1}{6}, \frac{1}{9}, \frac{1}{12}\right\}$; $Y = \{a, b, c, d\}$, $f = \left\{\left(\frac{1}{6}, b\right), \left(\frac{1}{9}, d\right), \left(\frac{1}{12}, a\right)\right\}$

(*iv*) $X = \{y_1, y_2, y_3\}$; $Y = \{k, l\}$, $f = \{(y_1, l), (y_2, l), (y_3, k)\}$.

Sol. (*i*) Neither one-to-one nor onto (*ii*) one-to-one onto

(*iii*) one-to-one (*iv*) onto.

Problem 4. *Consider the sets $X = \{k, l, m, n\}$ to $Y = \{7, 8, 9, 10\}$. Let $f : X \to Y$ such that*

(*i*) $f = \{(k, 10), (l, 7), (m, 8), (n, 9)\}$ (*ii*) $f = \{(k, 7), (l, 8), (m, 7), (n, 8)\}$

Determine whether f^{-1} is a function.

Sol. (*i*) The f^{-1} is $f^{-1} = \{(10, k), (7, l), (8, m), (9, n)\}$

The f^{-1} is also a function.

(*ii*) The f^{-1} is $f^{-1} = \{(7, k), (8, l), (7, m), (8, n)\}$

The domain (f) ≠ set Y. So f^{-1} is not a function.

Problem 5. *Let X = Y = Z = R, be the set of reals.*

Consider the functions f and g such that

$$f(x) = x + 2,\ x \in R,\ g(y) = y^2,\ y \in \text{R}.$$

Find the composition function gof.

Sol. The composition of f with g is given by

$$gof(x) = g(x + 2) = (x + 2)^2,\ x \in \text{R}.$$

Problem 6. *Consider the functions f, g : R → R difined by*

$$f(x) = x^2 + 3x + 1,\ g(x) = 2x - 3.$$

Find the composition functions

(*i*) *fof* (*ii*) *fog* (*iii*) *gof*.

Sol. (*i*)

$$(fof)(x) = f[f(x)] = f(x^2 + 3x + 1)$$
$$= (x^2 + 3x + 1)^2 + 3(x^2 + 3x + 1) + 1$$
$$= x^4 + 9x^2 + 1 + 6x^3 + 2x^2 + 6x + 3x^2 + 9x + 3 + 1$$
$$= x^4 + 6x^3 + 14x^2 + 15x + 5.$$

(*ii*)

$$(gof)(x) = g[f(x)] = g(x^2 + 3x + 1)$$
$$= g(x^2 + 3x + 1) - 3$$
$$= 2x^2 + 6x + 2 - 3$$
$$= 2x^2 + 6x - 1.$$

(*iii*)

$$(fog)(x) = f[g(x)] = f(2x - 3)$$
$$= (2x - 3)^2 + 3(2x - 3) + 1$$
$$= 4x^2 + 9 - 12x + 6x - 9 + 1$$
$$= 4x^2 - 6x + 1.$$

Problem 7. *Let f, g, h be functions from N to N, where N is the set of natural numbers so that*

$$f(n) = n + 1$$
$$g(n) = 2n$$
$$h(n) = \begin{cases} 0 & \text{when } n \text{ is even} \\ 1 & \text{when } n \text{ is odd}. \end{cases}$$

Determine fof, fog, gof, goh, hog, (fog)oh.

Sol. (*i*)

$$fof(n) = n + 1$$
$$f\,[f(n)] = n + 1$$
$$f(n + 1) = (n + 1) + 1 = n + 2.$$

(*ii*)

$$fog(n) = 2n$$
$$f\,[g(n)] = 2n$$
$$f\,[g(n)] = 2(2n) = 4n.$$

(*iii*)

$$gof(n) = n + 1$$
$$g[f(n)] = n + 1$$
$$g(n + 1) = 2(n + 1) = 2n + 2.$$

(*iv*)

$$goh(n) = 0\ ;\ n \text{ is even}$$
$$goh(n) = 1\ ;\ n \text{ is odd}$$
$$g(0) = 0 \text{ when } n \text{ is even}$$
$$g(1) = 2 \text{ when } n \text{ is odd.}$$

(*v*)

$$hog(n) = 2n$$
$$h[g(n)] = 2n$$
$$h[g(n)] = 1, \text{ as } n \text{ is even every time.}$$

(*vi*)

$$fogoh(n) = 0\ ;\ n \text{ is even}$$
$$fogoh(n) = 1\ ;\ n \text{ is odd}$$

$$fog[h(n)] = 0 \text{ and } fog[h(n)] = 1$$
$$fog(0) = 0 \text{ and } fog(1) = 2$$
$$f[g(0)] = 0 \text{ and } f[g(1)] = 2$$
$$f(0) = 1 \text{ and } f(2) = 2 + 1 = 3.$$

So, $(fog)oh = 1$ when n is even, $(fog)oh = 3$ when n is odd.

Problem 8. *Consider $A = B = C = R$ and let $f : A \to B$ and $g : B \to C$ be defined by $f(x) = x + 9$ and $g(y) = y^2 + 3$.*

Find the following composition functions :

(i) $(fof)(a)$ (ii) $(gog)(a)$

(iii) $(fog)(b)$ (iv) $(gof)(b)$

(v) $(gof)(4)$ (vi) $(fog)(-4)$.

Sol. (i) $(fof)(a) = f[f(a)] = f(a + 9) = (a + 9) + 9 = a + 18.$

(ii) $(gog)(a) = g[g(a)] = g(a^2 + 3) = (a^2 + 3)^2 + 3 = a^4 + 6a^2 + 12.$

(iii) $(fog)(b) = f[g(b)] = f(b^2 + 3) = (b^2 + 3) + 9 = b^2 + 12.$

(iv) $(gof)(b) = g[f(b)] = g(b + 9) = (b + 9)^2 + 3 = b^2 + 18b + 84.$

(v) $(gof)(4) = g[f(4)] = g(13) = (13)^2 + 3 = 172.$

(vi) $(fog)(-4) = f[g(-4)] = f(19) = 19 + 9 = 28.$

Problem 9. *Let $X = \{a, b, c\}$. Define $f : X \to X$ such that*

$$f = \{(a, b), (b, a), (c, c)\}$$

Find (i) f^{-1} (ii) f^2 (iii) f^3 (iv) f^4.

Sol. (i) The inverse function $f^{-1} = \{(b, a), (a, b), (c, c)\}$

(ii) The f^2 is fof (Fig. 20).

$$(fof)(a) = f[f(a)] = f(b) = a, \qquad (fof)(b) = f[f(b)] = f(a) = b$$

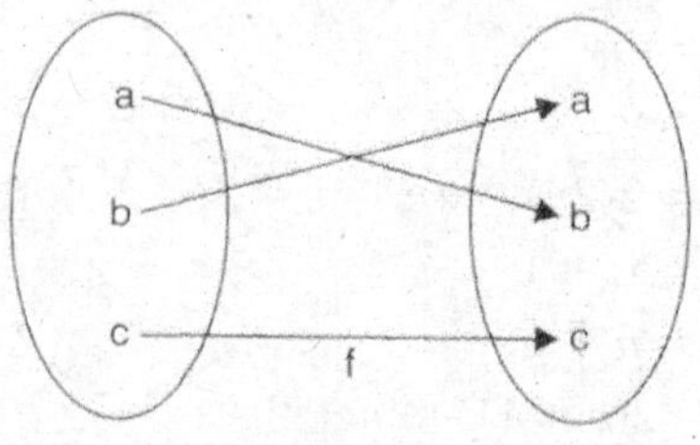

Fig. 20.

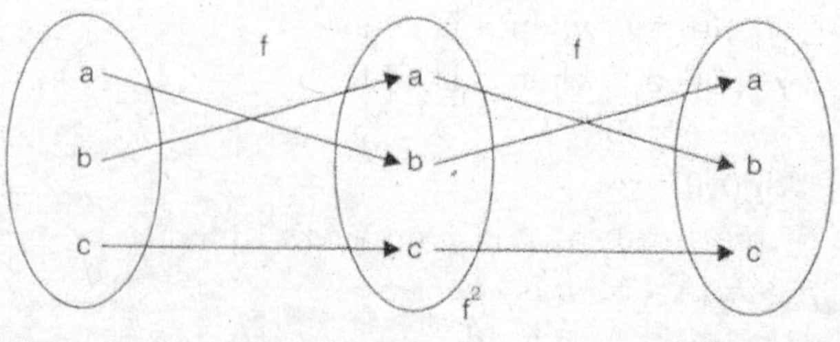

Fig. 21.

$$(fof)(c) = f[f(c)] = f(c) = c,$$

So, $$f^2 = \{(a, a), (b, b), (c, c)\}.$$

(*iii*) The f^3 is *fofof* *i.e.*, fof^2 (Fig. 22).

$$(fof)^2 (a) = f[f^2(a)] = f(a) = b, \qquad (fof^2)(b) = f[f^2(b)] = f(b) = a$$
$$(fof^2)(c) = f[f^2(c)] = f(c) = c$$

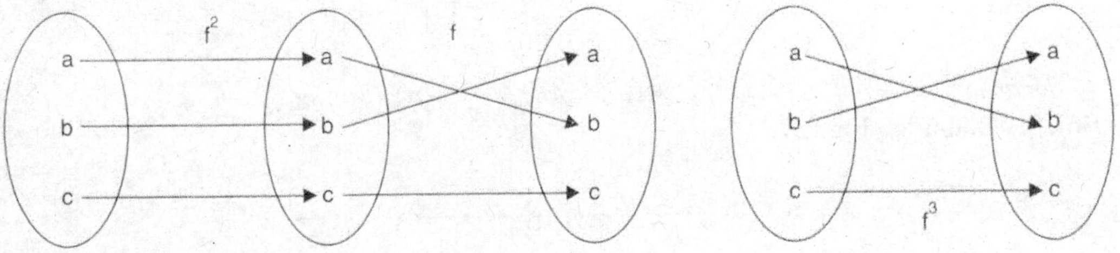

Fig. 22.

So, $$f^3 = \{(a, b), (b, a), (c, c)\}.$$

(*iv*) The f^4 is *fofofof* *i.e.*, fof^3 (Fig. 23).

$$(fof^3)(a) = f[f^3(a)] = f(b) = a, \qquad (fof^3)(b) = f[f^3(b)] = f(a) = b$$
$$(fof^3)(c) = f[f^3(c)] = f(c) = c$$

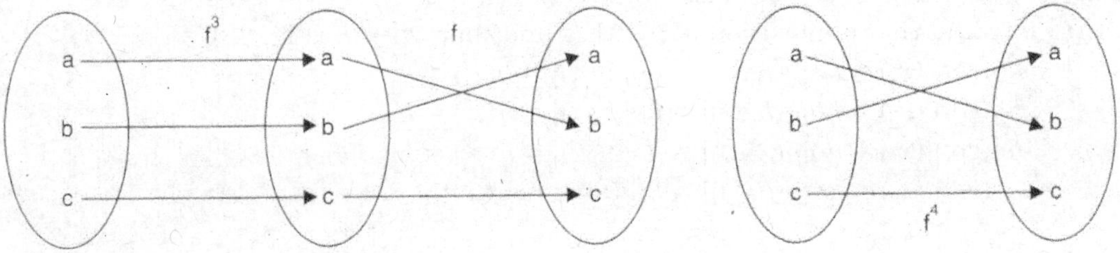

Fig. 23.

So, $$f^4 = \{(a, a), (b, b), (c, c)\}.$$

Problem 10. *Consider the functions f, g and h as in Figs. 24, 25 and 26.*

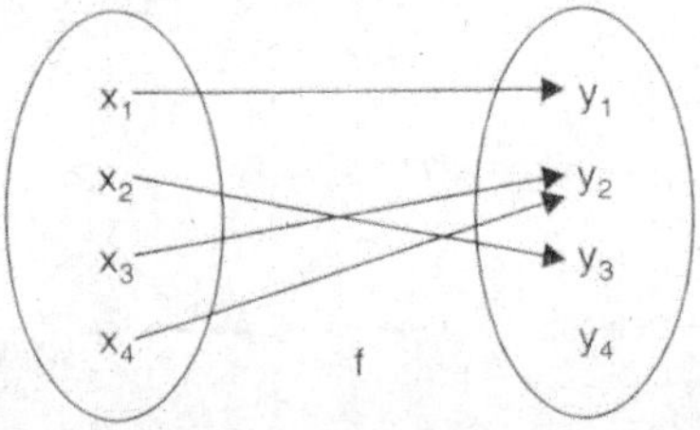

Fig. 24.

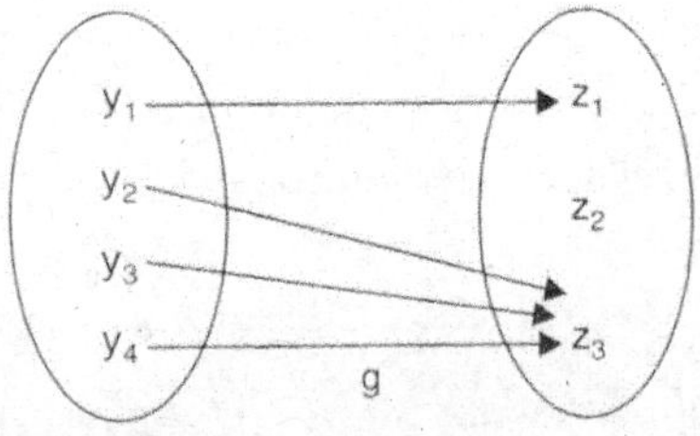

Fig. 25.

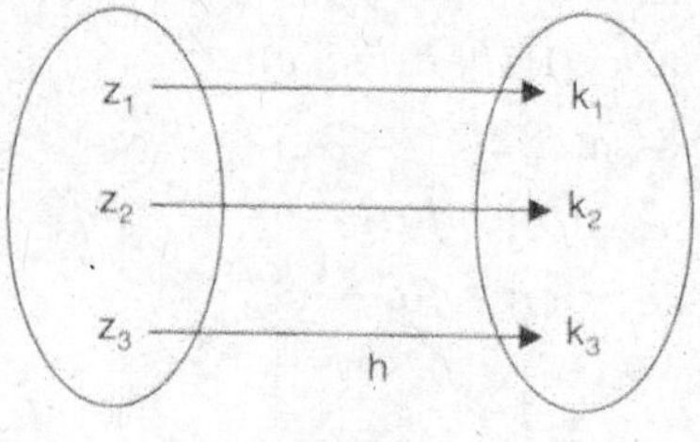

Fig. 26.

Determine (*i*) *gof* (*ii*) *ho(gof)* (*iii*) *(hog)of.*

Sol. (*i*) Consider Fig. 27.

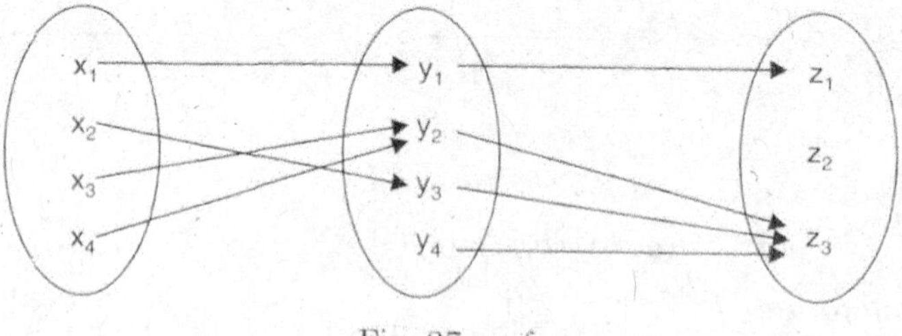

Fig. 27. *gof.*

$$(gof)(x_1) = g[f(x_1)] = g(y_1) = z_1, \qquad (gof)(x_2) = g[f(x_2)] = g(y_3) = z_3$$
$$(gof)(x_3) = g[f(x_3)] = g(y_2) = z_3, \qquad (gof)(x_4) = g[f(x_4)] = g(y_2) = z_3$$

So, $gof = \{(x_1, z_1), (x_2, z_3), (x_3, z_3), (x_4, z_3)\}$.

(*ii*) First find the composition of *f* with *g* and then with *h* (Fig. 28).

$$ho(gof)\ (x_1) = ho[g(f(x_1))] = ho[g(y_1)] = h(z_1) = k_1$$
$$ho(gof)\ (x_2) = ho[g(f(x_2))] = ho[g(y_3)] = h(z_3) = k_3$$
$$ho(gof)\ (x_3) = ho[g(f(x_3))] = ho[g(y_2)] = h(z_3) = k_3$$
$$ho(gof)\ (x_4) = ho[g(f(x_4))] = h[g(y_2)] = h(z_3) = k_3$$

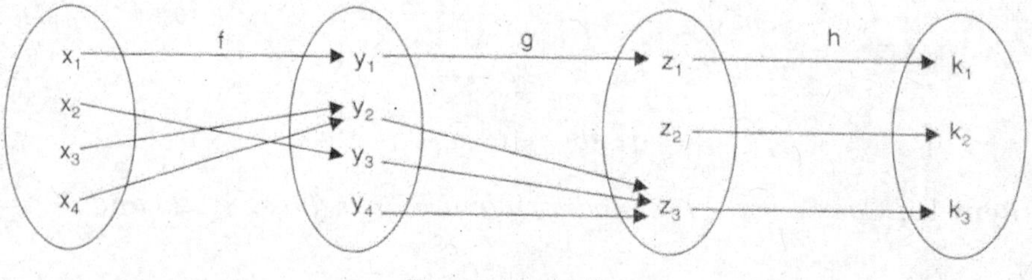

Fig. 28. *ho(gof).*

So, $ho(gof) = \{(x_1, k_1), (x_2, k_3), (x_3, k_3), (x_4, k_3)\}$.

(*iii*) First find the composition of *g* with *h* and then with *f*(Fig. 29).

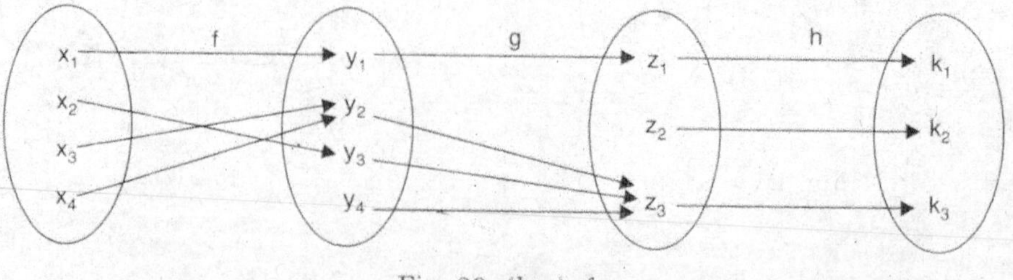

Fig. 29. *(hog)of.*

$(hog)(y_1) = h(g(y_1)) = h(z_1) = k_1,$ $\quad$ $(hog)(y_2) = h(g(y_2)) = h(z_3) = k_3$

$(hog)(y_3) = h(g(y_3)) = h(z_3) = k_3,$ $\quad$ $(hog)(y_4) = h(g(y_4)) = h(z_3) = k_3$

So, $\quad hog = \{(y_1, k_1), (y_2, k_2), (y_3, k_3), (y_4, k_4)\}$

Now,

$((hog)of)\ (x_1) = (hog)(f(x_1)) = hog(y_1) = k_1$

$((hog)of)\ (x_2) = (hog)(f(x_2)) = hog(y_3) = k_3$

$((hog)of)\ (x_3) = (hog)(f(x_3)) = hog(y_2) = k_3$

$((hog)of)\ (x_4) = (hog)(f(x_4)) = hog(y_2) = k_3$

So, $\quad (hog)of = \{(y_1, k_1), (y_2, k_3), (y_3, k_3), (y_4, k_3)\}.$

Problem 11. *Determine the value of each of the following floor functions :*

(i) $[8 . 2]$ *(ii)* $[19 . 1231]$ *(iii)* $[-9 . 6]$

(iv) $[8]$ *(v)* $[\sqrt{5}]$ *(vi)* $[-11 . 2]$.

Sol. *(i)* $[8 . 2] = 8$ *(ii)* $[19 . 1231] = 19$ *(iii)* $[-9 . 6] = -10$

(iv) $[8] = 8$ *(v)* $[\sqrt{15}] = 3$ *(vi)* $[-11 . 2] = -12$.

Problem 12. *Determine the value of each of the following ceiling functions :*

(i) $[8 . 3]$ *(ii)* $[19 . 1231]$ *(iii)* $[-9 . 7]$

(iv) $[9]$ *(v)* $[\sqrt{17}]$ *(vi)* $[-11 . 1]$.

Sol. *(i)* $[8 . 3] = 9$ *(ii)* $[19 . 1231] = 20$ *(iii)* $[-9 . 7] = -9$

(iv) $[9] = 9$ *(v)* $[\sqrt{17}] = 5$ *(vi)* $[-11 . 1] = -11$.

Problem 13. *Let $X = \{10, 20, 30\}$ and $U = \{10, 20, 30, ..., 100\}$. Compute the following function values :*

(i) $f \times (10)$ *(ii)* $f \times (80)$ *(iii)* $f \times (110)$

(iv) $f \times (30)$ *(v)* $f \times (40)$.

Sol. *(i)* $f \times (10) = 1$ *(ii)* $f \times (80) = 0$ *(iii)* $f \times (110) =$ Undefined

(iv) $f \times (30) = 1$ *(v)* $f \times (40) = 0$.

Problem 14. *Determine the value of the following functions :*

(i) 18(MOD 3) *(ii) 11(MOD 9)* *(iii) 40(MOD 7)*

Sol. *(i)* 18(MOD 3) = 0 *(ii)* 11(MOD 9) = 2 *(iii)* 40(MOD 7) = 5.

Problem 15. *Let f be the function from $X = \{0, 1, 2, 3, 4\}$ to X defined by $f(x) = 4x$(MOD 5). Write f as a set of ordered pairs. Is f one-to-one ?*

Sol. We have $\quad f(x) = 4x$(MOD 5) $\quad$ (given)

Now put the value of X in the function $f(x)$ one by one and obtain the corresponding second value of the pair

e.g., $\quad f(0) = 4 \times 0$(MOD 5) = 0(MOD 5) = 0

$f(1) = 4 \times 1$(MOD 5) = 4(MOD 5) = 4 and so on.

Thus, the required set of ordered pairs is

$$f = \{(0, 0), (1, 4), (2, 3), (3, 2), (4, 1)\}$$

Yes, f is one-to-one since there is exactly one value of $f(x)$ for each $x \in X$.

Problem 16. *Consider the following Boolean function :*

T(x, y) : x is greater than 10 and y is less than 10.

Determine the value of the Boolean functions

T(25, 8), T(5, 25) and T(10, 10).

Sol. The solution of the above functions is :

T(25, 8) is true

T(5, 25) is false

and T(10, 10) is false.

Problem 17. *Let P be a function defined by* $P(x, y) = (\sim x \vee y) \vee x$. *Determine each of the following :*

(*i*) *p(false, false)* (*ii*) *P(true, true)* (*iii*) *P(false, true)* (*iv*) *P(true, true).*

Sol. The values of the above functions is as follows :

(*i*) P(false, false) is false. (*ii*) P(true, true) is true.

(*iii*) P(false, true) is true. (*iv*) P(true, false) is true.

Problem 18. *Determine the value of the following exponential functions :*

(*i*) 10^{-3} (*ii*) 4^{-4} (*iii*) $8^{-3/2}$ (*iv*) $7^{1/2}$

(*v*) 3^6 (*vi*) 2^{10}.

Sol. (*i*) $10^{-3} = \frac{1}{10^3} = \frac{1}{1000}$ (*ii*) $4^{-4} = \frac{1}{4^4} = \frac{1}{4.4.4.4} = 0.0039$ (*iii*) $8^{-3/2} = 0.0441942$

(*iv*) $7^{1/2} = 2.64575$ (*v*) $3^6 = 729$ (*vi*) $2^{10} = 1024$.

Problem 19. *Determine the value of the following logarithmic functions :*

(*i*) $\log_2 24$ (*ii*) $\log_4 4$ (*iii*) $\log_2 1024$

(*iv*) $\log_{10} 1$ (*v*) $\log_2 0.0001$.

Sol. (*i*) $\log_2 24 = 4$ as $2^5 = 32$, which is greater.

(*ii*) $\log_4 4 = 1$ (*iii*) $\log_2 1024 = 10$ as $2^{10} = 1024$

(*iv*) $\log_{10} 1 = 0$ (*v*) $\log_2 0.0001 = -13$.

Problem 20. *For the following hash function, show how the data would be inserted in the order given in initially empty cells. Use the collision resolution policy.*

h(x) = x MOD 11 and cells indexed from 0 to 10 and data is given as :

53, 281, 743, 13, 20, 10, 796, 377.

Sol. The following figure shows the result of storing the above data in initially empty cells.

10		13	377	796		281	743		53	20
0	1	2	3	4	5	6	7	8	9	10

The collision occurred at the numbers 743, 20 and 10. One collision resolution policy is to find the next highest unoccupied cell. Thus $h(743) = 6$ but is stored at 7, the next highest cell.

Problem 21. *For the following two permutations*

$$f = \begin{pmatrix} 1 & 2 & 3 & 4 & 5 & 6 \\ 4 & 5 & 6 & 3 & 1 & 2 \end{pmatrix} \text{ and } g = \begin{pmatrix} 1 & 2 & 3 & 4 & 5 & 6 \\ 5 & 6 & 4 & 2 & 1 & 3 \end{pmatrix}$$

find the product fog and express f as a product of cycles and then as a product of transpositions.

Sol. The product of two permutations f and g is

$$fog = \begin{pmatrix} 1 & 2 & 3 & 4 & 5 & 6 \\ 4 & 5 & 6 & 3 & 1 & 2 \end{pmatrix} \circ \begin{pmatrix} 1 & 2 & 3 & 4 & 5 & 6 \\ 5 & 6 & 4 & 2 & 1 & 3 \end{pmatrix}$$

$$= \begin{pmatrix} 1 & 2 & 2 & 4 & 5 & 6 \\ 1 & 2 & 3 & 4 & 5 & 6 \end{pmatrix}$$

To write f as a product of cycles, we have to find different cycles in f. Since f contains a single cycle, hence f can be written as

$$f = (1, 4, 3, 6, 2, 5).$$

To write f as a product of transpositions, we have to find different transpositions from the product of cycles of f. Thus, f as a product of transpositions is as follows :

$$(1, 4, 3, 6, 2, 5) = (1, 5), (1, 2), (1, 6), (1, 3), (1, 4).$$

6

Graphs

BASIC TERMINOLOGY

The graphs consist of points or nodes called vertices which are connected to each other by way of lines called edges. These lines may be directed or undirected.

UNDIRECTED GRAPHS

An undirected graph G consists of a set of vertices, V and a set of edges E. The edge set contains the unordered pair of vertices. If $(u, v) \in E$ then we say u and v are connected by an edge where u and v are vertices in the set V.

Example 1. *Let* $V = \{1, 2, 3, 4\}$ *and* $E = \{(1, 2), (1, 4), (3, 4), (2, 3)\}$. *Draw the graph.*

Sol. The graph can be drawn in several ways.

Two of which are as follows (Fig. 1 and Fig. 2).

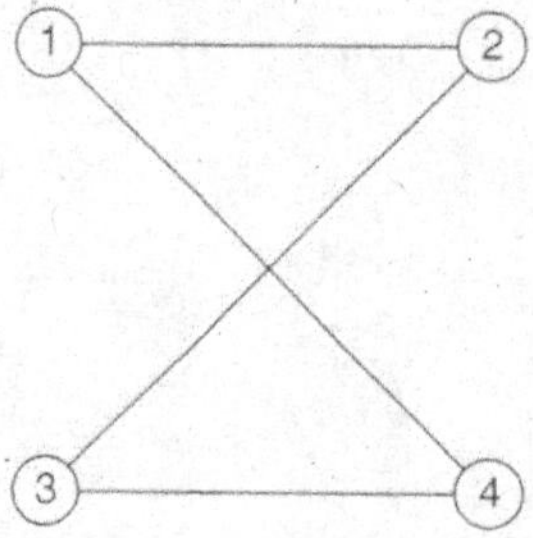

Fig. 1.

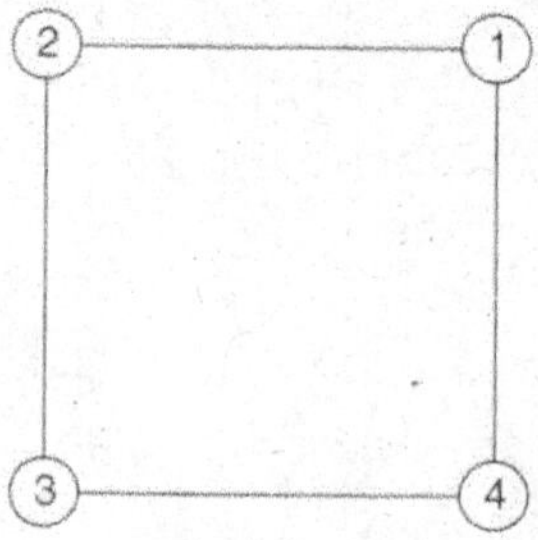

Fig. 2.

Example 2. *Consider the graph shown in Fig. 3. Determine the edge set and the vertex set of the graph.*

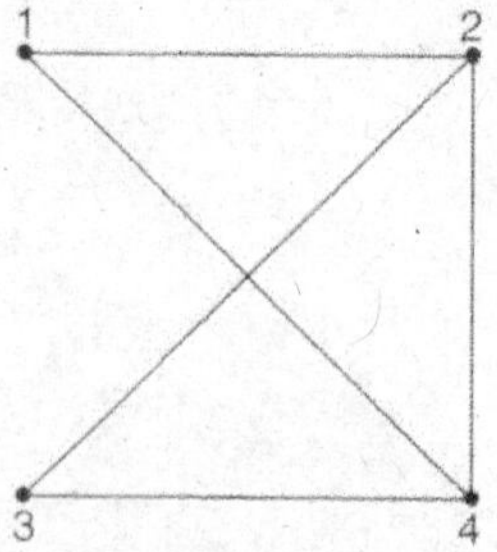

Fig. 3.

Sol. The edge set is

$$E = \{(1, 2), (1, 4), (2, 3), (2, 4), (3, 4)\}$$

The vertex set is

$$V = \{1, 2, 3, 4\}.$$

Degree of a Vertex. The degree of a vertex is the number of edges incident on a vertex v. The self-loop is counted twice. The degree of vertex is denoted by $d(v)$.

Example 3. *Consider the graph G shown in Fig. 4. Determine the degree of each vertex of graph.*

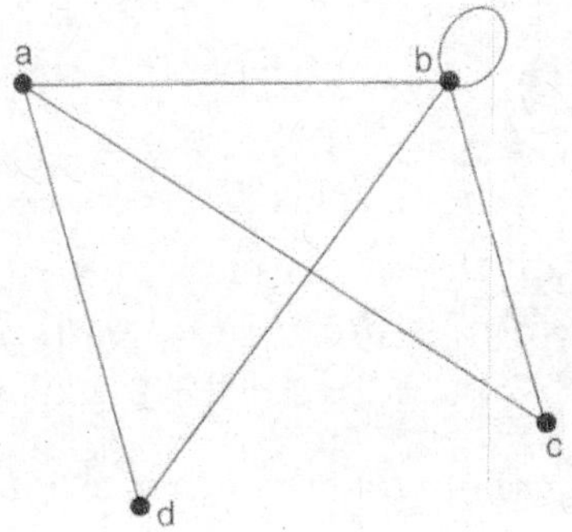

Fig. 4.

Sol. The degree of each vertex is as follows :

$$d(a) = 3 \; ; \; d(b) = 5 \; ; \; d(c) = 2 \; ; \; d(d) = 2.$$

Theorem I. *Show that the sum of degree of all the vertices in a graph G, is even.*

Proof. Each edge contribute two degrees in a graph. Also, each edge contributes one degree to each of the vertices on which it is incident.

Hence, if there are N edges in G, then we have

$$2N = d(v_1) + d(v_2) + ... + d(v_N)$$

Thus, 2N is always even.

Theorem II. *Prove that in any graph, there are an even number of vertices of odd degree.*

Proof. Consider a graph having vertices of degree even and odd. Now, make two groups of vertices. One with even degree of vertices $v_1, v_2, ..., v_k$ and other with odd degree of vertices $u_1, u_2, ..., u_n$.

Suppose,

$$V = d(v_1) + d(v_2) + ... + d(v_k)$$
$$U = d(u_1) + d(u_2) + ... + d(u_n).$$

Now, we know that sum of degree of all the vertices is even (Theorem I). So, V + U is even.

Since, V is the sum of K even numbers. Hence, it is even. But U is the sum of n odd numbers. So, to be U an even number, n must be even. Hence proved.

Example 4. *Verify that the sum of the degree of all the vertices is even for the graph shown in Fig. 5.*

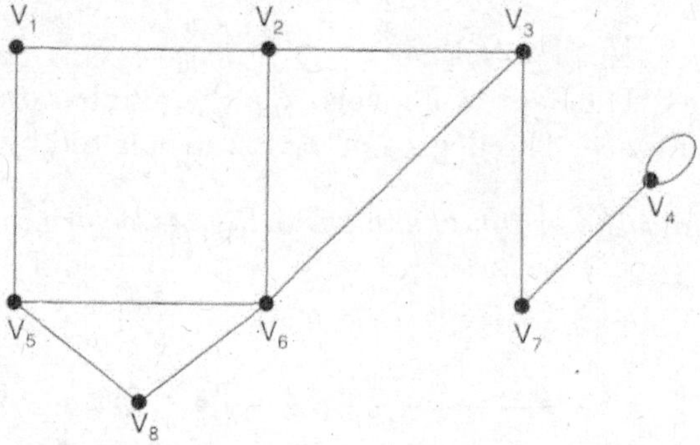

Fig. 5.

Sol. The sum of degree of all the vertices is

$$= d(v_1) + d(v_2) + d(v_3) + d(v_4) + d(v_5) + d(v_6) + d(v_7) + d(v_8)$$
$$= 2 + 3 + 3 + 3 + 3 + 4 + 2 + 2 = 22, \text{ which is even.}$$

Example 5. *Verify that there are an even number of vertices of odd degree in the graph shown in Fig. 6.*

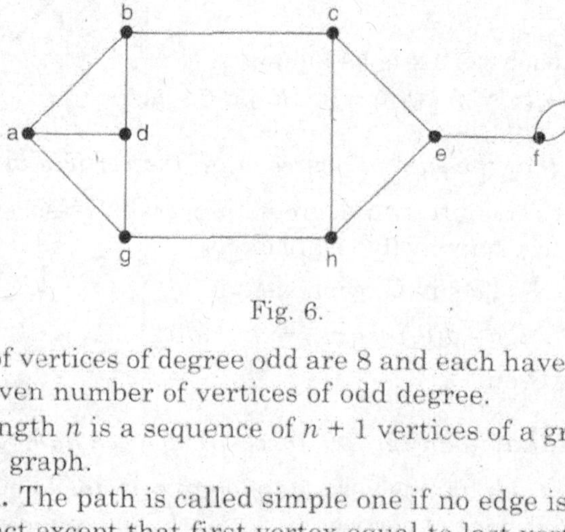

Fig. 6.

Sol. The number of vertices of degree odd are 8 and each have degree three in the above graph. Hence, we have even number of vertices of odd degree.

Path. A path of length n is a sequence of $n + 1$ vertices of a graph in which each pair of vertices is an edge of the graph.

1. **A Simple Path.** The path is called simple one if no edge is repeated in the path *i.e.*, all the vertices are distinct except that first vertex equal to last vertex.

2. **An Elementary Path.** The path is called elementary one if no vertex is repeated in the path *i.e.*, all the vertices are distinct.

3. **Circuit or Closed Path.** The circuit or closed path is a path which starts and ends at the same vertex *i.e.*, $v_o = v_n$.

4. **Simple Circuit Path.** The simple circuit is a simple path which is a circuit.

Theorem III. *Suppose a graph G contains two distinct paths from a vertex u to a vertex v. Show that G has a cycle.*

Proof. Consider two distinct paths from u to v be $P_1 = (e_1, e_2, e_3, \ldots\ldots, e_n)$ and $P_2 = (e_1', e_2', e_3', \ldots\ldots, e_n')$.

Now delete from the paths P_1 and P_2 all the initial edges which are identical *i,e.*, of we have $e_1 = e_1'$, $e_2 = e_2'$, $e_3 = e_3'$, ..., $e_k = e_k'$ but $e_{k+1} \neq e'_{k+1}$. We will delete all the first k edges of both the paths P_1 and P_2.

Now, after deleting the k edges both the paths start from the same vertex, (let u_1) and end at v.

Now, to construct a cycle, start from vertex u, and follow the left over path of P_1 until we first meet any vertex of the left over path of P_2.

If this vertex is u_2, then the remaining cycle is computed by following the left over path of P_2 which starts from u_2 and ends at v.

Example 6. *Consider the graph shown in Fig. 7. Give an example of the following :*

(*i*) *A simple path from V_1 to V_6.*

(*ii*) *An elementary path from V_1 to V_6.*

(*iii*) *A simple path which is not elementary from V_1 to V_6.*

(*iv*) *A path which is not simple and starting from V_2.*

(*v*) *A simple circuit starting from V_1.*

(*vi*) *A circuit which is not simple and starting from V_2.*

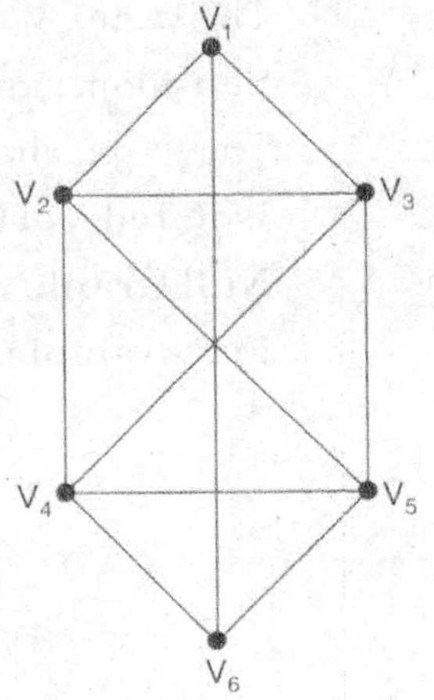

Fig.7.

Sol. (*i*) A simple path from V_1 to V_6 is

$$V_1, V_2, V_3, V_4, V_5, V_6.$$

(*ii*) An elementary path from V_1 to V_6 is

$$V_1, V_2, V_3, V_5, V_4, V_6.$$

(*iii*) A simple path which is not elementary from V_1 to V_6 is

$$V_1, V_2, V_3, V_5, V_2, V_4, V_6.$$

(*iv*) A path which is not simple and starting from V_2 is

$$V_2, V_3, V_4, V_5, V_3, V_4, V_6.$$

(*v*) A simple circuit starting from V_1 is

$$V_1, V_2, V_4, V_6, V_5, V_3, V_1.$$

(*vi*) A circuit which is not simple and starting from V_2 is

$$V_2, V_3, V_1, V_2, V_5, V_4, V_2.$$

Pendent Vertex. A vertex with degree one is called a pendent vertex.

Pendent Edge. The only edge which is incident with a pendent vertex is called the pendent edge.

Odd Vertex. A vertex having degree odd is called an odd vertex.

Even Vertex. A vertex having degree even is called an even vertex.

Incident Edge. An edge is called incident with the vertices is connects.

Adjacent Vertices. Two vertices are called adjacent if they are connected by an edge. If there is an edge (u, v), then we can say vertex u is adjacent to vertex v and vertex v is adjacent to vertex u.

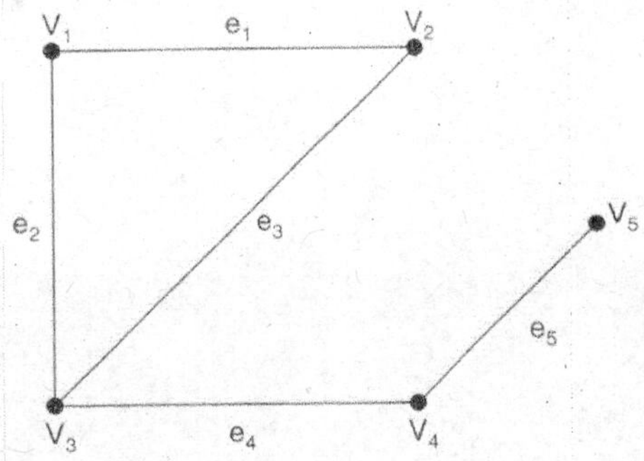

Fig.8.

Example 7. *Consider the graph as shown in Fig. 8. Determine the following :*

(*i*) *Pendent vertices*

(*ii*) *Pendent edges*

(*iii*) *Odd vertices*

(*iv*) *Even vertices*

(*v*) *Incident edges*

(*vi*) *Adjacent vertices.*

Sol. (*i*) The vertex V_5 is the pendent vertex.

(*ii*) The edge (V_4, V_5) or e_5 is the pendent edge.

(*iii*) The vertices V_3 and V_5 are odd vertices.

(*iv*) The vertices V_1, V_2 and V_4 are even vertices.

(*v*) The edge e_1 is incident on V_1 and V_2. The edge e_2 is incident on V_1 and V_3.
The edge e_3 is incident on V_2 and V_3. The edge e_4 is incident on V_3 and V_4.
The edge e_5 is incident on V_4 and V_5.

(*vi*) The vertex V_1 is adjacent to V_2 and V_3. The vertex V_2 is adjacent to V_1 and V_3.
The vertex V_3 is adjacent to V_1 and V_4. The vertex V_4 is adjacent to V_3 and V_5.
The vertex V_5 is adjacent to V_4.

Self-loop. A self-loop is an edge e if it has a same end point.

The graph shown in Fig. 4 contains the self-loop at vertex b *i.e.*, $e = (b, b)$.

Isolated Vertex. An isolated vertex is a vertex having no edges incident to it.

Null Graph. A null graph is defined as a graph which consists only the isolated vertices.

For example. The graph shown in Fig. 9 is a null graph and the vertices are isolated vertices.

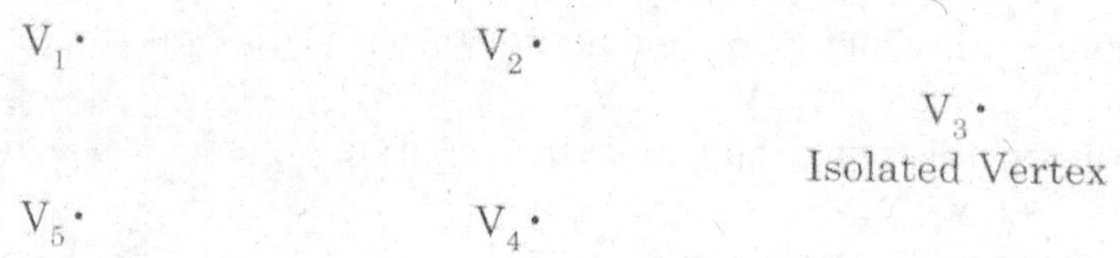

Fig. 9. Null Graph

Complementary Graph. The complement of a graph G is defined to be a graph which has the same number of vertices as in graph G and has two vertices connected if and only if they are not connected in graph.

Example 8. *Consider the graph G shown in Fig. 10. Find the complement of this graph.*

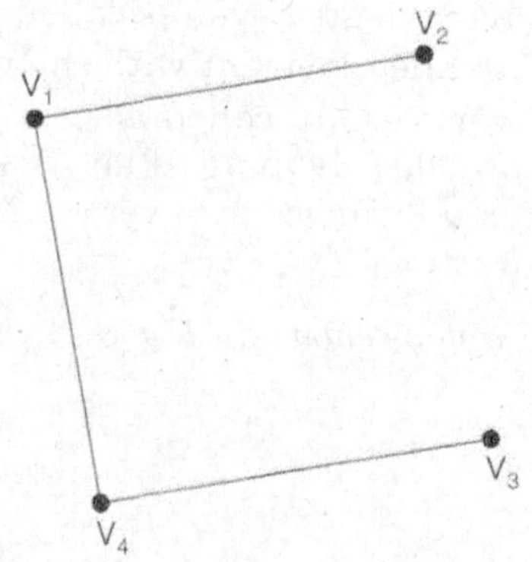

Fig. 10.

Sol. The complement of above graph is shown in Fig. 11.

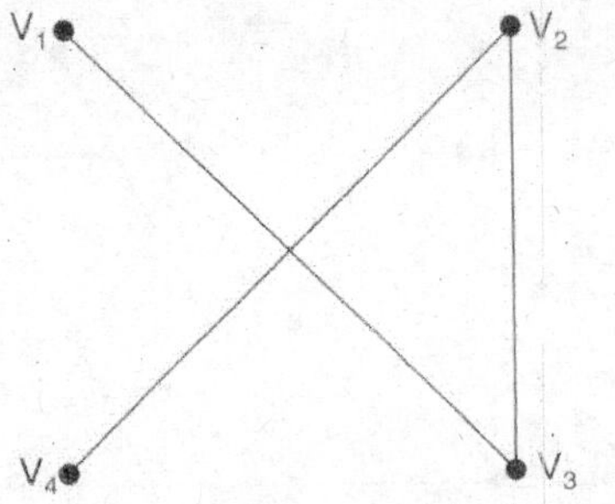

Fig. 11. Complement of Graph G.

Undirected Complete Graph. An undirected complete graph G = (V, E) of n vertices is a graph in which each vertex is connected to every other vertex *i.e.*, and edge exists between every pair of distinct vertices. It is denoted by K_n. A complete graph with n vertices will have $n(n-1)/2$ edges.

Example 9. *Draw undirected complete graphs* K_4 *and* K_6.

Sol. The undirected complete graph of K_4 is shown in Fig. 12 and that of K_6 is shown in Fig. 13.

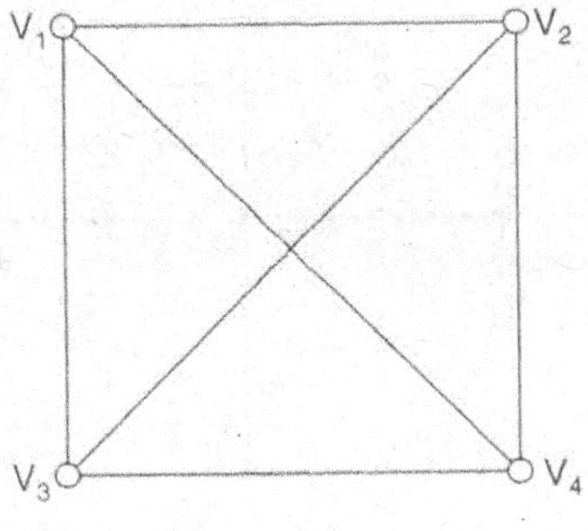

Fig. 12. K_4

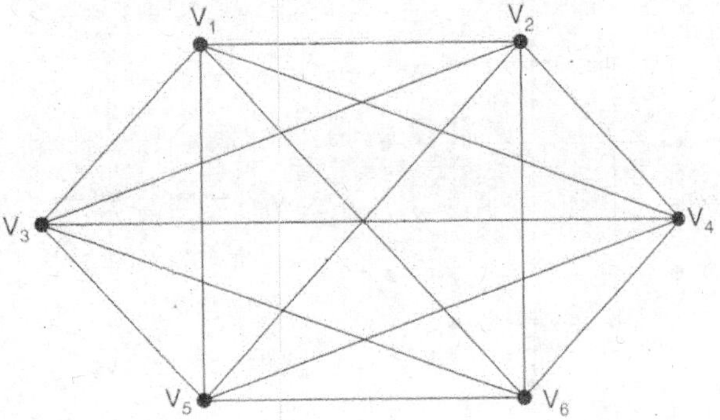

Fig. 13. K_6

Connected Graph. A graph is called connected if there is a path from any vertex u to v or vice-versa.

Disconnected Graph. A graph is called disconnected if there is no path between any two of its vertices.

Connected Component. A subgraph of graph G is called the connected component of G, if it is not contained in any bigger subgraph of G, which is connected. It is defined by listing its vertices.

Example 10. *Consider the graph shown in Fig. 14. Determine its connected components.*

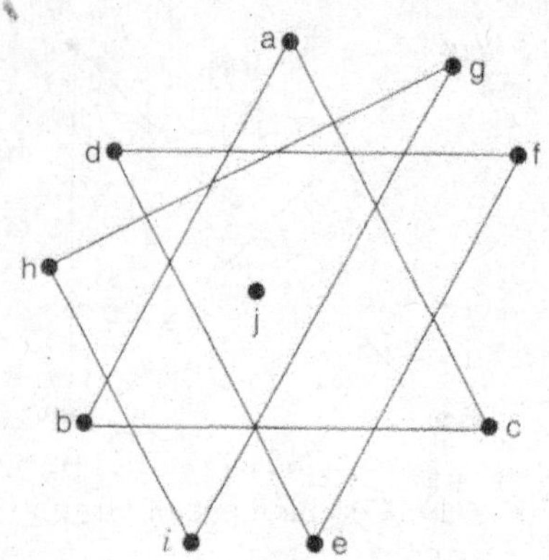

Fig. 14.

Sol. The connected components of this graph is {*a, b, c*}, {*d, e, f*}, {*g, h, i*} and {*j*}.

Example 11. *Consider the graphs shown in Figs. 15, 16 and 17. Determine whether the graphs are (a) Connected graphs or (b) Disconnected graphs.*

Also write their connected components.

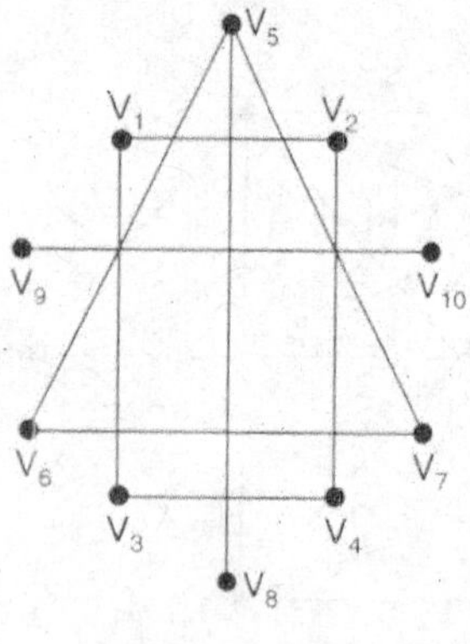

Fig. 15.

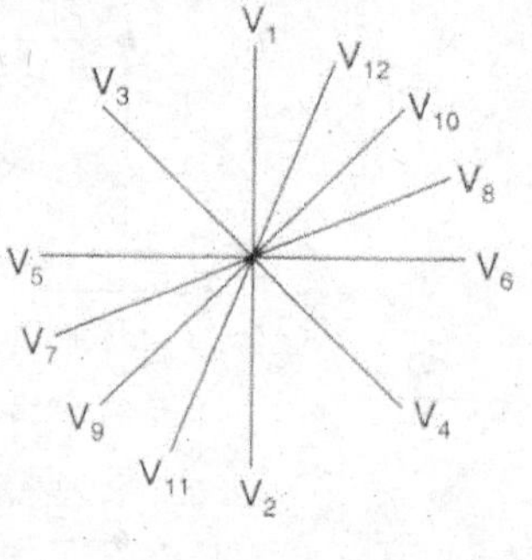

Fig. 16.

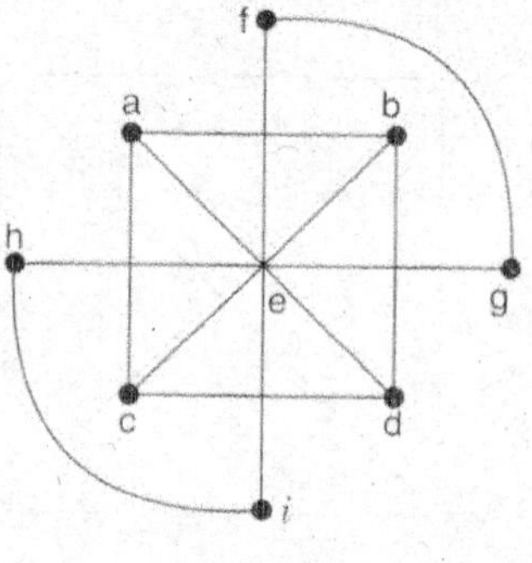

Fig. 17.

Sol. (*i*) The graph shown in Fig. 15 is a disconnected graph and its connected components are

$$\{V_1, V_2, V_3, V_4\}, \{V_5, V_6, V_7, V_8\} \quad \text{and} \quad \{V_9, V_{10}\}.$$

(*ii*) The graph shown in Fig. 16 is a disconnected graph and its connected components are

$$\{V_1, V_2\}, \{V_3, V_4\}, \{V_5, V_6\}, \{V_7, V_8\}, \{V_9, V_{10}\} \quad \text{and} \quad \{V_{11}, V_{12}\}.$$

(*iii*) The graph shown in Fig. 17 is a connected graph.

Subgraph. A subgraph of a graph G = (V, E) is a graph G′ = (V′, E′) in which V′ ⊆ V and E′ ⊆ E and each edge of G′ has the same end vertices in G′ as in graph G.

Note. A single vertex is a subgraph.

Example 12. *Consider the graph G shown in Fig. 18. Show the different subgraphs of this graph.*

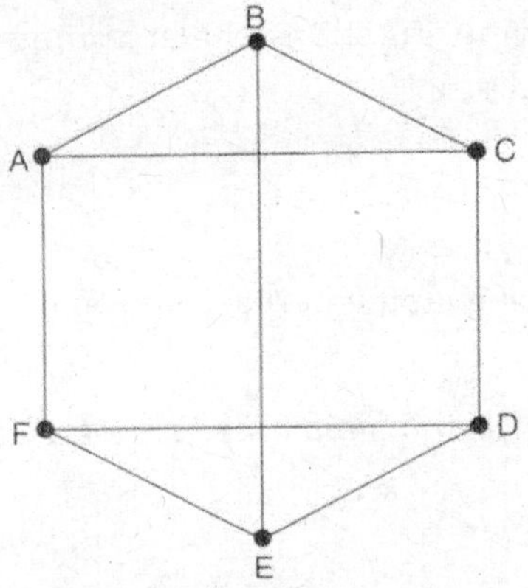

Fig. 18.

Sol. The following are all subgraphs of the above graph (shown in Figs. 19, 20, 21, 22).

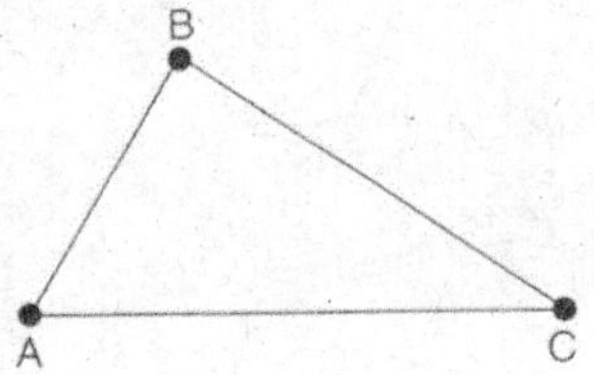

Fig. 19.

Fig. 20.

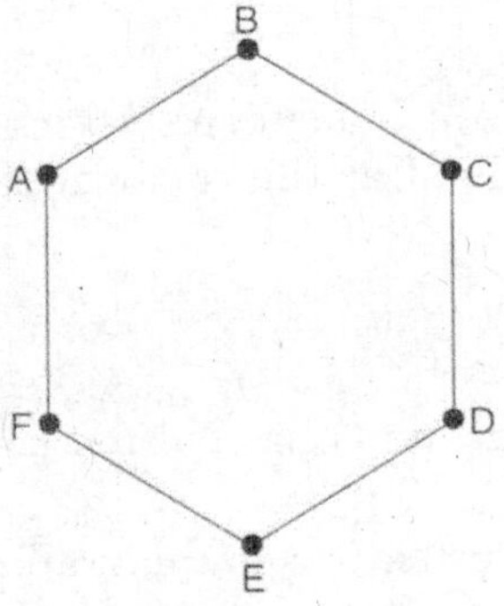

Fig. 21.

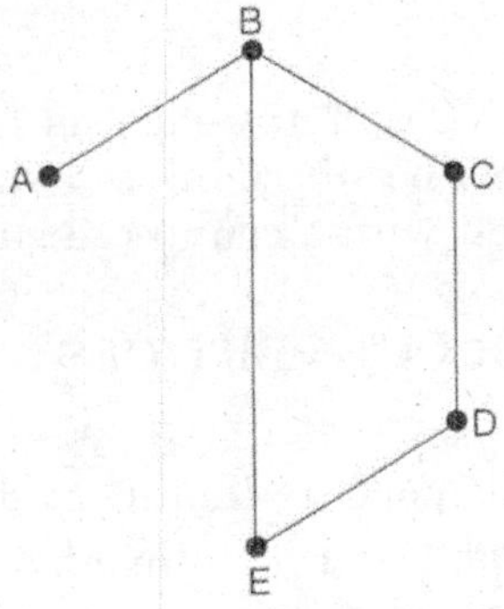

Fig. 22.

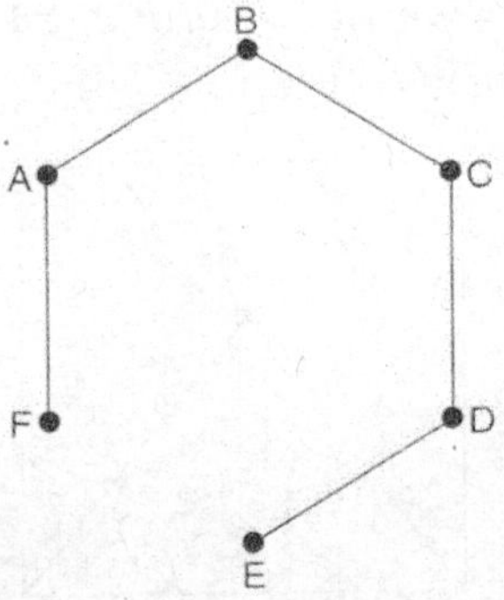

Fig. 23. Spanning Subgraph.

Spanning Subgraph. A graph G_1 is called a spanning subgraph of G if G_1 contains all the vertices of G.

For example : The following Fig. 23 is the spanning subgraph of the graph shown in Fig. 18.

CUT SET

Consider a connected graph G = (V, E). A cut set for G is a smallest set of edges such that removal of the set, disconnects the graph whereas the removal of any proper subset of this set, left a connected subgraph.

Example 13. *Consider the graph shown in Fig. 24. Determine the cut set for this graph.*

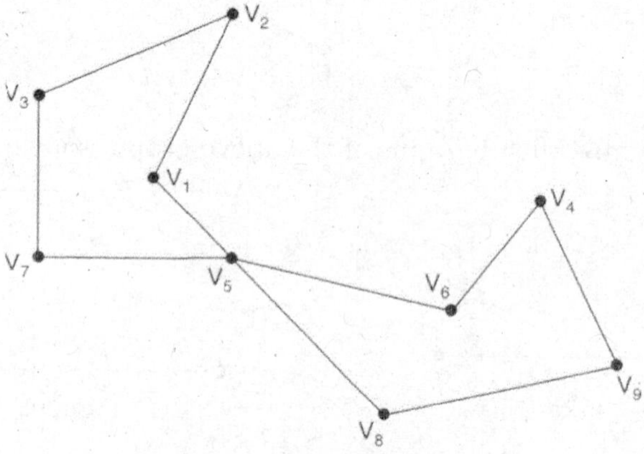

Fig. 24.

Sol. For this graph, the edge set $\{(V_1, V_5), (V_7, V_5)\}$ is a cut set. After the removal of this set, we have left with a disconnected subgraph. While after the removal of any of its proper subset, we have left with a connected subgraph.

CUT POINTS OR CUT VERTICES

Consider a graph G = (V, E). A cut point for a graph G, is a vertex v such that $G-v$ has more connected components than G or disconnected.

The subgraph $G-v$ is obtained by deleting the vertex v from the graph G and also deleting all the edges incident on v.

Example 14. *Consider the graph shown in Fig. 25. Determine the subgraphs*
(i) $G - v_1$ *(ii)* $G - v_3$ *(iii)* $G - v_5$.

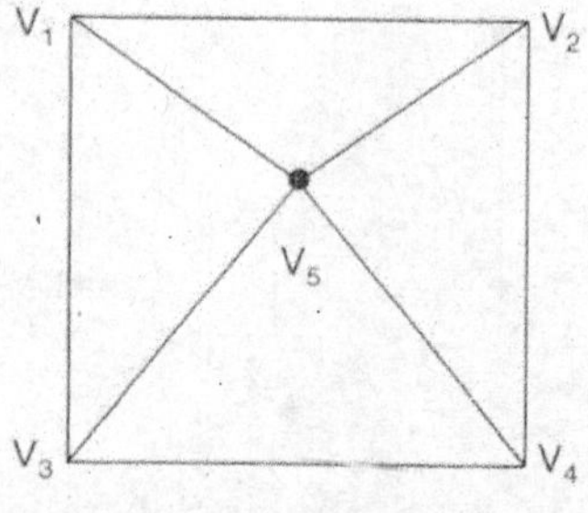

Fig. 25.

Sol. (*i*) The subgraph $G - v_1$ is shown in Fig. 26.

(*ii*) The subgraph $G – v_3$ is as shown in Fig. 27.

(*iii*) The subgraph $G – v_5$ is as shown in Fig. 28.

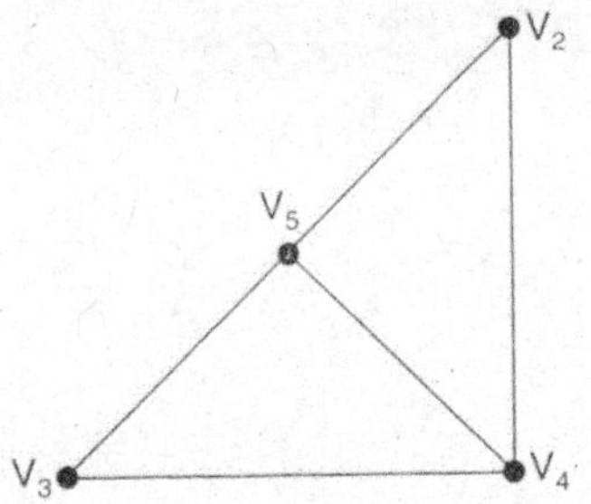

Fig. 26.

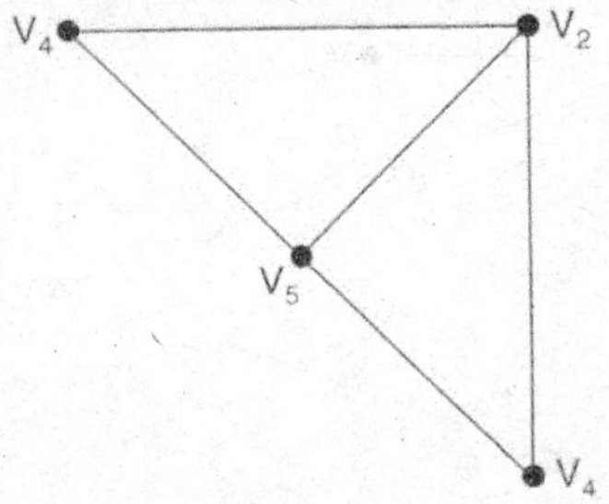

Fig. 27.

Fig. 28.

Example 15. *Consider the graph G shown in Fig. 29. Determine all the cut points of G.*

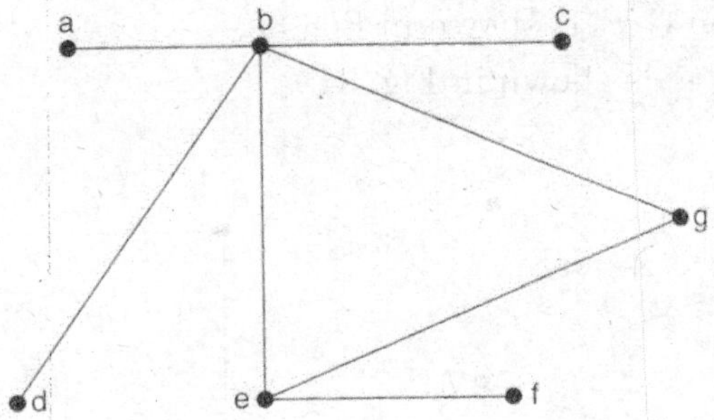

Fig. 29.

Sol. (*a*) The vertex b is cut point for G. Since, $G–b$ has more than one connected components as shown in Fig. 30.

(*b*) The vertex e is also a cut point for G. Since $G–e$ has more than one connected components as shown in Fig. 31.

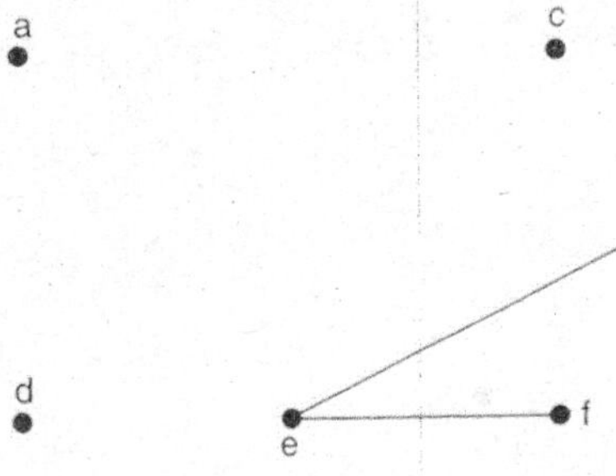

Fig. 30.

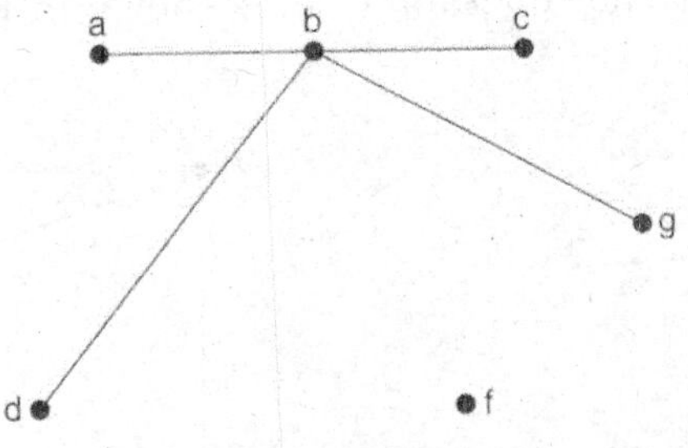

Fig. 31.

BRIDGE (CUT EDGES)

Consider a graph $G = (V, E)$. A bridge for a graph G, is an edge e such that G-e has more connected components than G or disconnected.

Example 16. *Consider the graph shown in Fig. 32. Determine the subgraphs*
(i) $G - e_1$ *(ii)* $G - e_3$ *(iii)* $G - e_4$.

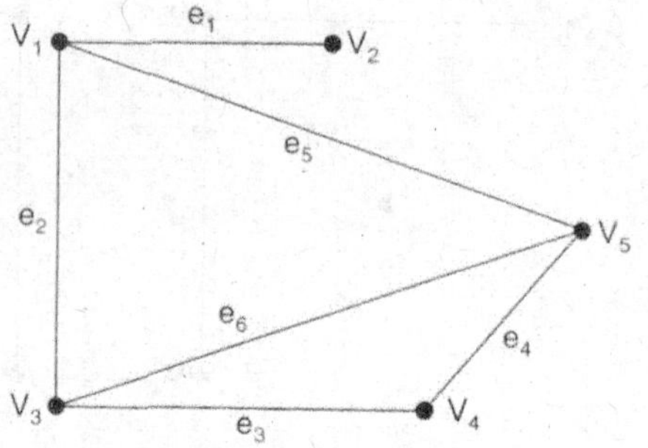

Fig. 32.

Sol. *(i)* The subgraph $G–e_1$ is shown in Fig. 33.
(ii) The Subgraph $G–e_3$ is shown in Fig. 34.

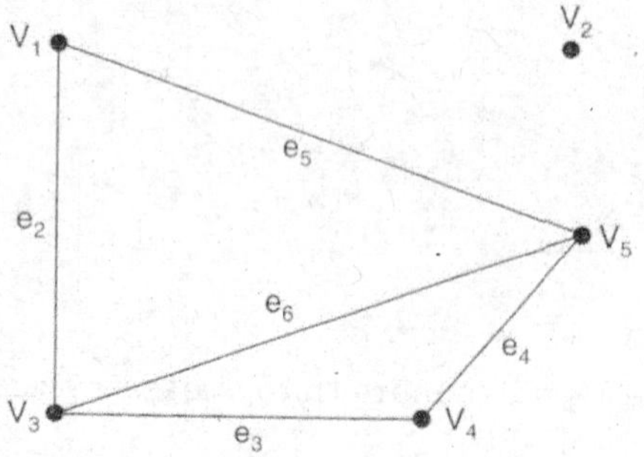

Fig. 33.

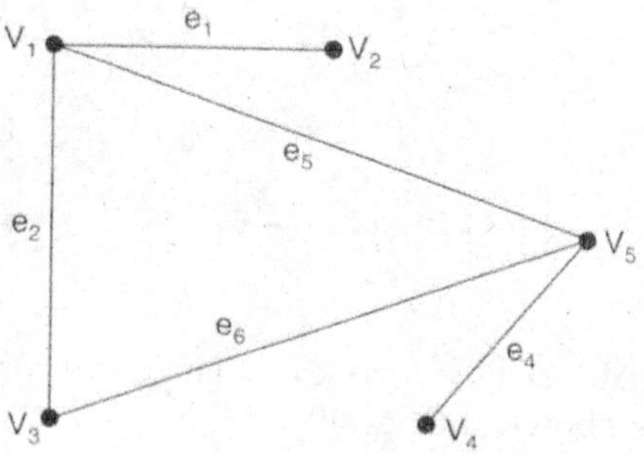

Fig. 34.

(iii) The subgraph $G–e_4$ is shown in Fig. 35.

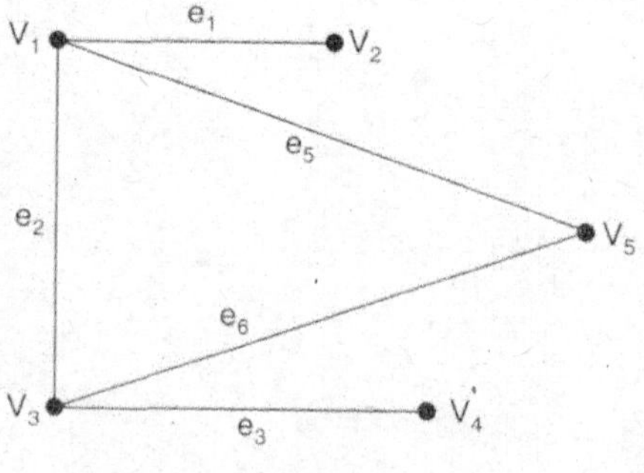

Fig. 35.

Example 17. *Consider the graph G shown in Fig. 36. Determine all the bridges of G.*

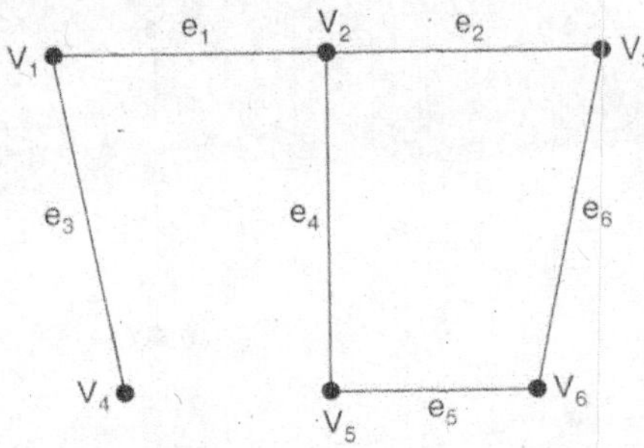

Fig. 36.

Sol. (*a*) The edge e_1 is a bridge for G. Since G–e_1 has more than one connected components as shown in Fig. 37.

(*b*) The edge e_3 is a bridge for G. Since G–e_3 has more than one connected components as shown in Fig. 38.

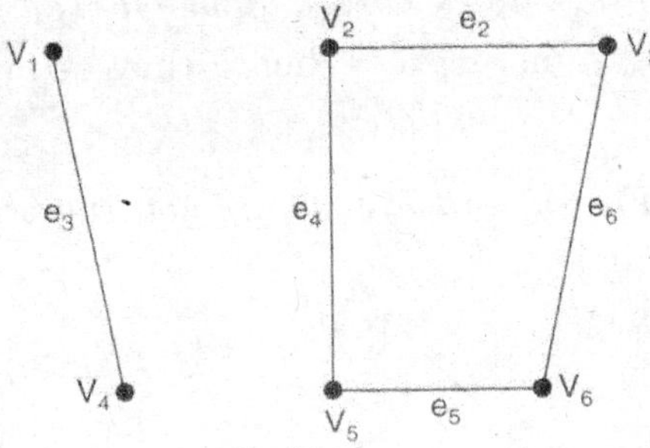

Fig. 37.

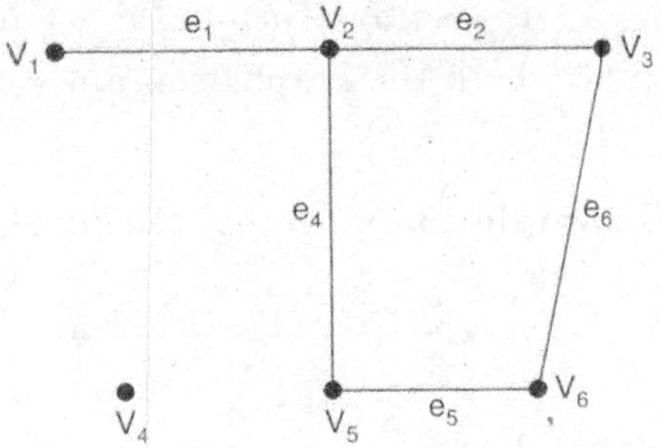

Fig. 38.

ISOMORPHIC GRAPHS

Two graphs G_1 and G_2 are called isomorphic graphs if there is a one-to-one correspondence between their vertices and between their edges *i.e.*, the graphs have identical representation except that the vertices may have different labels.

For example : Consider the following graphs shown in Fig. 39 and Fig. 40. They are isomorphic graphs.

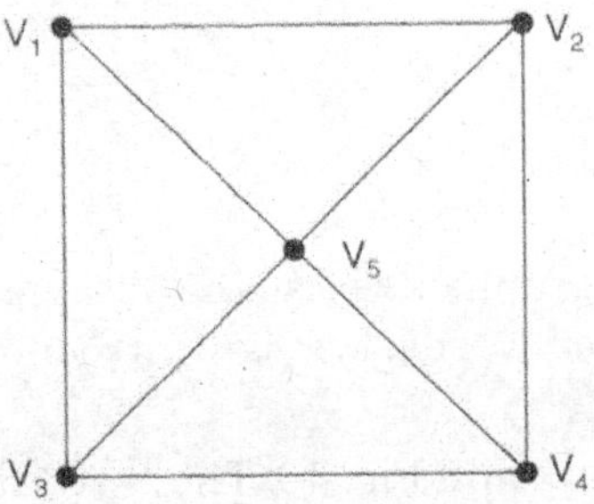

Fig. 39.

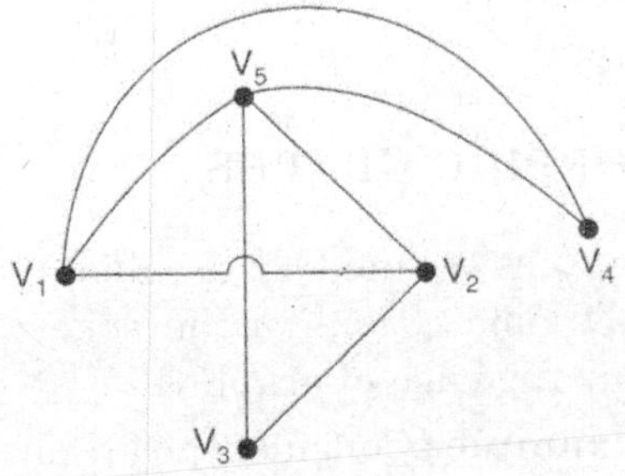

Fig. 40.

Example 18. *Show that the graphs shown in Fig. 41 and Fig. 42 are isomorphic.*

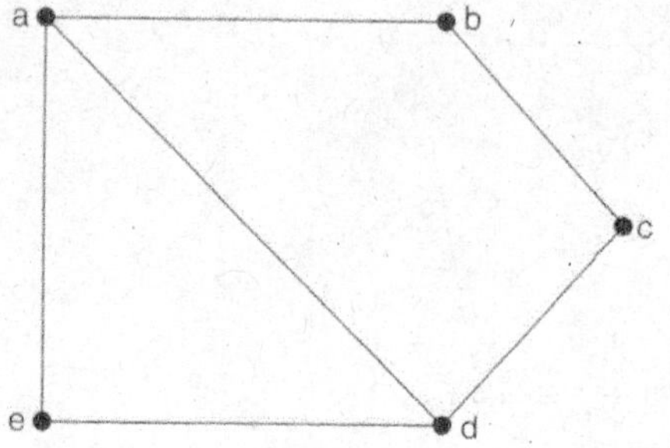

Fig. 41.

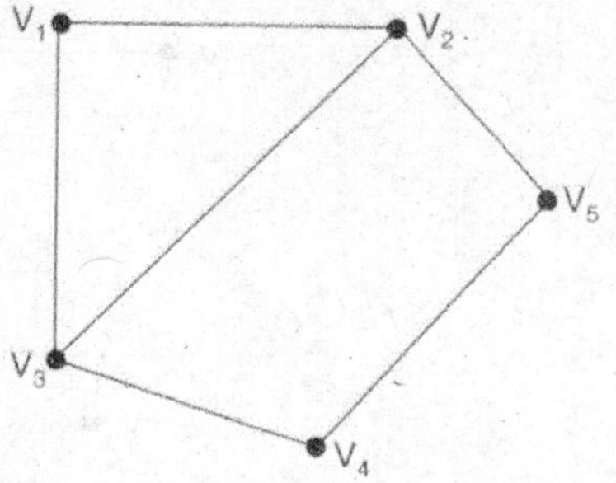

Fig. 42.

Sol. Compare the degrees of vertices of two graphs and find the vertices from both the graphs having same degrees and make the pairs of the vertices in decreasing order of degree. If both the graphs contain vertices having same degree, then they are isomorphic otherwise not. The pairs of vertices in decreasing order of degree are as follows :

$$d(a) \leftrightarrow d(v_2),\ d(d) \leftrightarrow d(v_3),\ d(b) \leftrightarrow d(v_1),\ d(e) \leftrightarrow d(v_4),\ d(c) \leftrightarrow d(v_5).$$

Since, both the graphs contain vertices having same degrees, hence they are isomorphic.

Example 19. *Show that the graphs shown in Fig. 43 and Fig. 44 are not isomorphic.*

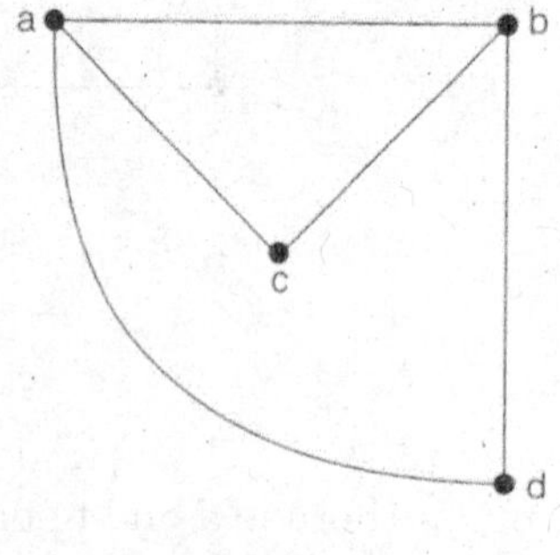

Fig. 43.

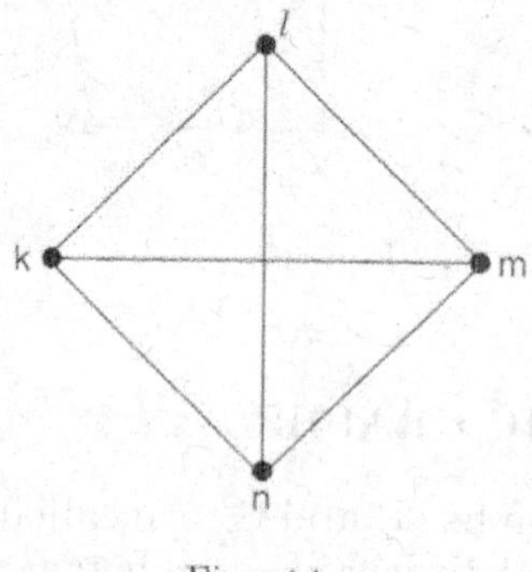

Fig. 44.

Sol. The graphs are not isomorphic because the vertices of the graph shown in Fig. 44 is having degree 3 but the graph shown in Fig. 43 contains two vertices having degree less than three.

HOMEOMORPHIC GRAPHS

Two graphs G_1 and G_2 are called homeomorphic graphs if G_2 can be obtained from G_1 by a sequence of subdivisions of the edges of G_1. In other words, we can introduce vertices of degree two in any edge of graph G_1.

For example : Consider the graph shown in Fig. 45 and Fig. 46. They are homeomorphic graphs.

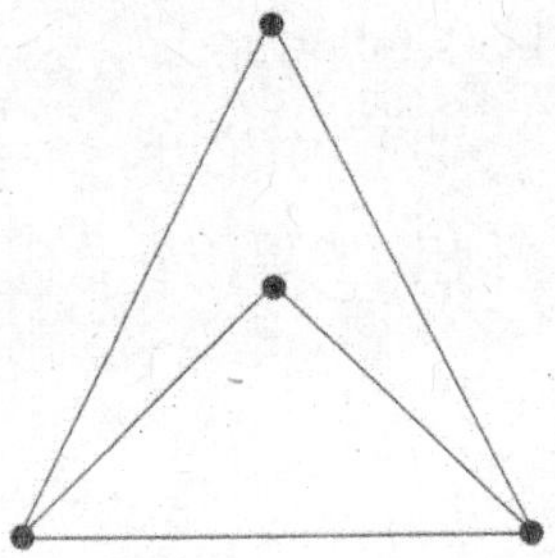

Fig. 45. G_1.

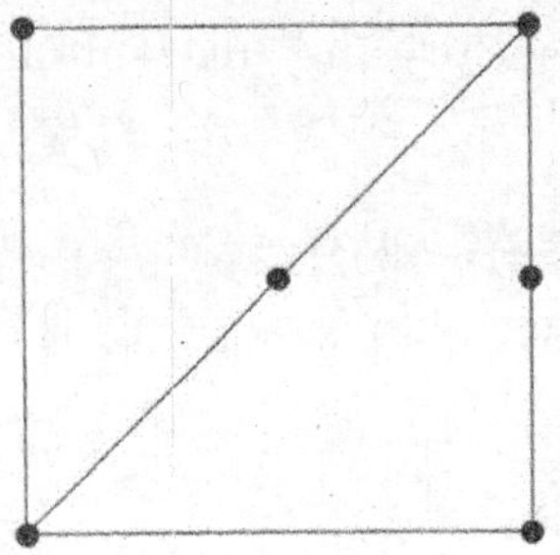

Fig. 46. G_2.

Example 20. *Show that the graphs shown in Figs. 47 and 48 are homeomorphic.*

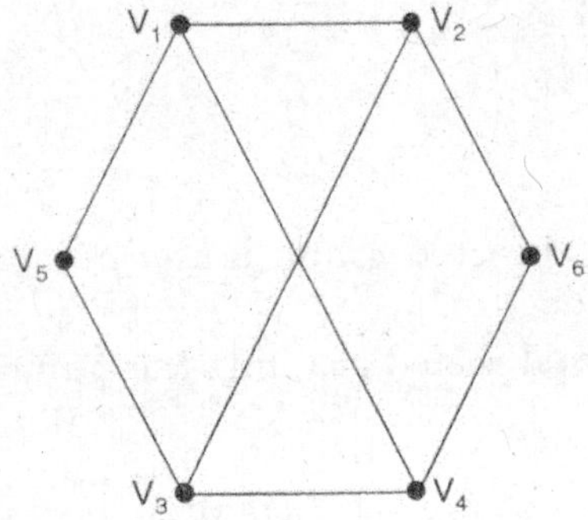

Fig. 47. G_1.

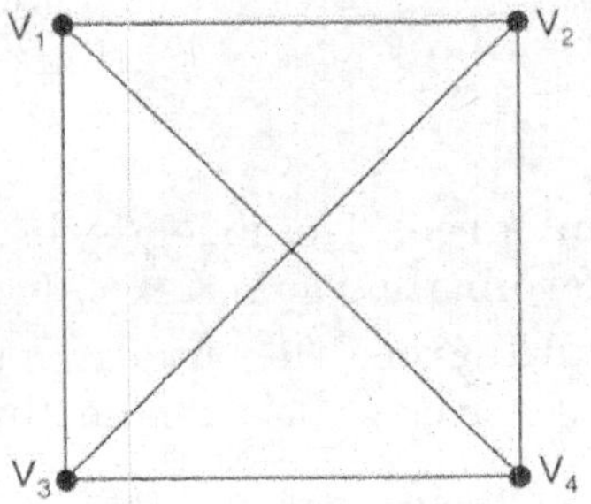

Fig. 48. G_2.

Sol. The two graphs are homeomorphic because G_1 can be obtained from G_2 by introducing vertices of degree 2 on edges (V_1, V_3) and (V_2, V_4).

DIRECTED GRAPHS

A directed graphs or digraph G is defined as an ordered pair (V, E), where V is the set of points called vertices and E is the set of edges. Each edge in the graph G is assigned a direction and is identified with an ordered pair (u, v), where u is the initial vertex and v is the end vertex.

Example 21. *Consider the directed graph G = (V, E) as shown in Fig. 49. Determine the vertex set and edge set of graph G.*

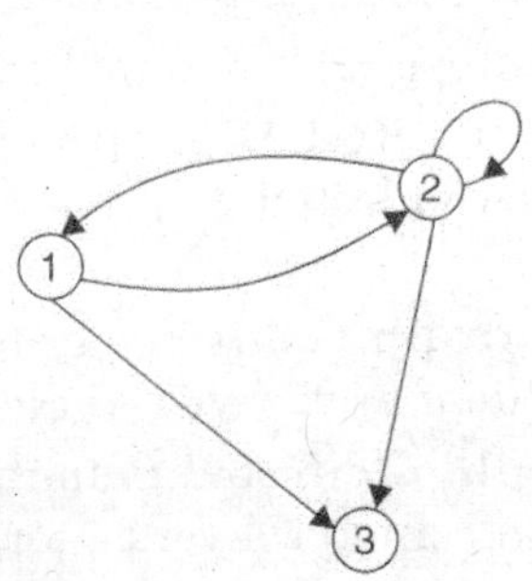

Fig. 49.

Sol. The vertex and edge set of graph G = (V, E) is as follows

$$G = \{\{1, 2, 3\}, \{(1, 2), (2, 1), (2, 2), (2, 3), (1, 3)\}\}.$$

Example 22. *Let G = {{a, b, c, d} {(a, b), (b, c), (c, c), (d, d), (d, a)}}. Draw the graph G.*

Sol. The graph of G = (V, E) is shown in Fig. 50.

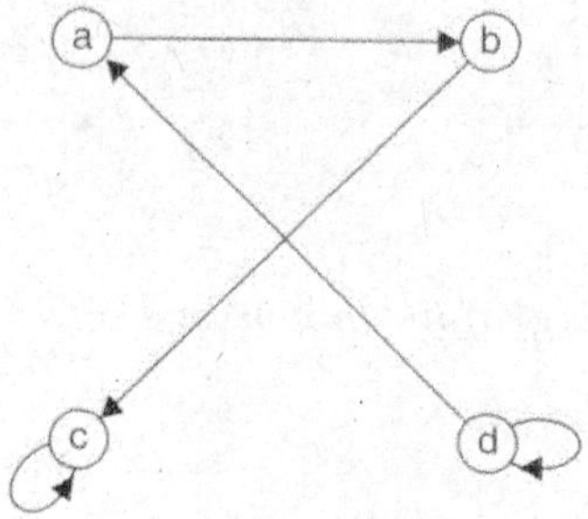

Fig. 50.

1. **Indegree.** The indegree of a vertex v in a directed graph is the number of edges ending or terminating at vertex v. It is denoted by indeg (v).

2. **Outdegree.** The outdegree of a vertex v in a directed graph is the number of edges beginning at vertex v. It is denoted by outdeg (v).

Example 23. *Consider the directed graph shown in Fig. 51. Determine the indegree and outdegree of each of vertices of the graph.*

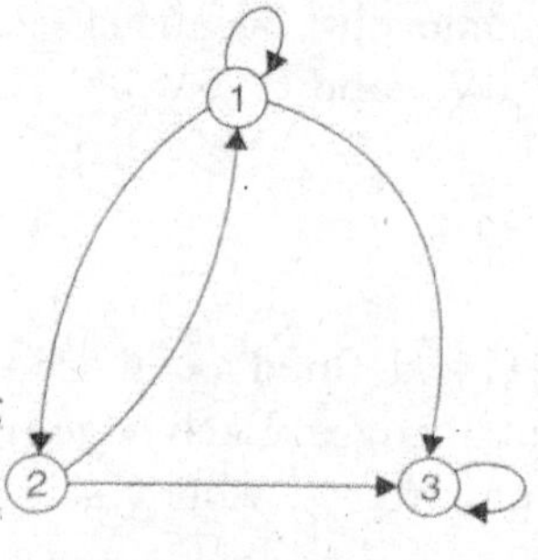

Fig. 51.

Sol. The indegree of digraph is indeg (1) = 2, indeg (2) = 1, indeg (3) = 3

The outdegree of digraph is outdeg (1) = 3, outdeg (2) = 2, outdeg (3) = 1.

Weakly Connected or Connected Digraph. A directed graph is called weakly connected if its undirected graph is connected *i.e.,* the graph obtained after neglecting the direction.

Unilaterally Connected Digraph. A directed graph is called unilaterally connected if there is a directed path from any node u to v or vice-versa, for any pair of nodes of the graph.

Strongly Connected Digraph. A directed graph is called strongly connected if there is a directed path from any node u to v and vice-versa, for any pair of nodes of the graph.

Disconnected Digraph. A directed graph is called disconnected if its undirected graph is disconnected.

Example 24. *Consider the graphs shown in Figs. 52, 53 and 54 which of the graphs are*

(i) Unilaterally connected digraph *(ii) Weakly connected digraph*

(iii) Strongly connected digraph

(iv) Disconnected digraph (also find its connected components).

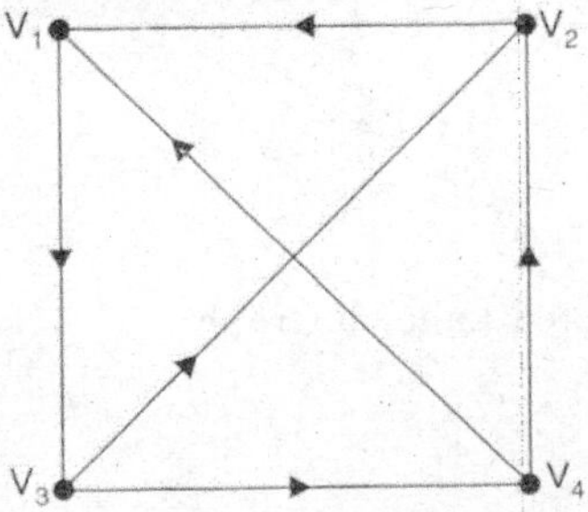

Fig. 52.

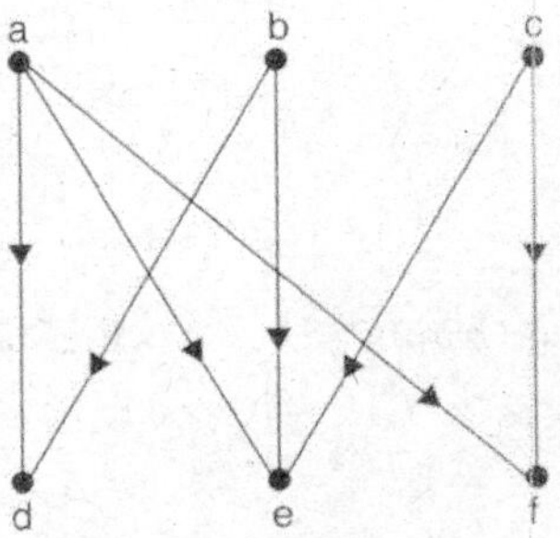

Fig. 53.

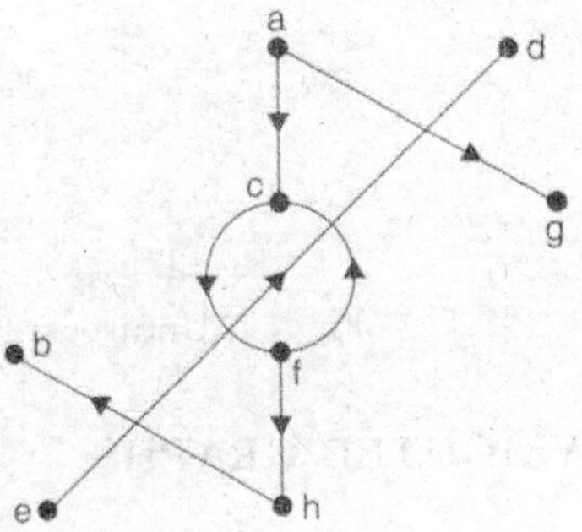

Fig. 54.

Sol. The graph shown in Fig. 52 is strongly connected because there is a path from every vertex u to v and also there is a path from v to u. It is also weakly and unilaterally connected because a strongly connected digraph is both weakly and unilaterally connected.

The graph shown in Fig. 53 is weakly connected but not unilaterally connected because there is no directed path from vertex a to b or a to c etc. but its undirected graph is connected.

The graph shown in Fig. 54 is a disconnected graph. The components of this graph are $\{g, a, c, f, h, b\}$ and $\{e, d\}$.

Directed Complete Graph. A directed complete graph G = (V, E) on n vertices is a graph in which each vertex is connected to every other vertex by an arrow. It is denoted by K_n.

Example 25. *Draw directed complete graphs K_3 and K_5.*

Sol. Place the number of vertices at appropriate place and then draw an arrow from each vertex to every other vertex as shown in Figs. 55 and 56.

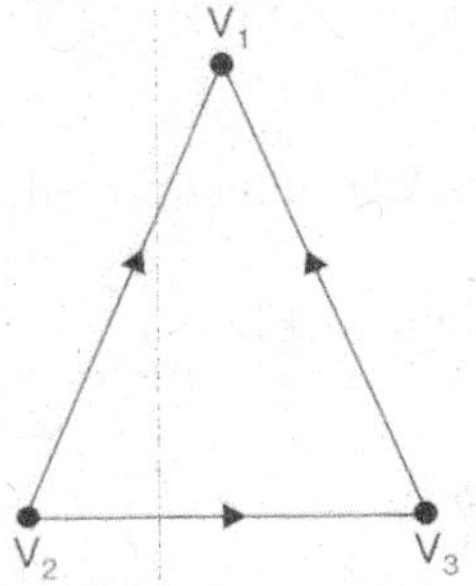

Fig. 55. K_3.

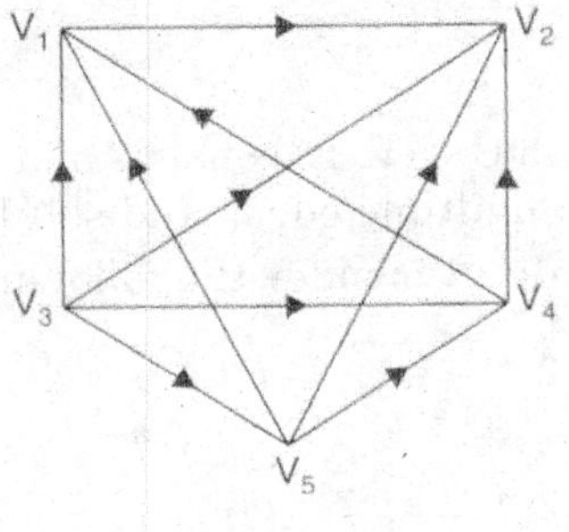

Fig. 56. K_5.

LABELED GRAPHS

A graph G = (V, E) is called a labeled graph if its edges are labeled with some name or data. So, we can write these labels in place of an ordered pair in the edge set.

For example : The graphs shown in Figs. 57 and 58 are labeled graphs.

$G = \{\{1, 2, 3, 4, 5\}, \{e_1, e_2, e_3, e_4, e_5\}\}$. $G = \{\{a, b, c, d\}, \{e_1, e_2, e_3, e_4\}\}$.

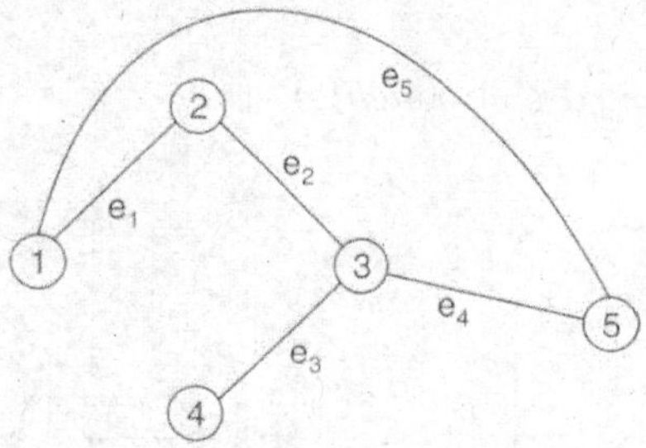

Fig. 57. Undirected Labeled Graph.

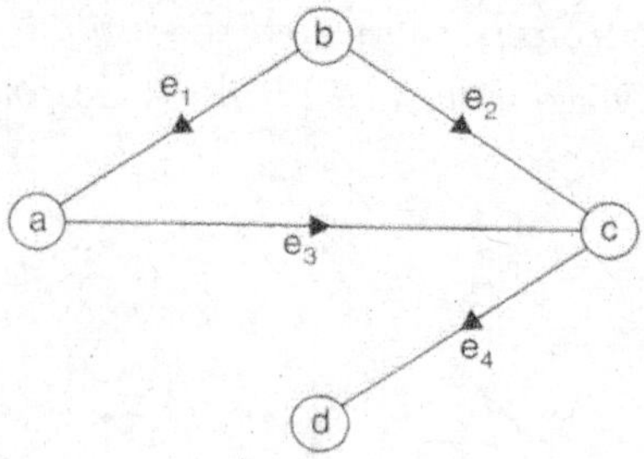

Fig. 58. Directed Labeled Graph.

WEIGHTED GRAPHS

A graph G = (V, E) is called a weighted graph if each edge of graph G is assigned a positive number *w* called the weight of the edge *e*.

For example : The graph shown in Figs. 59 and 60 is a weighted graph.

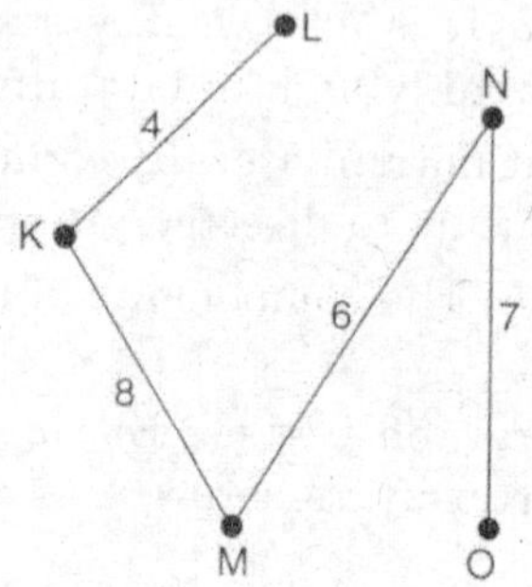

Fig. 59.

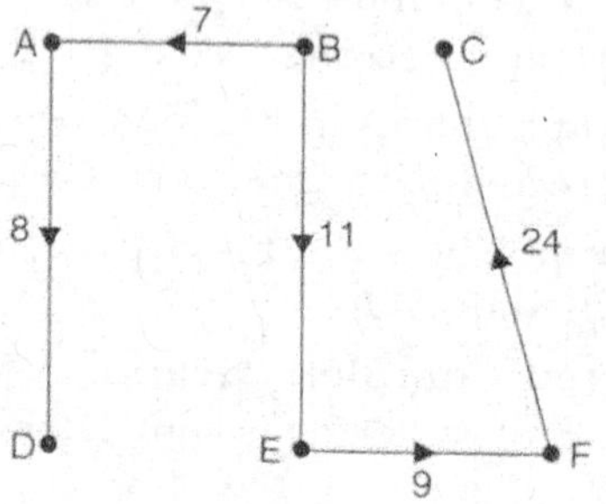

Fig. 60.

Multiple Edges. Two edges e_1 and e_1' which are distinct are said to be multiple edges if they connect the end points *i.e.*, if $e_1 = (u, v)$ and $e_1' = (u, v)$ then e_1 and e_1' are multiple edges.

MULTIGRAPH

A multigraph G = (V, E) consists of a set of vertices V and a set of edges E such that edge set E may contain multiple edges and self loops.

For example : Consider the following graph shown in Fig. 61.

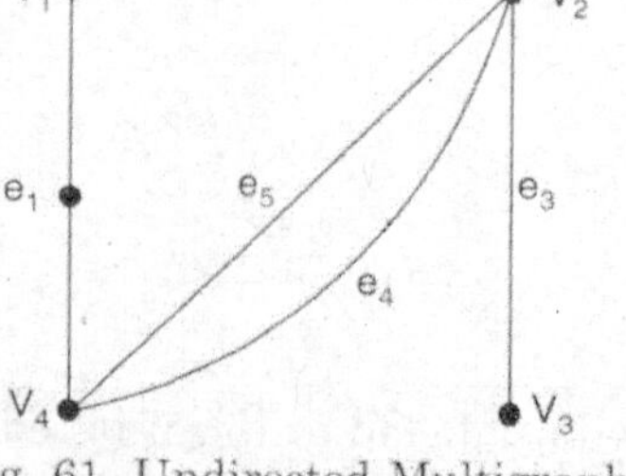

Fig. 61. Undirected Multigraph.

In the above Fig. 61, e_4 and e_5 are multiple edges, e_6 is a self-loop.

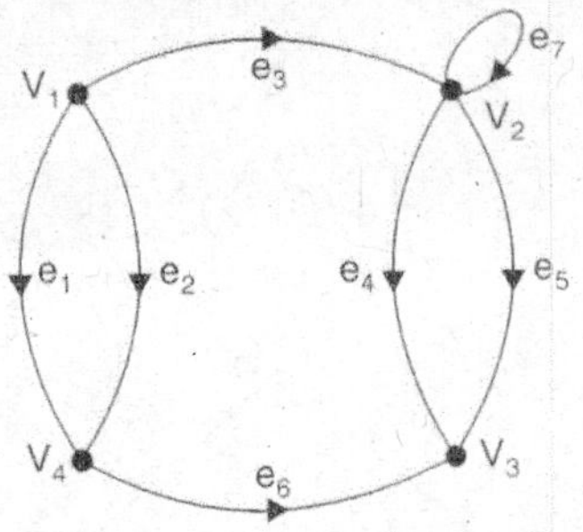

Fig. 62. Directed Multigraph.

In the graph shown in Fig. 62, the edges e_1, e_2 and e_4, e_5 are multiple edges. e_7 is a loop.

Traversable Multigraphs. Consider a multigraph G = (V, E). If the multigraph G consists of a path which includes all vertices and whose edge list contains each edge of graph exactly once. Then, the multigraph G is called a traversable multigraph.

The sufficient and necessary condition for a multigraph to be traversable is that it should be connected and have either zero or two vertices of odd degree.

For example : Consider the multigraph shown in Fig. 63.

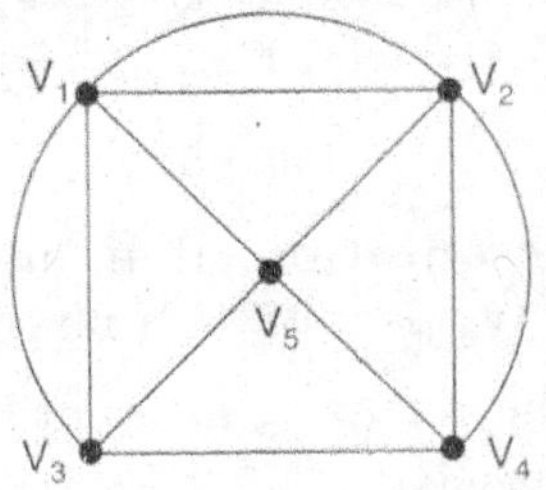

Fig. 63.

The multigraph has three even degree vertices *i.e.*, V_3, V_4 and V_5 and two odd degree vertices *i.e.*, V_1 and V_2. Hence it is a traversable multigraph.

REPRESENTATION OF GRAPHS

There are two important ways to represent a graph G with the matrices *i.e.*, adjacency matrix and incidence matrix representation.

(a) Representation of Undirected Graph

(*i*) **Adjacency matrix representation.** If an undirected graph G consists of n vertices then the adjacency matrix of graph is an $n \times n$ matrix $A = [a_{ij}]$ and defined by

$$a_{ij} = \begin{cases} 1, \text{ if } \{V_i, V_j\} \text{ is an edge } i.e., v_i \text{ is adjacent to } v_j \\ 0, \text{ if there is no edge between } v_i \text{ and } v_j \end{cases}$$

If there exists an edge between vertex v_i and v_j, where i is a row and j is a column then value of $a_{ij} = 1$.

If there is no edge between vertex v_i and v_j, then value of $a_{ij} = 0$.

Example 26. *Find the adjacency matrix M_A of graph G shown in Fig. 64.*

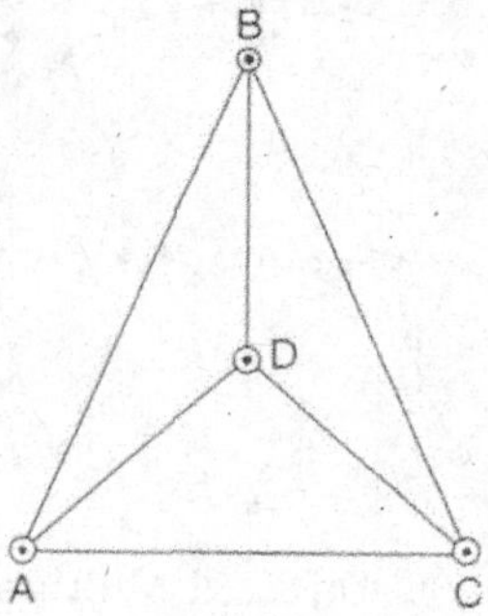

Fig. 64.

Sol. Since the graph G consists of four vertices. Therefore, the adjacency matrix will be a 4 × 4 matrix. The adjacency matrix is as follows in Fig. 65.

$$M_A = \begin{array}{c} \\ A \\ B \\ C \\ D \end{array}\begin{array}{c} \begin{array}{cccc} A & B & C & D \end{array} \\ \begin{bmatrix} 0 & 1 & 1 & 1 \\ 1 & 0 & 1 & 1 \\ 1 & 1 & 0 & 1 \\ 1 & 1 & 1 & 0 \end{bmatrix} \end{array}$$

Fig. 65.

(*ii*) **Incidence matrix representation.** If an undirected graph consists of n vertices and m edges, then the incidence matrix is an $n \times m$ matrix $C = [c_{ij}]$ defined by

$$c_{ij} = \begin{cases} 1, \text{ if the vertex } V_i \text{ incident by edge } e_j \\ 0, \text{ otherwise} \end{cases}$$

There is a row for every vertex and a column for every edge in the incidence matrix.

The number of one's in an incidence matrix of undirected graph (without loops) is equal to the sum of degrees of all the vertices of the graph.

Example 27. *Consider the undirected graph G as shown in Fig. 66. Find its incidence matrix M_I.*

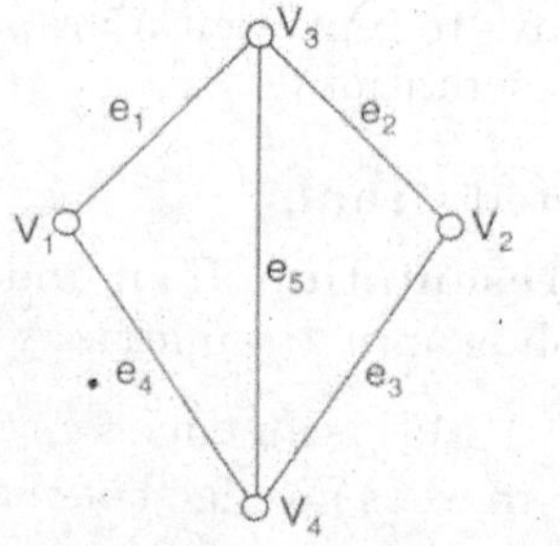

Fig. 66.

Sol. The undirected graph consists of four vertices and five edges. Therefore, the incidence matrix is an 4 × 5 matrix, which is shown in Fig. 67.

$$M_1 = \begin{array}{c} \\ V_1 \\ V_2 \\ V_3 \\ V_4 \end{array} \begin{array}{c} \begin{array}{ccccc} e_1 & e_2 & e_3 & e_4 & e_5 \end{array} \\ \begin{bmatrix} 1 & 0 & 0 & 1 & 0 \\ 0 & 1 & 1 & 0 & 0 \\ 1 & 1 & 0 & 0 & 1 \\ 0 & 0 & 1 & 1 & 1 \end{bmatrix} \end{array}$$

Fig. 67.

(b) Representation of Directed Graph

(*i*) **Adjacency matrix representation.** If a directed graph G consists of n vertices, then the adjacency matrix of graph is an $n \times n$ matrix $A = [a_{ij}]$ defined by

$$a_{ij} = \begin{cases} 1, \text{ if } V_i, V_j \text{ is an edge } i.e., \text{ if } V_i \text{ is initial vertex and } V_j \text{ is final vertex} \\ 0, \text{ if there is no edge between } V_i \text{ and } V_j \end{cases}$$

If there exists an edge between vertex V_i and V_j with V_i as initial vertex and V_j as final vertex, then value of $a_{ij} = 1$.

If there is no edge between vertex V_i and V_j then value of $a_{ij} = 0$.

The number of one's in the adjacency matrix of a directed graph is equal to the number of edges.

Example 28. *Consider the directed graph shown in Fig. 68. Determine its adjacency matrix M_A.*

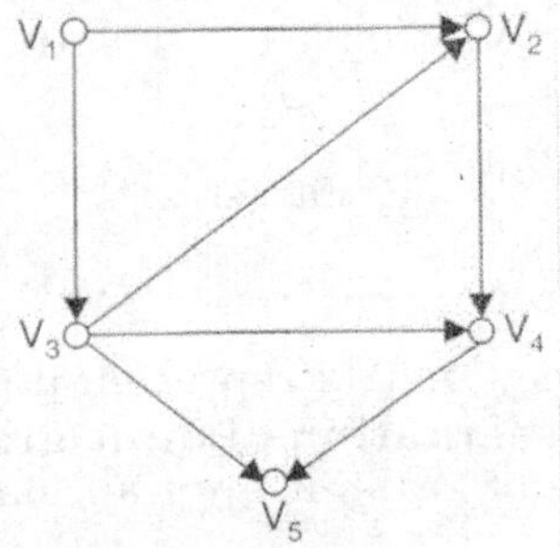

Fig. 68.

Sol. Since the directed graph G consists of five vertices. Therefore, the adjacency matrix will be a 5 × 5 matrix. The adjacency matrix of the directed graph is as follows in Fig. 69.

$$M_A = \begin{array}{c} \\ V_1 \\ V_2 \\ V_3 \\ V_4 \\ V_5 \end{array} \begin{array}{c} \begin{array}{ccccc} V_1 & V_2 & V_3 & V_4 & V_5 \end{array} \\ \begin{bmatrix} 0 & 1 & 1 & 0 & 0 \\ 0 & 0 & 0 & 1 & 0 \\ 0 & 1 & 0 & 1 & 1 \\ 0 & 0 & 0 & 0 & 1 \\ 0 & 0 & 0 & 0 & 0 \end{bmatrix} \end{array}$$

Fig. 69.

(*ii*) **Incidence matrix representation.** If a directed graph consists of n vertices and m edges then the incidence matrix is an $n \times m$ matrix $C = [c_{ij}]$, defined by

$$c_{ij} = \begin{cases} 1, \text{ if } V_i \text{ is initial vertex of edge } e_j \\ -1, \text{ if } V_i \text{ is final vertex of edge } e_j \\ 0, \text{ if } V_i \text{ is not incident on edge } e_j \end{cases}$$

The number of one's in the incidence matrix is equal to the number of edges in the graph.

Example 29. *Consider the directed graph G shown in Fig. 70. Find its incidence matrix M_I.*

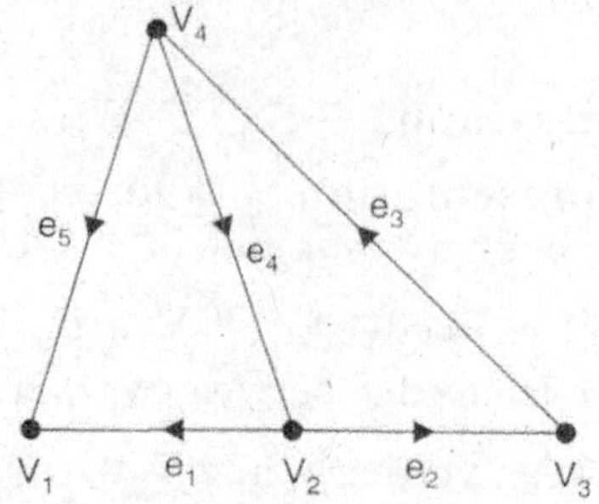

Fig. 70.

Sol. The directed graph consists of four vertices and five edges. Therefore, the incidence matrix is a 4 × 5 matrix which is shown in Fig. 71.

$$M_I = \begin{array}{c} \\ V_1 \\ V_2 \\ V_3 \\ V_4 \end{array}\begin{array}{c} \begin{array}{ccccc} e_1 & e_2 & e_3 & e_4 & e_5 \end{array} \\ \begin{bmatrix} -1 & 0 & 0 & 0 & -1 \\ 1 & 1 & 0 & -1 & 0 \\ 0 & -1 & 1 & 0 & 0 \\ 0 & 0 & -1 & 1 & 1 \end{bmatrix} \end{array}$$

Fig. 71.

(c) Representation of Multigraph

Represented only by adjacency matrix representation.

(*i*) **Adjacency matrix representation of multigraph.** If a multigraph G consists of n vertices, then the adjacency matrix of graph is an $n \times n$ matrix $A = [a_{ij}]$ and is defined by

$$a_{ij} = \begin{cases} \text{N, If there are one or more than one edges between vertex } v_i \text{ and } v_j, \text{ where} \\ \text{N is the number of edges.} \\ \text{0, otherwise.} \end{cases}$$

If there exists one or more than one edges between vertex v_i and v_j then $a_{ij} = N$, where N is the number of edges between v_i and v_j.

If there is no edge between vertex v_i and v_j then value of $a_{ij} = 0$.

Example 30. *Consider the multigraph shown in Fig. 72. Determine its adjacency matrix.*

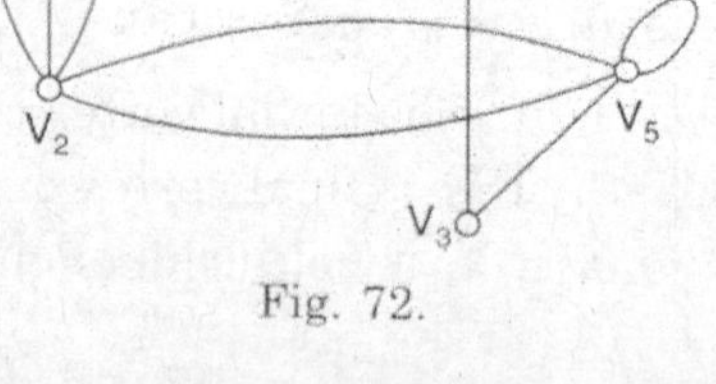

Fig. 72.

Sol. Since the multigraph consists of five vertices. Therefore, the adjacency matrix will be an 5 × 5 matrix. The adjacency matrix of the multigraph is as follows in Fig. 73.

$$M_A = \begin{array}{c} \\ V_1 \\ V_2 \\ V_3 \\ V_4 \\ V_5 \end{array} \begin{array}{c} \begin{array}{ccccc} V_1 & V_2 & V_3 & V_4 & V_5 \end{array} \\ \begin{bmatrix} 0 & 3 & 0 & 0 & 1 \\ 3 & 0 & 0 & 0 & 2 \\ 0 & 0 & 0 & 1 & 1 \\ 1 & 0 & 1 & 1 & 0 \\ 0 & 2 & 1 & 0 & 1 \end{bmatrix} \end{array}$$

Fig. 73.

OTHER IMPORTANT GRAPHS

(*a*) **Bipartite Graph.** A graph G = (V, E) is called a bipartite graph if its vertices V can be partitioned into two subsets V_1 and V_2 such that each edge of G connects a vertex of V_1 to a vertex of V_2. In other words, no edge joining two vertices in V_1 or two vertices in V_2. It is denoted by $K_{m,n}$, where m and n are number of vertices in V_1 and V_2 respectively.

(*b*) **Complete Bipartite Graph.** A graph G = (V, E) is called a complete bipartite graph if its vertices V can be partitioned into two subsets V_1 and V_2, such that each vertex of V_1 is connected to each vertex of V_2. The number of edges in a complete bipartite graph is $m.\ n$ as each of the m vertices is connected to each of the n vertices. It is denoted by $K_{m,n}$ and $m \le n$.

Example 31. *Draw the bipartite graphs $K_{2,4}$ and $K_{3,4}$. Assuming any number of edges.*

Sol. First draw the appropriate number of vertices on two parallel columns or rows and connect the vertices in one column or row with the vertices in other column or row. The bipartite graphs $K_{2,4}$ and $K_{3,4}$ are shown in Fig. 74 and Fig. 75 respectively.

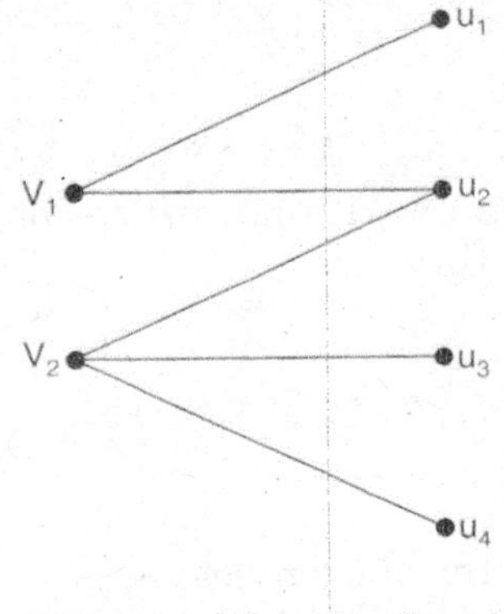

Fig. 74. Bipartite Graph $K_{2,4}$.

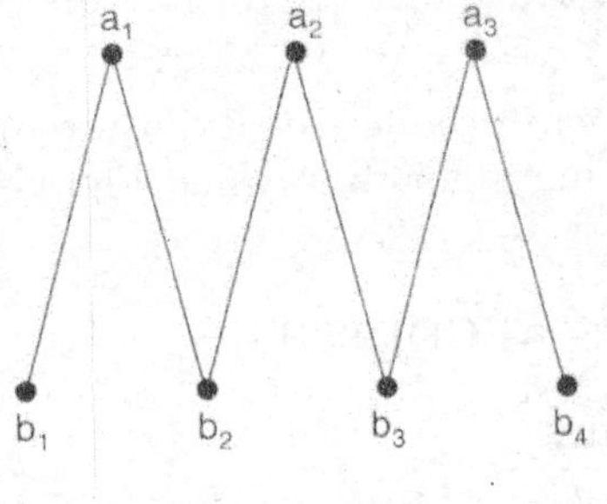

Fig. 75. Bipartite Graph $K_{3,4}$.

Example 32. *Draw the complete bipartite graphs $K_{3,4}$ and $K_{1,5}$.*

Sol. First draw the appropriate number of vertices in two parallel columns or rows and connect the vertices in first column or row with all the vertices in second column or row. The graphs $K_{3,4}$ and $K_{1,5}$ are shown in Fig. 76 and Fig. 77 respectively.

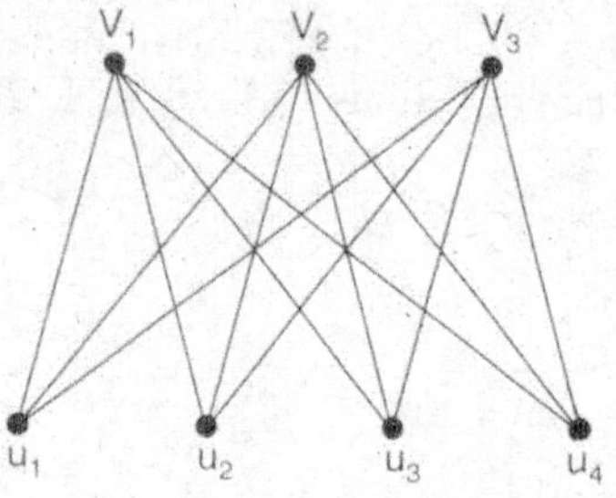

Fig. 76. $K_{3,4}$.

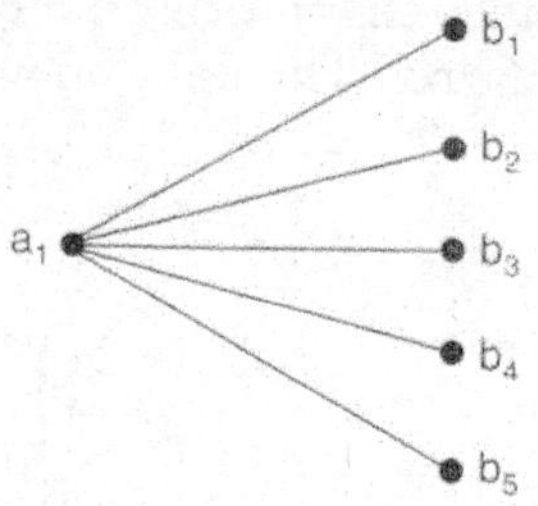

Fig. 77. $K_{1,5}$.

Euler Path. An Euler path through a graph is a path whose edge list contains each edge of the graph exactly once.

Euler Circuit. An Euler circuit is a path through a graph, in which the initial vertex appears second time as the terminal vertex.

(*c*) **Euler Graph.** An Euler graph is a graph that possesses an Euler circuit. An Euler circuit uses every edge exactly once but vertices may be repeated.

Example 33. *The graph shown in Fig. 78 is an Euler graph. Determine Euler circuit for this graph.*

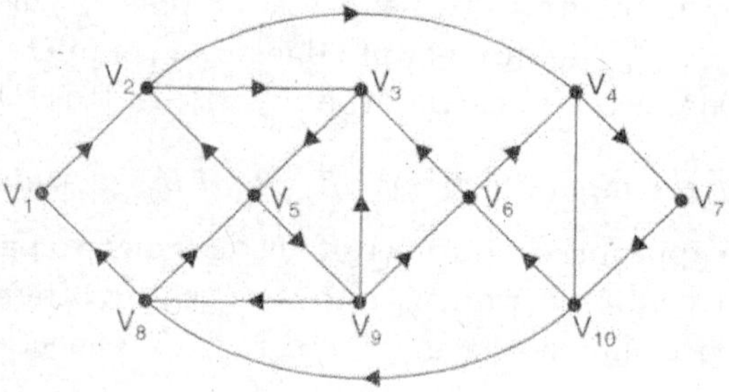

Fig. 78. Euler Graph.

Sol. The Euler circuit for this graph is

$$V_1, V_2, V_3, V_5, V_2, V_4, V_7, V_{10}, V_6, V_3, V_9, V_6, V_4, V_{10}, V_8, V_5, V_9, V_8, V_1.$$

We can produce an Euler circuit for a connected graph with no vertices of odd degree. The following algorithm called Fleury's algorithm can do this—

FLEURY'S ALGORITHM

Input

Let G = (V, E) be a connected graph with every vertex of even degree.

Step 1. Select a vertex v_0 of V as the starting vertex to construct the circuit and designate P : v_0 as the starting of the path to be constructed.

Step 2. Consider that P : $v_0, v_1, v_2, ..., v_k$ as been constructed so far. If at v_k there is only one edge say $\{v_k, v_{k+1}\}$, then extend P to P : $v_0, v_1, v_2 ..., v_{k+1}$. Remove $\{v_k, v_{k+1}\}$ from edge set E and v_{k+1} from V. If at v_k there are several edges, choose one that is not a bridge to the remaining graph, say $\{v_k, v_{k+1}\}$. Extend P to P : $v_0, v_1, v_2, ..., v_k, v_{k+1}$ and remove $\{v_k, v_{k+1}\}$ from edge set E.

Step 3. Repeat step 2 until no edges left in E.
END.

Example. *Use Fleury's algorithm to construct an Euler circuit for the graph shown in Fig. 78(a).*

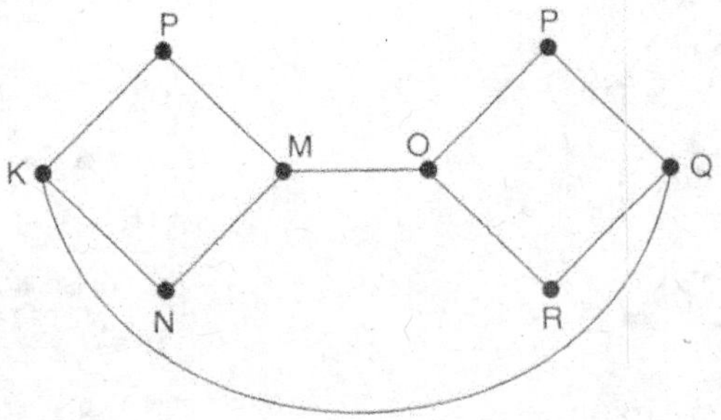

Fig. 78(*a*).

Sol. Start from any vertex, as per step 1. So, choose vertex *k* as the starting vertex. The summary of the results applying step 2 repeatedly is shown in the table—

Current path	*Next edge to traverse*	*Reason*
P : *k*	{*k*, *l*}	No edge from *k* is a bridge. Choose any one.
P : *k*, *l*	{*l*, *m*}	Only one edge from *l*.
P : *k*, *l*, *m*	{*m*, *n*}	No edge from *m* is a bridge. Choose any one.
P : *k*, *l*, *m*, *n*	{*n*, *k*}	Only one edge from *n*.
P : *k*, *l*, *m*, *n*, *k*	{*k*, *m*}	No edge from *k* is a bridge. Choose any one.
P : *k*, *l*, *m*, *n*, *k*, *m*	{*m*, *o*}	Only one edge from *m* remains.
P : *k*, *l*, *m*, *n*, *k*, *m*, *o*	{*o*, *p*}	No edge from *k* is a bridge. Choose any one.
P : *k*, *l*, *m*, *n*, *k*, *m*, *o*, *p*	{*p*, *q*}	Only one edge from *k* remains.
P : *k*, *l*, *m*, *n*, *k*, *m*, *o*, *p*, *q*	{*q*, *r*}	No edge from *k* is a bridge. Choose any one.
P : *k*, *l*, *m*, *n*, *k*, *m*, *o*, *p*, *q*, *r*	{*r*, *o*}	Only one edge from *r* remains.
P : *k*, *l*, *m*, *n*, *k*, *m*, *o*, *p*, *q*, *r*, *o*	{*o*, *q*}	Only one edge from *o* remains.
P : *k*, *l*, *m*, *n*, *k*, *m*, *o*, *p*, *q*, *r*, *o*, *q*	{*q*, *k*}	Only one edge from *q* remains.
P : *k*, *l*, *m*, *n*, *k*, *m*, *o*, *p*, *q*, *r*, *o*, *q*, *k*		

Hamiltonian Path. A Hamiltonian path through a graph is a path whose vertex list contain each vertex of the graph exactly once, except if path is a circuit.

Hamiltonian Circuit. A Hamiltonian circuit is a path in which the initial vertex appears a second time as the terminal vertex.

(*d*) **Hamiltonian Graph.** A Hamiltonian graph is a graph that possesses a Hamiltonian path. A Hamiltonian path uses each vertex exactly once but edges may not be included.

Example 34. *The graph shown in Fig. 79 is a Hamiltonian graph. Determine Hamiltonian circuit for this graph.*

Sol. The Hamiltonian circuit is shown in Fig. 80.

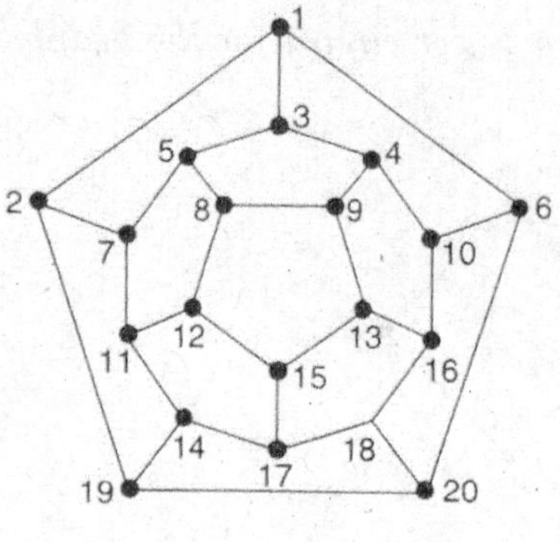

Fig. 79. Hamiltonian Graph.

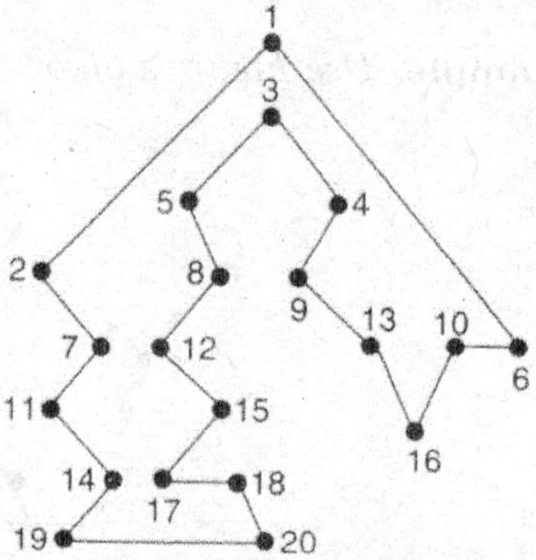

Fig. 80. Hamiltonian Circuit.

(*e*) **Regular Graph.** A graph is said to be regular or K-regular if all its vertices have the same degree K. A graph whose all vertices have degree 2 is called 2-regular graph. A complete graph K_n is regular of degree $n - 1$.

Example 35. *Draw regular graphs of degree 2 and 3.*

Sol. The regular graphs of degree 2 and 3 are shown in Figs. 81 and 82.

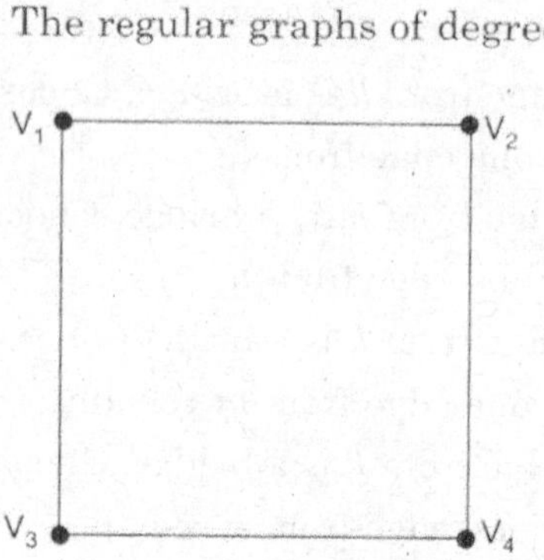

Fig. 81. 2-regular Graph.

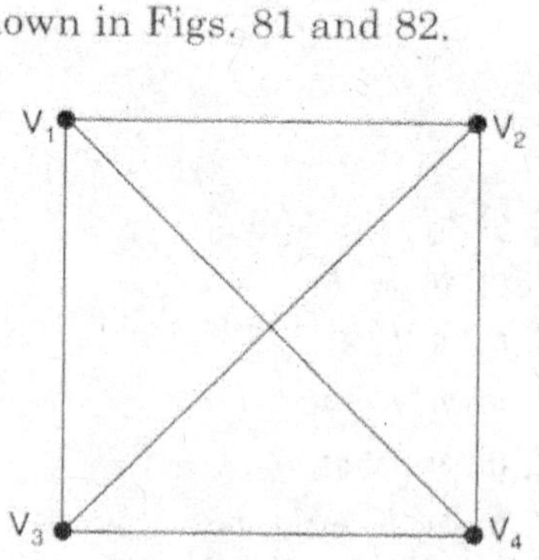

Fig. 82. Regular Graph.

Example 36. *Draw a 2-regular graph of five vertices.*

Sol. The 2-regular graph of five vertices is shown in Fig. 83.

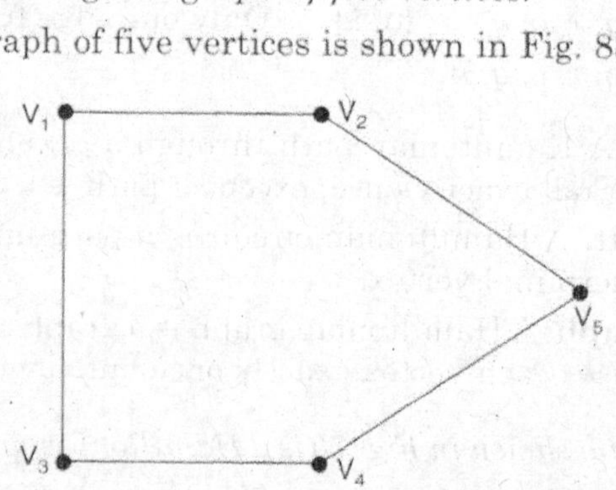

Fig. 83.

Example 37. *Draw a 3-regular graph of five vertices.*

Sol. It is not possible to draw 3-regular graph of five vertices. The 3-regular graph must have an even number of vertices.

Theorem IV. *Prove that K-regular graph must have even number of vertices when the value of K is odd.*

Proof. Consider a graph with n vertices. Let T is the sum of degrees of all the n vertices of a K-regular graph. Then, we have

$$T = K \cdot n$$

The sum T must be even (from the theorem I).

Now, suppose that K is odd, so the value of n must be even.

(*f*) **Planar Graph.** A graph is said to be planar if it can be drawn in a plane so that no edges cross.

For example : The graphs shown in Fig. 84 and Fig. 85 are planar graphs.

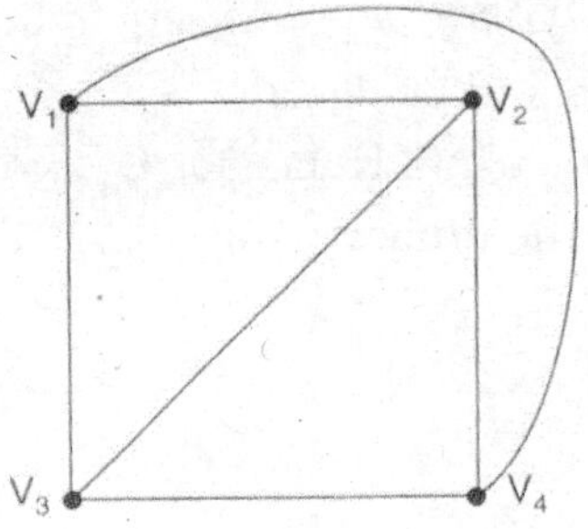

Fig. 84.

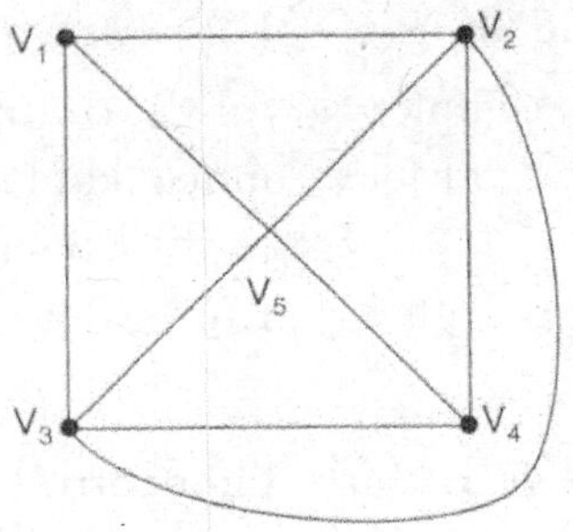

Fig. 85.

Region of a Graph. Consider a planar graph G = (V, E). A region is defined to be an area of the plane that is bounded by edges and cannot be further subdivided. A planar graph divides the plane into one or more regions. One of these regions will be infinite.

Finite Region. If the area of the region is finite, then that region is called finite region.

Infinite Region. If the area of the region is infinite, that region is called infinite region. A planar graph has only one infinite region.

Example 38. *Consider the graph shown in Fig. 86. Determine the number of regions, finite regions and an infinite region.*

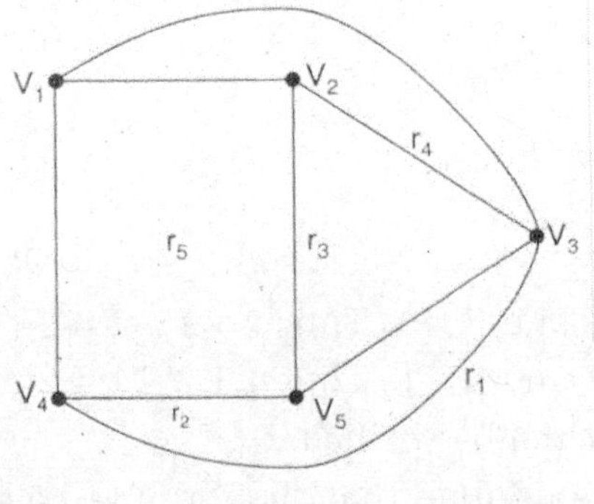

Fig. 86.

Sol. There are five regions in the above graph *i.e.*, r_1, r_2, r_3, r_4 and r_5.

There are four finite regions in the graph *i.e.*, r_2, r_3, r_4 and r_5.

There is only one infinite region *i.e.*, r_1.

Properties of Planar Graphs

1. If a connected planar graph G has e edges and r regions, then $r \leq \frac{2}{3}e$.

2. If a connected planar graph G has e edges, v vertices and r regions, then $v - e + r = 2$.
[Euler's rule]

3. If a connected planar graph G has e edges and v vertices, then $3v - e \geq 6$.

4. A complete graph K_n is planar if and only if $n < 5$.

5. A complete bipartite graph $K_{m, n}$ is planar if and only if $m < 3$ or $n > 3$.

Example 39. *Prove that complete graph K_4 is planar.*

Sol. The complete graph K_4 contains 4 vertices and 6 edges.

We know that for a connected planar graph $3v - e \geq 6$. Hence for K_4, we have

$$3 \times 4 - 6 = 6 \text{ which satisfies the property (3).}$$

Thus K_4 is a planar graph. Hence proved.

State and Prove Euler's Theorem

Statement. Consider any connected planar graph G = (V, E) having R regions, V vertices and E edges. Then

$$V + R - E = 2.$$

Proof. Use induction on the number of edges to prove this theorem.

Basis of Inductions. Assume that the edges $e = 1$. Then we have two cases, graphs of which are shown in Figs. 87 and 88.

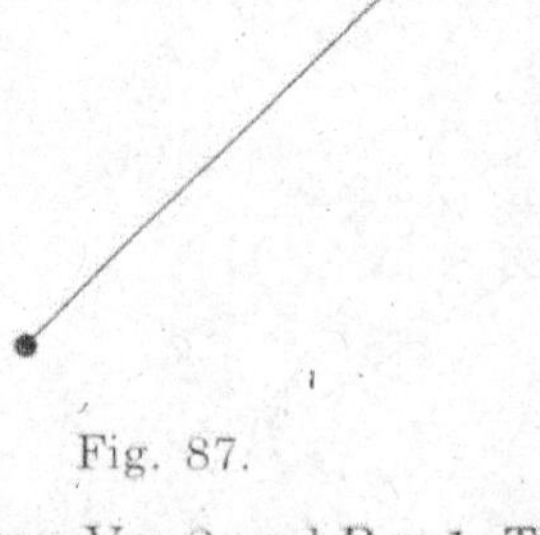

Fig. 87.

Fig. 88.

In Fig. 87 we have V = 2 and R = 1. Thus 2 + 1 – 1 = 2

In Fig. 88 we have V = 1 and R = 2. Thus 1 + 2 – 1 = 2.

Hence, the basis of induction is verified.

Induction Step. Let us assume that the formula holds for connected planar graphs with K edges.

Let G be a graph with K + 1 edges.

Firstly, we suppose that G contains no circuits. Now, take a vertex v and find a path starting at v. Since G is circuit free, whenever we find an edge, we have a new vertex. At last we will reach a vertex v with degree 1. So we cannot move further as shown in Fig. 89.

Now remove vertex v and the corresponding edge incident on v. So, we are left with a graph G* having K edges as shown in Fig. 90.

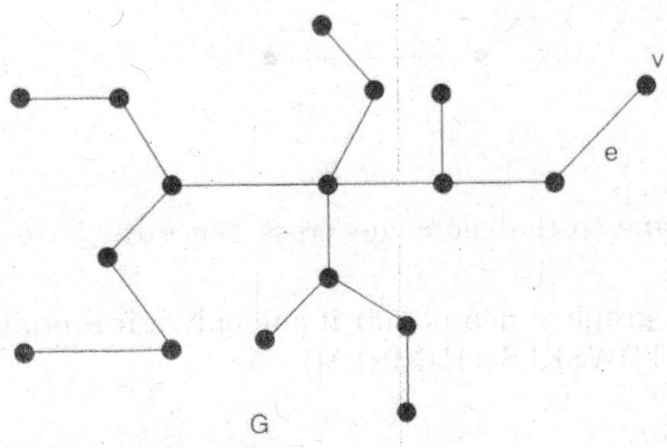

Fig. 89. G.

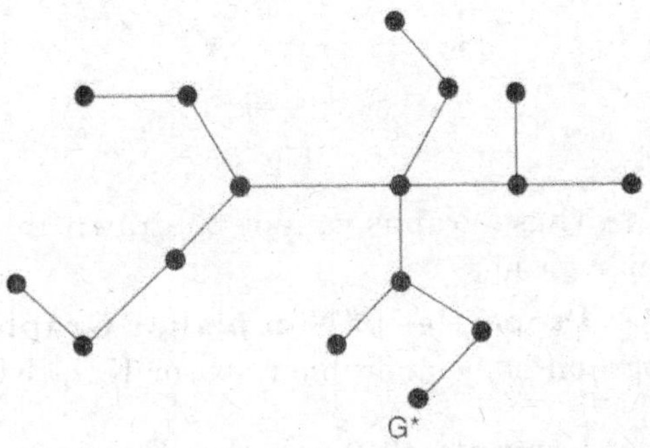

Fig. 90. G*.

Hence, by inductive assumption, Euler's formula holds for G*.

Now, since G has one more edge than G*, one more vertex than G* with same number of regions as in G*. Hence, the formula also holds for G.

Secondly, we assume that G contains a circuit and e is an edge in the circuit shown in Fig. 91.

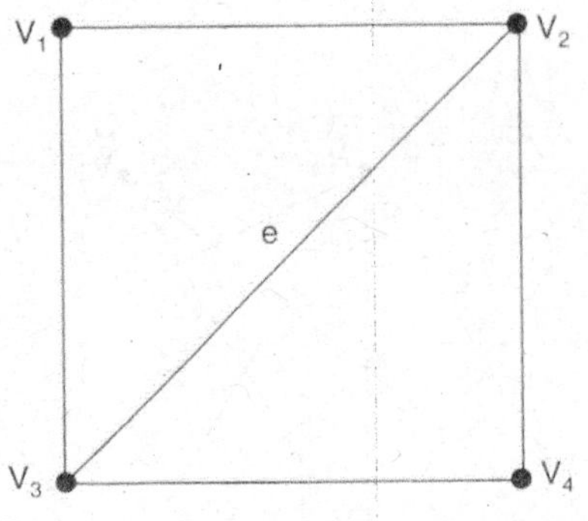

Fig. 91.

Fig. 92.

Now, as e is the part of a boundary for two regions. So, we only remove the edge and we are left with graph G* having K edges (Fig. 92).

Hence, by inductive assumption, Euler's formula holds for G*.

Now, since G has one more edge than G*, one more region than G* with same number of vertices as G*. Hence the formula also holds for G which, verifies the inductive step and hence proves the theorem.

(*g*) **Non Planar Graphs.** A graph is said to be non planar if it cannot be drawn in a plane so that no edges cross.

For example : The graphs shown in Figs. 93 and 94 are non planar graphs.

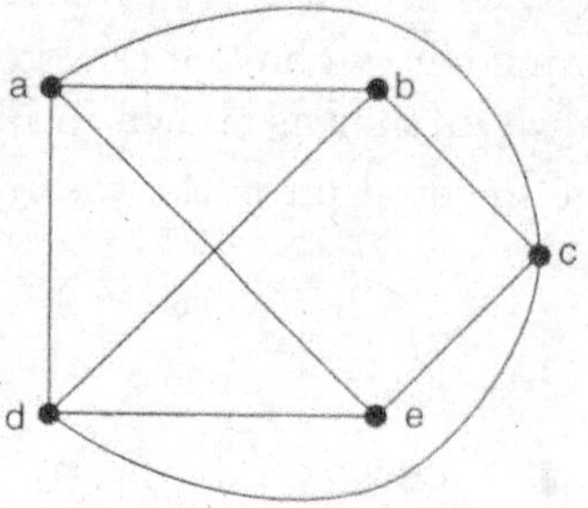

Fig. 93.

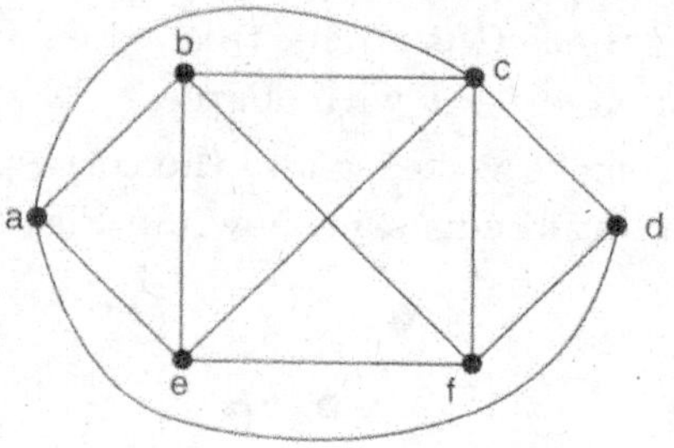

Fig. 94.

These graphs cannot be drawn in a plane so that no edges cross hence they are non planar graphs.

Properties of Non planar Graphs. A graph is non-planar if and only if it contains a subgraph homeomorphic to K_5 or $K_{3,3}$ [KURATOWSKI'S THEOREM].

Example 40. *Show that K_5 is non-planar.*

Sol. The complete graph K_5 contains 5 vertices and 10 edges.

Now, for a connected planar graph $3v - e \geq 6$.

Hence, for K_5 we have $3 \times 5 - 10 = 5$ (which does not satisfy property 3 because it must be greater than or equal to 6).

Thus, K_5 is a non planar graph.

Example 41. *Show that the graphs shown in Figs. 95 and 96 are non-planar by finding a subgraph homeomorphic to K_5 or $K_{3,3}$.*

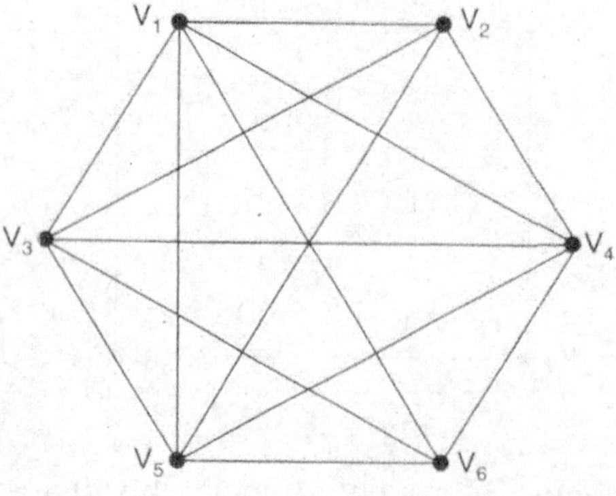

Fig. 95. G_1.

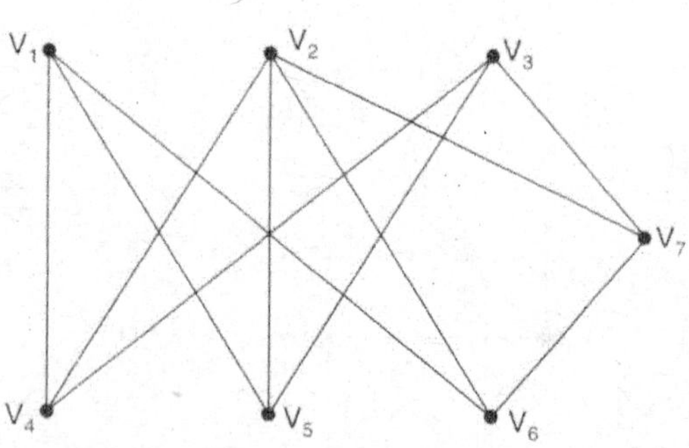

Fig. 96. G_2.

Sol. If we remove the edges (V_1, V_4), (V_3, V_4) and (V_5, V_4) the graph G_1, becomes homeomorphic to K_5. Hence it is non-planar.

If we remove the edge (V_2, V_7) the graph G_2 becomes homeomorphic to $K_{3,3}$. Hence it is non-planar.

Theorem V. *Prove that every planar graph has at least one vertex of degree 5 or less than 5.*

Proof. Consider a graph G, whose all vertices are of degree 6 or more, then the sum of the degrees of all the vertices would be greater than or equal to $6v$. We know that the sum of the degrees of the vertices is twice the number of edges. Therefore, we have

$$6v \leq 2e$$

or $$v \leq \frac{e}{3} \qquad \ldots(i)$$

But, any planar graph have the property,

$$r \leq \frac{2e}{3} \qquad \ldots(ii)$$

Also, from Euler's formula, we have

$$2 = v - e + r \qquad \ldots(iii)$$

Now, putting the value of v and r from (*i*) and (*ii*) in (*iii*), we have

$$2 \leq \frac{e}{3} - e + \frac{2e}{3} = 0$$

Since, the statement $2 \leq 0$ is not true, hence we conclude that there must exist some vertex in G with degree 5 or less than 5.

GRAPH COLOURING

Suppose that G = (V, E) is a graph with no multiple edges. A vertex colouring of G is an assignment of colours to the vertices of G such that adjacent vertices have different colours. A graph G is M-colourable if there exists a colouring of G which uses M-colours.

Proper Colouring. A colouring is proper if any two adjacent vertices u and v have different colours otherwise it is called improper colouring.

A graph can be coloured by assigning a different colour to each of its vertices. However, for most graphs a colouring can be found that uses fewer colours than the number of vertices in the graph.

Example 42. *Consider the following graph and colours C = {r, w, b, y}. Colour the graph properly using all colours or fewer colours.*

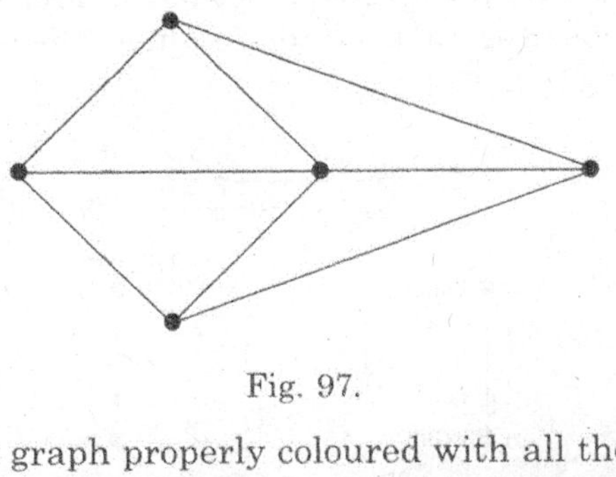

Fig. 97.

Sol. Fig. 98 shows the graph properly coloured with all the four colours.

Fig. 99 shows the graph properly coloured with three colours.

It is not possible to colour this graph property with two colours.

Chromatic number of G. The minimum number of colours needed to produce a proper colouring of a graph G is called the chromatic number of G and is denoted by $\chi(G)$.

The graph shown in Fig. 97 is minimum 3-colourable, hence $\chi(G) = 3$.

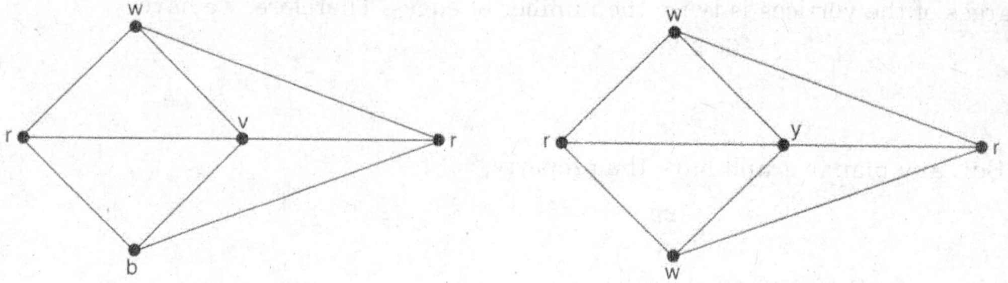

Fig. 98.

Example 43. *The chromatic number of* k_n *is n.*

Sol. A colouring of k_n can be constructed using n colours by assigning a different colour to each vertex. No two vertices can be assigned the same colour, since every two vertices of this graph are adjacent. Hence the chromatic number of $k_n = n$.

Example 44. *The chromatic number of complete bipartite graph* $k_{m,\, n}$, *where m and n are positive integers is two.*

Sol. The number of colours needed does not depend upon m and n. However, only two colours are needed colour the set of m vertices with one colour and the set of n vertices with a second colour. Since, edges connect only a vertex from the set of m vertices and a vertex from the set of n vertices, no two adjacent vertices have the same colour.

Note 1. Every connected bipartite simple graph has a chromatic number of 2 or 1.

2. Conversely, every graph with a chromatic number of 2 is bipartite.

Example 45. *The chromatic number of graph* c_n, *where* c_n *is the cycle with n vertices is either 2 ro 3.*

Sol. Two colours are needed to colour c_n, where c_n is even. To construct such a colouring, simply pick a vertex and colour it black. Then move around the graph in clockwise direction colouring the second vertex white, the third vertex black, and so on. The nth vertex can be coloured white since the two vertices adjacent to it, namely the $(n - 1)$th and the first are both coloured black as shown in Fig. 99.

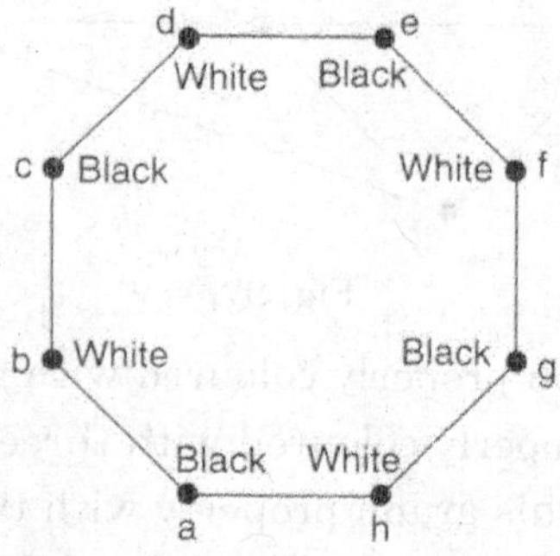

Fig. 99.

When n is odd and $n > 1$, the chromatic number of c_n is 3. To construct such a colouring, pick an initial vertex. First use only two colours and alternate colours as the graph is traversed in a clockwise direction. However, the nth vertex reached is adjacent to two vertices of different colours, the first and $(n - 1)$th. Hence, a third colour is needed.

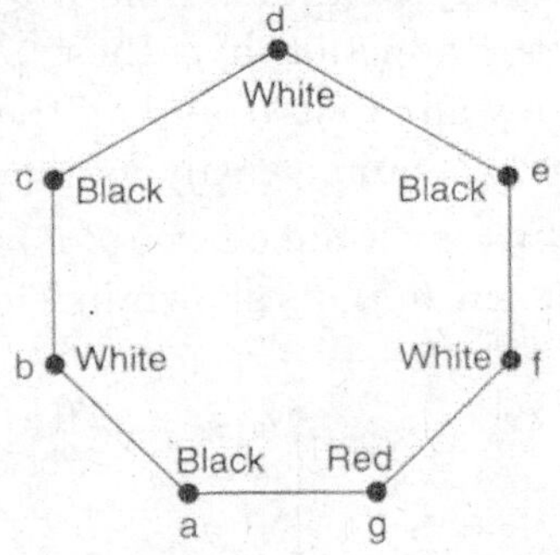

Fig. 100.

Theorem VI. *The following are equivalent for a graph G :*

(i) G is 2-colourable

(ii) G is bipartite

(iii) Every cycle of G has even length.

Proof. The proof of this theorem is directly obtained from the example 43, 44 and 45.

Example 46. *Determine the chromatic number of the graphs shown in Fig. 101.*

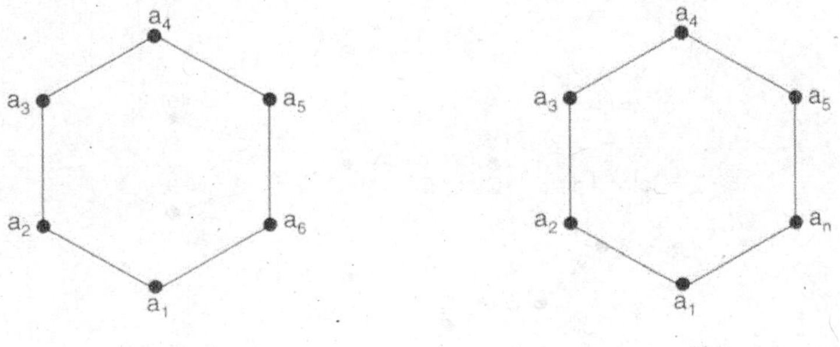

(*a*) (*b*)

Fig. 101.

Sol. The graphs shown in Fig. 101(*a*), has the chromatic number $\chi(G) = 2$.

The graph shown in Fig. 101(*b*) has the chromatic number

$\chi(G) = 2$, when n is an even number and

$\chi(G) = 3$, where n is odd.

Theorem VII. *The vertices of every planar graph can be properly coloured with five colours.*

Proof. We will prove this theorem by induction. All the graphs with 1, 2, 3, 4 or 5 vertices can be properly coloured with five colours. Now let us assume that every planar graph with $n - 1$ vertices can be properly coloured with five colours. Next, if we prove that any planar graph G with n vertices will require no more than five colours, we have done.

Consider the planar graph G with n vertices.

Since G is planar, it must have at least one vertex with degree five or less as shown in theorem V. Assume this vertex to be 'u'.

Let G_1 be a graph of $n - 1$ vertices obtained from G by deleting vertex 'u'. The G_1 graph requires no more than five colours (Induction hypothesis). Consider that the vertices in G_1 have been properly coloured and now add to it 'u' and all the edges incident on u. If the degree of u is 1, 2, 3, or 4, a proper colour to u can be easily assigned.

Now, we have one case left, in which the degree of u is 5, and all the 5 colours have been used in colouring the vertices adjacent to u, as shown in Fig. 102.

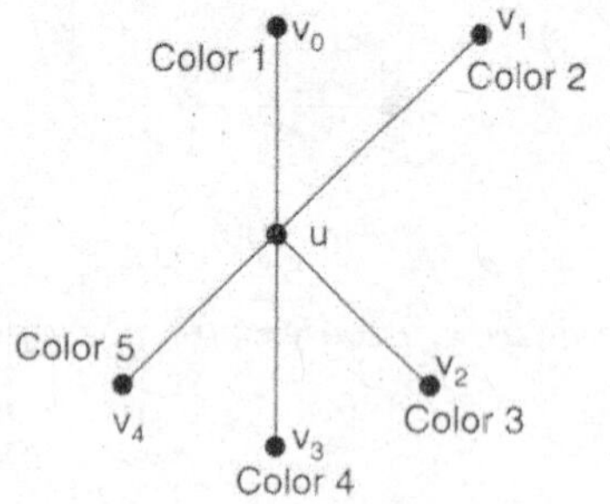

Fig. 102.

Suppose that there is a path in G_1 between vertices v_0 and v_3 coloured alternately with colours 1 and 4 as shown in Fig. 103.

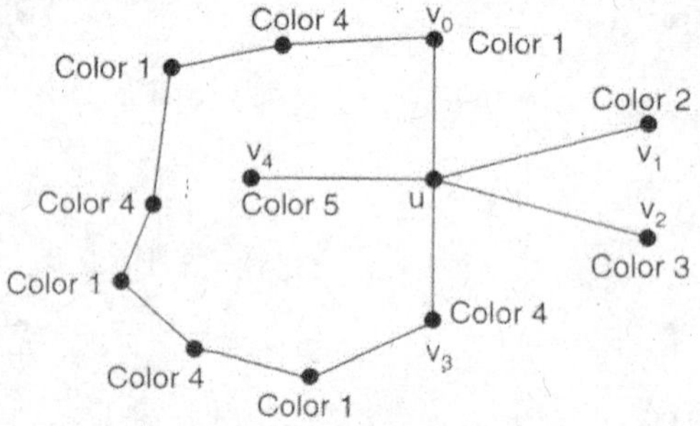

Fig. 103.

Then a similar path between v_4 and v_2, coloured alternately with colours 5 and 3, can not exist ; otherwise, these two paths will intersect and cause G to be non-planar.

Thus, if there is no path between v_4 and v_2 coloured alternately with colour 5 and 3 of all vertices connected to v_2 through vertices of alternating colours 5 and 3. This interchange will colour vertex v_2 with colour 5 and yet keep G, properly coloured. As vertex v_4 is still with colour 5, the colour 3 is left over with which to colour vertex u which proves the theorem.

SHORTEST PATH IN WEIGHTED GRAPHS

Weighted graphs can be used to represent highways connecting the different cities. The weighted edges represent the distance between different cities and the vertices represent the cities. A common problem with this type of graph is to find the shortest path from one city to another city. There are many ways to tackle this problem one of which is as follows :

Shortest Paths from Single Source. We will find shortest paths from a single vertex to all other vertices of the graph. The first algorithm was proposed by E. Dijkstra in 1959. Some common terms related with this algorithm are as follows :

Path Length. The length of a path is the sum of the weights of the edges on that path.

Source. The starting vertex of the graph from which we have to start to find the shortest path.

Destination. The terminal or last vertex upto which we have to find the path.

Dijkstra's Algorithm. This algorithm maintains a sets of vertices whose shortest path from source is already known. The graph is represented by its cost adjacency matrix, where cost being the weight of the edge. In the cost adjacency matrix of the graph, all the diagonal values are zero. If there is no path from source vertex V_s to any other vertex V_i then it is represented by + ∞. In this algorithm, we have assumed all weights are positive.

1. Initially there is no vertex in sets.

2. Include the source vertex V_s in S. Determine all the paths from V_s to all other vertices without going through any other vertex.

3. Now, include that vertex in S which is nearest to V_s and find shortest paths to all the vertices through this vertex and update the values.

4. Repeat the step 3 until $n - 1$ vertices are not included in S if there are n vertices in the graph.

After completion of the process, we got the shortest paths to all the vertices from the source vertex.

Example 47. *Find the shortest path between K and L in the graph shown in Fig. 104 by using Dijkstra's Algorithm.*

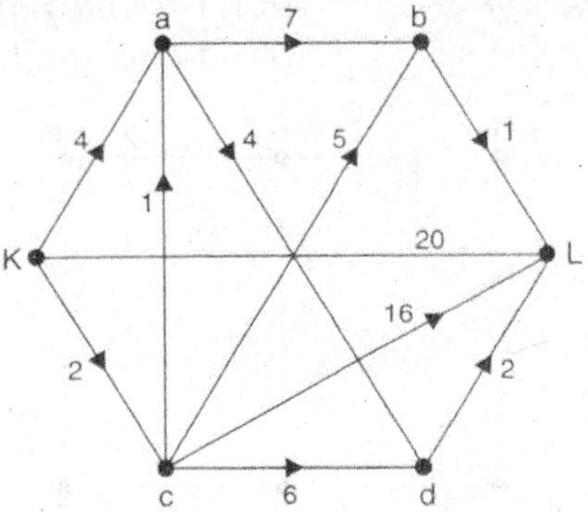

Fig. 104.

Sol. Step I. Include the vertex K is S and determine all the direct paths from K to all other vertices without going through any other vertex.

S	Distance to all other vertices					
	K	*a*	*b*	*c*	*d*	L
K	0	4(K)	∞	2(K)	∞	20(K)

Step II. Include the vertex in S which is nearest to K and determine shortest paths to all vertices through this vertex and update the values. The nearest vertex is *c*.

S	Distance to all other vertices					
	K	*a*	*b*	*c*	*d*	L
K, *c*	0	3(K, *c*)	7(K, *c*)	2(K)	8(K, *c*)	18(K, *c*)

Step III. The vertex which is 2nd nearest to K is 9, included in S.

S	Distance to all other vertices					
	K	*a*	*b*	*c*	*d*	L
K, *c*, *a*	0	3(K, *c*)	7(K, *c*)	2(K)	7(K, *c*, *a*)	18(K, *c*)

Step IV. The vertex which is 3rd nearest to K is *b*, is included in S.

S	Distance to all other vertices					
	K	*a*	*b*	*c*	*d*	L
K, *c*, *a*, *b*	0	3(K, *c*)	7(K, *c*)	2(K)	7(K, *c*, *a*)	8(K, *c*, *b*)

Step V. The vertex which is next nearest to K is *d*, is included in S.

S	Distance to all other vertices					
	K	*a*	*b*	*c*	*d*	L
K, *c*, *a*, *b*, *d*	0	3(K, *c*)	7(K, *c*)	2(K)	7(K, *c*, *a*)	8(K, *c*, *b*).

Since, $n - 1$ vertices included in S. Hence we have found the shortest distance from K to all other vertices. Thus, the shortest distance between K and L is 8 and the shortest path is K, *c*, *b*, L.

SOLVED PROBLEMS

Problem 1. *Consider the graph shown in Fig. 105. Give an example of the following :*

(*i*) *An elementary path*
(*ii*) *A simple path*
(*iii*) *A path which is not simple*
(*iv*) *A simple path which is not elementary*
(*v*) *A simple circuit*
(*vi*) *A circuit which is not simple.*

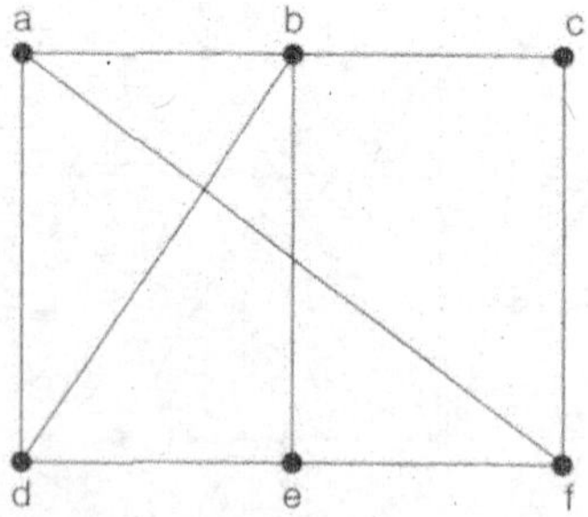

Fig. 105.

Sol. There are many solutions to the above problems, but we will take only one for each.

(*i*) An elementary path is *a*, *b*, *c*, *f*, *e*, *d*.

(*ii*) A simple path is *a*, *f*, *e*, *b*, *d*.

(*iii*) A path which is not simple is *a*, *b*, *e*, *e b*, *d*.

(*iv*) A simple path which is not elementary is *a*, *b*, *e*, *f*, *c*, *b*, *d*.

(*v*) A simple circuit is *a, b, c, f, a.*

(*vi*) A circuit which is not simple is *a, b, e, f, c, b, d, a.*

Problem 2. *Give an example of a graph with six vertices that has exactly two cut points.*

Sol. The graph with exactly two cut points is shown in Fig. 106.

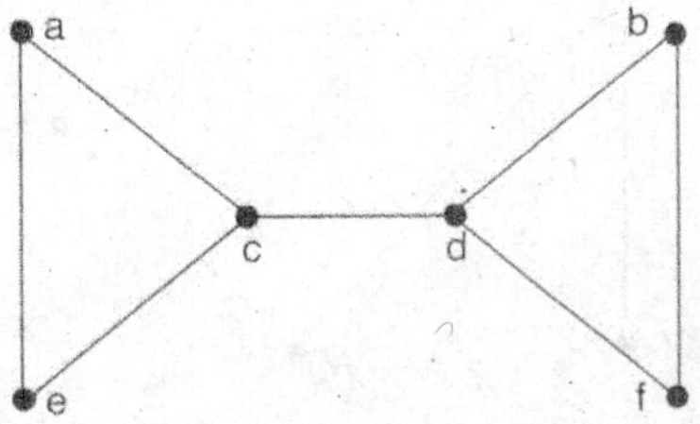

Fig. 106.

The two cut points in this graph are *c* and *d*.

Problem 3. *Give an example of a graph with six vertices that has no cut points.*

Sol. The graph with no cut points is shown in Fig. 107.

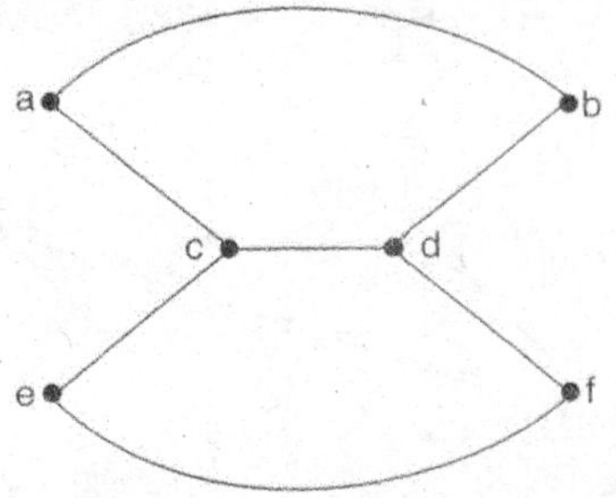

Fig. 107.

Problem 4. *Give an example of a graph with six vertices for the following :*

(*a*) *that has exactly two bridges.*

(*b*) *that has no bridges.*

Sol. (*a*) The graph that has exactly two bridges is shown in Fig. 108.

The two bridges are e_1 and e_2.

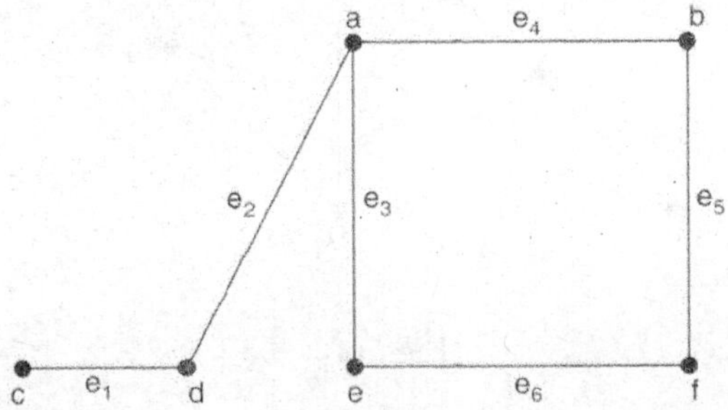

Fig. 108.

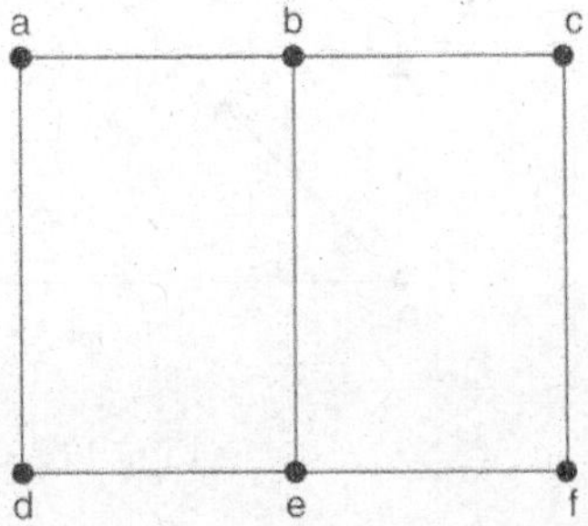

Fig. 109.

(*b*) The graph with no bridges is shown in Fig. 109.

Problem 5. *Draw a graph whose every edge is a bridge.*

Sol. The graph shown in figure is a graph whose every edge is a bridge because if any edge is removed from the graph *h*, we got two components or a disconnected graph.

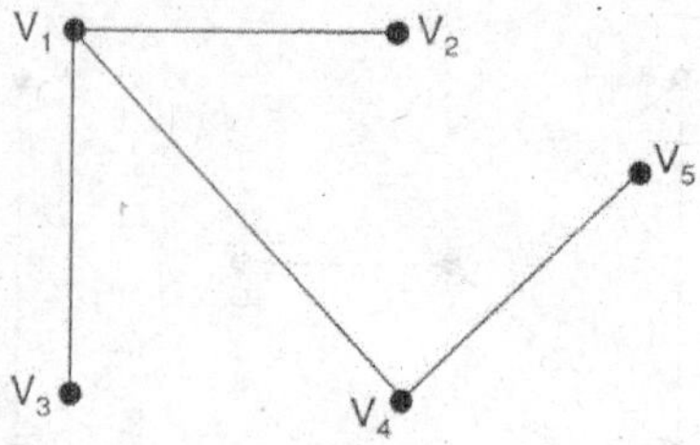

Fig. 110.

Problem 6. *Consider the directed graph as shown in Fig. 111. Show the four different subgraphs of this graph having at least four vertices.*

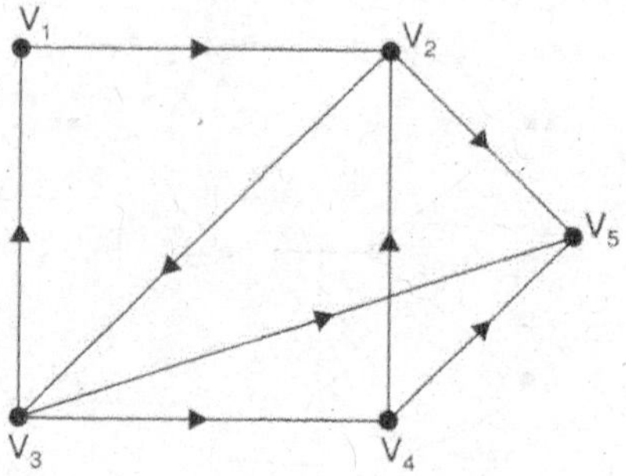

Fig. 111.

Sol. The subgraphs of the directed graph are shown in Figs. 112, 113, 114 and 115.

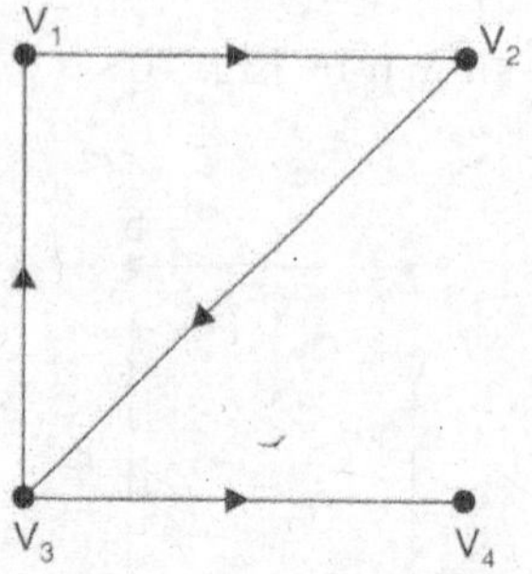

Fig. 112.

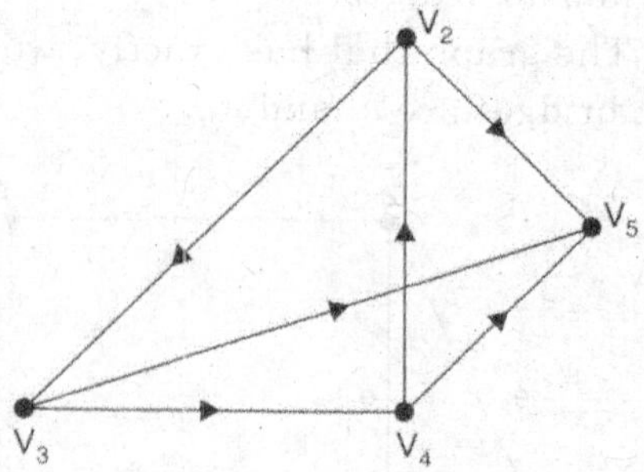

Fig. 113.

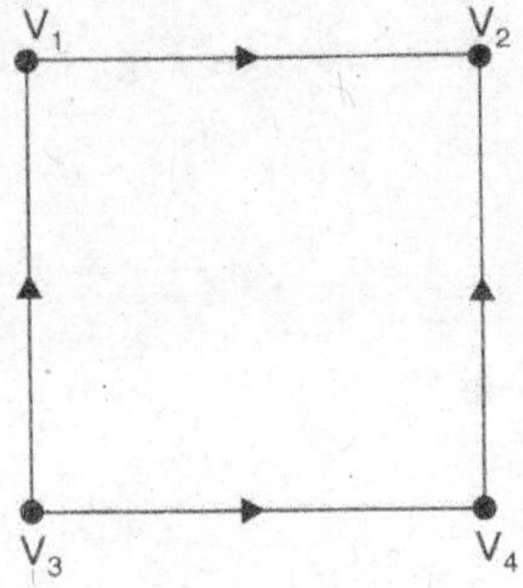

Fig. 114.

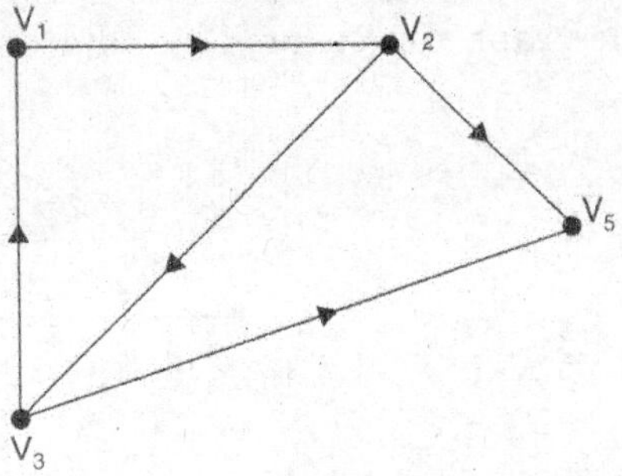

Fig. 115.

Problem 7. *Consider the multigraph shown in Fig. 116. Show two different subgraphs of this multigraph which are itself multigraphs.*

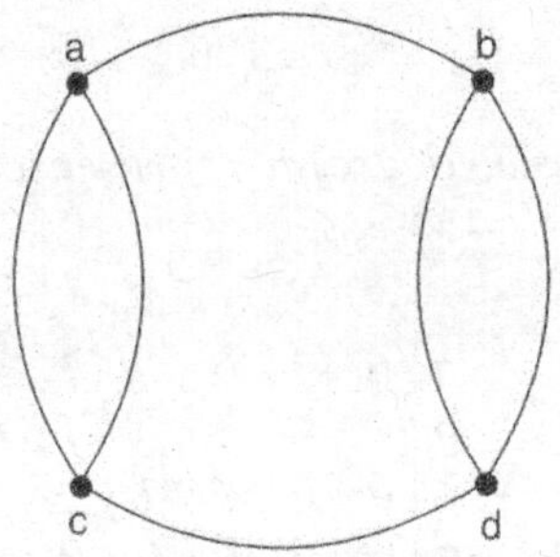

Fig. 116.

Sol. The two different subgraphs of this multigraph which are itself multigraphs are shown in Figs. 117 and 118.

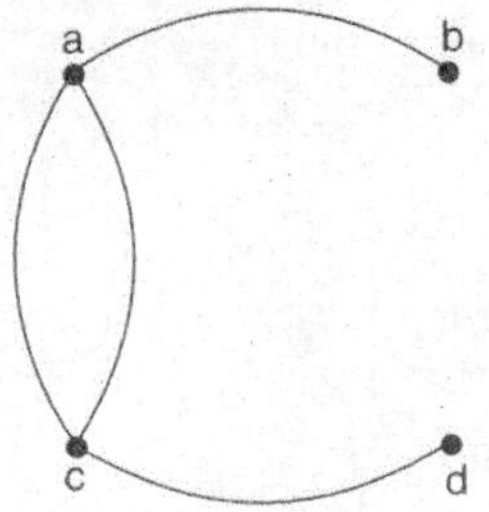

Fig. 117.

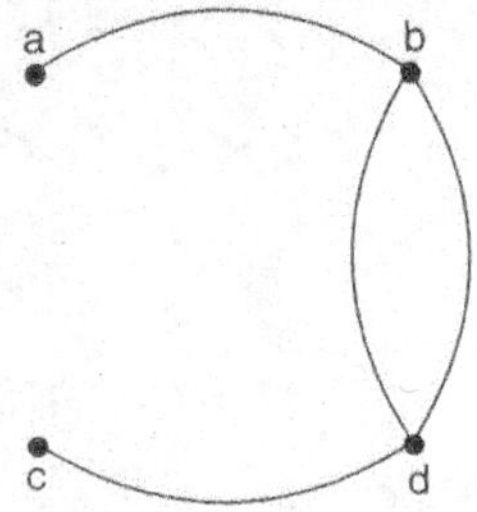

Fig. 118.

Problem 8. *Draw the undirected graph represented by adjacency matrix M_A shown in Fig. 119.*

$$M_A = \begin{array}{c} \\ V_1 \\ V_2 \\ V_3 \\ V_4 \\ V_5 \end{array} \begin{array}{c} \begin{array}{ccccc} V_1 & V_2 & V_3 & V_4 & V_5 \end{array} \\ \begin{bmatrix} 0 & 1 & 1 & 0 & 0 \\ 1 & 0 & 1 & 0 & 0 \\ 1 & 1 & 0 & 1 & 0 \\ 0 & 0 & 1 & 0 & 1 \\ 0 & 0 & 0 & 1 & 1 \end{bmatrix} \end{array}$$

Fig. 119.

Sol. The graph represented by adjacency matrix M_A is shown in Fig. 120.

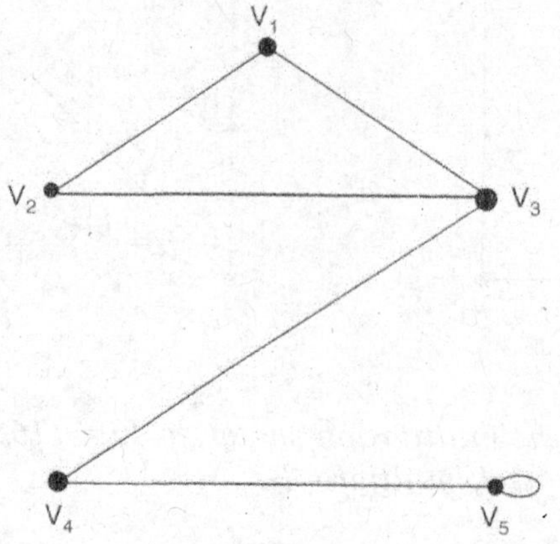

Fig. 120.

Problem 9. *Draw the undirected graph represented by incidence matrix M_I shown in Fig. 121.*

$$M_I = \begin{array}{c} \\ a \\ b \\ c \\ d \\ e \end{array} \begin{array}{c} \begin{array}{cccccc} e_1 & e_2 & e_3 & e_4 & e_5 & e_6 \end{array} \\ \begin{bmatrix} 0 & 1 & 0 & 0 & 1 & 1 \\ 0 & 1 & 1 & 0 & 1 & 0 \\ 0 & 0 & 0 & 1 & 0 & 1 \\ 1 & 0 & 0 & 0 & 0 & 0 \\ 1 & 0 & 0 & 1 & 0 & 0 \end{bmatrix} \end{array}$$

Fig. 121.

Sol. The graph represented by incidence matrix M_I is shown in Fig. 122.

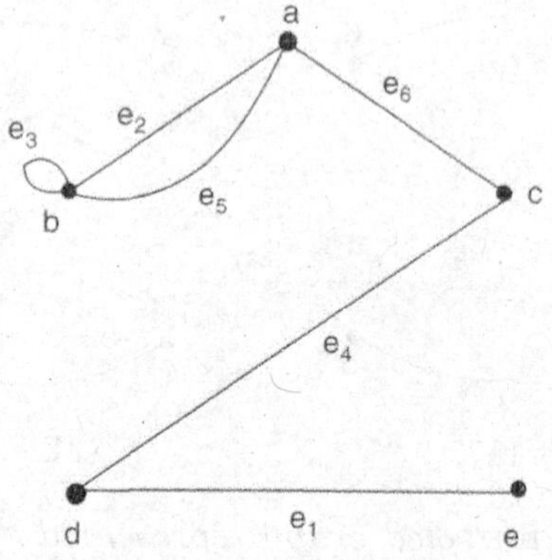

Fig. 122.

Problem 10. *Draw the multigraph G whose adjacency matrix M_A is shown in Fig. 123.*

$$M_A = \begin{bmatrix} 1 & 3 & 0 & 0 \\ 3 & 0 & 1 & 1 \\ 0 & 1 & 2 & 2 \\ 0 & 1 & 2 & 0 \end{bmatrix}$$

Fig. 123.

Sol. The multigraph corresponding to the adjacency matrix M_A is shown in Fig. 124.

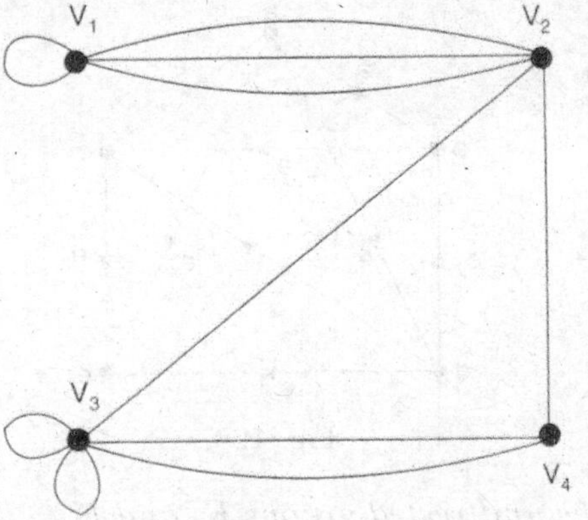

Fig. 124.

Problem 11. *Draw the directed graph G whose adjacency matrix M_A is shown in Fig. 125.*

$$M_A = \begin{array}{c} \\ a \\ b \\ c \\ d \\ e \\ f \end{array}\begin{array}{c} \begin{array}{cccccc} a & b & c & d & e & f \end{array} \\ \begin{bmatrix} 0 & 0 & 1 & 0 & 0 & 0 \\ 0 & 0 & 0 & 0 & 1 & 0 \\ 0 & 0 & 0 & 1 & 0 & 0 \\ 0 & 0 & 0 & 1 & 1 & 0 \\ 0 & 0 & 0 & 0 & 1 & 1 \\ 0 & 1 & 0 & 0 & 0 & 1 \end{bmatrix} \end{array}$$

Fig. 125.

Sol. The directed graph corresponding to the adjacency matrix M_A is shown in Fig. 126.

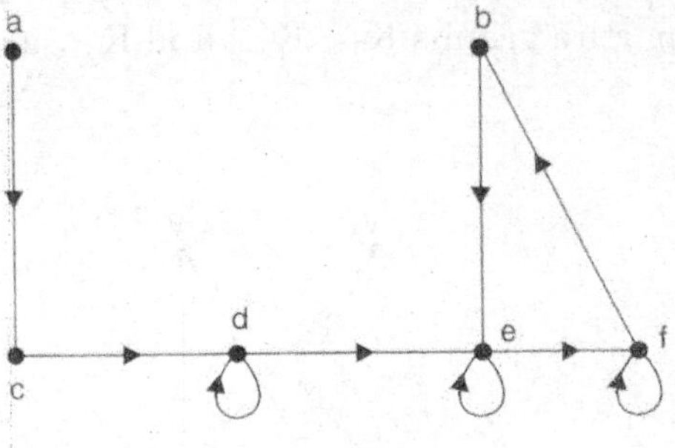

Fig. 126.

Problem 12. *Draw the directed graph G whose incidence matrix M_I is shown in Fig. 127.*

$$M_I = \begin{array}{c} \\ a \\ b \\ c \\ d \\ e \end{array}\begin{array}{c} \begin{array}{ccccccccc} e_1 & e_2 & e_3 & e_4 & e_5 & e_6 & e_7 & e_8 & e_9 \end{array} \\ \begin{bmatrix} -1 & -1 & 0 & 1 & 1 & 0 & 0 & 0 & 0 \\ 1 & 0 & 1 & 0 & 0 & 0 & -1 & 0 & 0 \\ 0 & +1 & -1 & 0 & 0 & -1 & 0 & 0 & 1 \\ 0 & 0 & 0 & -1 & 0 & 1 & 1 & 1 & 0 \\ 0 & 0 & 0 & 0 & -1 & 0 & 0 & -1 & -1 \end{bmatrix} \end{array}$$

Fig. 127.

Sol. The directed graph corresponding to the incidence matrix M_I is shown in Fig. 128.

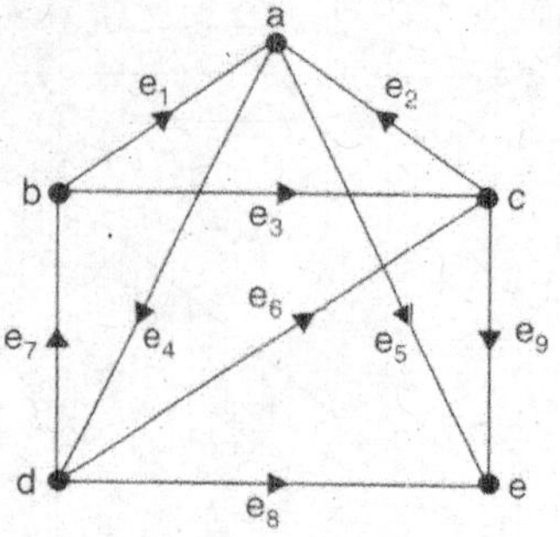

Fig. 128.

Problem 13. *Draw the undirected graphs K_3 and K_5.*

Sol. Undirected graphs K_3 and K_5 are shown in Figs. 129 and 130.

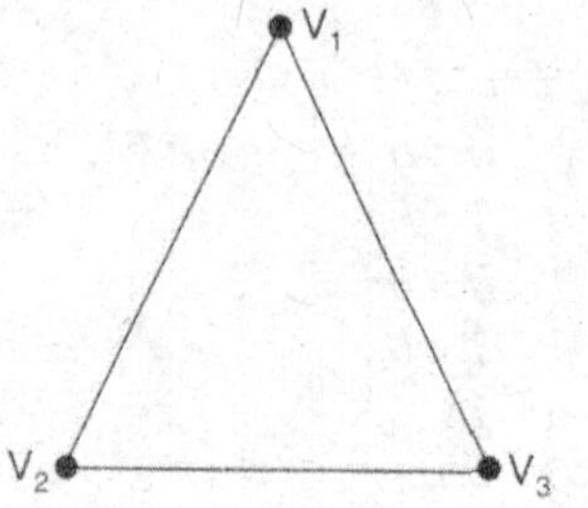

Fig. 129. K_3.

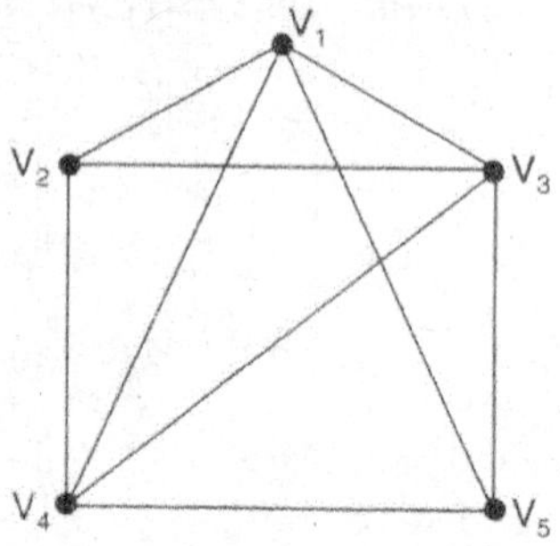

Fig. 130. K_5.

Problem 14. *Draw the complete bipartite graphs $K_{2,\,3}$, $K_{2,\,4}$ and $K_{2,\,5}$.*

Sol. The complete bipartite graphs $K_{2,3}$, $K_{2,\,4}$ and $K_{2,\,5}$ are shown in Figs. 131, 132 and 133 respectively.

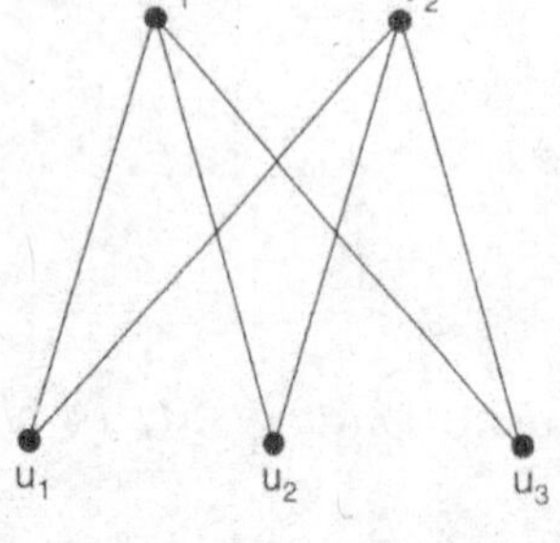

Fig. 131. $K_{2,\,3}$.

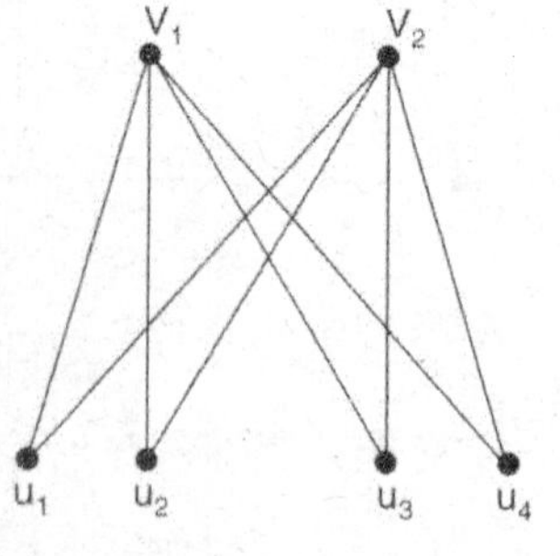

Fig. 132. $K_{2,\,4}$.

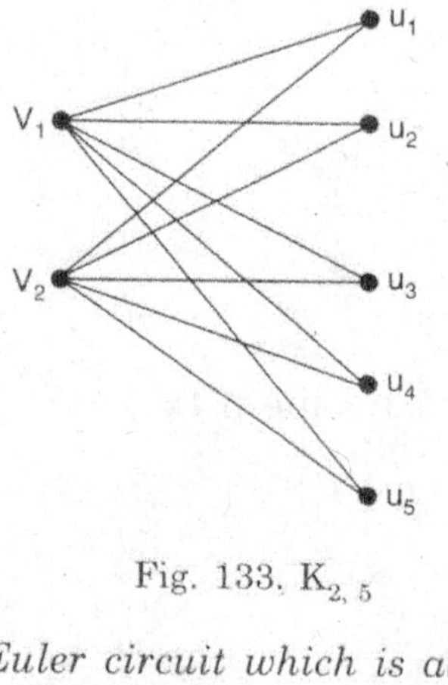

Fig. 133. $K_{2,\,5}$

Problem 15. *Give an example of a graph that has an Euler circuit which is also a Hamiltonian circuit.*

Sol. The graph having an Euler circuit which is also a Hamiltonian circuit is shown in Fig. 134.

In this graph V_1, V_2, V_3, V_4, V_1 is both an Euler circuit as well as Hamiltonian circuit.

Problem 16. *Give an example of a graph that has an Euler circuit and a Hamiltonian circuit, which are distinct.*

Fig. 134.

Sol. The graph having an Euler circuit and a Hamiltonian circuit which are distinct is shown in Fig. 135.

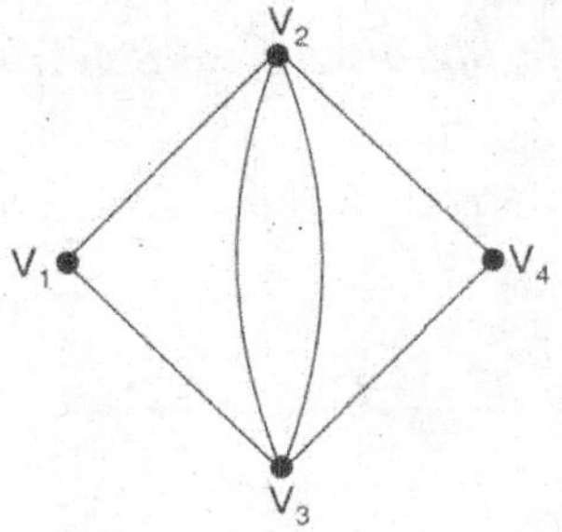

Fig. 135.

The Euler circuit is $V_1, V_5, V_2, V_5, V_4, V_2, V_1$.
The Hamiltonian circuit is V_1, V_2, V_4, V_5, V_1.

Problem 17. *Give an example of a graph which has an Euler circuit but not a Hamiltonian circuit.*

Sol. The graph having an Euler circuit but not a Hamiltonian circuit is shown in Fig. 136.

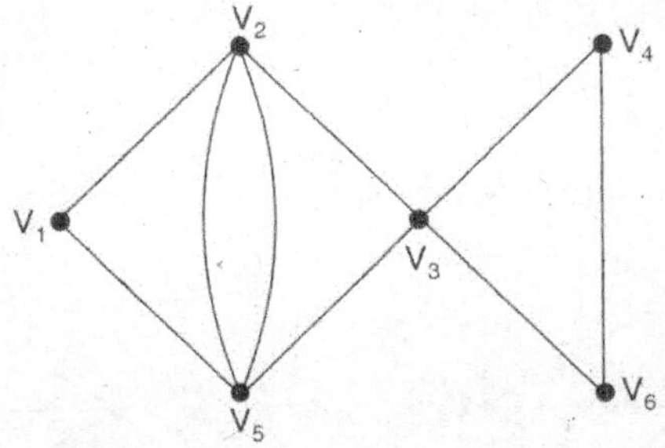

Fig. 136.

The Euler circuit is $V_1, V_5, V_2, V_5, V_3, V_4, V_6, V_3, V_2, V_1$.
There is no Hamiltonian circuit.

Problem 18. *Give an example of a graph which has a Hamiltonian circuit but not an Euler circuit.*

Sol. The graph having a Hamiltonian circuit but not an Euler circuit is shown in Fig. 137.

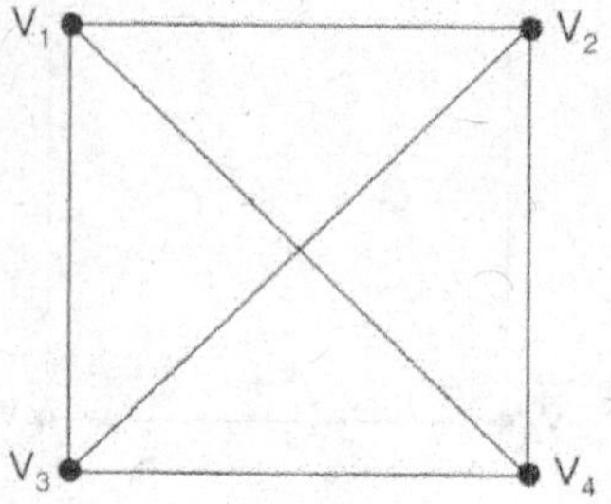

Fig. 137.

The Hamiltonian circuit is V_1, V_2, V_4, V_3, V_1.

There is no Euler circuit.

Problem 19. *Give an example of a graph that has neither an Euler circuit nor a Hamiltonian circuit.*

Sol. The graph having neither an Euler circuit nor a Hamiltonian circuit is shown in Fig. 138.

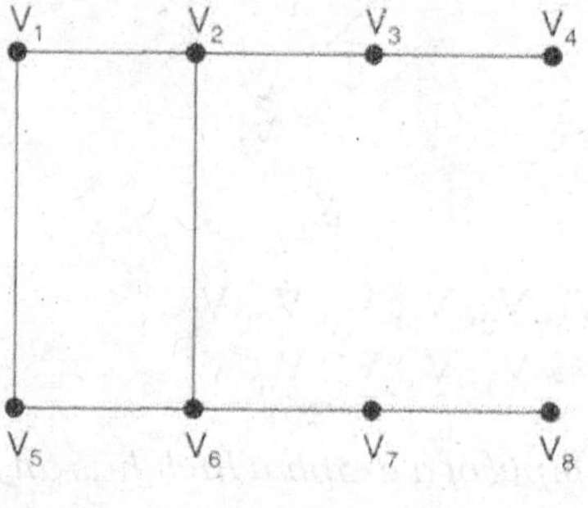

Fig. 138.

Problem 20. *Draw a planar representation of graphs shown in Figs. 139 and 140.*

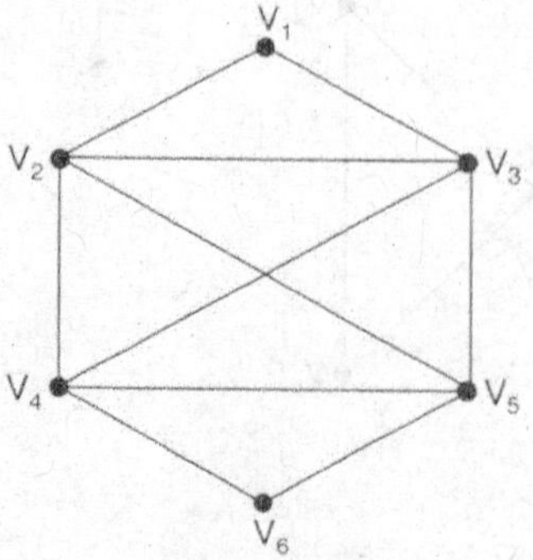

Fig. 139.

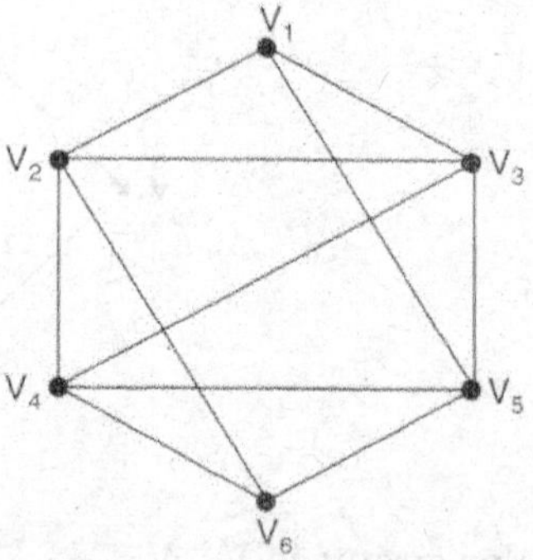

Fig. 140.

Sol. The planar representation of graph shown in Fig. 141 is shown in Fig. 142.

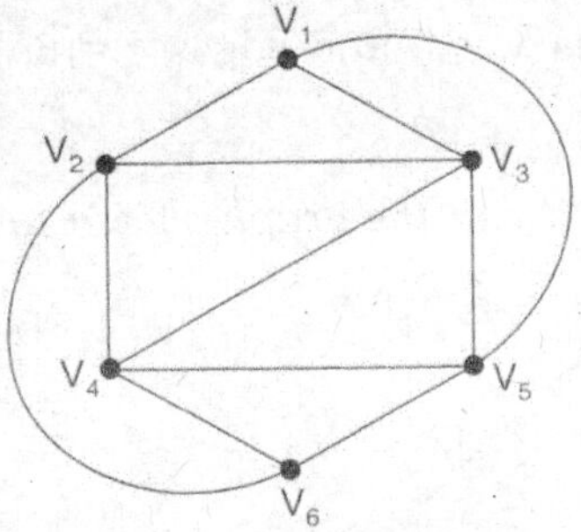

Fig. 141.

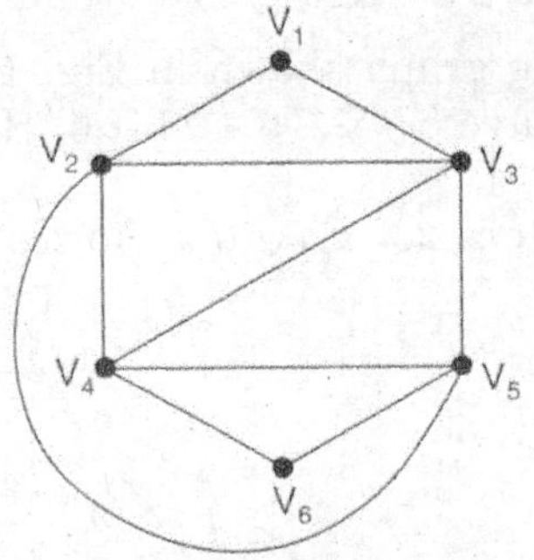

Fig. 142.

The planar representation of graph shown in Fig. 140 is shown in Fig. 141.

Problem 21. *Show that* $V - E + R = 2$ *for the connected planar graphs shown in Figs. 143. and 144.*

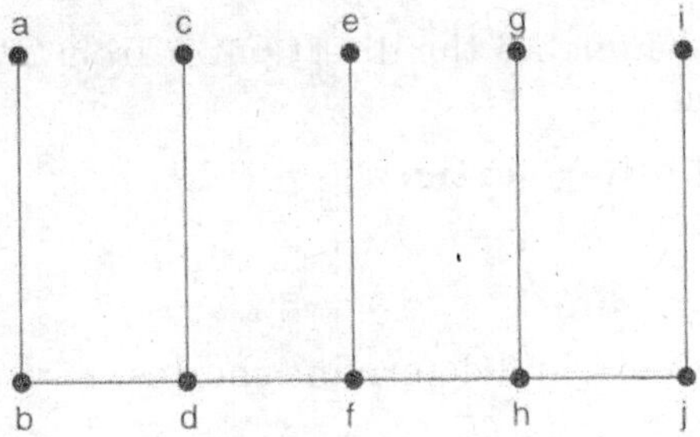

Fig. 143.

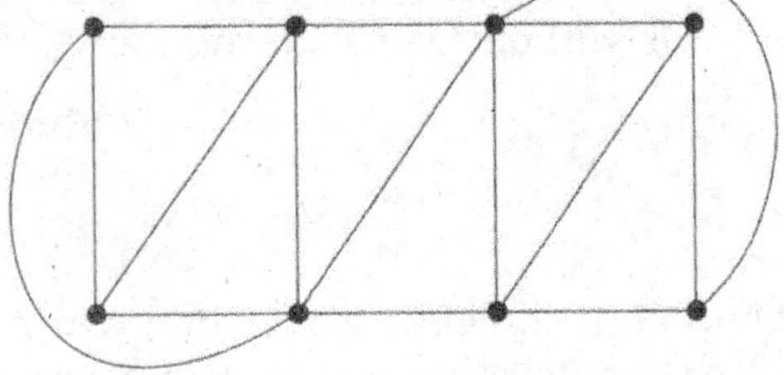

Fig. 144.

Sol. (*i*) The graph shown in Fig. 143 contains vertices V = 10, edges E = 9 and regions R = 1. Putting the values, we have 10 – 9 + 1 = 2. Hence proved.

(*ii*) The graph shown in Fig. 144 contains vertices V = 8, edges E = 15 and regions R = 9. Putting the values, we have 8 – 15 + 9 = 2. Hence proved.

Problem 22. *Show that* $e \geq 3V - 6$ *for the connected planar graphs shown in Figs. 145 and 146.*

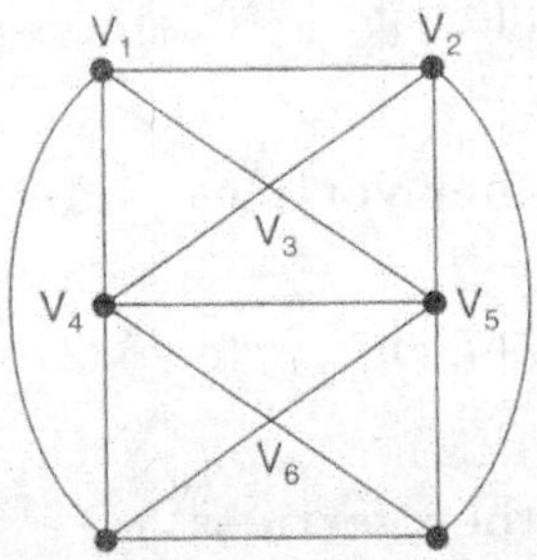

Fig. 145.

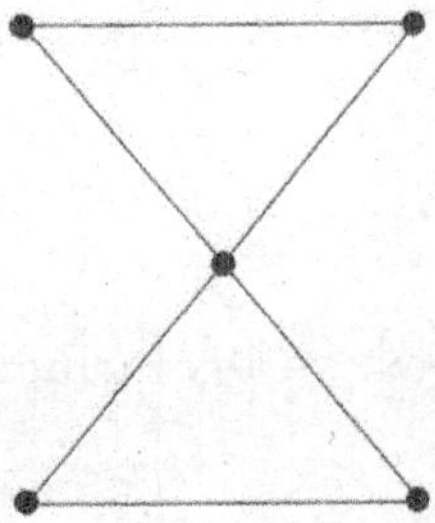

Fig. 146.

Sol. (*i*) The graph shown in Fig. 145 contains vertices V = 8 and edges e = 17. Putting the values we have $e = 3 \times 8 - 6 = 18 \geq 17$. Hence proved.

(*ii*) The graph shown in Fig. 146 contains vertices V = 5 and edges e = 6. Putting the values, we have $3 \times 5 - 6 = 11 > 6$. Hence proved.

Problem 23. *Find the shortest path between a and z in the graph shown in Fig. 147.*

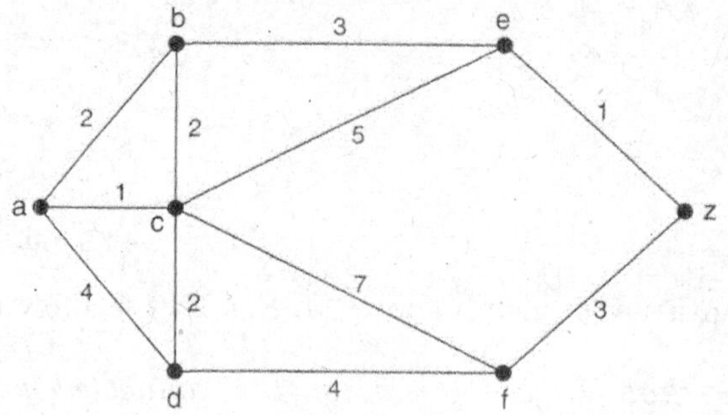

Fig. 147.

Sol. Step I. Include the vertex a in S and determine all the direct paths from a to all other vertices without going through any other vertices.

S	Distance to all other vertices						
	a	*b*	*c*	*d*	*e*	*f*	*z*
a	0	2(*a*)	1(*a*)	4(*a*)	∞	∞	∞

Step II. Include the vertex in S which is nearest to a and determine shortest path to all the vertices through this vertex. The nearest vertex is c.

S	Distance to all other vertices						
	a	*b*	*c*	*d*	*e*	*f*	*z*
a, *c*	0	2(*a*)	1(*a*)	3(*a*, *c*)	6(*a*, *c*)	8(*a*, *c*)	∞

Step III. Include the vertex in S which is 2nd nearest to S and determine shortest path to all the vertices through this vertex. The 2nd nearest vertex is b.

S	Distance to all other vertices						
	a	*b*	*c*	*d*	*e*	*f*	*z*
a, *c*, *b*	0	2(*a*)	1(*a*)	3(*a*, *c*)	5(*a*, *b*)	8(*a*, *c*)	∞

Step IV. Next vertex included in S is d.

S	Distance to all other vertices						
	a	*b*	*c*	*d*	*e*	*f*	*z*
a, *c*, *b*, *d*	0	2(*a*)	1(*a*)	3(*a*, *c*)	5(*a*, *b*)	7(*a*, *c*)	∞

Step V. Next vertex included in S is e.

S	Distance to all other vertices						
	a	*b*	*c*	*d*	*e*	*f*	*z*
a, *c*, *b*, *d*, *e*	0	2(*a*)	1(*a*)	3(*a*, *c*)	5(*a*, *b*)	7(*a*, *c*)	6(*a*, *b*, *e*)

Step VI. Next vertex included in S is *f*.

S	Distance to all over vertices						
a, c, b, d, e, f	*a*	*b*	*c*	*d*	*e*	*f*	*z*
	0	2(*a*)	1(*a*)	3(*a, c*)	5(*a, b*)	7(*a, c*)	6(*a, b, e*).

This ends our procedure as $n - 1$ vertices are included in S. Thus, the shortest distance between a and z is 6 and the shortest path is a, b, e, z.

7

Trees

GENERAL TREES

A graph which has no cycle is called an acyclic graph. A tree is an acyclic graph or graph having no cycles.

A tree or general tree is defined as a non-empty finite set of elements called vertices or nodes having the property that each node can have minimum degree 1 and maximum degree n. It can be partitioned into $n + 1$ disjoint subsets such that the first subset contains the root of the tree and the remaining n subsets contains the elements of the n subtree. (Fig. 1)

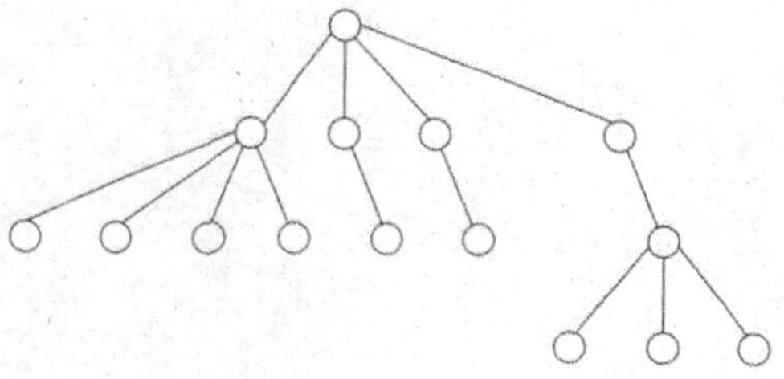

Fig. 1. General Tree.

DIRECTED TREES

A directed tree is an acyclic directed graph. It has one node with indegree 1, while all other nodes have indegree 1 as shown in Figs. 2 and 3.

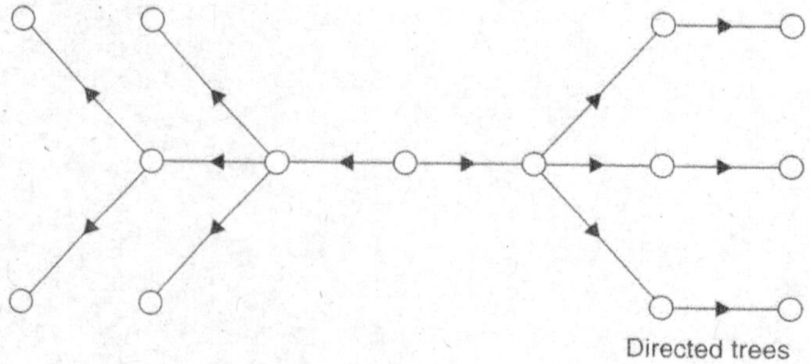

Fig. 2.

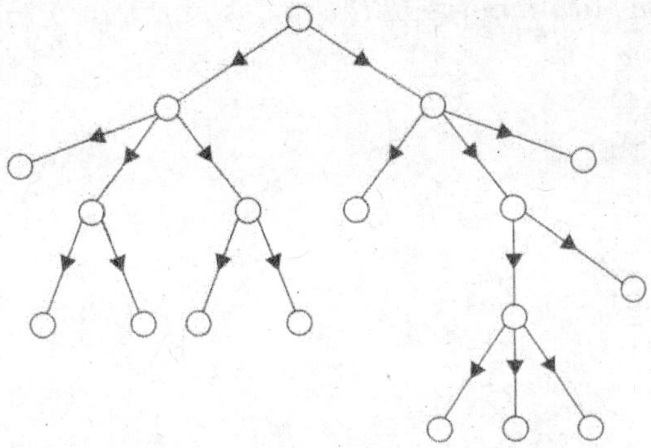

Fig. 3.

The node which has outdegree 0 is called an external node or a terminal node or a leaf. The nodes which has outdegree greater than or equal to one are called internal nodes or branch nodes.

ORDERED TREES

If in a tree at each level, an ordering is defined, then such a tree is called an ordered tree.

e.g., the trees shown in Figs. 4 and 5 represent the same tree but have different orders.

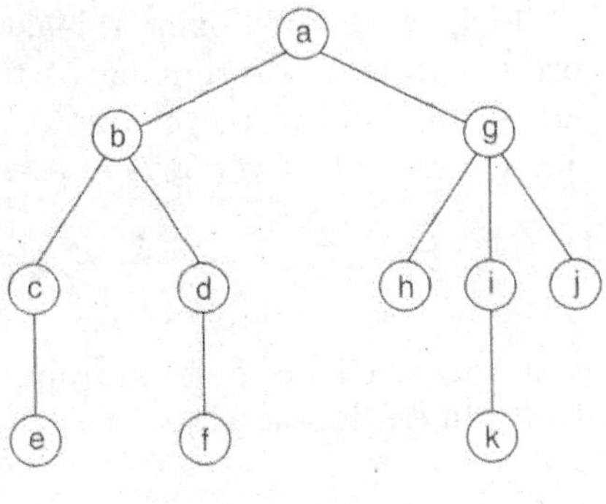

Fig. 4.

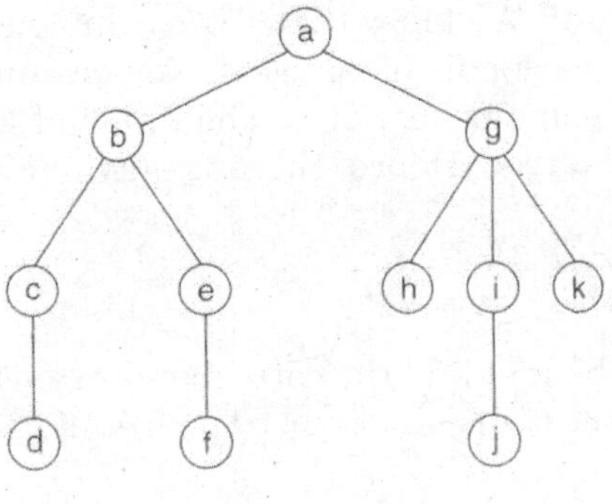

Fig. 5.

ROOTED TREES

If a directed tree has exactly one node or vertex called root whose incoming degree is 0 and all other vertices have incoming degree one, then the tree is called rooted tree.

* A tree with no nodes is a rooted tree (the empty tree).
* A single node with no children is a rooted tree.

PATH LENGTH OF A VERTEX

The path length of a vertex in a rooted tree is defined to be the number of edges in the path from the root to the vertex.

Example 1. *Find the path lengths of the nodes b, f, l, q in Fig. 6.*

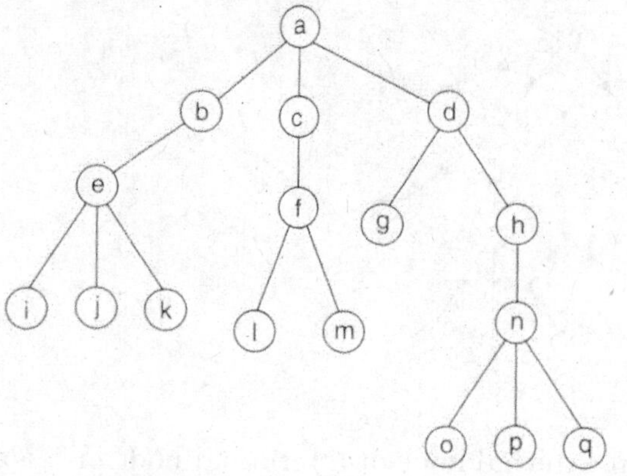

Fig. 6.

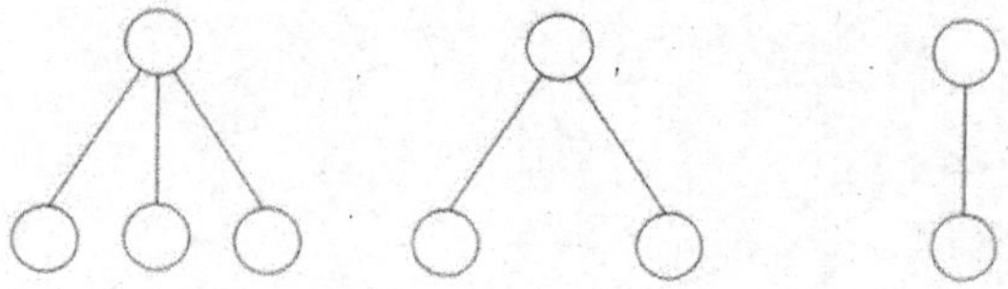

Fig. 7.

Sol. The path length of node *b* is one.

The path length of node *f* is two.

The path length of node *l* is three.

The path length of node *q* is four.

Theorem I. *Prove that there is one and only one path between every pair of vertices in a tree T.*

Proof. We know that T is a connected graph, in which there must exist at least one path between every pair of vertices. Now assume that there exists two different paths from some node *a* to some node *b* of T. The union of these two paths will contain a cycle and therefore T cannot be a tree. Hence, there is only one path between every pair of vertices in a tree.

FOREST

If the root and the corresponding edges connecting the nodes are deleted from a tree, we obtain a set of disjoint trees. This set of disjoint trees is called a forest. (Fig. 7)

BINARY TREE

If the outdegree of every node is less than or equal to 2, in a directed tree then the tree is called a binary tree. A tree consisting of no nodes (empty tree) is also a binary tree. A binary tree is shown in Figs. 8 and 9.

BASIC TERMINOLOGY

Root. A binary tree has a unique node called the root of the tree.

Left Child. The node to the left of the root is called its left child.

Right Child. The node to the right of the root is called its right child.

Parent. A node having left child or right child or both is called parent of the nodes.

Siblings. Two nodes having the same parent are called siblings.

Leaf. A node with no children is called a leaf. The number of leaves in a binary tree can vary from one (minimum) to half the number of vertices (maximum) in a tree.

Ancestor. If a node is the parent of another node, then it is called ancestor of that node. The root is an ancestor of every other node in the tree.

Descendent. A node is called descendent of another node if it is the child of the node or child of some other descendent of that node. All the nodes in the tree are descendents of the root.

Left Subtree. The subtree whose root is the left child of some node is called the left subtree of that node.

Example 2. *For the tree as shown in Fig. 8.*

(*i*) *Which node is the root ?*

(*ii*) *Which nodes are leaves ?*

(*iii*) *Name the parent node of each node.*

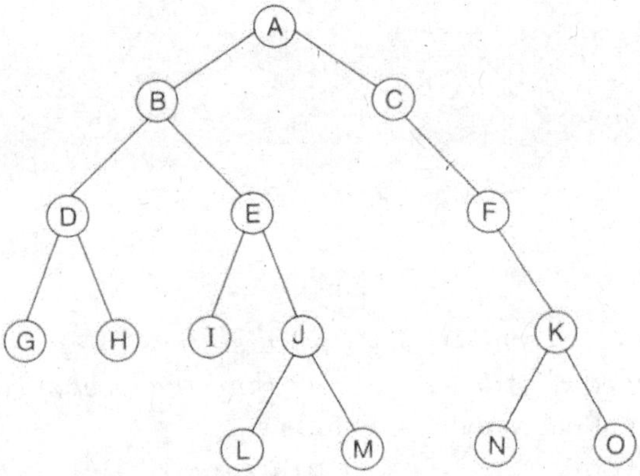

Fig. 8.

Sol. (*i*) The node A is the root node.

(*ii*) The nodes G, H, I, L, M, N, O are leaves.

(*iii*)

Nodes	Parent
B, C	A
D, E	B
F	C
G, H	D
I, J	E
K	F
L, M	J
N, O	K

Right Subtree. The subtree whose root is the right child of some node is called the right subtree of that node.

Level of a Node. The level of a node is its distance from the root. The level of root is defined as zero. The level of all other nodes is one more than its parent node. The maximum number of nodes at any level N is 2^N.

Depth or Height of a Tree. The depth or height of a tree is defined as the maximum number of nodes in a branch of tree. This is one more than the maximum level of the tree *i.e.*, the depth of root is one. The maximum number of nodes in a binary tree of depth d is $2^d - 1$, where $d \geq 1$.

External Nodes. The nodes which has no children are called external nodes or terminal nodes.

Internal Nodes. The nodes which has one or more than one children are called internal nodes or non-terminal nodes.

Example 3. *For the tree as shown in Fig. 9.*

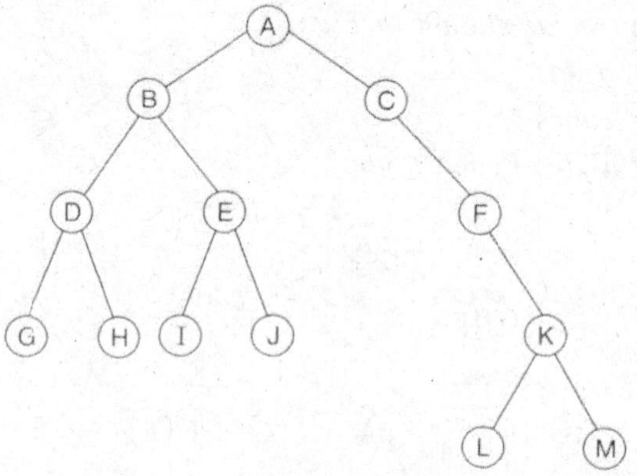

Fig. 9.

(*i*) *List the children of each node.* (*ii*) *List the siblings.*
(*iii*) *Find the depth of each node.* (*iv*) *Find the level of each node.*

Sol. (*i*) The children of each node is as follows :

Node	Children
A	B, C
B	D, E
C	F
D	G, H
E	I, J
F	K
K	L, M

(*ii*) The siblings are as follows :

Siblings
B and C
D and E
G and H
I and J
L and M are all siblings.

(*iii*)

Node	Depth or Height
A	1
B, C	2
D, E, F	3
G, H, I, J, K	4
L, M	5

(*iv*)

Node	Level
A	0
B, C	1
D, E, F	2
G, H, I, J, K	3
L, M	4

BINARY EXPRESSION TREES

Algebraic expression can be conveniently expressed by its expression tree. An expression having binary operators can be decomposed into

< left operand or expression > (operator) < right operand or expression >

depending upon precedence of evaluation.

The expression tree is a binary tree whose root contains the operator and whose left subtree contains the left expression and right subtree contains the right expression.

Example 4. *Construct the binary expression tree for the expression $(a + b) * (d/c)$.*

Sol. The binary expression tree for the expression $(a + b) * (d/c)$ is shown in Fig. 10.

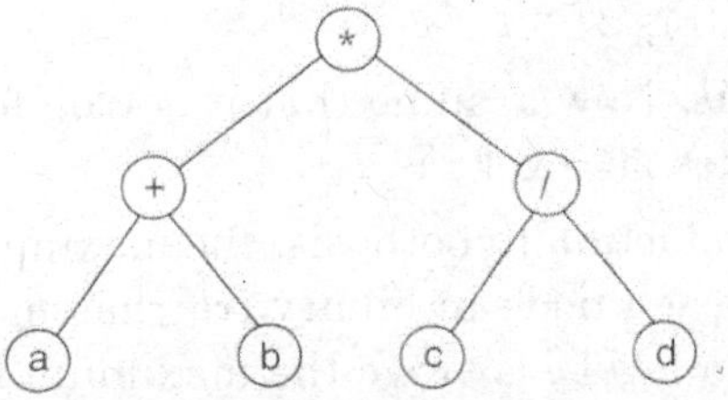

Fig. 10.

COMPLETE BINARY TREE

Complete binary tree is a binary tree if all its levels, except possibly the last, have the maximum number of possible nodes as for left as possible. The depth of complete binary tree having n nodes is $\log_2 n + 1$.

For example : The tree shown in Fig. 11 is a complete binary tree.

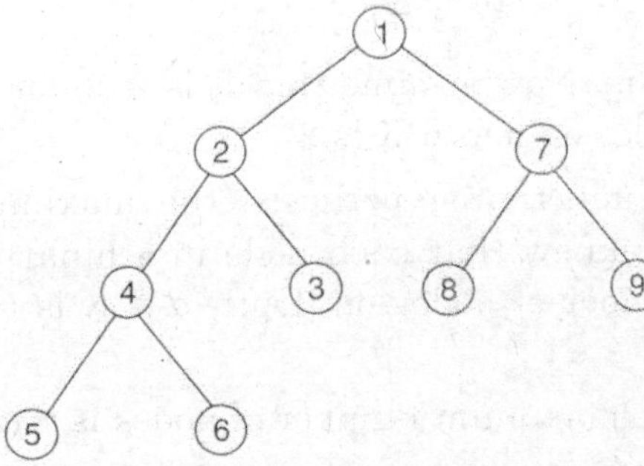

Fig. 11.

FULL BINARY TREE

Full binary tree is a binary tree in which all the leaves are on the same level and every non-leaf node has two children.

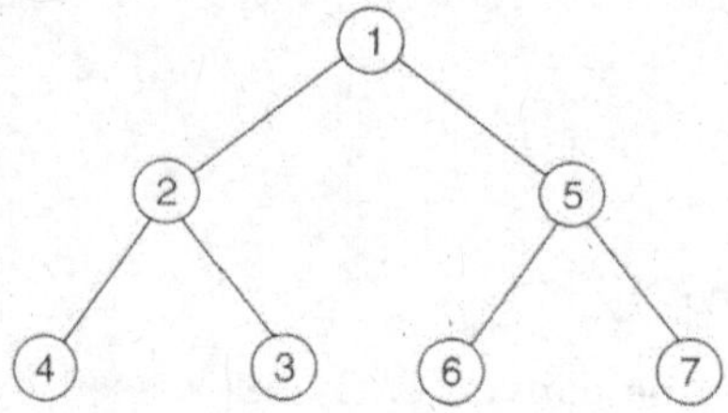

Fig. 12.

For example : The tree shown in Fig. 12 is a full binary tree.

Theorem II. *Prove that the maximum number of nodes on level n of a binary tree is* 2^n, *where* $n \geq 0$.

Proof. This can be proved by induction.

Basis of Induction. The only node at level $n = 0$ is the root node. Thus, the maximum number of nodes on level $n = 0$ is $2^0 = 1$.

Induction Hypothesis. Now assume that it is true for level j, where $n \geq j \geq 0$. Therefore, the maximum no. of nodes on level j is 2^j.

Induction Step. By induction hypothesis, the maximum number of nodes on level $j - 1$ is 2^{j-1}. Since, we know that each node in binary tree has maximum degree 2. Therefore, the maximum number of nodes on level j is twice the maximum number of level $j - 1$.

Hence, at level j, the maximum number of nodes is

$$= 2 \cdot 2^{j-1}$$

$$= 2^j. \text{ Hence proved.}$$

Theorem III. *Prove that the maximum number of nodes in a binary tree of depth d is* $2^d - 1$, *where* $d \geq 1$.

Proof. This can be proved by induction.

Basis of Induction. The only node at depth $d = 1$ is the root node. Thus, the maximum number of nodes on depth $d = 1$ is $2^1 - 1 = 1$.

Induction Hypothesis. Now assume that it is true for depth K, $d > k \geq 1$. Therefore, the maximum number of nodes on depth K is $2^k - 1$.

Induction Step. By induction hypothesis, the maximum number of nodes on depth K − 1 is $2^{k-1} - 1$. Since, we know that each node in a binary tree has maximum degree 2, therefore, the maximum number of nodes on depth d = K is twice the maximum number of nodes on depth K − 1.

So, at depth d = K, the maximum number of nodes is $= (2 \cdot 2^{K-1}) - 1$

$$= 2^{K-1+1} - 1 = 2^K - 1. \text{ Hence proved.}$$

Theorem IV. *Prove that in a binary tree, if n_E is the number of external nodes or leaves and n_I is the no. of internal nodes, then $n_E = n_I + 1$.*

Proof. Let n be the total number of nodes in the tree. Then, we may have three types of nodes in the tree.

n_E = the number of nodes having zero degree.

n_I = the number of nodes having two degree.

n_0 = the number of nodes having one degree.

So, we have $\quad n = n_E + n_I + n_0 \quad$ ...(1)

Let us assume that the number of edges of the tree is E. So, with these E edges we can connect E + 1 nodes.

Hence $\quad n = E + 1 \quad$ ...(2)

Since, all edges are either from a node of degree one or from a node of degree two,

therefore, $\quad E = n_0 + 2n_I \quad$ ...(3)

Put this value in the eq. (2), we have

$$n = n_0 + 2n_I + 1 \quad ...(4)$$

Subtract eq. (4) from eq. (1), we get

$$n_E = n_I + 1$$

Hence proved.

TRAVERSING BINARY TREES

Traversing means to visit all the nodes of the tree. There are three standard methods to traverse the binary trees. These are as follows :

1. Preorder traversal
2. Postorder traversal
3. Inorder traversal.

1. Preorder traversal. The preorder traversal of a binary tree is a recursive process. The preorder traversal of a tree is

(*i*) Visit the root of the tree.

(*ii*) Traverse the left subtree in preorder.

(*iii*) Traverse the right subtree in preorder.

2. Postorder traversal. The postorder traversal of a binary tree is a recursive process. The postorder traversal of a tree is

(*i*) Traverse the left subtree in postorder.

(*ii*) Traverse the right subtree in postorder.

(*iii*) Visit the root of the tree.

3. Inorder traversal. The inorder traversal of a binary tree is a recursive process. The inorder traversal of a tree is

(*i*) Traverse in inorder the left subtree.

(*ii*) Visit the root of the tree.

(*iii*) Traverse in inorder the right subtree.

Example 5. *Determine the preorder, postorder, and inorder traversal of the binary tree as shown in Fig. 13.*

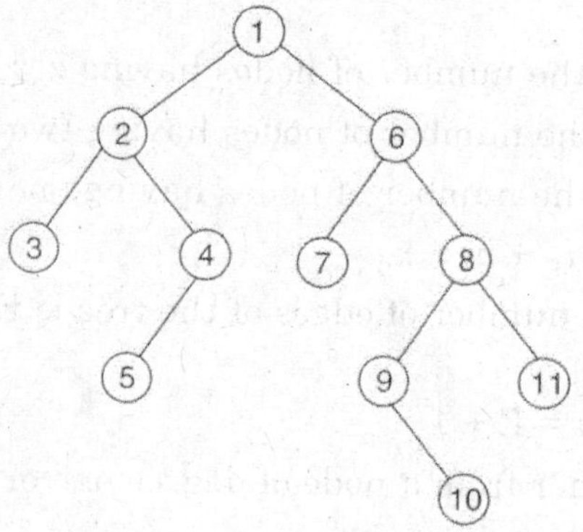

Fig. 13.

Sol. The preorder, postorder and inorder traversal of the tree is as follows :

Preorder	:	1	2	3	4	5	6	7	8	9	10	11
Postorder	:	3	5	4	2	7	10	9	11	8	6	1
Inorder	:	3	2	5	4	1	7	6	9	10	8	11.

Example 6. *Give the preorder, inorder and postorder traversals of the tree shown in Fig. 14.*

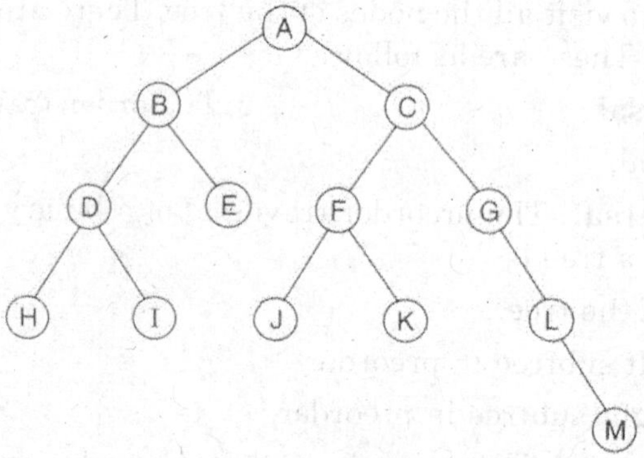

Fig. 14.

Sol.

Preorder	:	A,	B,	D,	H,	I,	E,	C,	F,	J,	K,	G,	L, M
Inorder	:	H,	D,	I,	B,	E,	A,	J,	F,	K,	C,	G,	L, M
Postorder	:	H,	I,	D,	E,	B,	J,	K,	F,	M,	L,	G,	C, A.

ALGORITHMS

(a) Algorithm to Draw a Unique Binary Tree When Inorder and Preorder Traversal of the Tree is Given

1. We know that the root of the binary tree is the first node in its preorder. Draw the root of the tree.

2. To find the left child of the root node, first use the inorder traversal to find the nodes in the left subtree of the binary tree. (All the nodes that are left to the root node in the inorder traversal are the nodes of the left subtree). After that the left child of the root is obtained by selecting the first node in the preorder traversal of the left subtree. Draw the left child.

3. In the same way, use the inorder traversal to find the nodes in the right subtree of the binary tree. Then the right child is obtained by selecting the first node in the preorder traversal of the right subtree. Draw the right child.

4. Repeat the steps 2 and 3 with each new node until every node is not visited in preorder. Finally, we obtain the unique tree.

Example 7. *Draw the unique binary tree when the inorder and preorder traversal is given as follows :*

Inorder	:	*B*	*A*	*D*	*C*	*F*	*E*	*J*	*H*	*K*	*G*	*I*
Preorder	:	*A*	*B*	*C*	*D*	*E*	*F*	*G*	*H*	*J*	*K*	*I.*

Sol. We know that the root of the binary tree is the first node in preorder traversal. Now check A, in the inorder traversal, all the nodes that are left of A, are nodes of left subtree and all the nodes that are right of A, are nodes of right subtree. Read the next node in preorder and check its position against the root node, if it is left of root node, then draw it as left child, otherwise draw it as right child. Repeat the above process for each new node until all the nodes of preorder traversal are read and finally we obtain the binary tree as shown in Fig. 15.

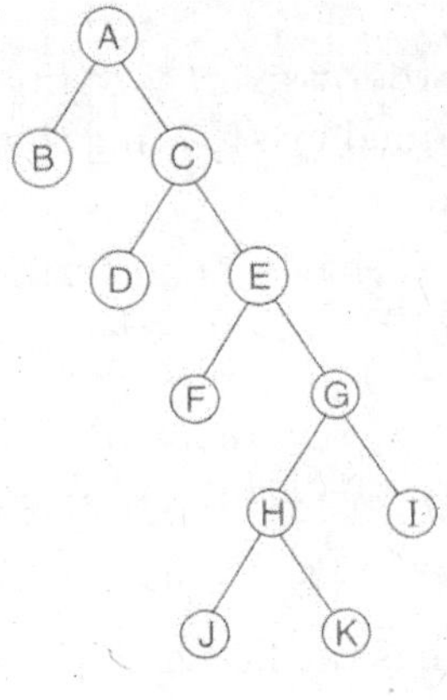

Fig. 15.

Example 8. *Draw the unique binary tree when the following is given :*

Inorder	:	*d*	*b*	*h*	*e*	*a*	*f*	*c*	*i*	*j*	*g*
Preorder	:	*a*	*b*	*d*	*e*	*h*	*c*	*f*	*g*	*i*	*j*

Sol. The first node in preorder is *a* and hence *a* is the root node. Unique binary tree is shown in Fig. 16.

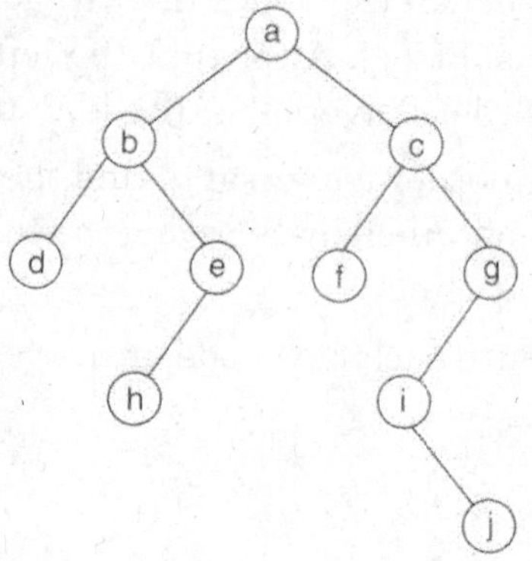

Fig. 16.

(b) Algorithm to Draw a Unique Binary Tree When Inorder and Postorder Traversal of the Tree is Given

1. We know that the root of the binary tree is the last node in its postorder. Draw the root of the tree.

2. To find the right child of the root node, first use the inorder traversal to find the nodes in the right subtree of the binary tree. (All the nodes that are right to the root node in the inorder traversal are the nodes of the right subtree). After that the right child of the root is obtained by selecting the last node in the postorder traversal of the right subtree. Draw the right child.

3. In the same way, use the inorder traversal to find the nodes in the left subtree of the binary tree. Then the left child is obtained by selecting the last node in the postorder traversal of the left subtree. Draw the left child.

4. Repeat the steps 2 and 3 with each new node until every node is not visited in postorder. After visiting the last node, we obtain the unique tree.

Example 9. *Draw the unique binary tree for the given Inorder and Postorder traversal*

Inorder	:	4	6	10	12	8	2	1	5	7	11	13	9	3
Postorder	:	12	10	8	6	4	2	13	11	9	7	5	3	1

Sol. We know that the root node is the last node in postorder traversal. Hence one is the root node.

Now check the inorder traversal, we know that root is at the centre, hence all the nodes that are left to the root node in inorder traversal are the nodes of left subtree and all that are right to the root node are the nodes of the right subtree.

Now, visit the next node from back in postorder traversal and check its position in inorder traversal, if it is on left of root then draw it as left child and if it is on right, then draw it as right child.

Repeat the above process for each new node and we obtain the binary tree as shown in Fig. 17.

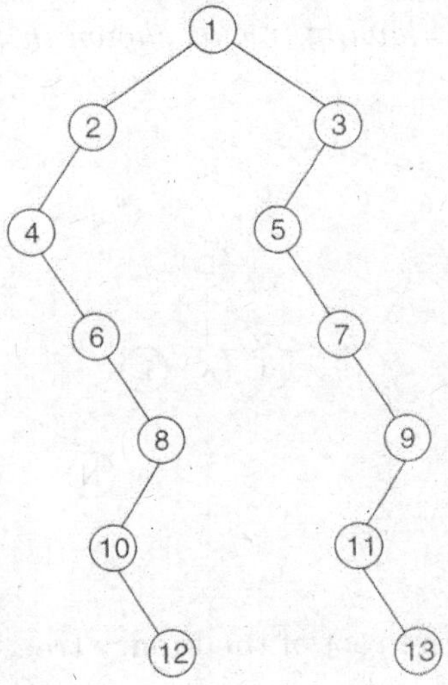

Fig. 17.

Example 10. *Draw the binary tree when Inorder and Postorder traversal is given :*

Inorder :	*m*	*k*	*n*	*j*	*o*	*l*	*u*	*s*	*v*	*q*	*t*	*p*	*r*
Postorder :	*m*	*n*	*k*	*o*	*u*	*v*	*s*	*t*	*q*	*r*	*p*	*l*	*j*.

Sol. We know that the last node in Postorder is the root node, hence *j* is the root. Now applying the algorithm as above, we obtain the tree shown in Fig. 18.

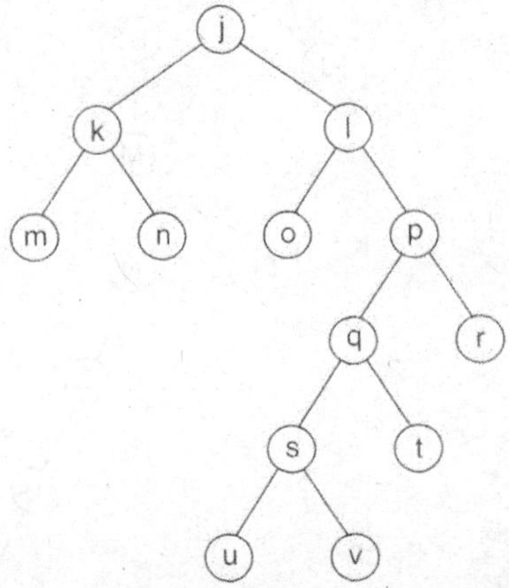

Fig. 18.

(*c*) Algorithm to Convert General Tree into the Binary Tree

1. Starting from the root node, the root of the tree is also the root of the binary tree.

2. The first child C_1 (from left) of the root node in the tree is the left child C_l of the root node in binary tree and the sibling of the C_1 is the right child of C_l and so on.

3. Repeat the step 2 for each new node.

Example 11. *Convert the following tree as shown in Fig. 19 into binary tree.*

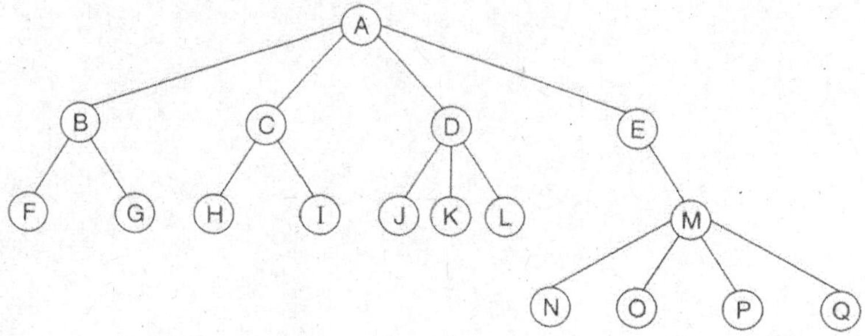

Fig. 19.

Sol. The root of the tree is the root of the binary tree. Hence A is the root of the binary tree. Now B becomes the left child of A in binary tree, C becomes the right child of B, D becomes right child of C, and E becomes right child of D in the binary tree and similarly applying the algorithm we obtain the binary tree as shown in Fig. 20.

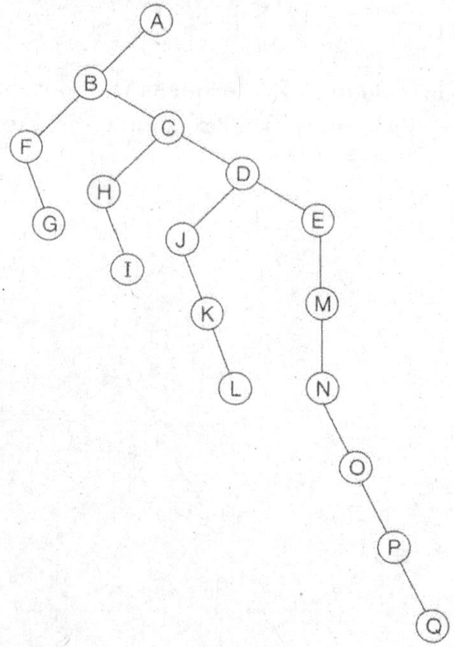

Fig. 20.

Example 12. *Convert the general tree as shown in Fig. 21 into binary tree.*

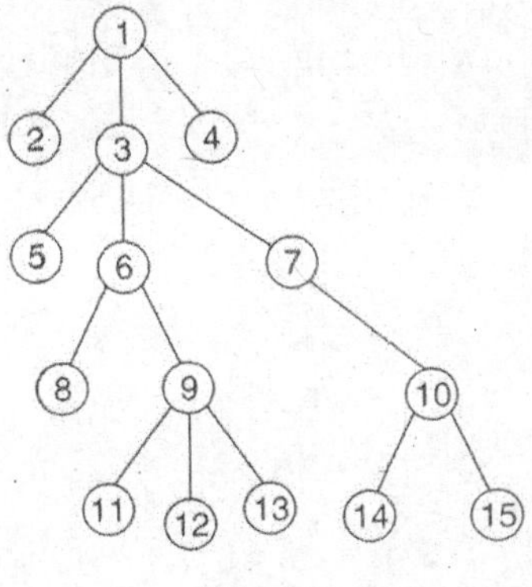

Fig. 21.

Sol. The root node 1 in general tree is the root node of the binary tree. Now applying the above algorithm we obtain the binary tree as shown in Fig. 22.

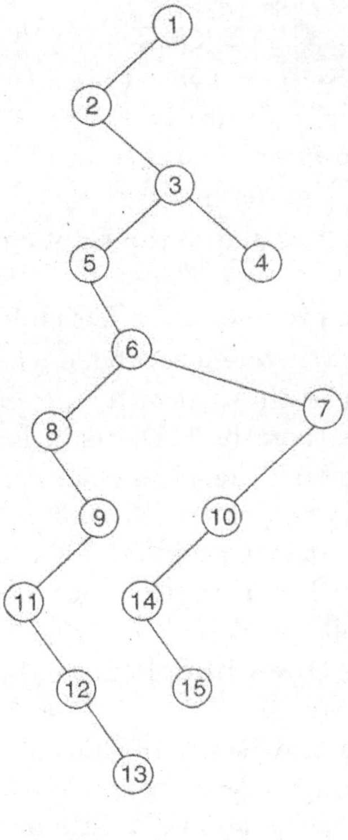

Fig. 22.

BINARY SEARCH TREES

Binary search trees has the property that the node to the left contains a smaller value than the node pointing to it and the node to the right contains a larger value than the node pointing to it.

It is not necessary that a node in a 'Binary Search Tree' point to the nodes whose values immediately precede and follow it.

For example : The tree shown in Fig. 23 is a binary search tree.

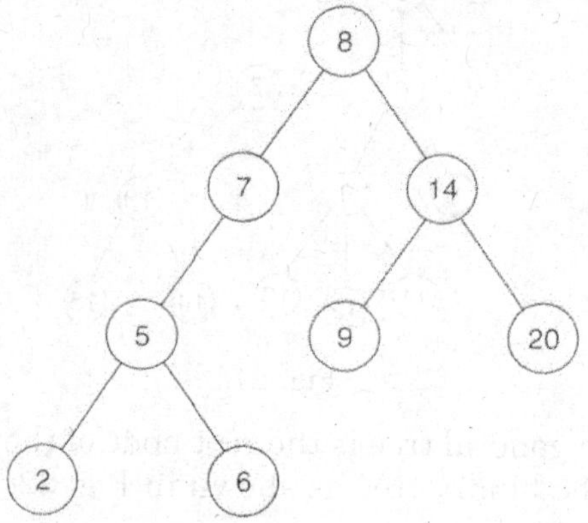

Fig. 23.

Inserting into a Binary Search Tree. Consider a binary search tree T. Suppose we have given an ITEM of information to insert in T. The ITEM is inserted as a leaf in the tree. The following steps explains a procedure to insert an ITEM in the binary search tree T.

1. Compare the ITEM with the root node.

2. If ITEM > ROOT NODE, proceed to the right child and it becomes root node for the right subtree.

3. If ITEM < ROOT NODE, proceed to the left child.

4. Repeat the above steps until we meet a node which has no left and right subtree.

5. Now if the ITEM is greater than node, then the ITEM is inserted as the right child and if the ITEM is less than node, then the ITEM is inserted as the left child.

Deleting in a Binary Search Tree. Consider a binary search tree T. Suppose we want to delete a given ITEM from binary search tree. To delete an ITEM from a binary search tree, we have three cases, depending upon the number of children of the deleted node.

1. Deleted Node has no Children. Deleting a node which has no children is very simple, as just replace the node with null.

2. Deleted Node has Only One Child. Replace the value of deleted node with the only child.

3. Deleted Node has Two Children. In this case, replace the deleted node with the node that is closest in the value to the deleted node. To find the closest value, we move once to the left and then to the right as far as possible. This node is called immediate predecessor. Now replace the value of deleted node with immediate predecessor and then delete the replaced node by using case 1 or 2.

Example 13. *Show the binary search tree after inserting 3, 1, 4, 6, 9, 2, 5, 7 into an initially empty binary search tree.*

Sol. The insertion of the above nodes in the empty binary search tree is shown in Fig. 24.

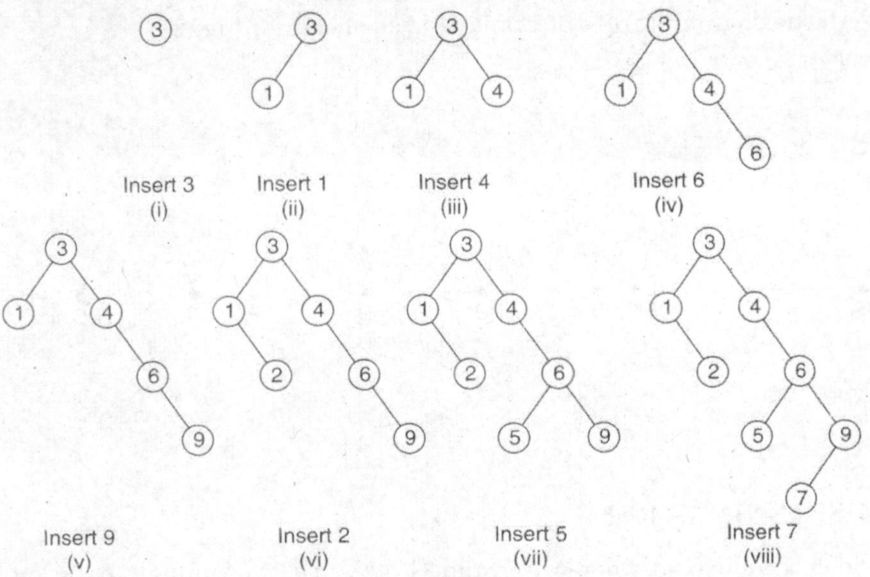

Fig. 24.

Example 14. *Show the binary tree shown in Fig. 24 (viii) after deleting the root node.*

Sol. To delete the root node, first replace the root node with the closest element of the root. For this, first move one step left and then to the right as far as possible to the node. Then delete the replaced node. The tree after deletion is shown in Fig. 25.

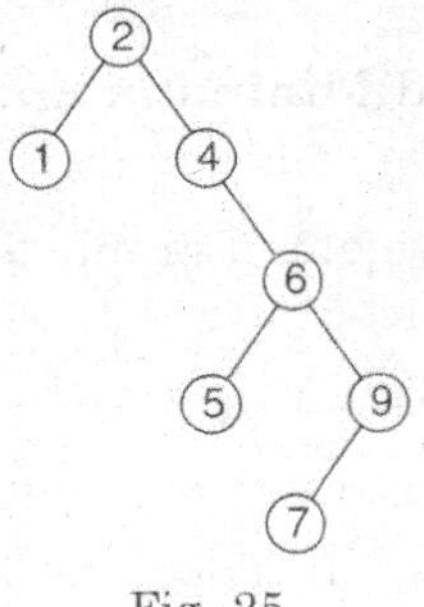

Fig. 25.

SPANNING TREE

Consider a connected graph G = (V, E). A spanning tree T is defined as a subgraph of G if T is a tree and T includes all the vertices of G.

Example 15. *Draw all the spanning trees of the graph G shown in Fig. 26.*

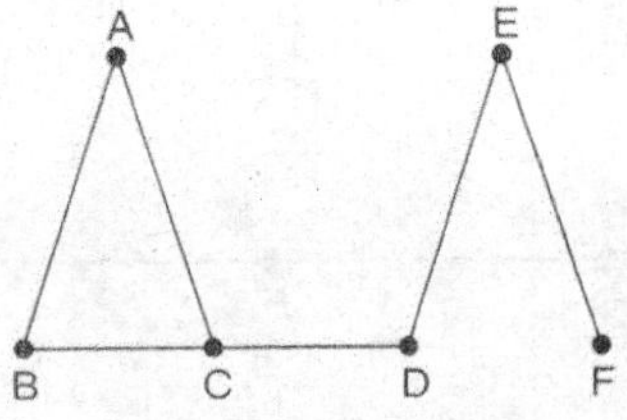

Fig. 26. Graph G.

Sol. All the spanning trees of graph G is as shown in Fig. 27.

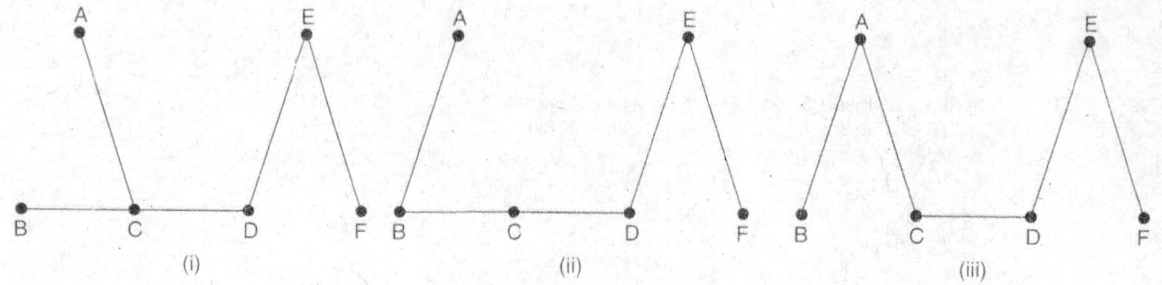

Fig. 27.

MINIMUM SPANNING TREE

Consider a connected weighted graph G = (V, E). A minimal spanning tree T of the graph G is a tree whose total weight is smallest among all the spanning trees of the graph G. The total weight of the spanning tree is the sum of the weights of the edges of the spanning trees.

The minimum weight of the spanning tree is unique but the spanning tree may not be unique because more than one spanning tree are possible when more than one edges exist having the same weight.

Kruskal's Algorithm to Find Minimum Spanning Tree. This algorithm finds the minimum spanning tree T of the given connected weighted graph G.

1. Input the given connected weighted graph G with n vertices whose minimum spanning tree T, we want to find.
2. Order all the edges of the graph G according to increasing weights.
3. Initialise T with all vertices but do not include any edge.
4. Add each of the graph G in T which does not form a cycle until $n - 1$ edges are added.

Example 16. *Determine the minimum spanning tree of the weighted graph shown in Fig. 28.*

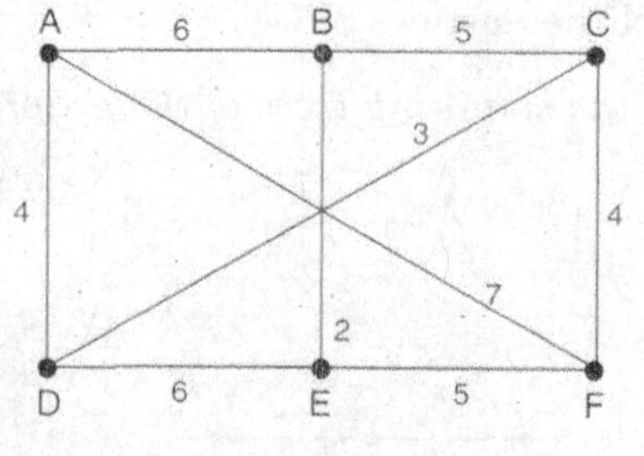

Fig. 28.

Sol. Using Kruskal's algorithm, arrange all the edges of the weighted graph in increasing order and initialise spanning tree T with all the six vertices of G. Now start adding the edges of G in T which do not form a cycle and having minimum weights until five edges are not added as there are six vertices. (Fig. 29).

Edges	Weights	Added or Not
(B, E)	2	Added
(C, D)	3	Added
(A, D)	4	Added
(C, F)	4	Added
(B, C)	5	Added
(E, F)	5	Not added
(A, B)	6	Not added
(D, E)	6	Not added
(A, F)	7	Not added.

Minimum Spanning Tree

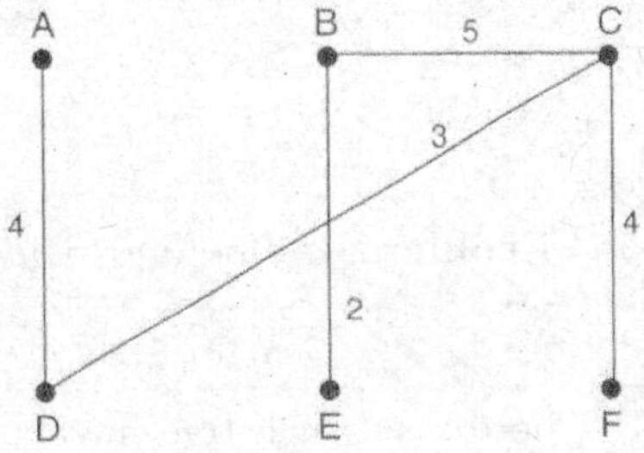

Fig. 29.

SOLVED PROBLEMS

Problem 1. *Show that if in a graph G there exists one and only one path between every pair of vertices, then G is a tree.*

Sol. The graph G is connected since there is a path between every pair of vertices. A cycle in a graph exists if there is at least one pair of vertices (v_1, v_2) such that there exist two distinct paths from v_1 to v_2. But the graph G has one and only one path between every pair of vertices. Thus, G contains no cycle. Hence, G is a tree.

Problem 2. *Draw two different binary trees with five nodes having only one leaf.*

Sol. The two trees out of many possible trees with five nodes having only one leaf is shown in Fig. 30.

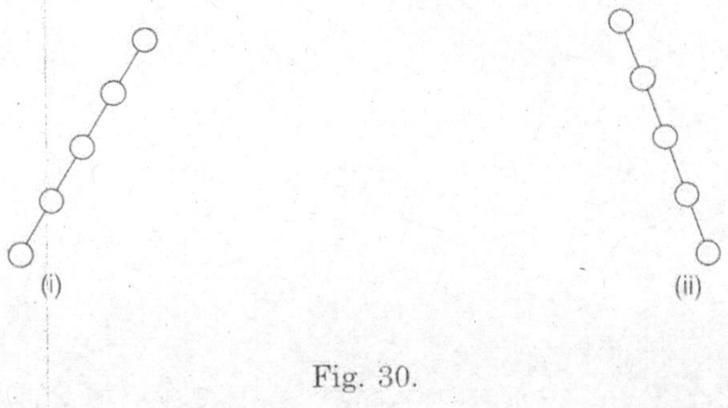

Fig. 30.

Problem 3. *Draw two different binary trees with five nodes having maximum number of leaves.*

Sol. There are many possible trees, out of which two different binary trees are shown in Fig. 31.

Fig. 31.

Problem 4. *How will you differentiate between a general tree and a binary tree ?*

General Tree	*Binary Tree*
1. There is no such tree having zero nodes or an empty general tree.	1. There may be an empty binary tree.
2. If some node has a child, then there is no such distinction.	2. If some nodes has a child, then it is distinguished as a left child or a right child.
3. The trees shown in Fig. 32. are same, when we consider them as general trees.	3. The trees shown in Fig. 32 are distinct, when we consider them as binary trees, because in (*i*), 4 is right child of 2 while in (*ii*), 4 is left child of 2.
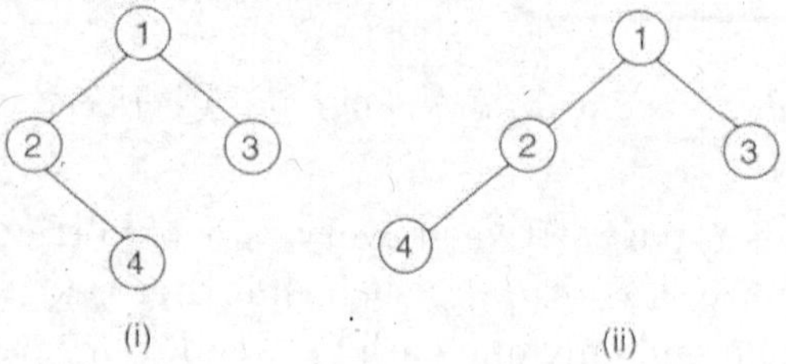 Fig. 32.	

Problem 5. *Determine the value of expression tree shown in Fig. 33.*

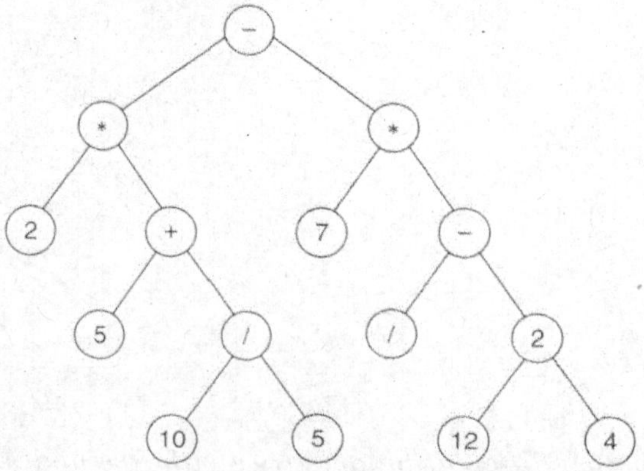

Fig. 33.

Sol. The value of expression tree is 2.

Problem 6. *Determine the value of expression tree shown in Fig. 34.*

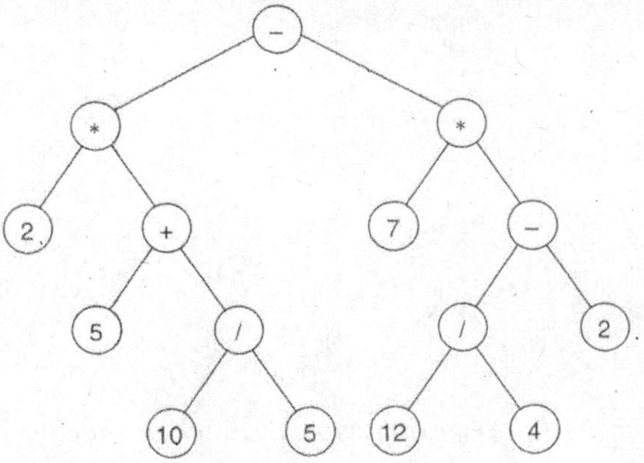

Fig. 34.

Sol. The value of expression tree is 7.

Problem 7. *Draw the unique binary tree when inorder and preorder traversal of tree is given as follows :*

Preorder	:	*10*	*9*	*8*	*7*	*6*	*5*	*4*	*3*	*2*	*1*
Inorder	:	*10*	*8*	*9*	*7*	*6*	*4*	*5*	*2*	*3*	*1*

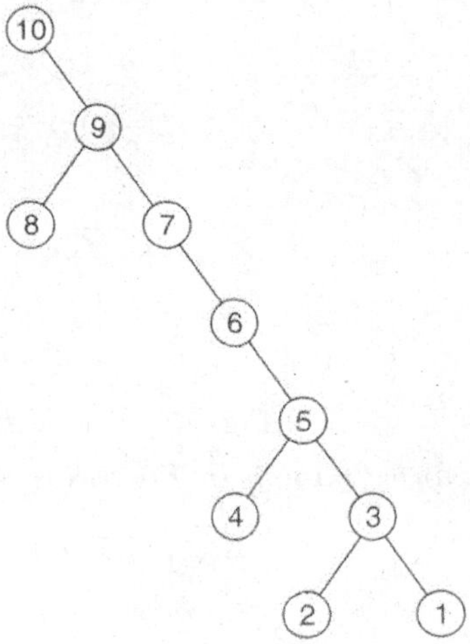

Fig. 35.

Sol. Unique binary tree is shown in Fig. 35.

Problem 8. *Draw the unique binary tree when inorder and preorder traversal of tree is given as follows :*

Inorder	:	*3*	*	*a*	+	*b*	*	*6*	*	c	+	*d*
Preorder	:	*	+	*	*3*	*a*	*b*	+	*	*6*	c	*d*

Sol. Unique binary tree is shown in Fig. 36.

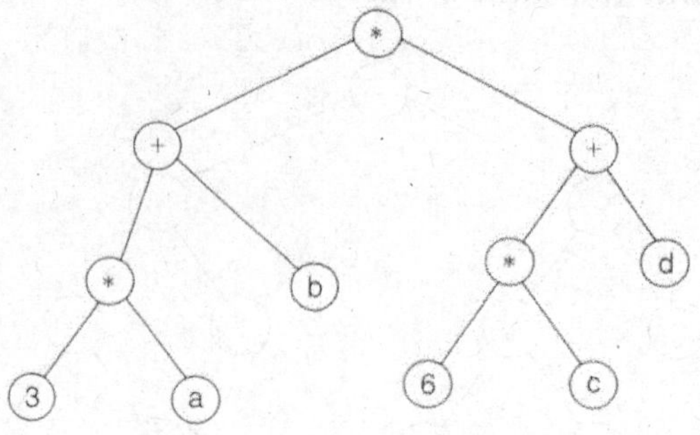

Fig. 36.

Problem 9. *Draw the binary expression tree, when inorder and postorder traversal of the tree is given as follows :*

Postorder :	*k*	*l*	− *m*	+	*n*	*p*	*q*	*r*	*t* − /	+	− *
Inorder :	*k*	−	*l* + *m*	*	*n*	−	*p*	+	*q* / *r*	−	*

Sol. Binary expression tree is shown in Fig. 37.

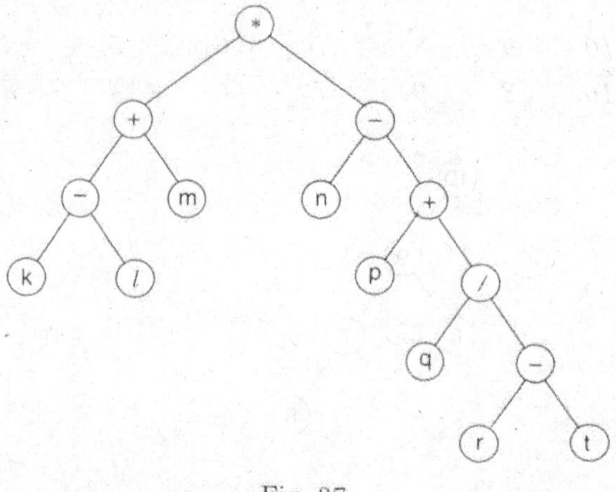

Fig. 37.

Problem 10. *Convert the forest shown in Fig. 38 into binary tree.*

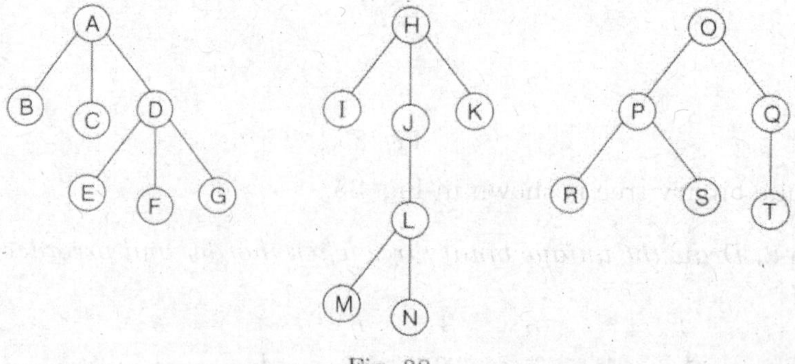

Fig. 38.

Sol. The root of the binary tree is the root node of the first tree (from left) and the root node of the second tree becomes the right son of the root node in binary tree and the root node

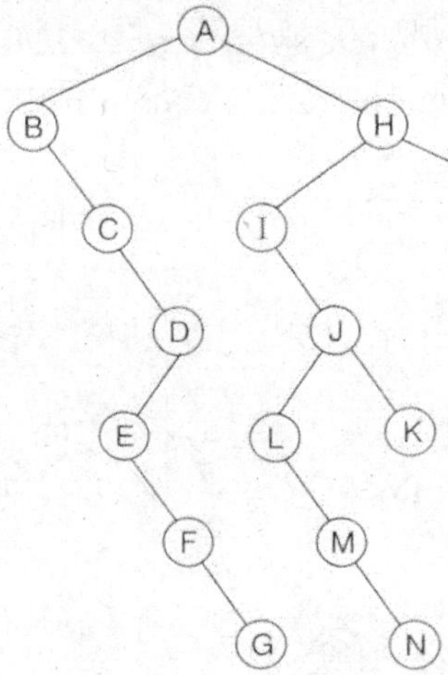

Fig. 39.

of the third tree becomes the right son of the right son in binary tree. Repeat this procedure for each level and we obtain the binary tree as shown in Fig. 39.

Problem 11. *A binary search tree contains the values 1, 2, 3, 4, 5, 6, 7, 8, 9. The tree is traversed in pre-order and the values are printed out. Determine the sequence of the print out values.*

Sol. First of all draw the binary search tree shown in Fig. 40.

Now traverse the tree in pre-order and get the output as below 1, 2, 3, 4, 5, 6, 7, 8, 9.

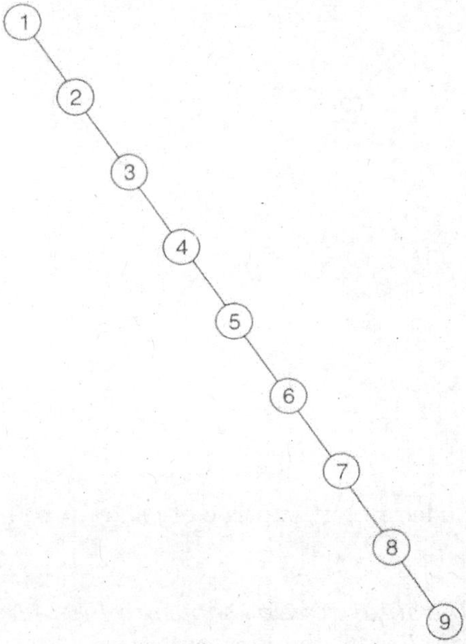

Fig. 40.

Problem 12. *A binary search tree is generated by inserting in order the following integers :*

50, 15, 62, 5, 20, 58, 91, 3, 8, 37, 60, 24.

Determine the number of nodes in the left subtree and right subtree of the root.

Sol. First of all draw the binary search tree as shown in Fig. 41.

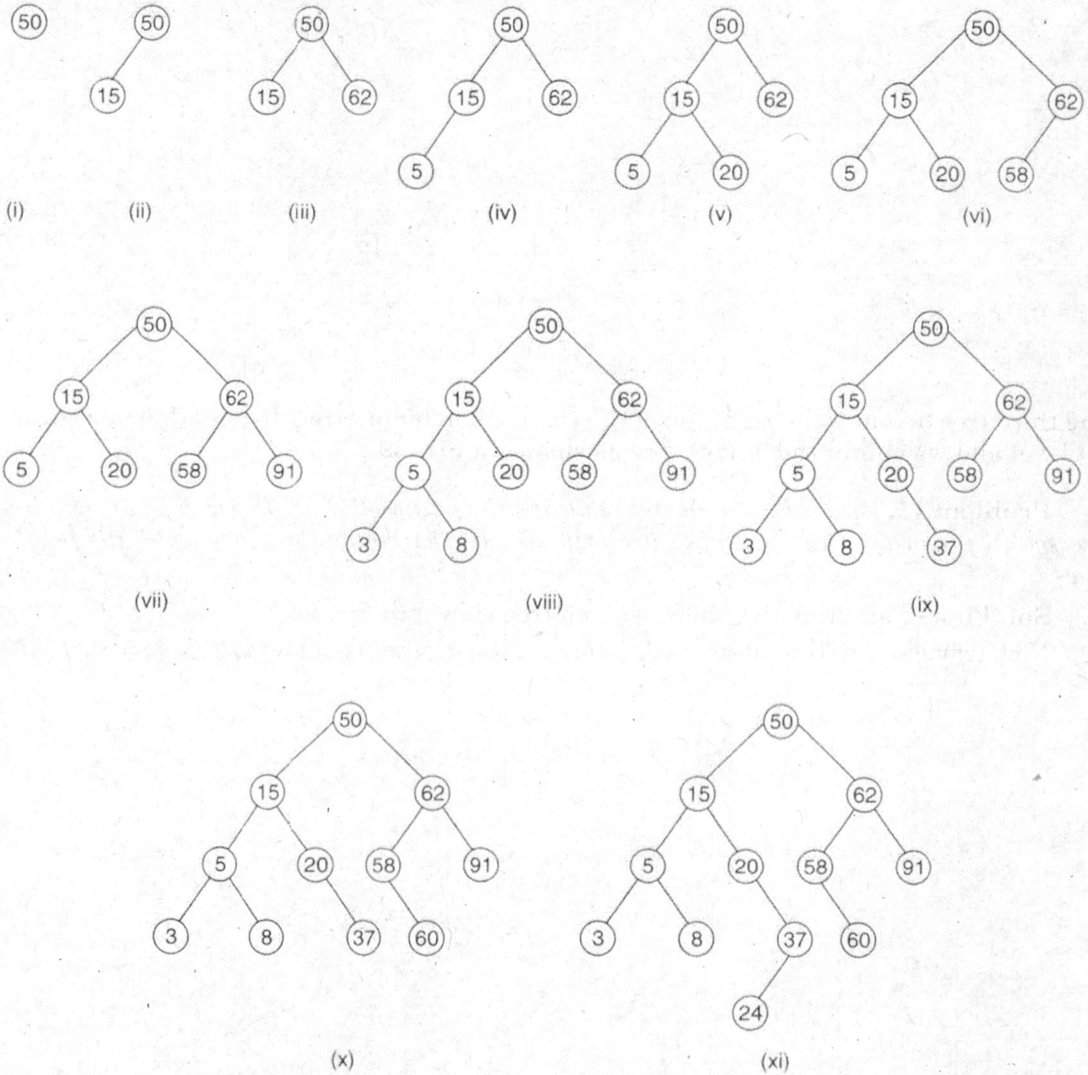

Fig. 41.

Thus, the number of nodes in left subtree of the root is 7 and right subtree of the root is 4.

Problem 13. *Consider the binary tree as shown in Fig. 42. Draw the binary tree for each of the following operations, if applied to the binary tree.*

(*i*) *Delete the node V* (*ii*) *Delete the node E*

(*iii*) *Delete the root node R.*

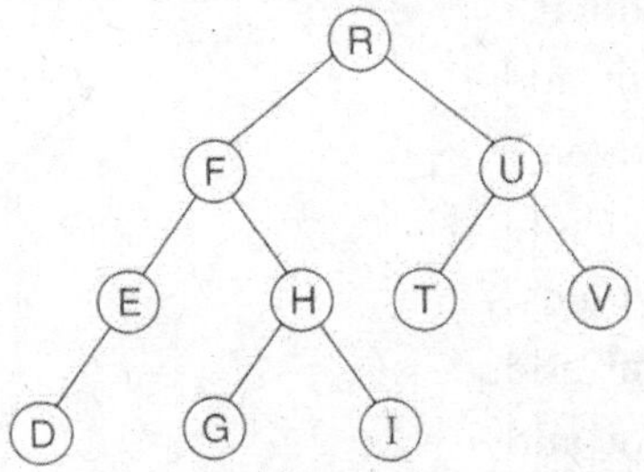

Fig. 42.

Sol. (*i*) The binary tree after deleting node V is shown in Fig. 43.

(*ii*) The binary tree after deleting node E is shown in Fig. 44.

(*iii*) The binary tree after deleting root node R is shown in Fig. 45.

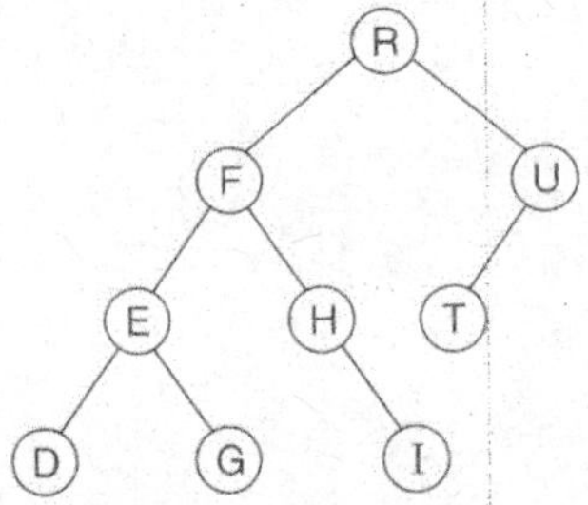

Fig. 43.

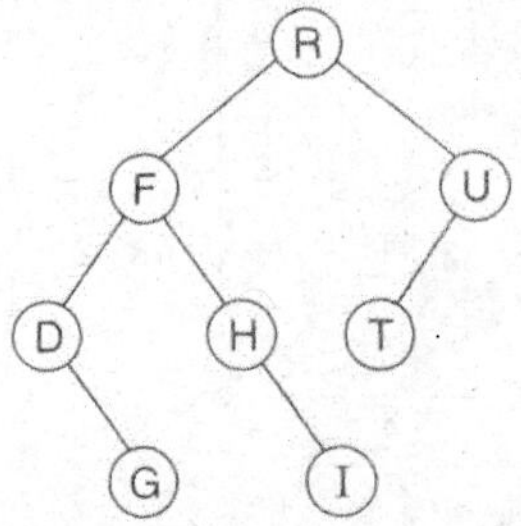

Fig. 44.

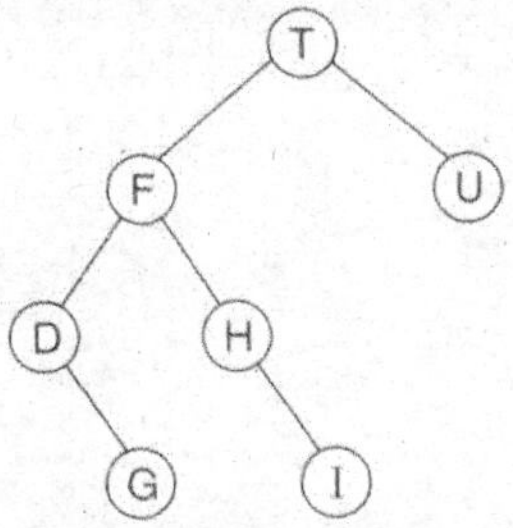

Fig. 45.

Problem 14. *Find a minimum spanning tree of the labelled connected graph shown in Fig. 46.*

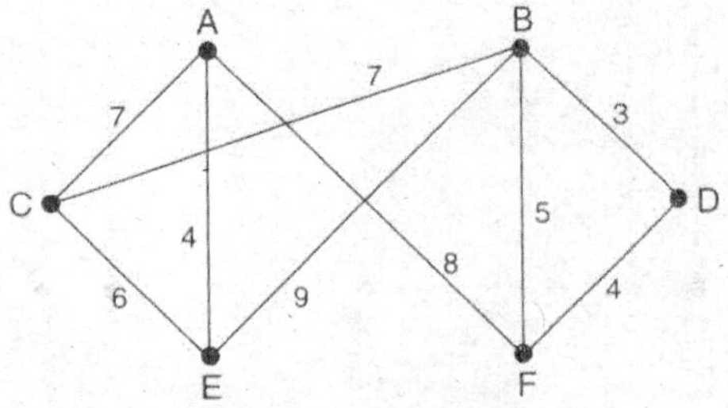

Fig. 46.

Sol. Using KRUSKAL'S ALGORITHM, arrange all the edges of the graph in increasing order and initialize spanning tree with all the vertices of G. Now, add the edges of G in T which do not form a cycle and have minimum weight until $n - 1$ edges are not added, where n is the number of vertices. The spanning tree is shown in Fig. 47.

Edges	Weights	Added or Not
(B, D)	3	Added
(A, E)	4	Added
(D, F)	4	Added
(B, F)	5	Not added
(C, E)	6	Added
(A, C)	7	Not added
(B, C)	7	Added
(A, F)	8	Not added
(E, B)	9	Not added

Minimum Spanning Tree

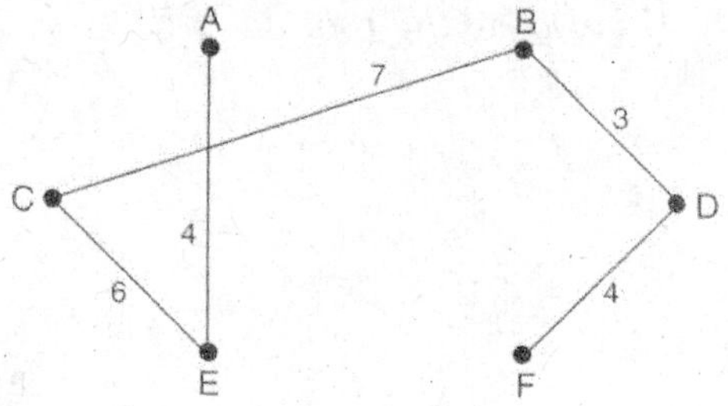

Fig. 47.

The minimum weight of spanning tree is = 24.

Problem 15. *Find all the spanning trees of graph G and find which is the minimal spanning tree of G shown in Fig. 48.*

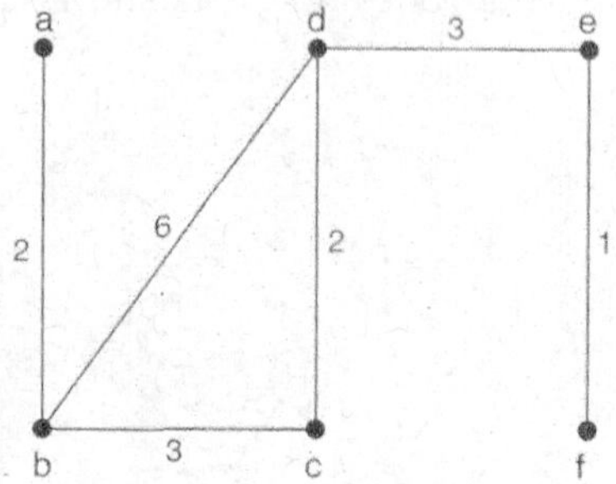

Fig. 48.

Sol. There are total three spanning trees of the graph G which are as shown in Fig. 49.

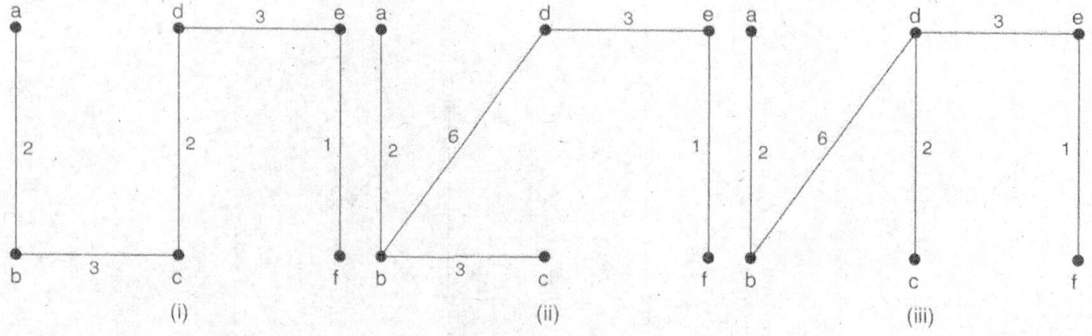

Fig. 49.

To find the minimal spanning tree, use the KRUSKAL'S ALGORITHM. The minimal spanning tree is shown in Fig. 50.

Edges	Weights	Added or Not
(E, F)	1	Added
(A, B)	2	Added
(C, D)	2	Added
(B, C)	3	Added
(D, E)	3	Added
(B, D)	6	Not added.

Minimal Spanning Tree

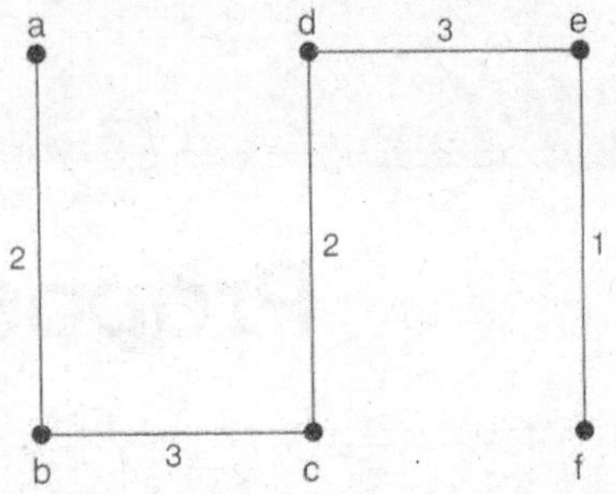

Fig. 50.

The first one is the minimal spanning having the minimum weight = 11.

8

Propositional Calculus

PROPOSITION

A proposition is a statement which is either true or false. It is a declarative sentence.

For example : The following statements are all propositions :

(*i*) Jawahar Lal Nehru is the first prime minister of India.

(*ii*) It rained yesterday.

(*iii*) If x is an integer, then x^2 is a + ve integer.

For example : The following statements are not propositions :

(*i*) Please report at 11 a.m. sharp

(*ii*) What is your name ?

(*iii*) $x^2 = 13$.

PROPOSITIONAL VARIABLES

The lower case letters starting from P onwards are used to represent propositions *e.g.*,

p : India is in Asia

$q : 2 + 2 = 4$.

Example 1. *Classify the following statements as propositions or non-propositions.*

(*i*) *The population of India goes upto 100 million in year 2000.*

(*ii*) $x + y = 30$

(*iii*) *Come here*

(*iv*) *The Intel Pentium-III is a 64-bit computer.*

Sol. (*i*) Proposition (*ii*) Not a proposition

(*iii*) Not a proposition (*iv*) Proposition.

COMBINATION OF PROPOSITIONS

We can combine the propositions to produce new propositions. There are three fundamental and three derived connectors to combine the propositions. These are explained as follows one by one.

(*a*) **Fundamental Connectors**

1. **Conjunction.** It means ANDing of two statements. Assume p and q be two propositions. Conjunction of p and q to be a proposition which is true when both p and q are true, otherwise false. It is denoted by $p \wedge q$. (Fig. 1)

Truth tables are used to determine the truth or falsity of the combined proposition.

p	q	$p \wedge q$
T	T	T
T	F	F
F	T	F
F	F	F

Fig. 1. Truth Table of $p \wedge q$.

2. **Disjunction.** It means ORing of two statements. Assume p and q be two propositions. Disjunction of p and q to be a proposition which is true when either one or both p and q are true and is false when both p and q are false. It is denoted by $p \vee q$. (Fig. 2)

p	q	$p \vee q$
T	T	T
T	F	T
F	T	T
F	F	F

Fig. 2. Truth Table of $p \vee q$.

3. **Negation.** It means opposite of original statement. Assume p be a proposition. Negation of p to be a proposition which is true when p is false, and is false when p is true. It is denoted by $\sim p$. (Fig. 3)

p	$\sim p$
T	F
F	T

Fig. 3. Truth Table of $\sim p$.

Example 2. *Consider the following :*

p : He is rich

q : He is Generous.

Write the proposition which combines the proposition p and q using conjunction ($\wedge$), disjunction ($\vee$), and negation ($\sim$).

Sol. Conjunction. He is rich and generous *i.e.*, $p \wedge q$.

Disjunction. He is rich or generous *i.e.*, $p \vee q$.

Negation. He is not rich *i.e.*, $\sim p$

He is not generous *i.e.*, $\sim q$.

It is false that he is rich or generous *i.e.*, $\sim (p \vee q)$.

He is neither rich nor generous *i.e.*, $\sim p \wedge \sim q$.

It is false that he is not rich *i.e.*, $\sim (\sim p)$.

Example 3. *Let p be "It is hot day" and q be "The temperature is 45°C". Write in simple sentences the meaning of following :*

(*i*) $\sim p$ (*ii*) $\sim (p \vee q)$ (*iii*) $\sim (p \wedge q)$

(*iv*) $\sim (\sim p)$ (*v*) $p \vee q$ (*vi*) $p \wedge q$

(*vii*) $\sim p \wedge \sim q$ (*viii*) $\sim (\sim p \vee \sim q)$.

Sol. (*i*) It is not a hot day.

(*ii*) It is false that it is hot day or temperature is 45°C.

(*iii*) It is not true that it is hot day and temperature is 45°C.

(*iv*) It is false that it is not a hot day.

(*v*) It is hot day or temperature is 45°C.

(*vi*) It is hot day and temperature is 45°C.

(*vii*) It is neither a hot day nor temperature is 45°C.

(*viii*) It is false that it is not a hot day or temperature is not 45°C.

Example 4. *Consider the following statements :*

p : He is coward.

q : He is lazy.

r : He is rich.

Write the following compound statements in the symbolic form.

(*i*) *He is either coward or poor.*

(*ii*) *He is neither coward nor lazy.*

(*iii*) *It is false that he is coward but not lazy.*

(*iv*) *He is coward or lazy but not rich.*

(*v*) *It is false that he is coward or lazy but not rich.*

(*vi*) *It is not true that he is not rich.*

(*vii*) *He is rich or else he is both coward and lazy.*

Sol. (*i*) $p \wedge \sim r$ (*ii*) $\sim p \wedge \sim q$ (*iii*) $\sim (p \wedge \sim q)$

(*iv*) $(p \vee q) \wedge \sim r$ (*v*) $\sim((p \vee q) \wedge \sim r)$ (*vi*) $\sim (\sim r)$

(*vii*) $r \vee (p \wedge q)$.

(*b*) **Derived Connectors**

1. **NAND.** It means negation after ANDing of two statements. Assume p and q be two propositions. Nanding of p and q to be a proposition which is false when both p and q are true, otherwise true. It is denoted by $p \uparrow q$. (Fig. 4)

p	q	$p \uparrow q$
T	T	F
T	F	T
F	T	T
F	F	T

Fig. 4.

2. **NOR or Joint Denial.** It means negation after ORing of two statements. Assume p and q be two propositions. NOring of p and q to be a proposition which is true when both p and q are false, otherwise false. It is denoted by $p \downarrow q$. (Fig. 5)

p	q	$p \downarrow q$
T	T	F
T	F	F
F	T	F
F	F	T

Fig. 5.

3. **XOR.** Assume p and q be two propositions. XORing of p and q is true if p is true or if q is true but not both and vice-versa. It is denoted by $p \oplus q$. (Fig. 6)

p	q	$p \oplus q$
T	T	F
T	F	T
F	T	T
F	F	F

Fig. 6.

Example 5. *Generate the truth table for following :*
(*i*) $A \oplus B \oplus C$ (*ii*) $A \uparrow B \uparrow C$.

Sol. The -1 truth table for above formulas are as shown in Figs. 7 and 8.

(*i*)

A	B	C	A ⊕ B	A ⊕ B ⊕ C
T	T	T	F	T
T	T	F	F	F
T	F	T	T	F
T	F	F	T	T
F	T	T	T	F
F	T	F	T	T
F	F	T	F	T
F	F	F	F	F

Fig. 7.

(*ii*) Truth table for (*ii*) is

A	B	C	A ↑ B	A ↑ B ↑ C
T	T	T	F	T
T	T	F	F	T
T	F	T	T	F
T	F	F	T	T
F	T	T	T	F
F	T	F	T	T
F	F	T	T	F
F	F	F	T	T

Fig. 8.

Example 6. *Prove that* $X \oplus Y \equiv (X \wedge \sim Y) \vee (\sim X \wedge Y)$.

Sol. Construct the truth table for both the propositions. (Fig. 9)

X	Y	X ⊕ Y	~ Y	~ X	X ∧ ~ Y	~ X ∧ Y	(X ∧ ~ Y) ∨ (~ X ∧ Y)
T	T	F	F	F	F	F	F
T	F	T	T	F	T	F	T
F	T	T	F	T	F	T	T
F	F	F	T	T	F	F	F

Fig. 9.

As the truth table for both the proposition are same.

$$X \oplus Y \equiv (X \wedge \sim Y) \vee (\sim X \wedge Y). \text{ Hence proved.}$$

Example 7. *Show that* $(p \oplus q) \vee (p \downarrow q)$ *is equivalent to* $p \uparrow q$.

Sol. Construct the truth table for both the propositions.

p	q	$(p \oplus q)$	$(p \downarrow q)$	$(p \oplus q) \vee (p \downarrow q)$	$p \uparrow q$
T	T	F	F	F	F
T	F	T	F	T	T
F	T	T	F	T	T
F	F	F	T	T	T

Fig. 10.

Since the values of $(p \oplus q) \vee (p \downarrow q)$ is same as $p \uparrow q$ as in Fig. 10. Hence, they are equivalent.

Example 8. *Show that* $(p \uparrow q) \oplus (p \uparrow q)$ *is equivalent to* $(p \vee q) \wedge (p \downarrow q)$.

Sol. Construct the truth table for both the propositions

p	q	$p \uparrow q$	$(p \uparrow q) \oplus (p \uparrow q)$	$p \vee q$	$p \downarrow q$	$(p \vee q) \wedge (p \downarrow q)$
T	T	F	F	T	F	F
T	F	T	F	T	F	F
F	T	T	F	T	F	F
F	F	T	F	F	T	F

Fig. 11.

Since, the values of $(p \uparrow q) \oplus (p \uparrow q)$ and $(p \vee q) \wedge (p \downarrow q)$ are same as in Fig. 11. Hence, they are equivalent.

(c) Some Other Connectors

1. **Conditional.** Statements of the form "If p then q" are called conditional statements. It is denoted as $p \to q$ and read as "p implies q" or "q is necessary for p" or "p is sufficient for q".

Conditional statement is true if both p and q are true or if p is false. It is false if p is true and q is false. The proposition p is called hypothesis and the proposition q is called conclusion. The truth table of conditional statement is

p	q	$p \to q$
T	T	T
T	F	F
F	T	T
F	F	T

Fig. 12. Truth Table of $p \to q$.

For example : The followings are conditional statements :

1. If $a = b$ and $b = c$, then $a = c$.
2. If I will get money, then I will purchase computer.

VARIATIONS IN CONDITIONAL STATEMENT

Contrapositive. The proposition $\sim q \to \sim p$ is called contrapositive of $p \to q$.

Converse. The proposition $q \to p$ is called the converse of $p \to q$.

Inverse. The proposition $\sim p \to \sim q$ is called the inverse of $p \to q$.

Example 9. *Show that $p \to q$ and its contrapositive $\sim q \to \sim p$ are logically equivalent.*

Sol. Construct truth table for both the propositions. (as in Fig. 13)

p	q	$\sim p$	$\sim q$	$p \to q$	$\sim q \to \sim p$
T	T	F	F	T	T
T	F	F	T	F	F
F	T	T	F	T	T
F	F	T	T	T	T

Fig. 13.

As, the values in both cases are same, hence both propositions are equivalent.

Example 10. *Show that proposition $q \to p$ and $\sim p \to \sim q$ is not equivalent to $p \to q$.*

Sol. Construct truth table for all the above propositions :

p	q	$\sim p$	$\sim q$	$p \to q$	$q \to p$	$\sim p \to \sim q$
T	T	F	F	T	T	T
T	F	F	T	F	T	T
F	T	T	F	T	F	F
F	F	T	T	T	T	T

Fig. 14.

As the values of $p \to q$ in table is not equal to $q \to p$ and $\sim p \to \sim q$ as in Fig. 14. So both of them are not equal to $p \to q$ but they are themselves logically equivalent.

Example 11. *Prove that the following propositions are equivalent to $p \to q$.*

(*i*) $\sim (p \wedge \sim q)$ (*ii*) $\sim p \vee q$ (*iii*) $\sim q \to \sim p$.

Sol. Construct the truth table for all the above propositions :

p	q	$\sim p$	$\sim q$	$\sim p \vee q$	$\sim q \to \sim p$	$(p \wedge \sim q)$	$\sim (p \wedge \sim q)$	$p \to q$
T	T	F	F	T	T	F	T	T
T	F	F	T	F	F	T	F	F
F	T	T	F	T	T	F	T	T
F	F	T	T	T	T	F	T	T

Fig. 15.

In the above table (Fig. 15) the values of $p \to q$ is equivalent to (*i*), (*ii*) and (*iii*), hence they are equivalent to $p \to q$. Hence proved.

2. **Biconditional.** Statements of the form "if and only if " are called biconditional statements.

It is denoted as $p \leftrightarrow q$ and read as "p if and only if q". The proposition $p \leftrightarrow q$ is true if p and q have the same truth values and is false if p and q do not have the same truth values. The name of biconditional comes from the fact that $p \leftrightarrow q$ is equivalent to $(p \to q) \wedge (q \to p)$.

The truth table of $p \leftrightarrow q$ is

p	q	$p \leftrightarrow q$
T	T	T
T	F	F
F	T	F
F	F	T

Fig. 16. Truth Table of $p \leftrightarrow q$.

For example : (*i*) Two lines are parallel if and only if they have same slope.

(*ii*) You will pass the exam if and only if you will work hard.

Example 12. *Prove that $p \leftrightarrow q$ is equivalent to $(p \to q) \wedge (q \to p)$.*

Sol. Construct the truth tables of both propositions :

p	q	$p \leftrightarrow q$
T	T	T
T	F	F
F	T	F
F	F	T

Fig. 17.

p	q	$p \to q$	$q \to p$	$(p \to q) \wedge (q \to p)$
T	T	T	T	T
T	F	F	T	F
F	T	T	F	F
F	F	T	T	T

Fig. 18.

Since, the truth tables are same, hence they are logically equivalent. (Fig. 17 and Fig. 18). Hence proved.

PRINCIPLE OF DUALITY

Two formulas A_1 and A_2 are said to be duals of each other if either one can be obtained from the other by replacing $\wedge$ (AND) by $\vee$ (OR) and $\vee$ (OR) by $\wedge$ (AND). Also if the formula contains T (True) or F (False), then we replace T by F and F by T to obtain the dual.

Note 1. The two connectives $\wedge$ and $\vee$ are called dual of each other.

2. Like AND and OR, $\uparrow$ (NAND) and $\downarrow$ (NOR) are dual of each other.

3. If any formula of proposition is valid, then its dual is also a valid formula.

Example 13. *Determine the dual of each of the following :*

(*a*) $p \wedge (q \wedge r)$ (*b*) $\sim p \vee \sim q$ (*c*) $(p \wedge \sim q) \vee (\sim p \wedge q)$

(*d*) $(p \uparrow q) \uparrow (p \uparrow q)$ (*e*) $((\sim p \vee q) \wedge (q \wedge \sim s)) \vee (p \vee F)$, *here F means false.*

Sol. To obtain the dual of all the above formulas, replace $\wedge$ by $\vee$ and $\vee$ by $\wedge$, and also replace T by F and F by T.

Also replace $\uparrow$ by $\downarrow$ and vice-versa.

(*a*) $p \wedge (q \wedge r) = p \vee (q \vee r)$ (*b*) $\sim p \vee \sim q = \sim p \wedge \sim q$

(*c*) $(p \wedge \sim q) \vee (\sim p \wedge q) = (p \vee \sim q) \wedge (\sim p \vee q)$

(*d*) $(p \downarrow q) \downarrow (p \downarrow q)$

(*e*) $((\sim p \vee q) \wedge (q \wedge \sim s)) \vee (p \vee \text{F}) = ((\sim p \wedge q) \vee (q \vee \sim s)) \wedge (p \wedge \text{T})$

EQUIVALENCE OF PROPOSITIONS

Two propositions are said to be logically equivalent if they have exactly the same truth values under all circumstances. The table 1 contains the fundamental logical equivalent expressions :

Table 1.

1. **De Morgan's Laws**
   $\sim(p \wedge q) \equiv \sim p \vee \sim q$
   $\sim(p \vee q) \equiv \sim p \wedge \sim q$
2. **Commutative Properties**
   $p \vee q \equiv q \vee p$; $p \wedge q \equiv q \wedge p$.
3. **Associative Properties**
   $(p \vee q) \vee r \equiv p \vee (q \vee r)$
   $(p \wedge q) \wedge r \equiv p \wedge (q \wedge r)$
4. **Distributive Properties**
   $p \wedge (q \vee r) \equiv (p \wedge q) \vee (p \wedge r)$
   $p \vee (q \wedge r) \equiv (p \vee q) \wedge (p \vee r)$
5. **Impotent Laws**
   $p \vee p \equiv p$ and $p \wedge p \equiv p$
6. **Complement Properties**
   $p \equiv \sim\sim p$
7. **Transposition**
   $(p \to q) \equiv (\sim q \to \sim p)$
8. **Material Implication**
   $(p \to q) \equiv (\sim p \vee q)$
9. **Material Equivalence**
   $(p \leftrightarrow q) \equiv [(p \to q) \wedge (q \to p)]$
   $(p \leftrightarrow q) \equiv [(p \to q) \vee (\sim p \wedge \sim q)]$
10. **Exportation**
   $[(p \wedge q) \to r] \equiv [p \to (q \to r)]$

Example 14. *Consider the following propositions*

$$\sim p \vee \sim q \text{ and } \sim(p \wedge q).$$

Are they equivalent ?

Sol. Construct the truth table for both (as in Fig. 19).

p	q	$\sim p$	$\sim q$	$\sim p \vee \sim q$	$p \wedge q$	$\sim(p \wedge q)$
T	T	F	F	F	T	F
T	F	F	T	T	F	T
F	T	T	F	T	F	T
F	F	T	T	T	F	T

Fig. 19.

Since, the final values of both the propositions are same, hence the two propositions are equivalent.

TAUTOLOGIES

A proposition P is a tautology if it is true under all circumstances. It means it contains only T in the final column of its truth table.

Example 15. *Prove that the statement* $(p \to q) \leftrightarrow (\sim q \to \sim p)$ *is a tautology.*

Sol. Make the truth table of above statement :

p	q	$p \to q$	$\sim q$	$\sim p$	$\sim q \to \sim p$	$(p \to q) \leftrightarrow (\sim q \to \sim p)$
T	T	T	F	F	T	T
T	F	F	T	F	F	T
F	T	T	F	T	T	T
F	F	T	T	T	T	T

Fig. 20.

As the final column contains all T's, so it is a tautology. (Fig. 20)

Example 16. *Prove that* $p \vee p \leftrightarrow p$ *is a tautology.*

Sol. Construct the truth table of the given statement :

p	$p \vee p$	$p \vee p \leftrightarrow p$
T	T	T
F	F	T

Fig. 21.

As the last column contains all T's, so it is a tautology. (Fig. 21)

CONTRADICTION

A statement that is always false is called a contradiction.

Example 17. *Show that the statement* $p \wedge \sim p$ *is a contradiction.*

p	$\sim p$	$p \wedge \sim p$
T	F	F
F	T	F

Fig. 22.

Sol. Construct the truth table of above statement.

Since, the last column contains all F's, so it is a contradiction. (Fig. 22)

CONTINGENCY

A statement that can be either true or false depending on the truth values of its variables, is called a contingency.

p	q	$p \rightarrow q$	$p \wedge q$	$(p \rightarrow q) \rightarrow (p \wedge q)$
T	T	T	T	T
T	F	F	F	T
F	T	T	F	F
F	F	T	F	F

Fig. 23.

Example 18. *Prove that the statement* $(p \rightarrow q) \rightarrow (p \wedge q)$ *is a contingency.*

Sol. Construct the truth table of above statement.

As, the value of final column depends on the truth value of the variables, so it is a contingency. (Fig. 23)

Example 19. *From the following formulae, find out tautology, contingency and contradiction.*

(*i*) $A \equiv A \wedge (A \vee B)$ (*ii*) $(p \wedge \sim q) \vee (\sim p \wedge q)$ (*iii*) $\sim (p \vee q) \vee (\sim p \vee \sim q)$.

Sol. (*i*) Construct the truth table for $A \to A \wedge (A \vee B)$.

A	B	$A \vee B$	$A \wedge (A \vee B)$	$A \to A \wedge (A \vee B)$
T	T	T	T	T
T	F	T	T	T
F	T	T	F	T
F	F	F	F	T

Fig. 24.

Since, the last column of the table contains all T's, hence it is a tautology. (Fig. 24)

(*ii*) Construct the truth table for $(p \wedge \sim q) \vee (\sim p \wedge q)$ as in Fig. 25.

p	q	$\sim p$	$\sim q$	$p \wedge \sim q$	$\sim p \wedge q$	$(p \wedge \sim q) \vee (\sim p \wedge q)$
T	T	F	F	F	F	F
T	F	F	T	T	F	T
F	T	T	F	F	T	T
F	F	T	T	F	F	F

Fig. 25.

Since, the value of the final column depends on the value of the different variables, hence it is a contingency.

(*iii*) Construct the truth table of the proposition $\sim (p \vee q) \vee (\sim p \vee \sim q)$ as in Fig. 26.

p	q	$\sim p$	$\sim q$	$p \wedge q$	$\sim (p \wedge q)$	$\sim p \vee \sim q$	$\sim (p \vee q) \vee (\sim p \vee \sim q)$
T	T	F	F	T	F	F	F
T	F	F	T	F	T	T	T
F	T	T	F	F	T	T	T
F	F	T	T	F	T	T	T

Fig. 26.

Since, the value of final column depends upon the value of different variables, hence it is a contingency.

Example 20. *Verify that proposition* $p \vee \sim (p \wedge q)$ *is tautology.*

Sol. Construct the truth table for above proposition. (Fig. 27)

p	q	$p \wedge q$	$\sim(p \wedge q)$	$p \vee \sim (p \wedge q)$
T	T	T	F	T
T	F	F	T	T
F	T	F	T	T
F	F	F	T	T

Fig. 27.

Since, the last column contains all T' s, hence it is a tautology.

Example 21. *Determine whether the following is a tautology, contingency and a contradiction :*

(*i*) $p \to (p \to q)$ (*ii*) $p \to (q \to p)$ (*iii*) $p \wedge \sim p$.

Sol. (*i*) Construct truth table for $p \to (p \to q)$ as in Fig. 28.

p	q	$p \to q$	$p \to (p \to q)$
T	T	T	T
T	F	F	F
F	T	T	T
F	F	T	T

Fig. 28.

Since, the value of last column depends on the value of different variables, hence it is a contingency.

(*ii*) Construct truth table for $p \to (q \to p)$ as in Fig. 29.

p	q	$q \to p$	$p \to (q \to p)$
T	T	T	T
T	F	T	T
F	T	F	T
F	F	T	T

Fig. 29.

Since, the last column contains all T's, hence it is a tautology.

(*iii*) Construct truth table for $p \wedge \sim p$ as in Fig. 30.

Since, the last column contain all F's, hence it is a contradiction.

p	$\sim p$	$p \wedge \sim p$
T	F	F
F	T	F

Fig. 30.

FUNCTIONALLY COMPLETE SETS OF CONNECTIVES

We have three basic and two conditional connectives *i.e.*, $\wedge$, $\vee$, $\sim$ $\Rightarrow$ and $\Leftrightarrow$. If we have given any formula containing all these connectives, we can write an equivalent formula with certain proper subsets of these connectives.

A set of connectives is called functionally complete if every formula can be expresses on terms of an equivalent formula containing the connectives from this set.

Example 22. *Write an equivalent formula for* $P \wedge (R \Leftrightarrow S) \vee (S \Leftrightarrow P)$ *which does not involve biconditional.*

Sol. We know that $\quad P \Leftrightarrow Q \equiv (P \Rightarrow Q) \wedge (Q \Rightarrow P)$ ...(*i*)

So, apply eqn. (*i*) to formula to obtain the formula equivalent to

$$[P \wedge (R \Leftrightarrow S) \vee (S \Leftrightarrow P)],$$

which does not involve biconditional.

$$= [P \wedge ((R \Rightarrow S) \wedge (S \Rightarrow R)) \vee ((S \Rightarrow P) \wedge (P \Rightarrow S))].$$

Example 23. *Write an equivalent formula for* $R \vee (S \Leftrightarrow T)$, *which does not involve biconditional as well as conditional.*

Sol. We know that

$$(P \Leftrightarrow Q) = (P \Rightarrow Q) \wedge (Q \Rightarrow P) \quad ...(i)$$

$$(P \Rightarrow Q) = \sim P \wedge Q \quad ...(ii)$$

So, applying the eqn. (*i*) and (*ii*) on the above formula, we can obtain an equivalent formula, which does not involve biconditional as well as conditional.

$$[R \vee (S \Leftrightarrow T)] = [R \vee ((S \Rightarrow T) \vee (T \Rightarrow S))]$$
$$= [R \vee ((\sim S \vee T) \vee (\sim T \vee S))].$$

Example 24. *Show that* $\{\sim, \wedge\}$ *is functionally complete.*

Sol. Take any formula which involve all the five connectives $\wedge, \vee, \sim, \Rightarrow$ and $\Leftrightarrow$. We can obtain an equivalence formula by first replacing biconditional and then replacing conditional and finally we can replace $\vee$.

As $\quad P \Leftrightarrow Q = (P \Rightarrow Q) \wedge (Q \Rightarrow P)$

$$= (\sim P \vee Q) \wedge (\sim Q \vee P) = (\sim (\sim\sim P \wedge \sim Q)) \wedge (\sim (\sim \sim Q \wedge \sim P))$$

Hence, $\{\sim, \wedge\}$ is functionally complete.

Similarly, we can show that $(\sim, \vee\}$ is functionally complete.

Example 25. *Show that* $\{\sim, \rightarrow\}$ *is functionally complete.*

Sol. We know that $\quad P \Rightarrow Q \equiv \sim P \vee Q.$

So, we have $\quad P \vee Q \equiv \sim P \Rightarrow Q.$

Since, $\{\sim, \vee\}$ is functionally complete. Hence from above, $\{\sim, \Rightarrow\}$ is also functionally complete.

i.e., given any formula which involve all the five connectives, we can obtain an equivalence formula using $\{\sim, \rightarrow\}$ by first replacing biconditional and then replacing $(\wedge)$ ANDing and finally replacing $\vee$.

Example 26. *Express* $P \Leftrightarrow Q$ *in terms of* $\{\sim, \wedge\}$ *only.*

Sol. $(P \Leftrightarrow Q) = (P \Rightarrow Q) \wedge (Q \Rightarrow P) \quad \because \quad (P \Leftrightarrow Q) = (P \Rightarrow Q) \wedge (Q \Rightarrow P)$

$= (\sim P \vee Q) \wedge (\sim Q \vee P) \quad \because \quad P \Rightarrow Q = \sim P \vee Q$

$= (\sim (\sim \sim P \wedge \sim Q)) \wedge (\sim (\sim \sim Q \wedge \sim P)) \quad \because \quad P \vee Q = \sim (\sim P \wedge \sim Q).$

Example 27. *Express* $(P \wedge \sim Q) \vee (\sim P \wedge \sim Q)$ *in terms of* $(\sim, \vee)$ *only.*

Sol. $(P \wedge \sim Q) \vee (\sim P \wedge \sim Q) = (\sim (\sim P \vee \sim \sim Q)) \vee (\sim (\sim \sim P \vee \sim \sim Q)) \because P \wedge Q = \sim (\sim P \vee \sim Q)$

ARGUMENT

An argument is an assertion ; that a group of propositions called premises, yields another proposition, called the conclusion. Let $P_1, P_2, P_3, ..., P_n$ is the group of propositions that yields the conclusion Q. Then, it is denoted as $P_1, P_2, P_n, ..., P_n \mid\!- Q$.

Conclusion. The conclusion of an argument is the proposition that is asserted on the basis of other proposition of the argument.

Premises. The propositions, which are assumed for accepting the conclusion, are called the premises of that argument.

(a) Valid Argument

An argument is called valid argument if the conclusion is true whenever all the premises are true.

The argument is also valid if and only if the ANDing of the group of propositions implies conclusion is a tautology *i.e.*, $P(p_1, p_2, p_3, ..., p_n) \to Q$ is a tautology. Where $P(p_1, p_2, p_3, ..., p_n)$ is the group of propositions and Q is the conclusion.

Some common valid argument forms are given in Table 2.

(b) Falacy Argument

An argument is called falacy or an invalid argument if it is not a valid argument.

Elementary Valid Argument Forms

Table 2.

1. **Modus ponens**

$p \to q$

p

$\therefore\ q$

2. **Modus tollens**

$p \to q$

$\sim q$

$\therefore\ \sim p$

3. **Hypothetical syllogism**

$p \to q$

$q \to r$

$\therefore\ p \to r$

4. **Disjunctive syllogism**

$p \vee q$

$\sim p$

$\therefore\ q$

5. **Constructive dilemma**

$(p \to q) \wedge (r \to s)$

$p \vee r$

$\therefore\ q \vee s$

6. **Absorption**

$p \to q$

$\therefore\ p \to (p \wedge q)$

7. **Simplification**

$p \wedge q$

$\therefore\ p$

8. **Conjunction**

p

q

$\therefore\ p \wedge q$

9. **Addition**

p

$\therefore\ p \vee q.$

Example 28. *Show that the following rule is valid :*

$$p \mid\!- p \vee q \quad \text{or} \quad p \ \therefore\ p \vee q.$$

Sol. We can prove this rule from the truth table (Fig. 31)

p	q	$p \vee q$
T	T	T
T	F	T
F	T	T
F	F	F

Fig. 31.

p is true in line 1 and 2 and $p \vee q$ is also true in line 1 and 2. Hence, argument is valid.

Example 29. *Show that the rule modus ponens is valid.*

$p \to q$

p

$\therefore \quad q$

Sol. The truth table of this rule is as follows : (Fig. 32)

p	q	$p \to q$
T	T	T
T	F	F
F	T	T
F	F	T

Fig. 32.

The p is the true in line 1 and 2 and $p \to q$ and p both are true in line 1 and $p, p \to q$ and q all are true in line 1. Hence, argument is valid.

Example 30. *Show that the rule of hypothetical syllogism is valid*

$p \to q$

$q \to r$

$\therefore \quad p \to r.$

Sol. The truth table of above rule is as in Fig. 33.

p	q	r	$p \to q$	$q \to r$	$p \to r$
T	T	T	T	T	T
T	T	F	T	F	F
T	F	F	F	T	F
T	F	T	F	T	T
F	T	T	T	T	T
F	T	F	T	F	T
F	F	T	T	T	T
F	F	F	T	T	T

Fig. 33.

$p \to q$ is true in lines 1, 2, 5, 6, 7, 8.

$q \to r$ is true in lines 1, 3, 4, 5, 7, 8.

Both $p \to q$ and $q \to r$ is true in lines 1, 5, 7, 8 $p \to r$ is also true in lines 1, 5, 7, 8. Hence, argument is valid.

Example 31. *Show that the rule of Modus Tollens is valid*

$p \to q$

$\sim q$

$\therefore \quad \sim p.$

Sol. The truth table of above propositions are (Fig. 34)

p	q	$\sim p$	$\sim q$	$p \to q$
T	T	F	F	T
T	F	F	T	F
F	T	T	F	T
F	F	T	T	T

Fig. 34.

$p \to q$ is true in line 1, 3 and 4. $\sim q$ is true in line 2 and 4. Both $p \to q$ and $\sim q$ are true in line 4. $\sim p$ is also true in line 4. Hence, the argument is valid.

Example 32. *Show that the rule of disjunctive syllogism is valid*

$$p \vee q$$
$$\sim p$$
$$\therefore \quad q.$$

Sol. The truth table of above rule is as follows : (Fig. 35)

p	q	$\sim p$	$p \vee q$
T	T	F	T
T	F	F	T
F	T	T	T
F	F	T	F

Fig. 35.

$p \vee q$ is true in line 1, 2 and 3. $\sim p$ is true in line 3. Both $p \vee q$ and $\sim p$ is true in line 3. As q is also true in line 3. Hence, argument is valid.

Example 33. *Show that the rule of simplification is valid*

$$p \wedge q$$
$$\therefore \quad p.$$

Sol. The truth table of above argument is as follows : (Fig. 36)

p	q	$p \wedge q$
T	T	T
T	F	F
F	T	F
F	F	F

Fig. 36.

$p \wedge q$ is true in line 1 and p is also true in line 1. Hence, the argument is valid.

Example 34. *Show that the rule of conjunction is valid.*

p

q

$\therefore \quad p \wedge q.$

Sol. The argument is valid if $p \wedge q \rightarrow p \wedge q$ is a tautology. The truth table for the above proposition is as follows : (Fig. 37)

p	q	$p \wedge q$	$p \wedge q \rightarrow p \wedge q$
T	T	T	T
T	F	F	T
F	T	F	T
F	F	F	T

Fig. 37.

As the proposition is a tautology. Hence, the argument is valid.

Example 35. *Show that the rule of absorption is valid*

$p \rightarrow q$

$\therefore \quad p \rightarrow (p \wedge q).$

Sol. We have to show that $(p \rightarrow q) \rightarrow [p \rightarrow (p \wedge q)]$ is tautology. The truth table of the above argument is as follows : (Fig. 38)

p	q	$p \wedge q$	$p \rightarrow q$	$p \rightarrow (p \wedge q)$	$(p \rightarrow q) \rightarrow [p \rightarrow (p \wedge q)]$
T	T	T	T	T	T
T	F	F	F	F	T
F	T	F	T	T	T
F	F	F	T	T	T

Fig. 38.

Since, the argument is a tautology. Hence, it is a valid argument.

PROOF OF VALIDITY

We can test the validity of any argument by constructing the truth tables. But as the no. of variable statements increases, the truth tables grow unwieldly. So, a more efficient method to test the validity of the argument is to deduce its conclusion from its premises by a sequence of elementary arguments each of which is known to be valid.

Example 36. *Prove that the argument $p \rightarrow \sim q, r \rightarrow q, r \vdash \sim p$ is valid without using truth tables.*

Sol. (*i*)	$p \rightarrow \sim q$	(Given)
(*ii*)	$r \rightarrow q$	(Given)
(*iii*)	$\sim q \rightarrow \sim r$	Contrapositive of (*ii*)
(*iv*)	$p \rightarrow \sim r$	Hypothetical syllogism using (*i*) and (*iii*)
(*v*)	$r \rightarrow \sim p$	Contrapositive of (*iv*)

(*vi*)	r is true	(Given)
(*vii*)	$\sim p$ is true	Modus ponens using (*v*) and (*vi*).

Example 37. *Prove that the argument* $p \to q, q \to r, r \to s, \sim s, p \vee t \vdash t$ *is valid without using truth tables.*

Sol. (*i*)	$p \to q$	(Given)
(*ii*)	$q \to r$	(Given)
(*iii*)	$r \to s$	(Given)
(*iv*)	$\sim s$	(Given)
(*v*)	$p \vee t$	(Given)
(*vi*)	$p \to r$	Hypothetical syllogism using (*i*) and (*ii*)
(*vii*)	$p \to s$	Hypothetical syllogism using (*vi*) and (*iii*)
(*viii*)	$\sim p$	Modus tollens using (*vii*) and (*iv*)
(*ix*)	t	Disjunctive syllogism using (*v*) and (*viii*).

Example 38. *Prove that the argument* $p, q \vdash (p \vee r) \wedge q$ *is valid without using truth tables.*

Sol. (*i*)	p	(Given)
(*ii*)	$p \vee r$	Rule of addition using (*i*)
(*iii*)	q	(Given)
(*iv*)	$(p \vee r) \wedge q$	Rule of conjunction using (*ii*) and (*iii*).

Example 39. *Prove that the argument* $p \to q, p \wedge r \vdash q$ *is valid without using truth table.*

Sol. (*i*)	$p \to q$	(Given)
(*ii*)	$p \wedge r$	(Given)
(*iii*)	p	Rule of simplification using (*i*)
(*iv*)	q	Modus ponens using (*i*) and (*iii*).

Example 40. *Prove that the argument* $(p \to q) \wedge (r \to s), (p \vee r) \wedge (q \vee r) \vdash q \vee s$

Sol. (*i*)	$(p \to q) \wedge (r \to s)$	(Given)
(*ii*)	$(p \vee r) \wedge (q \vee r)$	(Given)
(*iii*)	$(p \vee r)$	Simplification using (*iii*)
(*iv*)	$q \vee s$	Constructive dilemma using (*i*) and (*iii*).

Example 41. *Prove that the argument* $(p \wedge q) \vee (r \to s), t \to r, \sim (p \wedge q) \vdash t \to s$ *is valid without using truth tables.*

Sol. (*i*)	$(p \wedge q) \vee (r \to s)$	(Given)
(*ii*)	$t \to r$	(Given)
(*iii*)	$\sim (p \wedge q)$	(Given)
(*iv*)	$r \to s$	Disjunctive syllogism using (*i*) and (*iii*)
(*v*)	$t \to s$	Hypothetical syllogism using (*ii*) and (*iii*).

Example 42. *Prove that the argument* $(p \to q) \wedge (r \to s), q \to s, (q \to s) \to (p \vee r) \vdash q \vee s$ *is valid using deduction method.*

Sol. (*i*)	$(p \to q) \wedge (r \to s)$	(Given)
(*ii*)	$q \to s$	(Given)
(*iii*)	$(q \to s) \to (p \vee r)$	(Given)
(*iv*)	$p \vee r$	Modus ponens using (*iii*) and (*ii*)
(*v*)	$q \vee s$	Constructive dilemma using (*i*) and (*iv*).

Example 43. *Prove that the argument* $p \to (q \vee r), (s \wedge t) \to q, (q \vee r) \to (s \wedge t) \vdash p \to q$ *is valid without using truth table.*

Sol. (*i*)	$p \to (q \vee r)$	(Given)
(*ii*)	$(s \wedge t) \to q$	(Given)
(*iii*)	$(q \vee r) \to (s \wedge t)$	(Given)
(*iv*)	$p \to (s \wedge t)$	Hypothetical syllogism using (*i*) and (*iii*)
(*v*)	$p \to q$	Hypothetical syllogism using (*ii*) and (*iv*).

Example 44. *Prove that the argument* $p \vee (q \to p), \sim p \wedge r \vdash \sim q$ *is valid without using truth tables.*

Sol. (*i*)	$p \vee (q \to p)$	(Given)
(*ii*)	$\sim p \wedge r$	(Given)
(*iii*)	$\sim p$	Rule of simplification using (*ii*)
(*iv*)	$q \to p$	Disjunctive syllogism using (*i*) and (*iii*)
(*v*)	$\sim q$	Modus tollens using (*iv*) and (*iii*).

Example 45. *Test the validity of following argument. If I will select in IAS examination, then I will not be able to go to London. Since, I am going to London, I will not select in IAS examination.*

Sol. Let p be "I will select in IAS examination" and q be "I am going to London". Then the above argument can be written in symbolic form as follows :

$$p \to \sim q$$
$$q$$
$$\therefore \quad \sim p$$

Construct the truth table for above argument (Fig. 39)

p	q	$\sim p$	$\sim q$	$p \to \sim q$
T	T	F	F	F
T	F	F	T	T
F	T	T	F	T
F	F	T	T	T

Fig. 39.

$p \to \sim q$ is true in line 2, 3 and 4. q is true in line 1 and 4 and $\sim p$ is true in line 3 and 4. Hence, all three are true in line 4. So it is a valid statement.

Example 46. *Consider the following argument and determine whether it is valid.*

Either I will get good marks or I will not graduate. If I did not graduate I will go to Canada. I get good marks. Thus, I would not go to Canada.

Sol. Let p be "I will get good marks" and q be "I will graduate" and r be "I will go to Canada". Thus, the above argument can be written in symbolic form as follows :

$$p \vee \sim q$$
$$\sim q \to r$$
$$p$$
$$\therefore \quad \sim r$$

The truth table of above proposition is (Fig. 40)

p	q	r	$\sim q$	$\sim r$	$p \vee \sim q$	$\sim q \to r$
T	T	T	F	F	T	T
T	T	F	F	T	T	T
T	F	F	T	T	T	F
T	F	T	T	F	T	T
F	T	T	F	F	F	T
F	T	F	F	T	F	T
F	F	T	T	F	T	T
F	F	F	T	T	T	F

Fig. 40.

$p \vee \sim q$ is true in line 1, 2, 3, 4, 7, 8 and $\sim q \to r$ is true in line 1, 2, 4, 5, 6, 7 and p is true in line 1, 2, 3 and 4. $\sim r$ is true in 2, 3, 6 and 8. All the above are true in line 2. Hence, the argument is valid.

Example 47. *Determine the validity of the following argument without using truth tables.*

Either I will pass the examination, or, I will not graduate. If I do not graduate, I will go to Canada. I failed : Thus, I will go to Canada.

Sol. Let p be "I will pass the examination" and q be "I will graduate" and t be "I will go to Canada". Thus the above argument, in symbolic form can be written as

$$p \vee \sim q$$
$$\sim q \to t$$
$$\sim p$$
$$\therefore \quad t$$

Thus to prove the validity of the argument, use the standard results as follows :

(*i*)	$p \vee \sim q$	(Given)
(*ii*)	$\sim q \to t$	(Given)
(*iii*)	$\sim p$	(Given)
(*iv*)	$\sim q$	Disjunctive syllogism using (*i*) and (*iii*)
(*v*)	t	Modus ponens using (*ii*) and (*iv*)

Hence proved.

Example 48. *Determine the validity of the following argument using deduction method.*

If I study, then I will pass examination. If I do not go to picnic, then I will study. But I failed examination. Therefore, I went to picnic.

Sol. Let p be "I study" and q be "I will pass examination" and t be "I go to picnic". Then the above argument is written in symbolic form as follows :

$$p \to q$$
$$\sim t \to p$$
$$\sim p$$
$$\therefore \quad t$$

Thus to prove the validity of the argument use the rules of inference.

(*i*)	$p \to q$	(Given)
(*ii*)	$\sim t \to p$	(Given)
(*iii*)	$\sim p$	(Given)
(*iv*)	$\sim \sim t$	Modus tollens using (*ii*) and (*iii*)
(*v*)	t	Complement property using (*iv*)

Hence proved.

Example 49. *Prove the validity of the following argument using truth tables as well as deduction method.*

"If the market is free then there is no inflation. If there is no inflation then there are price controls. Since there are price controls, therefore, the market is free".

Sol. Let p be "The market is free" and q be "There is inflation" and r be "There are price controls". Then the above argument can be written in symbolic form as follows :

$$p \to \sim q$$
$$\sim q \to r$$
$$r$$
$$\therefore \quad p$$

Ist Method. By using truth tables

Construct the truth table of above argument (Fig. 41)

p	q	r	$\sim q$	$p \to \sim q$	$\sim q \to r$
T	T	T	F	F	T
T	T	F	F	F	T
T	F	F	T	T	F
T	F	T	T	T	T
F	T	T	F	T	T
F	T	F	F	T	T
F	F	T	T	T	T
F	F	F	T	T	F

Fig. 41.

$p \to \sim q$ is true in line 3, 4, 5, 6, 7 and 8 $\sim q \to r$ is true in line 1, 2, 4, 5, 6, 7 r is true in line 1, 4, 5, 7. All the above three are true in line 4 and 5. Also p is true in line 4. Hence the argument is valid.

IInd Method. Using deduction method

(*i*)	$p \to \sim q$	(Given)
(*ii*)	$\sim q \to r$	(Given)
(*iii*)	$p \to r$	Hypothetical syllogism using (*i*) and (*ii*)
(*iv*)	$\sim p \to \sim r$	Transposition using (*iii*)
(*v*)	r	(Given)
(*vi*)	$\sim \sim p$	Modus tollens using (*iv*) and (*v*)
(*vii*)	p	Complement of (*vi*).

EXISTENTIAL QUANTIFIER

If $p(x)$ is a proposition over the universe U. Then it is denoted as $\exists x\, p(x)$ and read as "There exists at least one value in the universe of variable x such that $p(x)$ is true. The quantifier $\exists$ is called the existential quantifier.

There are several ways to write a proposition, with an existential quantifier *i.e.,*

$(\exists x \in A)\, p(x)$ or $\exists x \in A$ such that $p(x)$ or $(\exists x)\, p(x)$ or $p(x)$ is true for some $x \in A$.

UNIVERSAL QUANTIFIER

If $p(x)$ is a proposition over the universe U. Then it is denoted as $\forall\, x, p(x)$ and read as "For every $x \in U$, $p(x)$ is true". The quantifier $\forall$ is called the universal quantifier.

There are several ways to write a proposition, with a universal quantifier.

$\forall x \in A, p(x)$ or $p(x), \forall x \in A$

or $\forall x, p(x)$ or $p(x)$ is true for all $x \in A$.

Example 50. *Let $A(x)$: x has a white colour, $B(x)$: x is a polar bear, $C(x)$: x is found in cold regions, over the universe of animals. Translate the following into simple sentences :*

(*i*) $\exists x\, (B(x) \wedge \sim A(x))$ (*ii*) $(\exists x)\, (\sim C(x))$

(*iii*) $(\forall x)\, (B(x) \wedge C(x) \to A(x))$.

Sol. (*i*) There exists a polar bear whose colour is not white.

(*ii*) There exists an animal that is not found in cold regions.

(*iii*) Every polar bear that is found in cold regions has a white colour.

Example 51. *Let $K(x)$: x is a two-wheeler, $L(x)$: x is a scooter. $M(x)$: x is manufactured by Bajaj. Express the following using quantifiers.*

(*i*) *Every two wheeler is a scooter.*

(*ii*) *There is a two wheeler that is not manufactured by Bajaj.*

(*iii*) *There is a two wheeler manufactured by Bajaj that is not a scooter.*

(*iv*) *Every two wheeler that is a scooter is manufactured by Bajaj.*

Sol. (*i*) $(\forall\, x)\, (K(x) \to L(x))$

(*ii*) $(\exists\, x)\, (K(x) \wedge M(x))$

(*iii*) $(\exists\, x)\, (K(x) \wedge M(x) \to \sim L(x))$

(*iv*) $(\forall\, x)\, (K(x) \wedge L(x) \to M(x))$

NEGATION OF QUANTIFIED PROPOSITIONS

When we negate a quantified proposition *i.e.*, when a universally quantified proposition is negated, we obtain an existentially quantified proposition and when an existentially quantified proposition is negated, we obtain a universally quantified proposition.

The two rules for negation of quantified proposition are as follows. These are also called De Morgan's law.

(*i*) $\sim \exists\, x\, p(x) \equiv \forall\, x \sim p(x)$ (*ii*) $\sim \forall\, x\, p(x) \equiv \exists\, x \sim p(x)$.

Example 52. *Negate each of the following propositions :*

(*i*) *All boys can run faster than all girls.*

(*ii*) *Some girls are more intelligent than all boys.*

(*iii*) *Some students do not live in hostel.*

(*iv*) *All students pass the semester exams.*

(*v*) *Some of the students are absent and the classroom's empty.*

Sol. (*i*) Some boys can run faster than some girls.

(*ii*) All girls are more intelligent than some boys.

(*iii*) All students live in hostel.

(*iv*) Some students do not pass the semester exams.

(*v*) All students are present and the class-room is full.

Example 53. *Negate each of the following propositions :*

(*i*) $\forall x\, P(x) \wedge \exists\, y\, q(y)$ (*ii*) $\forall\, x\, p(x) \wedge \forall\, y\, q(y)$

(*iii*) $\exists\, x\, p(x) \vee \forall\, y\, q(y)$ (*iv*) $\exists\, x\, p(x) \vee \exists\, y\, q(y)$

(*v*) $(\exists\, x \in U)\,(x + 6 = 25)$ (*vi*) $(\forall x \in U)\,(x < 25)$.

Sol. (*i*)
$$\sim (\forall\, x\, p(x) \wedge \exists\, y\, q(y))$$
$$\equiv \sim \forall\, x\, p(x) \vee \sim \exists\, y\, q(y) \qquad (\because\ \sim (p \wedge q) = \sim p \vee \sim q)$$
$$\equiv \exists\, x \sim p(x) \vee \forall\, y \sim q(y)$$

(*ii*)
$$\sim (\forall\, x\, p(x) \wedge \forall\, y\, q(y))$$
$$\equiv \sim \forall\, x\, p(x) \vee \sim \forall\, y\, q(y) \qquad (\because\ \sim (p \wedge q) = \sim p \sim q)$$
$$\equiv \exists\, x \sim p(x) \vee \exists\, y \sim q(y)$$

(*iii*)
$$\sim (\exists\, x\, p(x) \vee \forall\, y\, q(y))$$
$$\equiv \sim \exists\, x\, p(x) \wedge \sim \forall\, y\, q(y) \qquad (\because\ \sim (p \vee q) = \sim p \wedge \sim q)$$
$$\equiv \forall\, x \sim p(x) \wedge \exists\, y \sim q(y)$$

(*iv*)
$$\sim (\exists\, x\, p(x) \vee \exists\, y\, q(y))$$
$$\equiv \sim \exists\, x\, p(x) \wedge \sim \exists\, y\, q(y) \qquad (\because\ \sim (p \vee q) = \sim p \wedge \sim q)$$
$$\equiv \forall\, x \sim p(x) \wedge \forall\, y \sim q(y)$$

(*v*)
$$\sim (\exists\, x \in \mathrm{U})\,(x + 6) = 25$$
$$\equiv \forall x \in \mathrm{U} \sim (x + 6) = 25$$
$$\equiv (\forall x \in \mathrm{U})\,(x + 6) \neq 25$$

(*vi*)
$$\sim (\forall\, x \in \mathrm{U})\,(x < 25)$$
$$\equiv \exists\, x \in \mathrm{U} \sim (x < 25)$$
$$\equiv (\exists\, x \in \mathrm{U})\,(x \geq 25)$$

PROPOSITIONS WITH MULTIPLE QUANTIFIERS

The proposition having more than one variable can be quantified with multiple quantifiers. The multiple universal quantifiers can be arranged in any order without altering the meaning of the resulting proposition. Also the multiple existential quantifiers can be arranged in any order without altering the meaning of proposition.

The proposition which contains both universal and existential quantifiers, the order of these quantifiers can't be exchanged without altering the meaning of proposition *e.g.*, the proposition $\exists x \forall y\, p(x, y)$ means " There exists some x such that $p(x, y)$ is true for every y".

Example 54. *Write the negation for each of the following. Determine whether the resulting statement is true or false. Assume U = R.*

(i) $\forall x \exists m(x^2 < m)$ *(ii)* $\exists m \forall x(x^2 < m)$.

Sol. *(i)* Negation of $\forall x \exists m(x^2 < m)$ is $\exists x \forall m\ (x^2 \geq m)$. The meaning of $\exists x \forall m(x^2 \geq m)$ is that there exists some x such that $x^2 \geq m$, for every m. The statement is true as there is some greatest x such that $x^2 \geq m$, for every m.

(ii) Negation of $\exists m \forall x\ (x^2 < m)$ is $\forall m \exists x\ (x^2 \geq m)$. The meaning of $\forall m \exists x\ (x^2 \geq m)$ is that for every m, there exists some x such that $(x^2 \geq m)$. The statement is true as for every m, there exists some greatest x such that $x^2 \geq m$.

Example 55. *Check the validity of following formula under given interpretation.*

(a) $\forall x \exists y\, P(x, y)$ *under interpretation domain* = {1, 2} *and* P(1, 1) = T, P(1, 2) = F, P(2, 1) = T, P(2, 2) = T *where T and F refer to true and false respectively.*

(b) $\forall x\ (P(x) \rightarrow Q(a, f(x))$ *under interpretation domain* = {1, 2} *and* a = 1, f(1) = 2, f(2) = 1, P(1) = F, P(2) = T, Q(1, 1) = T, Q(1, 2) = T, Q(2, 1) = F *and* Q(2, 2) = T. *Here T and F refer to true and false respectively.*

Sol. *(a)* $\forall x \exists y$ P(x, y) means for every x, there exists some y such that P(x, y) is true.

Since P(1, 1) = T and P(2, 1) = T. Hence, the formula is valid under the domain = {1, 2}.

(b) $\forall x$ (P(x) $\rightarrow$ Q(a, F(x)) means for every x, whenever P(x) is true implies Q(a, $f(x)$) is also true. So, under the domain {1, 2}, the different values of $\forall x$ (P(x) $\rightarrow$ Q(a, $f(x)$)) are as follows :

x	P(x)	Q(a, $f(x)$)
1	F	T
2	T	T

So, when P(x) is true, Q(a, $f(x)$) is also true. Hence, the formula is valid.

SOLVED PROBLEMS

Problem 1. *Translate the following statements in proposition logic :*

(i) If you study, you will get good marks. If you do not study, you will enjoy. Therefore, either you will get good marks or you will enjoy.

(ii) If the catalogue is correct, then if the seeds are planted on April, flowers will bloom in July.

(iii) If John is elected class representative, then either Mary is elected treasurer or Alice is elected vice-treasurer.

(iv) Either taxes are increased or if expenditures rise, then the debt ceiling is raised.

Sol. (*i*) Let p be "You study" and q be "You will get good marks" and r be "You will enjoy". Then the proposition logic is

$$p \to q$$
$$\sim p \to r$$
$$\therefore \quad q \vee r$$

(*ii*) Let p be "The catalogue is correct" and q be "Seeds are planted in April" and r be "The flowers bloom in July". Then, proposition logic is

$$p \to (q \to r)$$

(*iii*) Let p be "John is elected class representative" and q be "Mary is elected treasurer" and r be "Alice is elected vice-treasurer". Then, proposition logic is

$$p \to q \vee r$$

(*iv*) Let p be "Taxes are increased" and q be "Expenditures rise" and r be "The debt ceiling is raised". Then proposition logic is

$$p \vee (q \to r).$$

Problem 2. *The meaning of proposition $p \to q$ is "If p then q" or "q is a necessary condition for p" or "p only if q" or "p is a sufficient condition for q".*

Write the following statements in terms of above.

1. *p : New Delhi is capital of India* *q : India is in Asia.*
2. $p : x^2 = 4$ $q : x = 2$
3. $p : x^2 = y^2$ $q : x = y.$
4. *p = he works hard q : he is a Gold Medalist.*

Sol. 1. (*i*) If New Delhi is capital of India, then India is in Asia.

(*ii*) India is in Asia is a necessary condition for New Delhi to be capital of India.

(*iii*) New Delhi is capital of India only if India is in Asia.

(*iv*) The fact that New Delhi is capital of India is a sufficient condition that India is in Asia.

2. (*i*) If $x^2 = 4$, then $x = 2$

(*ii*) $x = 2$ is necessary for $x^2 = 4$

(*iii*) $x^2 = 4$ only if $x = 2$

(*iv*) $x^2 = 4$ is sufficient for $x = 2$.

3. (*i*) If $x^2 = y^2$ then $x = y$

(*ii*) $x = y$ is necessary for $x^2 = y^2$

(*iii*) $x^2 = y^2$ only if $x = y$

(*iv*) $x^2 = y^2$ is sufficient for $x = y$.

4. (*i*) If he works hard then he is a Gold Medalist.

(*ii*) Gold medal is necessary for hard work.

(*iii*) He works hard only if he is a Gold Medalist.

(*iv*) Hard work is sufficient condition for Gold Medalist.

Problem 3. *Construct the truth table for the following statements*

(*i*) $(p \to p) \to (p \to \sim p)$ (*ii*) $(\sim q \to \sim p) \to (p \to q)$.

Sol. (*i*) The truth table for $(p \to p) \to (p \to \sim p)$ is as follows. (Fig. 42)

p	$\sim p$	$p \to p$	$p \to \sim p$	$(p \to p) \to (p \to \sim p)$
T	F	T	F	F
F	T	T	T	T

Fig. 42.

(*ii*) The truth table of statement $(\sim q \to \sim p) \to (p \to q)$ is as follows. (Fig. 43)

p	q	$\sim p$	$\sim q$	$\sim q \to \sim p$	$p \to q$	$(\sim q \to \sim p) \to (p \to q)$
T	T	F	F	T	T	T
T	F	F	T	F	F	T
F	T	T	F	T	T	T
F	F	T	T	T	T	T

Fig. 43.

Problem 4. *Construct the truth table for following statements*

(*i*) $(p \to (q \to r)) \to ((p \to q) \to (p \to r))$

(*ii*) $p \leftrightarrow (\sim p \vee \sim q)$

(*iii*) $(p \to p) \vee (p \to \sim p)$.

Sol. (*i*) Truth table for $(p \to (q \to r)) \to ((p \to q) \to (p \to r))$ is as follows. (Fig. 44)

p	q	r	$q \to r$	$p \to q$	$p \to r$	$\sim(p \to (q \to r))$	$((p \to q) \to (p \to r))$	$K \to L$
T	T	T	T	T	T	T	T	T
T	T	F	F	T	F	F	F	T
T	F	F	T	F	F	T	T	T
T	F	T	T	F	T	T	T	T
F	T	T	T	T	T	T	T	T
F	T	F	F	T	T	T	T	T
F	F	T	T	T	T	T	T	T
F	F	F	T	T	T	T	T	T

Fig. 44.

(*ii*) Truth table for $p \leftrightarrow (\sim p \vee \sim q)$ is as following. (Fig. 45)

p	q	$\sim p$	$\sim q$	$\sim p \vee \sim q$	$p \leftrightarrow (\sim p \vee \sim q)$
T	T	F	F	F	F
T	F	F	T	T	T
F	T	T	F	T	F
F	F	T	T	T	F

Fig. 45.

(*iii*) The truth table for $(p \to p) \vee (p \to \sim p)$ is as follows. (Fig. 46)

p	$\sim p$	$p \to p$	$p \to \sim p$	$(p \to p) \vee (p \to \sim p)$
T	F	T	F	T
F	T	T	T	T

Fig. 46.

Problem 5. *Assume the value of $p \to q$ is false. Determine the value of $(\sim p \vee \sim q) \to q$.*

Sol. Construct the truth table for both the statements and determine the value of $(\sim p \vee \sim q) \to q$ against the false values of $p \to q$. (Fig. 47)

p	q	$\sim p$	$\sim q$	$p \to q$	$\sim p \vee \sim q$	$(\sim p \vee \sim q) \to q$
T	T	F	F	T	F	T
T	F	F	T	F	T	F
F	T	T	F	T	T	T
F	F	T	T	T	T	F

Fig. 47.

So, when $p \to q$ is false, the value of $(\sim p \vee \sim q) \to q$ is also false.

Problem 6. *Given the value of $p \to q$ is true. Determine the value of $\sim p \vee (p \leftrightarrow q)$.*

Sol. Construct truth table for both statements. (Fig. 48)

p	q	$\sim p$	$p \to q$	$p \leftrightarrow q$	$\sim p \vee (p \leftrightarrow q)$
T	T	F	T	T	T
T	F	F	F	F	F
F	T	T	T	F	T
F	F	T	T	T	T

Fig. 48.

So, when the value of $p \to q$ is true *i.e.*, in line 1, 3 and 4. The value of $\sim p \vee (p \leftrightarrow q)$ is also true.

Problem 7. *Prove that the negation of conditional statement $\sim (p \to q)$ is equivalent of $p \wedge \sim q$.*

Sol. The truth table of the above propositions are as follows. (Fig. 49)

p	q	$\sim q$	$p \to q$	$\sim (p \to q)$	$p \wedge \sim q$
T	T	F	T	F	F
T	F	T	F	T	T
F	T	F	T	F	F
F	F	T	T	F	F

Fig. 49.

As the values of both the propositions are same, hence $\sim (p \to q) \equiv p \wedge \sim q$.

Problem 8. *Prove that the negation of biconditional statement $\sim (p \leftrightarrow q)$ is equivalent to $p \leftrightarrow \sim q$ or $\sim p \leftrightarrow q$.*

Sol. The truth table of all the above propositions is as follows. (Fig. 50)

p	q	$\sim p$	$\sim q$	$p \leftrightarrow q$	$\sim (p \leftrightarrow q)$	$p \leftrightarrow \sim q$	$\sim p \leftrightarrow q$
T	T	F	F	T	F	F	F
T	F	F	T	F	T	T	T
F	T	T	F	F	T	T	T
F	F	T	T	T	F	F	F

Fig. 50.

As the values of $\sim (p \leftrightarrow q)$ and $p \leftrightarrow \sim q$ and $\sim p \leftrightarrow q$ are same, hence they all are equivalent.

Problem 9. *From the following formulae find out tautology, contingency and contradiction.*

(*i*) $\sim (A \to B) \vee (\sim A \vee (A \wedge B))$ (*ii*) $(H \to (I \wedge J)) \to \sim (H \to I)$

(*iii*) $(p \leftrightarrow Q) \equiv (p \wedge Q) \vee (\sim p \wedge Q)$.

Sol. (*i*) Construct the truth table for $\sim (A \to B) \vee (\sim A \vee (A \wedge B))$ as in Fig. 51.

A	B	$\sim$ A	$\sim A \vee (A \wedge B)$	$A \to B$	$\sim (A \to B)$	$\sim (A \to B) \vee (\sim A \vee (A \wedge B))$
T	T	F	T	T	F	T
T	F	F	F	F	T	T
F	T	T	T	T	F	T
F	F	T	T	T	F	T

Fig. 51.

As the last column of the table contains all T's, hence it is a tautology.

(*ii*) Construct truth table for $(H \to I \wedge J) \to \sim (H \to I))$ as in (Fig. 52)

H	I	J	$(I \wedge J)$	$H \to (I \wedge J)$	$H \to I$	$\sim (H \to I)$	$(H \to I \wedge J) \to \sim (H \to I)$
T	T	T	T	T	T	F	F
T	T	F	F	F	T	F	T
T	F	F	F	F	F	T	T
T	F	T	F	F	F	T	T
F	T	T	T	T	T	F	F
F	F	T	F	T	T	F	F
F	T	F	F	T	T	F	F
F	F	F	F	T	T	F	F

Fig. 52.

As the value of last column depends upon the value of the variables, hence it is a contingency.

(*iii*) Construct the truth table for $(p \leftrightarrow q) \to (p \wedge q) \vee (\sim p \wedge q)$ as in (Fig. 53)

p	q	$\sim p$	$p \leftrightarrow q$	$p \wedge q$	$\sim p \wedge q$	$(p \wedge q) \vee (\sim p \wedge q)$	$(p \leftrightarrow q) \to (p \wedge q) \vee (\sim p \wedge q)$
T	T	F	T	T	F	T	T
T	F	F	F	F	T	T	T
F	T	T	F	F	T	T	T
F	F	T	T	F	F	F	T

Fig. 53.

As the last column contains all T's, hence the above formulae is a tautology.

Problem 10. *Prove that following is a tautology*

$$A \vee (\overline{B \wedge C}) \equiv (A \vee \overline{B}) \vee \overline{C}.$$

Sol. Construct the truth table for $A \vee (\overline{B \wedge C}) \to (A \vee \overline{B}) \vee \overline{C}$ as in (Fig. 54)

A	B	C	$\overline{B}$	$\overline{C}$	$\overline{B \wedge C}$	$A \vee (\overline{B \wedge C})$	$(A \vee \overline{B})$	$(A \vee \overline{B}) \vee \overline{C}$	$A \vee (\overline{B \wedge C}) \to (A \vee \overline{B}) \vee \overline{C}$
T	T	T	F	F	F	T	T	T	T
T	T	F	F	T	T	T	T	T	T
T	F	F	T	T	T	T	T	T	T
T	F	T	T	F	T	T	T	T	T
F	T	T	F	F	F	F	F	F	T
F	T	F	F	T	T	T	F	T	T
F	F	T	T	F	T	T	T	T	T
F	F	F	T	T	T	T	T	T	T

Fig. 54.

As the last column of the table contains all T's, hence it is a tautology.

Problem 11. *Determine whether the following are equivalent, using biconditional statement.*

(*i*) $p \leftrightarrow q \equiv (p \wedge q) \vee (\sim p \wedge \sim q)$

(*ii*) $(p \to q) \to t \equiv (p \wedge \sim q) \to t.$

Sol. To prove that the above pairs are equivalent. Prove that $(p \leftrightarrow q) \leftrightarrow ((p \wedge q) \vee (\sim p \wedge \sim q))$ and $((p \to q) \to t) \leftrightarrow ((p \wedge \sim q) \to t)$ are tautologies.

(*i*) Construct truth table for $(p \leftrightarrow q) \leftrightarrow ((p \wedge q) \vee (\sim p \wedge \sim q))$ as in (Fig. 55)

p	q	$\sim p$	$\sim q$	$p \leftrightarrow q$	$p \wedge q$	$\sim p \wedge \sim q$	$(p \wedge q) \vee (\sim p \wedge \sim q)$	$(p \leftrightarrow q) \leftrightarrow (p \wedge q) \vee (\sim p \wedge \sim q)$
T	T	F	F	T	T	F	T	T
T	F	F	T	F	F	F	F	T
F	T	T	F	F	F	F	F	T
F	F	T	T	T	F	T	T	T

Fig. 55.

As the proposition $(p \leftrightarrow q) \leftrightarrow ((p \wedge q) \vee (\sim p \wedge \sim q))$ is a tautology, hence they are equivalent.

(*ii*) Construct the truth table for $((p \to q) \to t) \leftrightarrow ((p \wedge \sim q) \to t)$ as in Fig. 56.

p	q	t	$\sim q$	$p \to q$	$(p \to q) \to t$	$p \wedge \sim q$	$p \wedge \sim q \to t$	$((p \to q) \to t) \leftrightarrow ((p \wedge \sim q) \to t))$
T	T	T	F	T	T	F	T	T
T	T	F	F	T	F	F	T	F
T	F	F	T	F	T	T	F	F
T	F	T	T	F	T	T	T	T
F	T	T	T	T	T	F	T	T
F	T	F	F	T	F	F	T	F
F	F	T	T	T	T	F	T	T
F	F	F	F	T	F	F	T	F

Fig. 56.

Since, the proposition is not a tautology, hence they are not equivalent.

Problem 12. *Determine the dual of each of the following :*

(a) $(P \wedge Q) \vee (\sim P \vee (\sim P \vee Q))$

(b) $\sim P \wedge (\sim Q \wedge R) \vee (Q \wedge R) \vee (P \wedge R)$

(c) $(P \downarrow Q) \uparrow R.$

Sol. The dual of above propositions are as follows :

(a) $(P \wedge Q) \vee (\sim \phi \vee (\sim P \vee Q)) = (P \vee Q) \wedge (\sim P \wedge (\sim P \wedge Q))$

(b) $\sim P \wedge (\sim Q \wedge R) \vee (Q \wedge R) \vee (P \wedge R) = \sim P \vee (\sim Q \vee R) \wedge (Q \vee R) \wedge (P \vee R)$

(c) $(P \downarrow Q) \uparrow R = (P \uparrow Q) \downarrow R.$

Problem 13. *Express $P \Rightarrow Q$ using Nand $\uparrow$ only.*

Sol. $P \Rightarrow Q \cong \sim P \vee Q \cong \sim (\sim \sim P \wedge \sim Q)$ $\quad \because \quad P \vee Q \cong \sim (\sim P \wedge \sim Q)$

$\cong \sim (P \wedge \sim Q)$ $\quad \because \quad \sim \sim P \equiv P$

$\cong P \uparrow \sim Q$ $\quad \because \quad P \uparrow Q \equiv \sim (P \wedge Q)$

$\cong P \uparrow Q \uparrow Q$ $\quad \because \quad \sim P = P \uparrow P.$

Problem 14. *Express $P \downarrow Q$ using $\uparrow$ only.*

Sol. $P \downarrow Q \cong \sim (P \vee Q) \cong (P \vee Q) \uparrow (P \vee Q)$

$\cong [(P \uparrow P) \uparrow (Q \uparrow Q)] \uparrow [(P \uparrow P) \uparrow (Q \uparrow Q)]$ $\quad \because \quad P \vee Q \cong (P \uparrow Q) \uparrow (P \uparrow Q)$

Problem 15. *Express the following formula using only $\sim$ and $\wedge$.*

$$(P \downarrow Q) \uparrow R$$

where $\downarrow$ denotes NOR and $\uparrow$ denotes NAND.

Sol. $(P \downarrow Q) \uparrow R \cong \sim (P \vee Q) \uparrow R$ $\quad \because \quad P \downarrow Q \cong \sim (P \vee Q)$

$\cong \sim ((\sim P \vee Q) \wedge R)$ $\quad \because \quad P \uparrow Q \cong \sim (P \wedge Q)$

$\cong \sim ((\sim \sim \sim (P \wedge \sim Q)) \wedge R)$ $\quad \because \quad P \vee Q \cong \sim (\sim P \wedge \sim Q)$

$\cong \sim ((\sim (P \wedge \sim Q)) \wedge R)$ $\quad \because \quad \sim \sim P = P$

Note. $\uparrow$ (NAND) and $\downarrow$ (NOR) are dual of each other.

Problem 16. *Show that the connective $\uparrow$ (Nand) is functionally complete.*

Sol. To show that the connective $\uparrow$ is functionally complete. We have to show that the set of connectives $(\wedge, \sim)$ and $(\vee, \sim)$ can be expressed in terms of $\uparrow$ alone which is shown as follows :

$$\sim P \cong \sim P \vee \sim P \cong \sim (P \wedge P) \cong P \uparrow P$$

$$P \vee Q \cong \sim (\sim P \wedge \sim Q) \cong \sim P \uparrow \sim Q \cong (P \uparrow P) \uparrow (Q \uparrow Q)$$

and $$P \wedge Q \cong \sim (P \uparrow Q) \cong (P \uparrow Q) \uparrow (P \uparrow Q)$$

Since the above expresses $\wedge$, $\vee$ and $\sim$ in terms of $\uparrow$ alone. Hence $\uparrow$ is functionally complete.

Problem 17. *Prove the validity of following arguments without using truth tables.*

1. $p \vee q, \sim p \mid\!- q$
2. $p, p \to q, q \to r \mid\!- r$
3. $p, q, (p \wedge q) \to r \mid\!- r$
4. $p, (p \wedge \sim q) \to \sim p \mid\!- p \to q.$

Sol. 1. (*i*) $p \vee q$ (Given)

(*ii*) $\sim p$ (Given)

(*iii*) q Disjunctive syllogism

2. (*i*)	p	(Given)
(*ii*)	$p \to q$	(Given)
(*iii*)	$q \to r$	(Given)
(*iv*)	$p \to r$	Hypothetical syllogism using (*ii*) and (*iii*)
(*v*)	r	Modus ponens using (*iv*) and (*i*)
3. (*i*)	p	(Given)
(*ii*)	q	(Given)
(*iii*)	$(p \wedge q) \to r$	(Given)
(*iv*)	$p \wedge q$	Rule of conjunction using (*i*) and (*ii*)
(*v*)	r	Modus ponens using (*iii*) and (*iv*)
4. (*i*)	p	(Given)
(*ii*)	$(p \wedge \sim q) \to \sim p$	(Given)
(*iii*)	$\sim (p \wedge \sim q)$	Modus tollens using (*ii*) and (*i*)
(*iv*)	$\sim\sim (p \to q)$	As $\sim (p \to q) \equiv (p \wedge \sim q)$
(*v*)	$p \to q$	Complement property using (*iv*).

Problem 18. *Prove that the argument* $(p \wedge q) \to r, p \to q \mid\!- p \to ((p \wedge q) \wedge r)$ *is valid without using truth table.*

Sol. (*i*)	$(p \wedge q) \to r$	(Given)
(*ii*)	$p \to q$	(Given)
(*iii*)	$p \to (p \wedge q)$	Rule of absorption using (*ii*)
(*iv*)	$(p \wedge q) \to ((p \wedge q) \wedge r)$	Rule of absorption using (*i*)
(*v*)	$p \to ((p \wedge q) \wedge r)$	Hypothetical syllogism using (*iii*) and (*iv*).

Problem 19. *Prove that the following arguments are valid without using truth tables.*

1. $(p \vee q) \to \sim r, r \vee t, p \mid\!- t$
2. $(p \wedge q) \to r, (r \to q), (r \to q) \to (q \wedge r) \mid\!- (p \wedge q) \to (q \wedge r)$

Sol. 1. (*i*)	$(p \vee q) \to \sim r$	(Given)
(*ii*)	$r \vee t$	(Given)
(*iii*)	p	(Given)
(*iv*)	$p \vee q$	Rule of addition using (*iii*)
(*v*)	$\sim r$	Modus ponens using (*i*) and (*iv*)
(*vi*)	t	Disjunctive syllogism using (*ii*) and (*v*)
2. (*i*)	$p \wedge q \to r$	(Given)
(*ii*)	$r \to q$	(Given)
(*iii*)	$(r \wedge q) \to (q \wedge r)$	(Given)
(*iv*)	$r \to (r \wedge q)$	Rule of absorption using (*ii*)
(*v*)	$(p \wedge q) \to (r \wedge q)$	Hypothetical syllogism using (*i*) and (*iv*)
(*vi*)	$(p \wedge q) \to (q \wedge r)$	Hypothetical syllogism using (*v*) and (*iii*)

Problem 20. *Prove the validity of following argument using deduction system.*

(1) $A \to B$ (2) $B \to \sim C$

(3) $\sim C \wedge D$ (4) $A \to D$

$\therefore B$.

Sol. (*i*)	$A \to B$	(Given)
(*ii*)	$B \to \sim C$	(Given)
(*iii*)	$\sim C \wedge D$	(Given)
(*iv*)	$A \to D$	(Given)
(*v*)	$A \to \sim C$	Hypothetical syllogism using (*i*) and (*ii*)
(*vi*)	$\sim C$	Simplification of (*iii*)
(*vii*)	$\sim A \to C$	Transposition of (*iv*)
(*viii*)	A	Modus tollens using (*vii*) and (*vi*)
(*ix*)	B	Modus ponens using (*i*) and (*viii*)

Problem 21. *Prove the validity of following by deduction method.*

(A) (i) $P \to Q$
(ii) $\sim Q \vee R$
(iii) $\sim (R \wedge \sim S)$
(iv) P
$\therefore S$

(B) (i) $P \vee Q$
(ii) $Q \to R$
(iii) $R \wedge S$
(iv) $P \to S$
(v) P
$\therefore S$

(C) (i) $(Q \to R) \wedge (S \to T)$
(ii) $(U \to V) \wedge (W \to X)$
(iii) $Q \vee U$
$\therefore R \vee V$

Sol. (A) (*i*)	$P \to Q$	(Given)
(*ii*)	$\sim Q \vee R$	(Given)
(*iii*)	$\sim (R \wedge \sim S)$	(Given)
(*iv*)	P	(Given)
(*v*)	Q	Modus ponens using (*i*) and (*iv*)
(*vi*)	$R \vee \sim Q$	Commutative property using (*ii*)
(*vii*)	$\sim R$	Rule of Modus tollens using (*vi*) and (*v*)
(*viii*)	$\sim R \vee S$	De Morgan's Law using (*iii*)
(*ix*)	S	Modus ponens using (*viii*) and (*vii*)
(B) (*i*)	$P \vee Q$	(Given)
(*ii*)	$Q \to R$	(Given)
(*iii*)	$R \wedge S$	(Given)
(*iv*)	$P \to S$	(Given)
(*v*)	P	(Given)
(*vi*)	S	Modus ponens using (*iv*) and (*v*)
(C) (*i*)	$(Q \to R) \wedge (S \to T)$	(Given)
(*ii*)	$(U \to V) \wedge (W \to X)$	(Given)
(*iii*)	$Q \vee U$	(Given)
(*iv*)	$(Q \to R)$	Simplification using (*i*)
(*v*)	$(U \to V)$	Simplification using (*ii*)
(*vi*)	$(Q \to R) \wedge (U \to V)$	Conjunction using (*iv*) and (*v*)
(*vii*)	$R \vee V$	Constructive dilemma using (*vi*) and (*iii*)

Problem 22. *Prove the validity of following argument using truth tables.*

"If it rains then it will be cold. If it is cold then I shall stay at home. Since it rains therefore, I shall stay at home".

Sol. Let p be "It rains" and q be "It will be cold" and r be "I shall stay at home". Then the above argument in symbolic form is

$$p \to q$$
$$q \to r$$
$$p$$
$$\therefore \quad r$$

Construct the truth table of $(p \to q) \wedge (q \to r) \wedge p \to r$ as in (Fig. 57).

p	q	r	$p \to q$	$q \to r$	$(p \to q) \wedge (q \to r)$	$(p \to q) \wedge (q \to r) \wedge p$	$(p \to q) \wedge (q \to r) \wedge p \to r$
T	T	T	T	T	T	T	T
T	T	F	T	F	F	F	T
T	F	F	F	T	F	F	T
T	F	T	F	T	F	F	T
F	T	T	T	T	T	F	T
F	T	F	T	F	F	F	T
F	F	T	T	T	T	F	T
F	F	F	T	T	T	F	T

Fig. 57.

As the last column contain all T's, hence the argument is valid.

Problem 23. *Translate the following into symbolic form and test the validity of the argument.*

If 6 is even then 2 does not divide 7. Either 5 is not prime or 2 divides 7. But 5 is prime, therefore, 6 is odd.

Sol. Let p be "6 is even" and q be "2 divide 7" and r be "5 is primed". Thus, the above argument in symbolic form can be written as

$$p \to \sim q$$
$$\sim r \vee q$$
$$r$$
$$\therefore \quad \sim p$$

Construct the truth table to test the validity. (Fig. 58)

p	q	r	$\sim p$	$\sim q$	$\sim r$	$p \to \sim q$	$\sim r \vee q$
T	T	T	F	F	F	F	T
T	T	F	F	F	T	F	T
T	F	F	F	T	T	T	F
T	F	T	F	T	F	T	T
F	T	T	T	F	F	T	T
F	T	F	T	F	T	T	T
F	F	T	T	T	F	T	F
F	F	F	T	T	T	T	F

Fig. 58.

$p \to \sim q$ is true in line 3, 4, 5, 6, 7 and 8, $\sim r \vee q$ is true in line 1, 2, 3, 4, 5 and r is true in line 1, 4, 5 and 7. All the three are true in line 4 and 5 and $\sim p$ is true in line 5. Hence, argument is valid.

Problem 24. *Determine the negation of the following statements*

(i) $\forall x \, \forall y \, \forall z, \, p(x, y, z)$ (ii) $\forall x \, \exists_y, \, p(x, y)$

(iii) $\exists_y \, \forall_x \, \forall_z, \, p(x, y, z)$.

Sol. (i) $\sim (\forall_x \forall_y \forall_z, \, p(x, y, z)) \equiv \exists_x \exists_y \exists_z \sim p(x, y, z)$

(ii) $\sim (\forall_x \exists_y, \, p(x, y)) \equiv \exists_x \forall_y, \sim p(x, y)$

(iii) $\sim (\exists_y \forall_x \forall_z, \, p(x, y, z)) \equiv \forall_y \exists_x \exists_z, \sim p(x, y, z)$.

Problem 25. *Determine the negation of the following statements*

(i) $\exists_x \forall_y \, (p(x) \vee q(y))$ (ii) $\forall_x \exists_y \, (p(x, y) \to q(x, y))$

(iii) $\forall_x \forall_y \, (p(x) \wedge q(y))$.

Sol. (i) $\sim (\exists x \, \forall_y, \, (p(x) \vee q(y)) \equiv \forall_x \exists_y \sim (p(x) \vee q(y))$

$\equiv \forall_x \exists_y \, (\sim p(x) \wedge \sim q(y))$

(ii) $\sim \forall_x \exists_y \, (p(x, y) \to q(x, y)) \equiv \exists_x \forall_y \sim (p(x, y) \to q(x, y))$

(iii) $\equiv \forall_x \forall_y \, (p \, (x) \wedge p(y)) \equiv \exists_x \exists_y \sim (p(x) \wedge p(y)) \equiv \exists_x \exists_y \, (\sim p(x) \vee \sim p(y))$.

Problem 26. *Let $U = Q$. Use quantifiers to express the following statements*

(i) $\sqrt{5}$ *is not rational.*

(ii) *Subtraction of any two rational numbers is rational.*

Sol. (i) $\sim (\exists \, x) \, (x^2 = 5)$

(ii) $(\forall x) \, (\forall y) \, (x - y$ is a rational$)$.

9

Probability Theory

The word 'probability' means the chance of occurring of a certain event. It is generally possible to predict the future of an event quantitatively with a certain probability of being correct. The probability is used in such cases where the outcome of trial is uncertain.

Before actually defining the probability, we will discuss some terms related with probability.

IMPORTANT TERMS RELATED WITH PROBABILITY

Trial. The performance of an experiment is called a trial.

Experiment. It is a well defined process that leads to a well defined outcome.

For example : Tossing a coin is an experiment. It may consist of one or more trials.

Sample Space. The set of all possible outcomes of an experiment is called sample space and is denoted by S.

e.g., when a die is thrown, sample space is $S = \{1, 2, 3, 4, 5, 6\}$.

It consists of six outcomes 1, 2, 3, 4, 5 and 6.

Event. The possible outcomes of a trial are called events. In other words, event is a subset of the sample space.

e.g., in the throw of a single die, getting a number less than four is an event. If this is denoted by L, then $L = \{1, 2, 3\}$ which is a subset of S.

Complement of Event. The set of all outcomes which are in sample space but not in event is called the complement of event.

e.g., in case of above example, the complement of event is $\overline{L} = \{4, 5, 6\}$.

Equally Likely Events. Events are said to be equally likely if one of them can not be expected to occur in preference to others. In other words, it means each outcome is as likely to occur as any other outcome. *e.g.,* when a die is throw, all the six faces *i.e.,* 1, 2, 3, 4, 5 and 6 are equally likely to occur.

Mutually Exclusive or Disjoint Events. Events are called mutually exclusive if they can not occur simultaneously.

e.g, suppose a card is drawn from a pack of cards, then the events getting a jack and getting a king are mutually exclusive because they can not occur simultaneously.

Exhaustive Events. The total number of all possible outcomes of an experiment are called exhaustive events. *e.g.,* in the tossing of a coin, either head or tail may turn up. Therefore, there are two possible outcomes. Hence, there are two exhaustive events in tossing a coin.

Independent and Dependent Events. Events are said to be independent if occurrence of any one event does not depend on occurrence of any other event.

Events are said to be dependent if occurrence of one effect the occurrence of other events *e.g.*, the outcome of throwing a die is independent of outcome of drawing a king from pack of cards.

PROBABILITY DEFINITION

The probability of happening of an event A, denoted by P(A), is defined as

$$P(A) = \frac{\text{number of cases favourable to A}}{\text{number of possible outcomes}}.$$

Thus, if an event an happen in m ways and fails to happen in n ways and $m + n$ ways is equally likely to occur then the probability of happening of the event A is given by

$$P(A) = \frac{m}{m+n}$$

and the probability of non-happening of A is

$$P(\bar{A}) = \frac{n}{m+n}.$$

Points to be Noted

1. The probability of an event which is certain to occur is one.
2. The probability of an event which is impossible is zero.
3. If the probability of happening of an events P(A) and that of not happening is $P(\bar{A})$, then

$$P(A) + P(\bar{A}) = 1,\ 0 \le P(A) \le 1,\ 0 \le P(\bar{A}) \le 1.$$

ADDITION THEOREM

Theorem I. *If A and B are two mutually exclusive events, then*

$$P(A \cup B) = P(A) + P(B).$$

Proof. Let n = total number of exhaustive cases

n_1 = number of cases favourable to A.

n_2 = number of cases favourable to B.

Now, we have A and B two mutually exclusive events. Therefore, $n_1 + n_2$ is the number of cases favourable to A or B.

$$\therefore \quad P(A \cup B) = \frac{\text{favourable cases}}{\text{total number of exhaustive cases}} = \frac{n_1 + n_2}{n} = \frac{n_1}{n} + \frac{n_2}{n}$$

But we have, $\quad P(A) = \frac{n_1}{n}$ and $P(B) = \frac{n_2}{n}$

Hence $\quad P(A \cup B) = P(A) + P(B).$

Example 1. *In a single throw of two dice find the probability of having 8 or 11.*

Sol. Number of favourable cases for throwing 8 are {(6, 2), (5, 3), (4, 4), (3, 5), (2, 6)} = 5

Number of total cases = 36

$\therefore$ Probability of having, $8 = \dfrac{5}{36}$

Number of favourable cases for throwing, 11

$$\{(6, 5), (5, 6)\} = 2$$

$\therefore$ Probability of throwing, $11 = \dfrac{2}{36}$

So, probability of throwing 8 or 11 $= \dfrac{5}{36} + \dfrac{2}{36} = \dfrac{7}{36}$.

Example 2. *A card is drawn at random from a well-shuffled pack of 52 cards. Find probability of getting*

(i) a jack, a queen and a king *(ii) a two of heart or two of diamond.*

Sol. *(i)* In a pack of 52 cards, we have 4 jacks, 4 queens and 4 kings.

$$\therefore \quad P(\text{a jack}) = \frac{{}^4C_1}{{}^{52}C_1} = \frac{4}{52} = \frac{1}{13}$$

$$P(\text{a queen}) = \frac{{}^4C_1}{{}^{52}C_1} = \frac{4}{52} = \frac{1}{13}$$

$$P(\text{a king}) = \frac{{}^4C_1}{{}^{52}C_1} = \frac{4}{52} = \frac{1}{13}.$$

By addition theorem of probability

$$P(\text{a jack or a queen or a king}) = \frac{1}{13} + \frac{1}{13} + \frac{1}{13} = \frac{3}{13}$$

(ii) P(two of heart or two of diamonds) = P(two of heart) + P(two of diamonds)

$$= \frac{1}{52} + \frac{1}{52} + \frac{2}{52} = \frac{1}{26}.$$

Example 3. *Two dice are tossed once. Find the probability of getting an even number on first dice or a total of 8.*

Sol. An even number can be got on a die in 3 ways because any one of 2, 4, 6, can come. The other die can have any number. This can happen in 6 ways.

$$\therefore P(\text{an even number on Ist die}) = \frac{3 \times 6}{36} = \frac{18}{36} = \frac{1}{2}$$

A total of 8 can be obtained in following cases $\{(2, 6), (3, 5), (4, 4), (5, 3), (6, 2)\}$

$$\therefore \quad P(\text{a total of } 8) = \frac{5}{36}$$

$$\therefore \quad \text{Total probability} = \frac{1}{2} + \frac{5}{36} = \frac{23}{36}.$$

Example 4. *Find the probability of getting a sum of 5 or 7 in toss of two dice.*

Sol. Chances for sum of 5 is $\{(1, 4), (2, 3), (4, 1), (3, 2)\}$

$$\therefore \quad P(\text{sum of } 5) = \frac{4}{36} = \frac{1}{9}$$

Chances for sum of 7 is {(1, 6), (6, 1), (2, 5), (5, 2), (3, 4), (4, 3)}

$$\therefore \quad P(\text{sum of } 7) = \frac{6}{36} = \frac{1}{6}$$

$$\therefore \quad \text{Required probability} = \frac{1}{9} + \frac{1}{6} = \frac{5}{18}.$$

Theorem II. *If A and B are two events that are not mutually exclusive, then*

$$P(A \cup B) = P(A) + P(B) - P(A \cap B).$$

Proof. Let n = total number of exhaustive cases

n_1 = number of cases favourable to A

n_2 = number of cases favourable to B

n_3 = number of cases favourable both A and B

But, A and B are not mutually exclusive. Therefore, A and B can occur simultaneously. So, $n_1 + n_2 - n_3$ is the number of cases favourable to A or B.

Therefore, $$P(A \cup B) = \frac{n_1 + n_2 - n_3}{n} = \frac{n_1}{n} + \frac{n_2}{n} - \frac{n_3}{n}$$

But we have, $$P(A) = \frac{n_1}{n},\ P(B) = \frac{n_2}{n} \text{ and } P(A \cap B) = \frac{n_3}{n}$$

Hence, $$P(A \cup B) = P(A) + P(B) - P(A \cap B).$$

Example 5. *A card is drawn from a deck of cards. Find the probability of getting an ace or a spade card.*

Sol. Let A be event that a card drawn is a spade and B be event that a card drawn is an ace

$$\therefore \quad P(A) = \frac{13}{52},\ P(B) = \frac{4}{52} \text{ and } P(A \cap B) = \frac{1}{52}$$

$$\therefore \quad P(A \cup B) = P(A) + P(B) - P(A \cap B) = \frac{13}{52} + \frac{4}{52} - \frac{1}{52} = \frac{16}{52} = \frac{4}{13}.$$

Example 6. *Two dice are tossed once. Find probability of getting an even number on first die or a total of 8.*

Sol. P(even number on Ist die or a total of 8) = P(even number on Ist die) + P(total of 8) P(even number on Ist die and total of 8)

Now, $P(\text{even number on Ist die}) = \frac{18}{36}$

Ordered pairs showing a total of 8 = {(6, 2), (5, 3), (4, 4), (3, 5), (2, 6)} = 5

$$\therefore \quad \text{Probability ; } P(\text{total of } 8) = \frac{5}{36}$$

$$P(\text{even number of Ist die and total of } 8) = \frac{3}{36}$$

$$\therefore \quad \text{Required probability} = \frac{18}{36} + \frac{5}{36} - \frac{3}{36} = \frac{20}{36} = \frac{5}{9}.$$

Example 7. *A dice is thrown twice. What is the probability that at least one of two throws comes up with number 3.*

Sol. Sample space contains 36 events.

Let 'A' be the event "first throw comes up with number 3" and B be the event "second throw comes up with number 3".

∴ Favourable events of

$$A = \{(3, 1), (3, 2), (3, 3), (3, 4), (3, 5), (3, 6)\} = 6$$

$$\therefore \quad P(A) = \frac{6}{36} = \frac{1}{6}$$

Favourable events of $B = \{(1, 3), (2, 3), (3, 3), (4, 3), (5, 3), (6, 3)\} = 6$

$$\therefore \quad P(B) = \frac{6}{36} = \frac{1}{6}$$

Also $\quad P(A \cap B) = \frac{1}{36}$ $\quad$ [∵ Only one event will be (3, 3)]

$$\therefore \text{ Required probability} = \frac{1}{6} + \frac{1}{6} - \frac{1}{36} = \frac{11}{36}.$$

Example 8. *What is the probability of getting a number greater than 2 or an even number in a single throw of a fair die.*

Sol. Let E_1 be event of getting a number greater than 2 or E_2 be the event of getting an even number.

$$\therefore \quad P(E_1) = \frac{4}{6} = \frac{2}{3}, \; P(E_2) = \frac{3}{6} = \frac{1}{2} \quad \text{and} \quad P(E_1 \cap E_2) = \frac{1}{3}.$$

$$\therefore \quad P(E_1 \cup E_2) = P(E_1) + P(E_2) - P(E_1 \cap E_2) = \frac{2}{3} + \frac{1}{2} - \frac{1}{3} = \frac{5}{6}.$$

Example 9. *A coin is tossed n times. What is probability that the head will present itself an odd number of times ?*

Sol. Each time tossing of coin may result in 2 ways. Hence, n throws may result in 2^n ways. The event will happen if among the result of n throws, head occupies either 1 place, 3 place, 5 place ... etc. For these, the number of ways are ${}^nC_1$, ${}^nC_3$, ${}^nC_5$,... .

Favourable number of ways $= {}^nC_1 + {}^nC_3 + {}^nC_5 + ... = 2^{n-1}$

$$P(\text{head occurs odd number of times}) = \frac{2^{n-1}}{2^n} = \frac{1}{2}.$$

Example 10. *A card is drawn from a deck of cards. Find probability of getting a king or a heart or a red card.*

Sol. Let A be event that a card drawn is a king

Let B be event that a card drawn is a heart

Let C be event that a card drawn is a red card

$$P(A) = \frac{4}{52}, \quad P(B) = \frac{13}{52}, \quad P(C) = \frac{26}{52}$$

$$P(A \cap B) = \frac{1}{52}, \quad P(B \cap C) = \frac{13}{52}$$

$$P(A \cap C) = \frac{2}{52}, \quad P(A \cap B \cap C) = \frac{1}{52}$$

$$P(A \cup B \cup C) = P(A) + P(B) + P(C) - P(A \cap B) - P(B \cap C) - P(A \cap C) + P(A \cap B \cap C)$$

$$= \frac{4}{52} + \frac{13}{52} + \frac{26}{52} - \frac{1}{52} - \frac{13}{52} - \frac{2}{52} + \frac{1}{52} = \frac{28}{52} = \frac{7}{13}.$$

Example 11. *From 30 tickets with the first 30 natural number one is drawn at random. Find chance that is a multiple of 3 or 7.*

Sol. $P(\text{a multiple of 3}) = \frac{10}{30}$; $P(\text{a multiple of 7}) = \frac{4}{30}$

$P(\text{multiple of both}) = \frac{1}{30}$

$\therefore$ Required probability $= \frac{10}{30} + \frac{4}{30} - \frac{1}{30} = \frac{13}{30}$.

Example 12. *Two events A and B have probability 0.25 and 0.50. The probability that both A and B occur simultaneously is 0.12. Find the probability that neither A nor B occurs.*

Sol. $P(A \cup B) = P(A) + P(B) - P(A \cap B) = 0.25 + 0.50 - 0.12 = 0.63$

$P(\overline{A \cup B}) = 1 - P(A \cup B) = 1 - 0.63 = 0.37.$

MULTIPLICATION THEOREM

Theorem III. *If A and B are two independent events, then the probability that both will occur is equal to product of their individual probabilities i.e., $P(A \cap B) = P(A) \times P(B)$.*

Proof. Let event

A can happen is n_1 ways of which p are successful

B can happen in n_2 ways of which q are successful

Now, combine successful event of A with successful event of B.

Thus, total number of successful cases $= p \times q$

We have, total number of cases $= n_1 \times n_2$.

Therefore, from definition of probability

$$P(A \text{ and } B) = P(A \cap B) = \frac{p \times q}{n_1 \times n_2} = \frac{p}{n_1} \times \frac{q}{n_2}.$$

We have $\quad P(A) = \frac{p}{n_1}, \; P(B) = \frac{q}{n_2}$

So, $\quad P(A \cap B) = P(A) \times P(B)$

If, there are three independent events A, B and C, then

$$P(A \cap B \cap C) = P((A \cap B) \cap C) = P(A \cap B) \times P(C)$$
$$= P(A) \times P(B) \times P(C).$$

In general, if there are n independent events, then

$$P(\bigcap_{i=1}^{n} A_i) = \prod_{i=1}^{n} P(A_i).$$

Example 13. *Two cards are drawn from a pack of cards in succession (with replacement). Find the probability that both are jacks.*

Sol. Probability that Ist car is a jack $= \frac{4}{52}$

Probability that IInd card is a jack $= \frac{4}{52}$

Probability that both cards are jacks $= \frac{4}{52} \times \frac{4}{52} = \frac{1}{169}$.

Example 14. *A die is tossed twice. Find probability of having a number greater than 3 on each toss.*

Sol. Probability of having number greater than 3 in Ist toss $= \frac{3}{6}$

Probability of having number greater than 3 in IInd toss $= \frac{3}{6}$.

Probability of combined event $= \frac{3}{6} \times \frac{3}{6} = \frac{1}{4}$.

Example 15. *A bag contains 5 white and 3 black balls and four balls are successively drawn out and not replaced. What is probability that they are alternately of different colours.*

Sol. The event is W, B, W, B and B, W, B, W, where W is for white and B for black ball. Probability of having W, B, W, B in that order

$$= \frac{5}{8} \times \frac{3}{7} \times \frac{4}{6} \times \frac{2}{5} = \frac{1}{14} \qquad \text{(Multiplication theorem)}$$

Probability of finding B, W, B, W in that order

$$= \frac{3}{8} \times \frac{5}{7} \times \frac{2}{6} \times \frac{4}{5} = \frac{1}{14}$$

∴ Probability of having balls of different colours alternately

$$= \frac{1}{14} + \frac{1}{14} = \frac{2}{14} = \frac{1}{7}. \qquad \text{(Addition theorem)}$$

Example 16. *From a bag containing 4 black and 5 white balls, a man draws four balls at random. Find the probability that there are 2 black and 2 white balls.*

Sol. Out of four black balls, 2 black balls can be drawn in ${}^4C_2$ ways.

Out of 5 white balls, 2 white balls can be drawn in ${}^5C_2$ ways.

∴ Number of favourable events $= {}^4C_2 \times {}^5C_2$

Out of 9 balls, 4 balls can be drawn in ${}^9C_4$ ways

$$\therefore \text{ Required probability} = \frac{{}^4C_2 \times {}^5C_2}{{}^9C_4} = \frac{4 \times 3}{2 \times 1} \times \frac{5 \times 4}{2 \times 1} \times \frac{4 \times 3 \times 2 \times 1}{9 \times 8 \times 7 \times 6} = \frac{10}{21}.$$

Example 17. *A bag contains 5 green and 7 red balls. Two balls are drawn. Find probability that one is green and other is red.*

Sol. P(A) = P(a green ball) $= \frac{1}{5}$; P(B) = P(a red ball) $= \frac{1}{7}$

By multiplication theorem

$$P(\text{A and B}) = P(A) \times P(B) = \frac{1}{5} \times \frac{1}{7} = \frac{1}{35}.$$

Example 18. *Two dices are thrown. Find probability of getting an odd number on one and a multiple of three on other.*

Sol. $P(A) = P(\text{an odd number}) = \frac{3}{6} = \frac{1}{2}$ [∵ 1, 3, 5 will be odd number]

$P(B) = P(\text{a multiple of 3}) = \frac{2}{6} = \frac{1}{3}$ [∵ 3 and 6 will be multiple of 3]

∴ Required probability $= \frac{1}{2} \times \frac{1}{3} = \frac{1}{6}.$

Example 19. *A card is drawn from a well shuffled deck of 52 cards. The outcome is noted, the card is replaced and the deck reshuffled. Another card is then drawn from the deck. Find the probability of getting first card a jack and the second card a black queen.*

Sol. $P(\text{a jack}) = \frac{4}{52} = \frac{1}{13}$

$P(\text{a black queen}) = \frac{2}{52} = \frac{1}{26}$

P(first card is jack and second card is a black queen) = P(a jack) × P(a black queen)

$$= \frac{1}{13} \times \frac{1}{26} = \frac{1}{318}.$$

Example 20. *From a pack of cards, two cards are drawn at random one by one with replacement. Find the probability that first is heart and second is king.*

Sol. Probability of getting heart $= \frac{13}{52}$

Probability of getting king $= \frac{4}{52}$

Probability that Ist is heart and IInd is king $= \frac{13}{52} \times \frac{4}{52} = \frac{1}{52}.$

Example 21. *An urn contains 25 balls numbered through 1 to 25. Two balls are drawn from urn with replacement. Find probability of getting one odd and one even number.*

Sol. Let E and O represent events of drawing even and odd numbered balls. There are 12 balls having even numbers and 13 balls having odd number.

∴ $P(E) = \frac{12}{25}, P(O) = \frac{13}{25}$

The total number of such cases consists of (first odd and second even) + (first even and second odd).

∴ P(one odd and one even) = P(O) × P(E) + P(E) × P(O)

$$= \frac{13}{25} \times \frac{12}{25} + \frac{12}{25} \times \frac{13}{25} = \frac{312}{625}.$$

CONDITIONAL PROBABILITY

Theorem IV. *If A and B are two dependent events then the probability of occurrence of A given that B has already occurred and is denoted by P(A/B) is given by*

$$P\left(\frac{A}{B}\right) = \frac{P(A \cap B)}{P(B)} \; ; P(A \cap B) = P(B) \times P\left(\frac{A}{B}\right).$$

Similarly, the probability of occurrence of B given that A has already occurred is given by

$$P\left(\frac{B}{A}\right) = \frac{P(A \cap B)}{P(A)} \; ; P(A \cap B) = P(A) \times P\left(\frac{B}{A}\right).$$

Proof. Let S be the sample space.

Then, we have

$$P(A \cap B) = \frac{P(A \cap B)}{P(S)} = \frac{P(A \cap B)}{P(A)} \times \frac{P(A)}{P(S)}$$

$$= \frac{P(A)}{P(S)} \times \frac{P(A \cap B)}{P(A)} \qquad \because \frac{P(A \cap B)}{P(A)} = P\left(\frac{B}{A}\right)$$

$$= P(A) \times P\left(\frac{B}{A}\right) \qquad \ldots(i)$$

Interchange A and B in eq. (*i*), we get

$$= P(B) \times P\left(\frac{A}{B}\right).$$

Example 22. *Find the probability of drawing a heart on each of two consecutive draws from a well shuffled pack of cards if the card is not replaced after the draw.*

Sol. Let event A is heart on the first draw and event B is heart on the second draw.

Then, $P(A \cap B) = P(A) \times P\left(\frac{B}{A}\right)$

Now, $P(A) = \frac{13}{52}$

When we get a heart on the first draw, the second draw has 51 outcomes and 12 are favourable.

$\therefore \quad P\left(\frac{B}{A}\right) = \frac{12}{51} = \frac{4}{17}$

$\therefore \quad P(A \cap B) = \frac{13}{52} \times \frac{4}{17} = \frac{1}{17}.$

Example 23. *From a deck of cards, three cards are drawn one by one without replacement. Find the probability that each time it is a card of spade.*

Sol. Let event A = first card is spade

Event B = second card is spade

Event C = third card is spade

$$P(A \cap B \cap C) = P(A) \times P\left(\frac{B}{A}\right) \times P\left(\frac{C}{B \cap A}\right)$$

$$= \frac{13}{52} \times \frac{12}{51} \times \frac{11}{50} = \frac{11}{850}.$$

Example 24. *A bag contains 5 white, 7 red and 8 black balls. If four balls are drawn one by one without replacement, what is the probability that all are white.*

Sol. Let events A, B, C and D represent Ist, IInd, IIIrd and IVth white ball drawn.

$$P(A) = \frac{5}{20},\ P\left(\frac{B}{A}\right) = \frac{4}{19}$$

$$P\left(\frac{C}{A \cap B}\right) = \frac{3}{18} \text{ and } P\left(\frac{D}{A \cap B \cap C}\right) = \frac{2}{17}$$

$$\therefore \quad P(A \cap B \cap C \cap D) = P(A) \times P\left(\frac{B}{A}\right) \times P\left(\frac{C}{A \cap B}\right) \times P\left(\frac{D}{A \cap B \cap C}\right)$$

$$= \frac{5}{20} \times \frac{4}{19} \times \frac{3}{18} \times \frac{2}{17} = \frac{1}{969}.$$

Example 25. *Two cards are drawn from a pack of 52 cards, one after the other without replacement. Find the chance that one of these cards is an ace and other is queen of opposite shade.*

Sol. There are two possibilities

(*i*) The first card drawn is an ace and second is a queen of opposite shade probability of this event

$$= \frac{4}{52} \times \frac{2}{51} = \frac{2}{663}.$$

(*ii*) The first card drawn is a queen and second is an ace of opposite colour. The probability of this event

$$= \frac{4}{52} \times \frac{2}{51} = \frac{2}{663}$$

Hence, required probability $= \frac{2}{663} + \frac{2}{663} = \frac{4}{663}$.

Example 26. *A bag contains 5 white and 3 red balls. Balls are drawn in succession and are not replaced. Show that the chance that Ist red ball will appear on fifth draw is* $\frac{3}{56}$.

Sol. Let W_i = event that white ball is drawn at *i*th draw

R_i = event that red ball is drawn at *i*th draw. We want that first ball should appear on fifth draw. This will happen in event W_1, W_2, W_3, W_4, R_5.

$$\therefore \quad P(W_1 \cap W_2 \cap W_3 \cap W_4 \cap R_5) = P(W_1) \times P\left(\frac{W_2}{W_1}\right) \times P\left(\frac{W_3}{W_2 \cap W_1}\right) \times P\left(\frac{W_4}{W_3 \cap W_2 \cap W_1}\right)$$

$$\times P\left(\frac{R_5}{W_4 \cap W_3 \cap W_2 \cap W_1}\right) = \frac{5}{8} \times \frac{4}{7} \times \frac{3}{6} \times \frac{2}{5} \times \frac{3}{4} = \frac{3}{56}.$$

Example 27. *There are two groups of subjects, one of which consists of 5 science and 3 engineering subjects and the other consists of 3 science and 5 engg. subjects. An unbiased die is cast. If the number 3 or number 5 turns up, a subject is selected at random from the first group. Otherwise the subject is selected at random from the second group. Find the probability that an engg. subject is selected ultimately.*

Sol. Let E_1 = event that subject is selected from the first group.

E_2 = Event that subject is selected from the second group.

Now, the probability of turning up the number 3 or 5 in the throw of an unbiased die is $\frac{2}{6}=\frac{1}{3}$.

$$\therefore \quad P(E_1)=\frac{1}{3} \quad \text{and} \quad P(E_2)=\frac{2}{3}$$

Let E is event that an engg. subject is selected.

$$\therefore \quad P\left(\frac{E}{E_1}\right)=\frac{3}{8} \quad \text{and} \quad P\left(\frac{E}{E_2}\right)=\frac{5}{8}$$

$$\therefore \quad P(E)=P(E_1)\times P\left(\frac{E}{E_1}\right)+P(E_2)\times P\left(\frac{E}{E_2}\right)=\frac{1}{3}\times\frac{3}{8}+\frac{2}{3}\times\frac{5}{8}=\frac{13}{24}.$$

Theorem V. *If $A_1, A_2, A_3, \ldots, A_n$ are n independent events and the probabilities of happening of these events is $P_1, P_2, P_3, \ldots, P_n$; then the probability that at least one of these events must happen is*

$$1-(1-p_1)(1-p_2)(1-p_3)\ldots(1-p_n).$$

Proof. The probability that at least one of $A_1, A_2, A_3, \ldots, A_n$ will occur is

P(at least one of $A_1, A_2, A_3, \ldots, A_n$ will occur)

$= 1 - P(\text{none of } A_1, A_2, A_3, \ldots, A_n \text{ will occur}) \quad (\because \; P(A)+P(\overline{A})=1)$

$= 1 - P[A_1 \text{ will not occur}, A_2 \text{ will not occur}, \ldots, A_n \text{ will not occur}]$

$= 1 - P[\overline{A}_1 \cap \overline{A}_2 \cap \overline{A}_3 \ldots\ldots \cap \overline{A}_n]$

$= 1 - P(\overline{A}_1)\times P(\overline{A}_2)\times P(\overline{A}_3)\ldots\ldots\times P(\overline{A}_n)$

[$\because$ $A_1, A_2, A_3, \ldots, A_n$ are independent events]

$= 1-(1-p_1)(1-p_2)\ldots(1-p_n)$

$\because$ $P(\overline{A}_i)=1-P(A_i)=1-p$ for 1, 2, 3, ..., n.

Example 28. *Shanu speaks truth in 60% cases and Nanu in 70% cases. In what %age of cases are they likely to contradict each other in stating same fact.*

Sol. They will contradict each other if one of them speaks the truth and other tells a lie.

Probability that Shanu speaks the truth and Nanu tells a lie

$$=\frac{60}{100}\times\left(1-\frac{70}{100}\right)=\frac{60}{100}\times\frac{30}{100}=\frac{9}{50}$$

Probability that Shanu tells a lie and Nanu speaks the truth

$$=\left(1-\frac{60}{100}\right)\times\frac{70}{100}=\frac{40}{100}\times\frac{70}{100}=\frac{7}{25}$$

The percentage of cases in which they contradict each other

$$=\left(\frac{9}{50}+\frac{7}{25}\right)\times 100=\frac{23}{50}\times 100=46\%.$$

Example 29. *The probability of A, B, C solving a problem are 1/2, 1/3, 1/4. Find the probability that the problem will be solved.*

Sol. We shall find the probability that the problem is not solved.

Probability that A, B, C do not solve the problem are

$$1-\frac{1}{2},\ 1-\frac{1}{3},\ 1-\frac{1}{4} \quad \text{or} \quad \frac{1}{2}, \frac{2}{3}, \frac{3}{4}$$

By the law of multiplication

Probability that the problem is not solved $= \frac{1}{2}\times\frac{2}{3}\times\frac{3}{4}=\frac{1}{4}$

Probability that the problem is solved $= 1-\frac{1}{4}=\frac{3}{4}$.

Example 30. *Suppose it is 11 to 6 against a person A who is now 38 years of age living till he is 73 and 6 to 3 against B now 43 living till he is 78. Find the chance that at least one of them will be alive 35 years hence.*

Sol. The chance that A will die within 35 years $= \frac{11}{17}$.

The chance that B will die within 35 years $= \frac{6}{9}$.

As the events are independent

Probability that both will die within 35 years $= \frac{11}{17}\times\frac{6}{9}=\frac{66}{153}$

$\therefore$ The chances that at least one of them will be alive $= 1-\frac{66}{153}=\frac{87}{153}$.

Example 31. *Certain questions are given to the three independent students A, B and C to solve. A solves 25%, B solves 50%, C solves 40% of question given to them. What is probability that the question will be solved ?*

Sol. The probability that the question will not be solved = Probability that A, B and C all three will not solve it.

$$= \left(1-\frac{25}{100}\right)\left(1-\frac{50}{100}\right)\left(1-\frac{40}{100}\right)=\frac{75}{100}\times\frac{50}{100}\times\frac{60}{100}=\frac{9}{40}$$

Probability that question will be solved $= 1-\frac{9}{40}=\frac{31}{40}$.

Example 32. *An anti-aircraft gun can take a maximum of four shots at an enemy plane moving away from it. The probability of hitting the plane at first, second, third and fourth shots are 0.4, 0.3, 0.2 and 0.1. What is the probability that the gun hits the plane ?*

Sol. The probability of hitting the plane at Ist, IInd, IIIrd and IVth shots are 0.4, 0.3, 0.2 and 0.1.

Probability of not hitting the plane at Ist, IInd, IIIrd and IVth shots are (1 – 0.4), (1 – 0.3), (1 – 0.2), (1 – 0.1) *i.e.*, 0.6, 0.7, 0.8 and 0.9 respectively.

The probability of not hitting the plane at all *i.e.*, probability that none of four shots hits the plane

$$= 0.6 \times 0.7 \times 0.8 \times 0.9 = 0.3024$$

Hence, probability of hitting the plane

$$= 1 - \text{probability of not hitting the plane} = 1 - 0.3024 = 0.6976.$$

SOLVED PROBLEM

Problem 1. *A bag contains 8 blue and 4 red balls. Two balls are drawn at random with replacement. Find the probability of getting one blue and one red ball.*

Sol. P(a blue ball) = P(B) = $\frac{1}{8}$; P(a red ball) = P(R) = $\frac{1}{4}$

P(a blue and a red ball) = P(B) × P(R) + P(R) × P(B)

$$= \frac{1}{8} \times \frac{1}{4} + \frac{1}{4} \times \frac{1}{8} = \frac{2}{32} = \frac{1}{16}.$$

Problem 2. *An urn contains 25 balls numbered through 25. Two balls are drawn from the urn with replacement. Find probability of getting (i) both odd numbers (ii) one odd and one even number (iii) at least one odd number (iv) no odd number.*

Sol. Let E = Event of drawing even numbered ball and O = event of drawing odd numbered ball.

There are 13 balls numbered odd and 12 balls numbered even.

$$P(E) = \frac{12}{25} \quad \text{and} \quad P(O) = \frac{13}{25}$$

(*i*) Since balls are replaced back in the urn

∴ P(both odd numbers) = P(O) × P(O) = $\frac{13}{25} \times \frac{13}{25} = \frac{169}{625}$

(*ii*) P(one odd and one even number) = P(O) × P(E) + P(E) × P(O)

$$= \frac{13}{25} \times \frac{12}{25} + \frac{12}{25} \times \frac{13}{25} = \frac{312}{625}.$$

(*iii*) Probability of getting both the balls even numbered

$$P(EE) = P(E) \,.\, P(E) = \frac{12}{25} \times \frac{12}{25} = \frac{144}{625}$$

∴ P(atleast one odd numbered ball) = 1 – P(EE) = 1 – $\frac{144}{625} = \frac{481}{625}$

(*iv*) P(no odd number) = P(both even numbers) = $\frac{144}{625}$.

Problem 3. *A bag contains 3 white, 3 black and 2 red balls. One by one, three balls are drawn without replacing them. Find the probability that the third ball is red.*

Sol. The third ball is red, can be in following 3 cases.

(*i*) Ist red, IInd non-red, IIIrd red = $\frac{2}{8} \times \frac{6}{7} \times \frac{1}{7}$

(*ii*) Ist non-red, IInd red, IIIrd red = $\frac{6}{8} \times \frac{2}{7} \times \frac{1}{6}$

(*iii*) Ist non-red, IInd non-red, IIIrd red = $\frac{6}{8} \times \frac{5}{7} \times \frac{2}{6}$

$\therefore$ Required probability $= \frac{2}{8} \times \frac{6}{7} \times \frac{1}{7} + \frac{6}{8} \times \frac{2}{7} \times \frac{1}{6} + \frac{6}{8} \times \frac{5}{7} \times \frac{2}{6} = \frac{1}{4}.$

Problem 4. *A bag contains 6 red, 4 white and 5 black balls. A person draws four balls from the box at random. Find the probability that among the balls drawn there is at least one ball of each colour ?*

Sol. The four balls may have at least one ball of each colour in the three following mutually exclusive events.

E_1 = 1 red, 1 white, 2 black balls are drawn.

E_2 = 2 red, 1 white, 1 black balls are drawn.

E_3 = 1 red, 2 white, 1 black balls are drawn.

$$\text{Total probability} = \frac{{}^6C_1 \times {}^4C_1 \times {}^5C_2}{{}^{15}C_4} + \frac{{}^6C_2 \times {}^4C_1 \times {}^5C_1}{{}^{15}C_4} + \frac{{}^6C_1 \times {}^4C_2 \times {}^5C_1}{{}^{15}C_4}$$

$$= \left\{\frac{(6 \times 4 \times 10) + (15 \times 4 \times 5) + (6 \times 6 \times 5)}{{}^{15}C_4}\right\} = 0.5275.$$

Problem 5. *A dice a thrown. Consider two events A = {1, 2, 3, 4} and B = {4, 5, 6}. Prove that the events A and B are*

(i) not independent *(ii) not mutually exclusive*

(iii) exhaustive.

Sol. $P(A) = \frac{4}{6} = \frac{2}{3}, \; P(B) = \frac{3}{6} = \frac{1}{2}; \quad P(AB) = \frac{1}{6}, \; P(A) \,.\, P(B) = \frac{1}{3}$

(*i*) As $P(AB) \neq P(A) \,.\, P(B)$

So, A and B are not independent *i.e.*, occurrence of one do not effect the occurrence of other.

(*ii*) $P(A \cap B) \neq$ null set

i.e., they are not mutually exclusive.

(*iii*) $A \cup B = \{1, 2, 3, 4, 5, 6\}$ *i.e.*, all the possible outcomes are existing. So, events A and B are exhaustive.

Problem 6. *If A and B are two events such that P(A + B) = 5/6, P($\overline{A}$) = 1/4, P(B) = 1/3. Show that A and B are (i) independent (ii) not exclusive.*

Sol. (*i*) $P(A \cup B) = P(A) + P(B) - P(A \cap B)$

or $P(A \cap B) = P(A) + P(B) - P(A \cup B)$

$$= \left[\left(1 - \frac{1}{4}\right)\right] + \frac{1}{3} - \frac{5}{6} = \frac{3}{4} + \frac{1}{3} - \frac{5}{6} = \frac{3}{12} = \frac{1}{4}$$

$$P(A) \,.\, P(B) = \frac{1}{3} \cdot \frac{3}{4} = \frac{1}{4}$$

As $P(A \cap B) = P(A) \,.\, P(B)$

So, events are independent.

(*ii*) $P(A \cap B) \neq 0$

So, events are not exclusive.

Problem 7. *Show that the sum of the probability of two independent events is one (certain).*

Sol. Let the probability of success of first event be p_1.

The probability of failure of first event be q_1.

Similarly, the probability of success of second event be p_2.

The probability of failure of second event be q_2.

Now, the probability of success of the first event and success of the second event $= p_1 \times p_2$.

Probability of success of first event of failure of second event $= p_1 \times q_2$.

Probability of failure of first event and success of second event $= p_2 \times q_1$.

Probability of failure of both the events $= q_1 \times q_2$.

So, there are these four possibilities.

Therefore, the sum of probabilities

$$= p_1 \times p_2 + p_1 \times q_2 + p_2 \times q_1 + q_1 \times q_2$$

$$= p_1 (p_2 + q_2) + q_1 (p_2 + q_2) = (p_1 + q_1)(p_2 + q_2)$$

$$= 1 \qquad [\because p_1 + q_1 = 1 \quad p_2 + q_2 = 1]$$

Problem 8. *If $P(A_1 + A_2) = \frac{5}{6}$, $P(A_1 A_2) = \frac{1}{3}$ and $P(\overline{A_2}) = \frac{1}{2}$. Show that the events A_1, A_2 are not mutually exclusive.*

Sol. $$P(\overline{A_2}) = \frac{1}{2} \quad \text{or} \quad P(A_2) = 1 - \frac{1}{2} = \frac{1}{2}\ ; P(A_1 A_2) = \frac{1}{3}$$

$$P(A_1) \cdot P(A_2) = \frac{1}{3} \Rightarrow P(A_1) \cdot \frac{1}{2} = \frac{1}{3} \text{ or } P(A_1) = \frac{2}{3}$$

$$P(A_1) + P(A_2) - P(A_1 A_2) = \frac{2}{3} + \frac{1}{2} - \frac{1}{3} = \frac{5}{6} = P(A_1 + A_2)$$

So, A_1 and A_2 are independent events and hence cannot be mutually exclusive.

Problem 9. *The UPSC has a list of 150 persons. Out of these 50 are womens and 100 men. 125 of them know Hindi and remaining do not know Hindi. 90 of them are teachers and remaining are not teachers. What is the probability of selecting a Hindi knowing woman teacher as examiner ?*

Sol. Probability of selecting a woman $= \frac{50}{150} = \frac{1}{3}$.

Probability of selecting a teacher $= \frac{90}{150} = \frac{3}{5}$.

Probability of selecting a Hindi knowing candidate $= \frac{125}{150} = \frac{5}{6}$.

As the events are independent, the probability of board selecting a Hindi knowing woman teacher

$$= \frac{1}{3} \times \frac{3}{5} \times \frac{5}{6} = \frac{1}{6}.$$

Problem 10. *In a box, there are 6 balls of which 3 are white and 3 are black. They are drawn successively. (i) without replacement (ii) with replacement. What is the chance that colour are alternate ?*

Sol. (*i*) Without replacement P(that 6 balls drawn follow pattern WBWBWB)

$$= \frac{3}{6} \times \frac{3}{5} \times \frac{2}{4} \times \frac{2}{3} \times \frac{1}{2} \times \frac{1}{1} = \frac{1}{20}.$$

P(that six balls follow pattern BWBWBW)

$$= \frac{3}{6} \times \frac{3}{5} \times \frac{2}{4} \times \frac{2}{3} \times \frac{1}{2} \times \frac{1}{1} = \frac{1}{20}.$$

∴ The probability that the colours of balls are alternate = probability that either the pattern WBWBWB or BWBWBW is obtained $= \frac{1}{20} + \frac{1}{20} = \frac{1}{10}$.

(*ii*) With replacement. The probability of getting pattern WBWBWB

$$= \frac{3}{6} \times \frac{3}{6} \times \frac{3}{6} \times \frac{3}{6} \times \frac{3}{6} \times \frac{3}{6} = \left(\frac{1}{2}\right)^6 = \frac{1}{64}.$$

The probability of getting the pattern BWBWBW

$$= \left(\frac{3}{6}\right)^6 = \left(\frac{1}{2}\right)^6 = \frac{1}{64}$$

Required probability = P(that the colour of balls are alternate)

$$= \frac{1}{64} + \frac{1}{64} = \frac{1}{32}.$$

Problem 11. *A box 'A' contains 2 white and 4 black balls. Another box 'B' contains 5 white and 7 black balls. A ball is transferred from the box A to box B. Then a ball is transferred from box B to box A. Find the probability that it will be white.*

Sol. We have two cases.

(*i*) When the white ball is transferred

Probability of drawing white ball from box A $= \frac{2}{6} = \frac{1}{3}$.

Now, box B contains 6 white and 7 black balls.

∴ Probability of drawing a white ball from it $= \frac{6}{13}$.

∴ Probability of compound event $= \frac{1}{3} \times \frac{6}{13} = \frac{2}{13}$.

(*ii*) When black ball is transferred

Probability of drawing black ball from box A $= \frac{4}{6} = \frac{2}{3}$.

Now, box B contains 5 white and 8 black balls.

∴ Probability of drawing a white ball from it $= \frac{5}{13}$.

∴ Probability of compound event $= \frac{2}{3} \times \frac{5}{13} = \frac{10}{39}$.

These two events are mutually exclusive.

∴ Probability of transferring a ball from box A and then drawing a white ball from box B $= \frac{2}{13} + \frac{10}{39} = \frac{16}{39}$.

Problem 12. *A and B are two candidates seeking admission in PMT. Probability that A is selected is 0.5 and the probability that both A and B are selected is almost 0.3. Is it possible that the probability of B getting selected is 0.9.*

Sol. $P(A) = 0.5,\ P(A \cap B) \le 0.3$

$$P(A \cup B) = P(A) + P(B) - P(A \cap B)$$

or $$P(B) = P(A \cup B) + P(A \cap B) - P(A) \le 1 + 0.3 - 0.5 = 0.8$$

$$[\because P(A \cup B) \le 1]$$

Hence, it is not possible that probability of B getting selected is 0.9.

Problem 13. *What is the probability of getting 53 Sundays in a leap year ?*

Sol. A leap year contains 366 days. Therefore, it contains 52 Sundays and two days extra.

These two extra days have the following equally likely cases

(*i*) Saturday and Sunday

(*ii*) Sunday and Monday

(*iii*) Monday and Tuesday

(*iv*) Tuesday and Wednesday

(*v*) Wednesday and Thursday

(*vi*) Thursday and Friday

(*vii*) Friday and Saturday

Hence, there are seven exhaustive cases and two favourable cases.

$\because$ Probability of getting 53 Sundays $= \dfrac{2}{7}$.

10

Counting Techniques

FIRST COUNTING PRINCIPLE

If an event can occur in r different steps, and

Step 1 can occur in n_1 ways.

Step 2 can occur in n_2 ways.

....

....

Step r can occur in n_r ways.

Then the number of possible events that can occur is $= n_1 . n_2 . n_3 ... n_r$.

This is the fundamental principle of counting.

Example 1. *A child has four hats, three pair of gloves and five pair of socks. Determine different possible triplets he can wear ?*

Sol. A hat can be selected in four ways.

A pair of gloves can be selected in three ways.

A pair of socks can be selected in five ways.

$\therefore$ By principle of counting.

Total number of possible triplets the child can wear are $= 4 \times 3 \times 5 = 60$ ways.

Example 2. *A person has to arrange five books on a shelf. In how many ways can he do so ?*

Sol. The first book can be arranged in 5 ways.

The second book can be arranged in 4 ways.

The third book can be arranged in 3 ways.

The fourth book can be arranged in 2 ways.

The fifth book can be arranged in 1 way.

Thus, by principle of counting,

Total number of ways five books can be arranged is $= 5 \times 4 \times 3 \times 2 \times 1 = 120$ ways.

Theorem I. *Prove that a set containing n elements has 2^n subsets. Use the first principle of counting.*

Proof. As we have n elements in the set, a subset can be constructed in n different steps *i.e.*,

Take or do not take first element.

Take or do not take second element.

....

....

Take or do not take nth element.

So, each step can be done in two different ways.

Hence the possible number of subsets is = 2 . 2 . 2 . 2 ... n times = 2^n.

Example 3. *How many different 8-bit strings are there that begin and end with one.*

Sol. A 8-bit string that begins and end with 1 can be constructed in 6 steps *i.e.*,

By selecting IInd bit, IIIrd bit, IVth bit, Vth bit, VIth bit and VIIth bit and each bit can be selected in 2 ways.

Hence, the total number of 8-bit strings that begins and end with 1 is

$$= 2 . 2 . 2 . 2 . 2 . 2 = 2^6.$$

Example 4. *How many different 2-digit numbers can be made from the digits 1, 2, 3, 4, 5, 6, 7, 8, 9, 0 ? When repetition is allowed ? When repetition is not allowed ?*

Sol. When repetition is allowed

The tens place can be filled by 10 ways and the units place can be filled by 10 ways.

∴ The total number of 2 digit numbers = 10 × 10 = 100.

When repetition is not allowed

The tens place can be filled by 10 ways and the units place can be filled by 9 ways.

∴ The total number of 2-digit numbers = 10 × 9 = 90.

Example 5. *A five person committee having members Ankit, Arjit, Sonu, Monu and Nonu is to select a president, vice-president and secretary.*

(*a*) *How many selections exclude Nonu ?*

(*b*) *How many selections include Sonu and Monu ?*

(*c*) *How many selections exclude Sonu and Monu ?*

(*d*) *How many selections are there is which Ankit is president ?*

Sol. (*a*) After excluding Nonu, we have to select three persons from the remaining four. Therefore, president can be selected in 4 ways, vice-president can be selected in 3 ways, and secretary can be selected in 2 ways.

Hence, the total number of selections that exclude Nonu is = 4 × 3 × 2 = 24.

(*b*) We have 3 ways to assign any post to Sonu. After selecting Sonu, there are 2 ways to assign any post to Monu. After selecting Sonu and Monu, we can assign the remaining post to any of the three persons.

Hence, the total number of selections that include Sonu and Monu is = 3 × 2 × 3 = 18.

(*c*) After excluding Sonu and Monu, we have to select three persons from the remaining three. Therefore, president can be selected in 3 ways, vice-president can be selected in 2 ways, and secretary can be selected in 1 way.

Hence, the total number of selections that exclude Sonu and Monu is = 3 × 2 × 1 = 6.

(*d*) When Ankit is selected as president, then we have to select vice-president and secretary from the remaining four. Therefore,

Vice-president can be selected in 4 ways and secretary can be selected in 3 ways.

Hence, the total number of selection in which Ankit is president is $= 4 \times 3 = 12$.

Example 6. *Ram has five different 'Data Structure Books', four different 'Discrete Structure Books' and five different 'Programming Language Books'.*

(*a*) *In how many ways Ram can arrange these books on a shelf ?*

(*b*) *In how many ways can these books be arranged on a shelf if all five programming language books are on the right ?*

(*c*) *In how many ways can these books be arranged on a shelf if all five data structure books are on the left and all five programming language books are on the right ?*

Sol. We have total $= 5 + 4 + 5 = 14$ books.

(*a*) The first book can be arranged in 14 ways.

The second book can be arranged in 13 ways.

The third book can be arranged in 12 ways.

....

....

The fourteenth book can be arranged in 1 way.

Hence, the total number of ways the book can be placed on a shelf

$$= 14 \times 13 \times 12 \times 11 \times 10 \times 9 \times 8 \times 7 \times 6 \times 5 \times 4 \times 3 \times 2 \times 1$$

$$= 8.717 \times 10^{10}.$$

(*b*) We have to place all the five programming language books on the right. Thus, the remaining books to be arranged are $= 14 - 5 = 9$.

The first book can be arranged in 9 ways.

The second book can be arranged in 8 ways.

....

....

The ninth book can be arranged in 1 way.

Hence, the total number of ways arranging remaining books is $= 9 \times 8 \times 7 \times 6 \times 5 \times 4 \times 3 \times 2 \times 1 = 362880$.

The five programming language books can be arranged in following ways :

The first one can be arranged in 5 ways.

The second one can be arranged in 4 ways.

....

....

The fifth one can be arranged in 1 way.

The total number of ways arranged programming language books is

$$= 5 \times 4 \times 3 \times 2 \times 1 = 120.$$

The total number of ways to arrange the books on a shelf when all five programming language books are on right is

$$= 362880 \times 120 = 43545600.$$

(*c*) As explained earlier, we have 120 ways to place five programming language books on the shelf. Also we have 120 ways to arrange five data structure books on the shelf. Similarly, we have 24 ways to arrange four discrete structure books on the shelf.

Hence, the total number of ways the book can be arranged on a shelf if all five data structure books are on the left and all five programming language books are on the right is

$$= 120 \times 120 \times 24 = 345600.$$

SECOND COUNTING PRINCIPLE

Consider that $\{D_1, D_2, D_3, \ldots, D_r\}$ is a pairwise disjoint family of sets and some set D_j has n_j elements. Then the number of possible selection of elements from the sets D_1 or D_2 or D_3 or D_4 or ... D_r is

$$n_1 + n_2 + n_3 + n_4 + \ldots + n_r.$$

We can also define this principle in another way, consider an event A_1 can occur in n_1 ways and another event A_2 can occur in n_2 ways and A_1 and A_2 are mutually exclusive, then A_1 or A_2 can occur in $(n_1 + n_2)$ ways. It is applicable for any number of events.

Example 7. *A five person committee having members Ankit, Arjit, Sonu, Monu and Nonu is to select a president, vice-president and secretary.*

(*a*) *In how many ways can this occur if either Sonu or Monu must be president ?*

(*b*) *How many selections are there in which either Nonu is a secretary or he is excluded ?*

(*c*) *How many selections exclude Ankit or Arjit ?*

Sol. (*a*) If Sonu is president, then vice-president can be selected in 4 ways and secretary can be selected in 3 ways.

Hence, the total number of ways to select the remaining is $= 4 \times 3 = 12$.

Similarly, if Monu is president, then the remaining persons can be selected in 12 ways as shown above.

As these are mutually exclusive events, hence the total number of ways if either Sonu or Monu must be president is

$$= 12 + 12 = 24.$$

(*b*) If Nonu is a secretary, then the remaining two posts can be filled in 12 ways as discussed earlier. The number of selections in which Nonu is excluded is 24.

As these are mutually exclusive events, hence the total number of selections in which either Nonu is a secretary or he is excluded at all is

$$= 12 + 24 = 36.$$

(*c*) The number of selections in which Ankit is excluded :

Since president can be selected in from the remaining 4 persons.

Vice-president can be selected from remaining 3 persons after setting first.

Secretary can be selected from remaining 2 persons after setting second.

Hence, total number of selections $= 4 \times 3 \times 2 = 24$.

Similarly, the number of selections in which Arjit is excluded is 24.

These two sets of selections are disjoint, hence the total number of selections in which Ankit or Arjit are excluded is

$$= 24 + 24 = 48.$$

DEFINE FACTORIAL *N*

The product of first n natural number is called factorial n. It is denoted by $n\,!$ or $\underline{|n}$. The factorial n can also be written as

$$n\,! = n(n-1)(n-2)(n-3)\ \ldots\ldots\ldots\ 1.$$

We have, $\quad 1\,! = 1$ and $0\,! = 1.$

Example 8. *Find the value of 5 !.*

Sol. $\quad 5\,! = 5 \times (5-1)(5-2)(5-3)(5-4) = 5 \times 4 \times 3 \times 2 \times 1 = 120.$

Example 9. *Find the value of* $\dfrac{10\,!}{8\,!}$.

Sol. $\quad \dfrac{10!}{8!} = \dfrac{10 \times 9 \times 8!}{8!} = 10 \times 9 = 90.$

Example 10. *Determine the value of* $\dfrac{n\,!}{(n-1)\,!}$.

Sol. $\quad \dfrac{n!}{(n-1)!} = \dfrac{n(n-1)\,!}{(n-1)\,!} = n.$

Example 11. *Find the value of* $\dfrac{n\,!}{r\,!(n-r)\,!}$, *when* $n = 6$, $r = 4$.

Sol. $\dfrac{n\,!}{r\,!(n-r)\,!}$. Substitute the value of n and r.

We have $\quad \dfrac{6!}{4!\,(6-4)\,!} = \dfrac{6!}{4!\,2!} = \dfrac{6 \times 5 \times 4\,!}{4\,! \times 2} = 15.$

Example 12. *Find the value of z, if* $\dfrac{1}{4\,!} + \dfrac{1}{5\,!} = \dfrac{z}{6\,!}$.

Sol. We have $\quad \dfrac{1}{4!} + \dfrac{1}{5!} = \dfrac{z}{6!} = \dfrac{5+1}{5!} = \dfrac{z}{6!}$

$$\dfrac{6}{5!} = \dfrac{z}{6!}\ ;\ z = \dfrac{6 \times 6!}{5!}\ ;\ z = \dfrac{6 \times 6 \times 5!}{5!}\ ;\ z = 36.$$

Example 13. *Show that* $3\,! + 4\,! \neq (3+4)\,!$.

Sol. $\quad 3\,! + 4\,! = (3 \times 2 \times 1) + (4 \times 3 \times 2 \times 1)$

$\quad = 6 + 24 = 30$

and $\quad (3+4)\,! = 7\,! = 7 \times 6 \times 5 \times 4 \times 3 \times 2 \times 1 = 5040$

Hence $\quad 3\,! + 4\,! \neq (3+4)\,!$.

Example 14. *Show that* $10\,! - 8\,! \neq (10-8)\,!$.

Sol. $\quad 10\,! - 8\,! = 3628800 - 40320 = 3588480$

and $\quad (10-8)\,! = (2)\,! = 2$

Hence $\quad 10\,! - 8\,! \neq (10-8)\,!$.

PERMUTATION

A permutation is an arrangement of a no. of objects in some definite order taken some or all at a time. The total number of permutations of n distinct objects taken r at a time is denoted by ${}^nP_r$ or $P(n, r)$, where $1 \le r \le n$.

Theorem II. *Prove that the number of different permutations of n distinct objects taken r at a time, $r \le n$ is given by*

$${}^nP_r = \frac{n!}{(n-r)!} = n\,(n-1)(n-2)\,...(n-r+1).$$

Proof. The number of permutations of n distinct objects taken r at a time is like filling of r places with n objects.

The first place can be filled in by any one of the n objects. So, this can be done in n ways $= {}^nP_1 = n$.

The second place can be filled in by any one of the $n - 1$ objects because after filling first place. We are left with $(n - 1)$ objects. Thus, the first two places can be filled in $n(n - 1)$ ways.

$$\therefore \qquad {}^nP_2 = n(n-1)$$

Similarly, the third place can be filled in by any one of the remaining $(n - 2)$ objects.

Therefore, the first three places can be filled in $n(n - 1)\,(n - 2)$ ways.

Proceeding in this way, we have the number of permutations of n different objects taken r at a time

$$= n(n-1)(n-2)\,...\,r$$

$$= n(n-1)(n-2)\,...(n-\overline{r-1})$$

$${}^nP_r = n(n-1)(n-2)\,...\,(n-r+1).$$

Theorem III. *Prove that the number of permutations of n things taken all at a time is n !.*

Proof. We know that

$${}^nP_n = \frac{n!}{(n-n)!} = \frac{n!}{0!} = \frac{n!}{1} \qquad [\because 0! = 1]$$

$$= n!$$

Hence proved.

Example 15. *Determine the value of the following*

(i) ${}^4P_2$ *(ii)* ${}^9P_3$ *(iii)* ${}^{20}P_2$ *(iv)* ${}^{52}P_4$.

Sol. *(i)* ${}^4P_2 = \dfrac{4!}{(4-2)!} = \dfrac{4 \times 3 \times 2!}{2!} = 12$

(ii) ${}^9P_3 = \dfrac{9!}{(9-3)!} = \dfrac{9 \times 8 \times 7 \times 6!}{6!} = 504$

(iii) ${}^{20}P_2 = \dfrac{20!}{(20-2)!} = \dfrac{20 \times 19 \times 18!}{18!} = 380$

(iv) ${}^{52}P_4 = \dfrac{52!}{(52-4)!} = \dfrac{52 \times 51 \times 50 \times 49 \times 48!}{48!} = 6497400.$

Example 16. *Determine the value of n if*

(i) $4 \times {}^nP_3 = {}^{n+1}P_3$. *(ii)* $6 \times {}^nP_3 = 3 \times {}^{n+1}P_3$. *(iii)* $3 \times {}^nP_4 = 7 \times {}^{n-1}P_4$.

Sol. *(i)*

$$4 \times \frac{n!}{(n-3)!} = \frac{(n+1)!}{(n+1-3)!} \qquad \left[\because {}^nP_r = \frac{n!}{(n-r)!}\right]$$

$$\frac{4 \times n!}{(n-3)!} = \frac{(n+1) \times n!}{(n-2)(n-3)!}$$

$$4(n-2) = (n+1)$$

$$4n - 8 = n + 1$$

$$3n = 9$$

$$n = 3.$$

(ii)

$$6 \times {}^nP_3 = 3 \times {}^{n+1}P_3$$

$$6 \times \frac{n!}{(n-3)!} = 3 \times \frac{(n+1)!}{(n+1-3)!} \qquad \left[\because {}^nP_r = \frac{n!}{(n-r)!}\right]$$

$$\frac{6 \times n!}{(n-3)!} = \frac{3(n+1)(n!)}{(n-2)(n-3)!}$$

$$6(n-2) = 3 \times (n+1)$$

$$6n - 12 = 3n + 3$$

$$6n - 3n = 12 + 3$$

$$3n = 15$$

$$n = 5.$$

(iii)

$$3 \times {}^nP_4 = 7 \times {}^{n-1}P_4$$

$$3 \times \frac{n!}{(n-4)!} = 7 \times \frac{(n-1)!}{(n-1-4)!} \qquad \left[\because {}^nP_r = \frac{n!}{(n-r)!}\right]$$

$$\frac{3 \times n \times (n-1)!}{(n-4)(n-5)!} = \frac{7 \times (n-1)!}{(n-5)!}$$

$$3n = 7(n-4)$$

$$3n = 7n - 28$$

$$3n - 7n = -28$$

$$-4n = -28$$

$$n = 7.$$

Example 17. *How many variable names of 8 letters can be formed from the letters a, b, c, d, e, f, g, h, i if no letter is repeated.*

Sol. There are 9 letters and 8 are to be selected.

$\therefore$ Total number of variable names of 8 letters is $= {}^9P_8 = \frac{9!}{(9-8)!} = \frac{9!}{1!} = 9!$.

Example 18. *There are 10 persons called on an interview. Each one is capable to be selected for the job. How many permutation are there to select 4 from the 10.*

Sol. There are 10 persons and 4 are to be selected.

$\therefore$ Total number of permutations to select 4 persons is = ${}^{10}P_4$

$$= \frac{10!}{(10-4)!} = \frac{10 \times 9 \times 8 \times 7 \times 6!}{6!} = 5040.$$

Example 19. *How many 6-digit numbers can be formed from the digits 0, 1, 2, 3, 4, 5, 6, 7, if no digit is repeated.*

Sol. There are 8 numbers are 6 are to be selected.

$$\therefore \quad \text{Total number of 6-digit numbers} = {}^{8}P_6 = \frac{8!}{(8-6)!} = \frac{8!}{2!}$$

$$= \frac{8 \times 7 \times 6 \times 5 \times 4 \times 3 \times 2!}{2!} = 22560.$$

Permutation with Restrictions. *The number of permutations of n different objects taken r at a time in which p particular objects do not occur is

$$^{n-p}P_r$$

*The number of permutations of n different objects taken r at a time in which p particular objects are present is

$$^{n-p}P_{r-p} \times {}^{r}P_p.$$

Example 20. *How many 6-digit numbers can be formed by using the digits 0, 1, 2, 3, 4, 5, 6, 7, 8 if every number is to start with '30' with no digit repeated.*

Sol. All the numbers begin with '30'. So, we have to choose 4-digits from the remaining 7-digits.

$\therefore$ Total number of numbers that begins with '30' is

$$^{7}P_4 = \frac{7!}{(7-4)!} = \frac{7 \times 6 \times 5 \times 4 \times 3!}{3!} = 840.$$

Example 21. *In how many ways 5 different microprocessor books and 4 different digital electronics books be arranged in a shelf so that all the four digital electronics books are together ?*

Sol. Consider the four digital electronics books as one unit. Thus, we have 6 units that can be arranged in 6 ! ways.

For each of these arrangements, 4 digital electronics books can be arranged among themselves in 4 ! ways.

$\therefore$ Total number of arrangements in which all the four digital electronics books are together is

$$= 6! \times 4! = 720 \times 24 = 17280.$$

Example 22. *How many permutations can be made out of the letter of word "COMPUTER" ? How many of these*

(i) begin with C ? *(ii) end with R ?*

(iii) begin with C and end with R ? *(iv) C and R occupy the end places ?*

Sol. There are 8 letters in the word 'COMPUTER' and all are distinct.

$\therefore$ The total number of permutations of these letters is 8 ! = 40320.

(*i*) Permutations which begin with C.

The first position can be filled in only one way *i.e.*, C and the remaining 7 letters can be arranged in 7 ! ways.

∴ Total number of permutations starting with C are

$$= 1 \times 7\,! = 5040.$$

(ii) Permutations which end with R.

The last position can be filled in only one way *i.e.*, R and the remaining 7 letters can be arranged in 7 ! ways.

∴ The total number of permutations ending with R are $= 7\,! \times 1 = 5040$.

(iii) Permutations begin with C and end with R.

The first position can be filled in only one way *i.e.*, C and the last place can also be filled in only one way *i.e.*, R and the remaining 6 letters can be arranged in 6 ! ways.

∴ The total number of permutations begin with C and end with R is

$$= 1 \times 6\,! \times 1 = 720.$$

(iv) Permutations is which C and R occupy end places.

C and R occupy end positions in 2 ! ways *i.e.*, C, R and R, C and the remaining 6 letters can be arranged in 6 ! ways.

∴ The total number of permutations in which C and R occupy end places is

$$= 2\,! \times 6\,! = 1440.$$

PERMUTATIONS WHEN ALL OF THE OBJECTS ARE NOT DISTINCT

Theorem IV. *The number of permutations of n objects, of which n_1 objects are of one kind and n_2 objects of another kind, when all are taken at a time is* $\dfrac{n\,!}{n_1\,!\,n_2\,!}$.

Proof. Let us assume that the number of required permutations be K. Now consider a single particular permutation of these K permutations, in which n_1 objects of one kind is followed by n_2 objects of other kind.

Also, assume that all n_1 object are distinct from all n_2 objects.

So, number of permutations of n_1 objects taken all at a time is $= {}^{n_1}P_{n_1} = n_1\,!$

Also, the number of permutations of n_2 objects taken all at a time is $= {}^{n_2}P_{n_2} = n_2\,!$

By the fundamental principle of counting, these K permutations will give rise to $n_1\,!\,n_2\,!$ permutations by arranging the objects of one kind within the places occupied by them.

Therefore, K permutations will give rise to K. $n_1\,!\,n_2\,!$ permutations.

For n distinct objects, the number of permutations is $= {}^nP_n = n\,!$

Therefore, $\quad K \times n_1\,!\,n_2\,! = n\,!$

$$K = \frac{n\,!}{n_1\,!\,n_2\,!}$$

This result can be generalised as follows :

If n_1 objects are of one kind, n_2 objects are of second kind, n_3 objects are of third kind, and so on upto n_t objects are of tth type is given by

$$\frac{n\,!}{n_1\,!\,n_2\,!\,n_3\,!\ldots n_t\,!} \qquad [\text{Here } n_1 + n_2 + n_3 + \ldots + n_t = n]$$

Example 23. *Determine the number of permutations that can be made out of the letters of the word 'PROGRAMMING'.*

Sol. There are 11 letters in the word 'PROGRAMMING' out of which G's and M's and R's are two each.

$\therefore$ The total number of permutations is

$$= \frac{11!}{2! \times 2! \times 2!} = \frac{11 \times 10 \times 9 \times 8 \times 7 \times 6 \times 5 \times 4 \times 3 \times 2!}{2 \times 1 \times 2 \times 1 \times 2!}$$

$$= 4989600.$$

Example 24. *There are 4 blue, 3 red and 2 black pens in a box. These are drawn one by one. Determine all the different permutations.*

Sol. There are total 9 pens in the box out of which 4 are blue, 3 are red and 2 are black.

$\therefore$ The total number of permutations is

$$= \frac{9!}{4! \times 3! \times 2!} = \frac{9 \times 8 \times 7 \times 6 \times 5 \times 4!}{4! \times 3 \times 2 \times 1 \times 2 \times 1} = 1080.$$

Example 25. *How many different variable names can be formed by using the letters a, a, a, b, b, b, b, c, c, c ?*

Sol. There are total 10 letters out of which 3 are a's, 4 are b's and there are 3 c's.

$\therefore$ Total number of permutations is

$$= \frac{10!}{3! \times 4! \times 3!} = \frac{10 \times 9 \times 8 \times 7 \times 6 \times 5 \times 4!}{3 \times 2 \times 1 \times 4! \times 3 \times 2 \times 1}$$

$$= 10 \times 3 \times 4 \times 7 \times 5 = 4200.$$

Example 26. *How many 7-digits numbers can be formed using digits 1, 7, 2, 7, 6, 7, 6 ?*

Sol. There are total 7-digits out of which 3 are 7's and 2 are 6's.

$\therefore$ Total number of permutations is $= \frac{7!}{3! \times 2!} = 420.$

PERMUTATIONS WITH REPEATED OBJECTS

Theorem V. *Prove that the number of different permutations of n distinct objects taken r at a time when every object is allowed to repeat any number of times is given by n^r.*

Proof. Assume that with n objects we have to fill r place when repetition of objects is allowed.

Therefore, the number of ways of filling the first place is $= n$

The number of ways of filling second place $= n$

....

....

The number of ways of filling rth place $= n$

Thus, the total number of ways of filling r places with n elements is

$$= n.\ n.\ n.\ n.\ \ldots\ r \text{ times} = n^r.$$

Example 27. *How many 4-digits numbers can be formed by using the digits 2, 4, 6, 8 when repetition of digits is allowed.*

Sol. We have 4-digits.

So, number of ways of filling unit's place = 4.

Number of ways of filling ten's place = 4.

Number of ways of filling hundred's place = 4.

Number of ways of filling thousand's place = 4.

Therefore, the total number of 4-digits numbers is

$$= 4 \times 4 \times 4 \times 4 = 256.$$

Example 28. *How many 2-digits even numbers can be formed by using the digits 1, 3, 4, 6, 8 when repetition of digits is allowed.*

Sol. We have three even numbers and two odd number.

Thus, number of ways of filling unit's place = 3.

Number of ways of filling ten's place = 5.

$\therefore$ Total number of two digits even numbers = $3 \times 5 = 15$.

Example 29. *In how many ways can 5 software projects be allotted to 6 final year students when all the 5 projects are not allotted to the same student.*

Sol. We have 5 projects and 6 students.

Each projects can be allotted in 6 ways.

Thus, the number of ways of alloting 5 projects is = $6 \times 6 \times 6 \times 6 \times 6 = 6^5$.

Number of ways in which all projects allotted to same student is = 6.

Therefore, total number of ways to allocate 5 projects to 6 students is = $6^5 - 6 = 7770$.

CIRCULAR PERMUTATIONS

The circular permutations are the permutations of the objects placed in a circle. Consider the letters *k, l, m, n, o* placed along the circle as shown in Fig. 1.

If we place letters linearly, there are five different permutations *i.e.*, *k, l, m, n, o* ; *l, m, n, o, k* ; *m, n, o, k, l* ; *n, o, k, l, m* ; *o, k, l, m, n,* but there is only one circular permutation *k, l, m, n, o*. Therefore, there is no starting and ending in circular permutation. We only consider the relative positions.

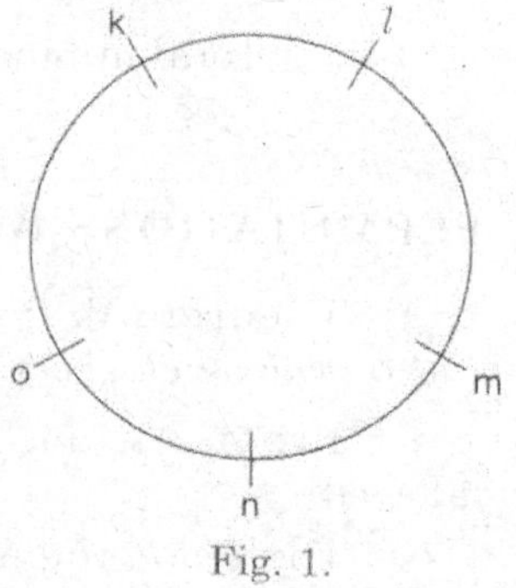

Fig. 1.

Theorem VI. *Prove that the number of circular permutations of n different objects is* $(n - 1)!$.

Proof. Let us consider that K be number of permutations required.

For each such circular permutation of K, there are n corresponding linear permutations. As shown earlier, we start from every object of n object in the circular permutation. Thus, for K circular permutations, we have K. n linear permutations.

Therefore, $\quad K.\, n = n!$ or $K = \dfrac{n!}{n}$

or
$$K = \frac{n \times (n-1)!}{n}$$
$$K = (n-1)!$$

Hence proved.

Example 30. *In how many ways can these letters a, b, c, d, e, f be arranged in a circle ?*

Sol. There are 6 letters and hence the number of ways to arrange these 6 letters in a circle is

$$= (6-1)! = 5! = 120.$$

Example 31. *In how many ways 10 programmers can sit on a round table to discuss the project so that project leader and a particular programmer always sit together.*

Sol. There are total 10 programmers but project leader and a particular programmer always sit together. So, both become a single unit and hence there are $(10 - 2 + 1) = 9$ remains.

Thus, these 9 units can be arranged on round table in $(9 - 1)!$ ways.

The two programmers *i.e.,* project leader and a particular programmer can be arranged in $2!$ ways.

Therefore, the total number of ways in which 10 programmers can sit on a round table is

$$= (9-1)! \times 2! = 8! \times 2! = 80640.$$

Example 32. *Determine the number of ways in which 5 software engineers and 6 electronics engineers can be sitted at a round table so that no two software engineers can sit together.*

Sol. There are 6 electronics engineers that can be arranged round a table in $(6 - 1)!$ ways. There are 5 software engineers and they are not to sit together so we have six places for software engineers and can be placed in $6!$ ways as shown in Fig. 2.

Therefore, total number of ways to arrange the engineers on a round table is

$$= (6-1)! \times 6! = 5! \times 6!$$
$$= 120 \times 720$$
$$= 86400.$$

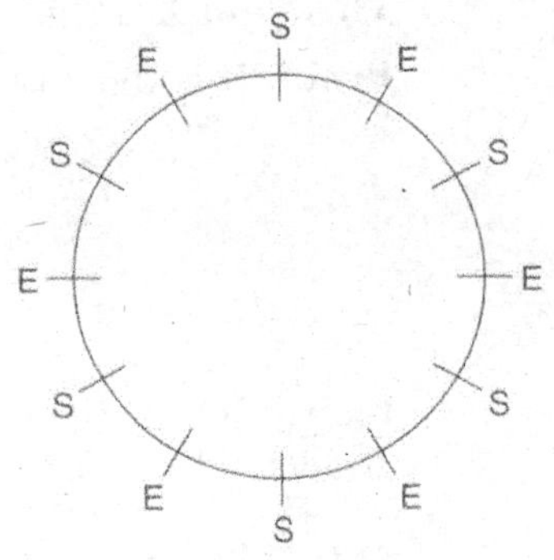

Fig. 2.

COMBINATION

A combination is a selection of some or all, objects from a set of given objects, where order of the objects does not matter. The number of combinations of n objects, taken r at a time is represented by ^nC_r or $C(n, r)$.

Theorem VII. *The number of combinations of n different things, taken r at a time is given by*

$$^nC_r = \frac{n!}{r!(n-r)!}, \quad n \geq r \geq 1.$$

Proof. The number of permutations of n different things, taken r at a time is given by

$$^nP_r = \frac{n!}{(n-r)!}$$

As there is no matter about the order of arrangement of the objects, therefore, to every combination of r things, there are $r!$ arrangements *i.e.,*

$$^nP_r = r!\,^nC_r \quad \text{or} \quad ^nC_r = \frac{^nP_r}{r!} = \frac{n!}{(n-r)!\,r!}, \; n \geq r$$

Thus, $$^nC_r = \frac{n!}{(n-r)!\,r!}.$$

Theorem VIII. *Prove that the number of combinations of n things taken all at a time is one.*

Proof. We know that

$$^nC_n = \frac{n!}{(n-n)!\,n!} = \frac{n!}{0!\,n!} = 1 \qquad [\because \; 0! = 1]$$

Theorem IX. *Prove that the number of combinations of n things taken none at a time is one.*

Proof. We know that

$$^nC_0 = \frac{n!}{(n-0)!\,0!} = \frac{n!}{n!\,0!} = \frac{n!}{n!} = 1 \qquad [\because 0! = 1]$$

Theorem X. *Prove that* $^nC_{n-r} = {}^nC_r$, $n \geq r \geq 1$.

Proof. We know that

$$^nC_{n-r} = \frac{n!}{n-(n-r)!\,(n-r)!} = \frac{n!}{(n-n+r)!\,(n-r)!}$$

$$= \frac{n!}{r!\,(n-r)!} = {}^nC_r.$$

Example 33. *Determine the value of following*

(*i*) $^{10}C_6$ (*ii*) $^{50}C_{45}$ (*iii*) $^{52}C_4$ (*iv*) $^{20}C_{10}$.

Sol. (*i*) $$^{10}C_6 = \frac{10!}{(6)!\times(10-6)!} = \frac{10\times9\times8\times7\times6!}{6!\times4\times3\times2\times1} = 10\times3\times7 = 210.$$

(*ii*) $$^{50}C_{45} = \frac{50!}{45!\times(50-45)!} = \frac{50\times49\times48\times47\times46\times45!}{45!\times5\times4\times3\times2\times1} = 2118760.$$

(*iii*) $$^{52}C_4 = \frac{52!}{4!\times(52-4)!} = \frac{52\times51\times50\times49\times48!}{4\times3\times2\times1\times48!} = 270725.$$

(*iv*) $$^{20}C_{10} = \frac{20!}{10!\times(20-10)!} = \frac{20\times19\times18\times17\times16\times15\times14\times13\times12\times11\times10!}{10!\times10!}$$
$$= 184756.$$

Example 34. *Determine the value of n if*

(*i*) $^nC_4 = {}^nC_3$ (*ii*) $^nC_{n-2} = 10$ (*iii*) $^{20}C_{n+2} = {}^{20}C_{2n-1}$.

Sol. *(i)* $${}^nC_4 = {}^nC_3$$

$$\frac{n!}{4! \times (n-4)!} = \frac{n!}{3!(n-3)!}$$

$$\frac{n!}{n!} = \frac{4! \times (n-4)!}{3! \times (n-3)!} = \frac{4 \times 3! \times (n-4)!}{3! \times (n-3) \times (n-4)!}$$

$$1 = \frac{4}{n-3}$$

$$n - 3 = 4$$

$$n = 7.$$

(ii) $${}^nC_{n-2} = 10$$

Thus $$\frac{n!}{(n-2)![n-(n-2)]!} = 10 \quad \text{or} \quad \frac{n!}{(n-2)! \times 2!} = 10$$

or $$\frac{n \times (n-1) \times (n-2)!}{(n-2)! \times 2!} = 10$$

$$n \times (n-1) = 10 \times 2!$$

$$n^2 - n - 20 = 0$$

$n = -4, 5$. Since -4 is not possible, hence $n = 5$.

(iii) $${}^{20}C_{n+2} = {}^{20}C_{2n-1}$$

Therefore, we have either

$$n + 2 = 2n - 1 \quad \text{or} \quad (n+2) + (2n-1) = 25$$

$$-n = -3 \quad \text{or} \quad 3n = 24$$

$$n = 3 \qquad n = 8$$

So $$n = 3, 8.$$

Example 35. *How many 16-bit strings are there containing exactly five 0's ?*

Sol. A 16-bit string having exactly five 0's is determined if we tell which bits are 0's. This can be done in ${}^{16}C_5$ ways.

Therefore, the total number of 16-bit strings is

$$= {}^{16}C_5 = \frac{16!}{5! \times (16-5)!}$$

$$= \frac{16 \times 15 \times 14 \times 13 \times 12 \times 11!}{5 \times 4 \times 3 \times 2 \times 1 \times 11!} = 4368.$$

Example 36. *How many ways can we select a software development group of 1 project leader, 5 programmers and 6 data entry operators from a group of 5 project leaders, 20 programmers and 25 data entry operators ?*

Sol. There are 5 project leaders out of which one can be selected in ${}^5C_1$ ways.

There are 20 programmers out of which five can be selected in ${}^{20}C_5$ ways.

There are 25 data entry operators out of which six can be selected in ${}^{25}C_6$ ways.

Therefore, the total number of ways to select the software development group is

$$= {}^5C_1 \times {}^{20}C_5 \times {}^{25}C_6 = 96101544000.$$

Example 37. *From 10 programmers in how many ways can 5 be selected when*

(a) A particular programmer is included every time.

(b) A particular programmer is not included at all.

Sol. We have to select 5 programmers from the 10 programmers. So, the number of ways to select them in $^{10}C_5$

$$= \frac{10!}{5! \times (10-5)!} = \frac{10 \times 9 \times 8 \times 7 \times 6 \times 5!}{5 \times 4 \times 3 \times 2 \times 1 \times 5!}$$
$$= 252.$$

(*a*) When a particular programmer is included every time then the remaining = 5 – 1 = 4 programmers can be selected from the remaining = 10 – 1 = 9 programmers. This can be done in $^9C_4$ ways

$$= \frac{9!}{4!(9-4)!} = \frac{9 \times 8 \times 7 \times 6 \times 5!}{4 \times 3 \times 2 \times 1 \times 5!} = 126.$$

(*b*) When a particular programmer is not included at all, then the five programmers can be selected from the remaining = 10 – 1 = 9 programmers.

This can be done in $^{11}C_5$ ways

$$= \frac{9!}{5!(9-5)!} = \frac{9 \times 8 \times 7 \times 6 \times 5 \times 4!}{5 \times 4 \times 3 \times 2 \times 1 \times 4!} = 126.$$

THE PIGEONHOLE PRINCIPLE

Theorem XI. *Show that if n pigeons are assigned to m pigeonholes and $m < n$, then there is at least one pigeonhole that contains two or more pigeons.*

Proof. Let us label the n pigeons with the numbers 1 through n and the m pigeonholes with the numbers 1 through m. Now starting with pigeon 1 and Pigeonhole 1, assign each pigeon in order to the pigeonhole with the same number. So we can assign as many pigeons as possible to distinct pigeonholes, but as we know that pigeonholes are less than pigeons *i.e.*, $m < n$. Thus, there remains $n - m$ pigeons that have not yet been assigned to a pigeonhole. Hence, there is at least one pigeonhole that will be assigned a second pigeon.

Example 38. *Show that if any four numbers from 1 to 6 are chosen, then two of them will add to 7.*

Sol. Make three sets containing two numbers whose sum is 7.

A = {1, 6}, B = {2, 5}, C = {3, 4}. The four numbers that will be chosen assigned to the set that contains it.

As there are only three sets, two numbers that are chosen is from the same set whose sum is 7.

Example 39. *Show that at least two people must have their birthday in the same month if 13 people are assembled in a room.*

Sol. We assigned each person the month of the year on which he was born. Since there are 12 months in a year.

So, according to pigeonhole principle, there must be at least two people assigned to the same month.

Example 40. *Show that if any eight + ve integers are chosen, two of them will have same remainder when divided by 7.*

Sol. Take any eight +ve integers. When these are divided by 7 each have some remainder. Since there are eight integers and only seven distinct remainders because number 7 can generate only 7 remainders, so two +ve integers must have same remainder.

EXTENDED PIGEONHOLE PRINCIPLE

It states that if n pigeons are assigned to m pigeonholes (The number of pigeons is very large than the number of pigeonholes), then one of the pigeonholes must contain at least $[(n-1)/m]+1$ pigeons.

Theorem XII. *Prove that extended pigeonhole principle.*

Proof. We can prove this by the method of contradiction. Assume that each pigeonhole does not contain more than $[(n-1)/m]$ pigeons. Then, there will be at most $m[(n-1)/m] \le m(n-1)/m = n-1$ pigeons in all. This is in contradiction to our assumptions. Hence, for given m pigeonholes, one of these must contain at least $[(n-1)/m]+1$ pigeons.

Example 41. *Show that if 9 colours are used to paint 100 houses, at least 12 houses will be of the same colour.*

Sol. Let us assume the colours be the pigeonholes and the houses the pigeons. Now 100 pigeons are to be assigned to 9 pigeonholes. Using the extended pigeonhole principle, $[(n-1)/m]+1$, where $n=100$ and $m=9$, we have $[(100-1)/9]+1=12$. Thus, there are 12 houses of the same colour.

SOLVED PROBLEMS

Problem 1. *How many different 8-bit strings are there that end with 0111 ?*

Sol. An 8-bit strings that end with 0111 can be constructed in 4 steps *i.e.*,

By selecting Ist bit, IInd bit, IIIrd bit and IVth bit and each bit can be selected in 2 ways.

Hence, the total no. of 8-bit strings that end with 0111 is

$$= 2 \,.\, 2 \,.\, 2 \,.\, 2 = 2^4.$$

Problem 2. *How many 2-digits numbers greater than 40 can be formed by using the digits 1, 2, 3, 4, 6, 7*

(a) When repetition is allowed (b) When repetition is not allowed.

Sol. (*a*) When repetition is allowed

We have to find the numbers greater than 40. Therefore,

Ten's place can be filled up by 3 ways.

Unit's place can be filled up by 6 ways.

∴ The total number of 2-digits numbers greater than 40 is $= 3 \times 6 = 18$.

(*b*) When repetition is not allowed

Ten's place can be filled up by 3 ways.

Unit's place can be filled up by 5 ways.

∴ The total number of 2-digits numbers greater than 40 is $= 3 \times 5 = 15$.

Problem 3. *How many words can be constructed of three English alphabets ?*

(a) When repetition of alphabets is allowed

(b) When repetition is not allowed.

Sol. There are 26 alphabets in English. Therefore,

(*a*) When repetition is allowed

First alphabet of word can be selected in 26 ways.

Second alphabet of word can be selected in 26 ways.

Third alphabet of word can be selected in 26 ways.

Hence, total number of words of three alphabets constructed is

$$= 26 \times 26 \times 26 = 17576.$$

(*b*) When repetition is not allowed

First alphabet of word can be selected in 26 ways.

Second alphabet of word can be selected in 25 ways.

Third alphabet of word can be selected in 24 ways.

Hence, the total number of words of three distinct alphabets is $= 26 \times 25 \times 24 = 15600$.

Problem 4. *Show that* $0\,! = 1$.

Sol. We have

$$^{n}P_{r} = \frac{n\,!}{(n-r)\,!} \qquad \ldots(i)$$

Now put $r = n$ in equation (*i*), we have

$$^{n}P_{n} = \frac{n\,!}{(n-n)\,!}$$

$$n\,! = \frac{n\,!}{0\,!}$$

$$0\,! = \frac{n\,!}{n\,!}$$

Hence $0\,! = 1$.

Problem 5. *There are n objects out of which r objects are to be arranged. Find the total number of permutations when*

(a) four particular objects always occur.

(b) four particular objects never occur.

Sol. (*a*) Number of ways to arrange first object $= r$

Number of ways to arrange second object $= r - 1$

Number of ways to arrange third object $= r - 2$

Number of ways to arrange fourth object $= r - 3$

Number of ways to arrange remaining $n - 4$ objects taking $r - 4$ at a time $= {}^{n-4}P_{r-4}$.

Therefore, the total number of permutation when four particular objects always occur is

$$= r(r-1)(r-2)(r-3)\,{}^{n-4}P_{r-4}. \qquad \text{[Using first principle of counting]}$$

(*b*) There are four particular objects which never occur in any arrangement. Hence set aside these four particular objects.

Thus, we have to find the number of arrangements of $n - 4$ objects taking r at a time.

$\therefore$ The total number of arrangements is $= {}^{n-4}P_r$.

Problem 6. *How many permutations can be made out of the letters of the word "Basic"? How many of these*

(i) begin with B? *(ii) end with C?*

(iii) B and C occupy the end places?

Sol. There are 5 letters in the word 'Basic' and all are distinct.

$\therefore$ The number of permutations of these letters is

$$= 5\,! = 5 \times 4 \times 3 \times 2 \times 1 = 120.$$

(*i*) Permutations which begin with B

The first position can be filled in only one way *i.e.*, B and the remaining 4 letters can be arranged in 4 ! ways.

$\therefore$ Total number of permutations starting with B is $= 1 \times 4\,! = 24$.

(*ii*) Permutations which end with C

The first position can be filled in only one way *i.e.*, C and the remaining 4 letters can be arranged in 4 ! ways.

$\therefore$ Total number of permutations ending with C is

$$= 4\,! \times 1 = 24.$$

(*iii*) Permutations in which B and C occupy end places

B and C occupy end positions in 2 ! ways *i.e.*, B, C and C, B and the remaining 3 letters can be arranged in 3 ! ways.

$\therefore$ Total number of permutations in which B and C occupy end places in

$$= 2\,! \times 3\,! = 12.$$

Problem 7. *Show that* ${}^nC_r + {}^nC_{r-1} = {}^{n+1}C_r$, *where* $n \geq r \geq 1$ *and n and r are natural numbers.*

Sol. Take L.H.S. of equation *i.e.*,

$$ {}^nC_r + {}^nC_{r-1} = \frac{n\,!}{r\,! \times (n-r)\,!} + \frac{n\,!}{(r-1)\,! \times (n-r+1)\,!}$$

$$= \frac{n\,!}{r(r-1)\,! \times (n-r)\,!} + \frac{n\,!}{(r-1)\,! \times (n-r+1)(n-r)\,!}$$

$$= \frac{n\,! \times (n-r+1) + n\,! \times r}{r(r-1)\,! \times (n-r)\,! \times (n-r+1)} = \frac{n\,! \times (n-r+1+r)}{r\,! \times (n-r+1)\,!}$$

$$= \frac{n\,! \times (n+1)}{r\,! \times (n-r+1)\,!} = {}^{n+1}C_r$$

Hence proved.

Problem 8. *In the 'Discrete Structures Paper' there are 8 questions. In how many ways can an examiner select five questions in all if first question is compulsory.*

Sol. Since the first question is compulsory, the examiner has to select 4 questions from the remaining 7 questions.

Therefore, the number of ways to select 5 questions is = $^7C_4$

$$= \frac{7!}{4! \times (7-4)!} = \frac{7 \times 6 \times 5 \times 4!}{4! \times 3 \times 2 \times 1} = 35.$$

Problem 9. *Determine the number of triangles that are formed by selecting points from a set of 15 points out of which 8 are collinear.*

Sol. When we take all the 15 points, the number of triangles formed is $^{15}C_3$.

As 8 points lie on the same line, they do not form any triangle. Thus, $^8C_3$ triangles are lost.

∴ The total number of triangles produced is

$$^{15}C_3 - {}^8C_3 = \frac{15!}{3 \times (15-3)!} - \frac{8!}{3!(8-3)!} = \frac{15 \times 14 \times 13 \times 12!}{3 \times 12!} - \frac{8 \times 7 \times 6 \times 5!}{3 \times 2 \times 1 \times 5!}$$

$$= \frac{15 \times 14 \times 13}{3} - \frac{8 \times 7 \times 6}{3 \times 2 \times 1} = 910 - 56 = 854.$$

Problem 10. *How many lines can be drawn through 10 points on a circle ?*

Sol. As all the points on the circle are not collinear. Thus, no lines will lost.

∴ The total number of lines drawn through a circle is = $^{10}C_2$

$$= \frac{10!}{2! \times (10-2)!} = \frac{10 \times 9 \times 8!}{2 \times 1 \times 8!} = 45.$$

Problem 11. *Determine the number of diagonals that can be drawn by joining the nodes of octagon.*

Sol. The number of lines that can be formed by joining 2 out of 8 points is = $^8C_2$

$$= \frac{8 \times 7}{2} = 28$$

Out of these 28 lines, the 8 are sides of the octagon.

∴ The number of diagonals is = 28 – 8 = 20.

Problem 12. *In a shipment, there are 40 floppy disks of which 5 are defective. Determine*

(a) in how many ways we can select five floppy disks ?

(b) in how many ways we can select five non-defective floppy disks ?

(c) in how many ways we can select five floppy disks containing exactly three defective floppy disks ?

(d) in how many ways we can select five floppy disks containing at least 1 defective floppy disks ?

Sol. (*a*) There are 40 floppy disks out of which we have to select 5 floppy disks. These can be done in $^{40}C_5$ ways *i.e.*,

$$= \frac{40!}{5!(40-5)!} = \frac{40 \times 39 \times 38 \times 37 \times 36 \times 35!}{5! \times 35!} = 658008.$$

(*b*) There are 40 – 5 = 35 non-defective floppy disks out of which we have to select 5. This can be done in $^{35}C_5$ ways.

$$= \frac{35!}{5!(35-5)!} = \frac{35 \times 34 \times 33 \times 32 \times 31 \times 30!}{5 \times 4 \times 3 \times 2 \times 1 \times 30!}$$

$$= 324632.$$

(*c*) To select exactly three defective floppy disks out of total 5 we have $^5C_3$ ways and the remaining 2 floppy disks can be selected in $^{35}C_2$ ways.

Therefore, the total number of ways to select 5 floppy disks out of which exactly 3 are defective is $= {}^5C_3 \times {}^{35}C_2$

$$= \frac{5!}{3!(5-3)!} \times \frac{35!}{2!\times(35-2)!} = \frac{5\times4\times3!}{3!\times2\times1} \times \frac{35\times34\times33!}{2\times1\times33!}$$

$$= 5950.$$

(*d*) There are five defective floppy disks out of which at least 1 must be selected. We know that the total number of ways to select 5 floppy disks out of total 40 disks is $= {}^{40}C_5$.

Also, the number of ways to select 5 floppy disks with number one defective is

$$= {}^{35}C_5.$$

Therefore, the total number of ways to select 5 floppy disks out of which at least one is defective is

$$= {}^{40}C_5 - {}^{35}C_5 = 611625.$$

Problem 13. *Seven members of a family have total Rs. 2886 in their pockets. Show that at least one of them must have at least Rs. 416 in his pocket.*

Sol. Let us assume the members be the pigeonholes and the Rupees the pigeons. Now 2886 pigeons are to be assigned to 7 pigeonholes. Using the extended pigeonhole principle, where $n = 2886$ and $m = 7$, we have $[(2886 - 1)/7] + 1 = 416$. Hence, there are 416 Rupees in one member's pocket.

Problem 14. *How many people must you have to guarantee that at least 9 of them will have birthdays in the same day of the week.*

Sol. Let us assume the days of week the pigeonholes and the people the pigeons. Now we have 7 pigeonholes and we have to find pigeons. Using the extended pigeonhole principle, we have

$$[(n-1)/7] + 1 = 9$$
$$[(n-1)/7] = 9 - 1 = 8$$
$$n - 1 = 8 \times 7$$
$$n = 56 + 1 = 57.$$

Thus, there must be 57 people to guarantee that at least 9 of them will have birthdays in the same day of the week.

11

Recurrence Relations and Generating Functions

DEFINITION

A recurrence relation is a functional relation between the independent variable x, dependent variable $f(x)$ and the differences of various order of $f(x)$. A recurrence relation is also called a difference equation and we will use these two terms interchangeably.

For example : The equation $f(x + 3h) + 3f(x + 2h) + 6f(x + h) + 9f(x) = 0$ is a recurrence relation.

It can also be written as

$$a_{r+3} + 3a_{r+2} + 6a_{r+1} + 9a_r = 0$$

or

$$y_{k+3} + 3y_{k+2} + 6y_{k+1} + 9y_k = 0.$$

For example : The Fibonacci sequence is defined by the recurrence relation $a_r = a_{r-2} + a_{r-1}$, $r \geq 2$, with the initial conditions $a_0 = 1$ and $a_1 = 1$.

ORDER OF THE RECURRENCE RELATION

The order of the recurrence relation or difference equation is defined to be the difference between the highest and lowest subscripts of $f(x)$ or a_r or y_k.

For example : The equation $13a_r + 20a_{r-1} = 0$ is a first order recurrence relation.

For example : The equation $8f(x) + 4f(x + 1) + 8f(x + 2) = k(x)$ is a second order difference equation.

DEGREE OF THE DIFFERENCE EQUATION

The degree of a difference equation is defined to be the highest power of $f(x)$ or a_r or y_k.

For example : The equation $y_{k+3}^3 + 2y_{k+2}^2 + 2y_{k+1} = 0$ has the degree 3, as the highest power of y_k is 3.

For example : The equation $a_r^4 + 3a_{r-1}^3 + 6a_{r-2}^2 + 4a_{r-3} = 0$ has the degree 4, as the highest power of a_r is 4.

For example : The equation $y_{k+3} + 2y_{k+2} + 4y_{k+1} + 2y_k = k(x)$ has the degree 1, because the highest power of y_k is 1 and its order is 3.

For example : The equation $f(x + 2h) - 4f(x + h) + 2f(x) = 0$ has the degree 1 and its order is 2.

LINEAR RECURRENCE RELATIONS WITH CONSTANT COEFFICIENTS

A recurrence relation is called linear if its degree is one.

The general form of linear recurrence relation with constant coefficient is

$$C_0 y_{n+r} + C_1 y_{n+r-1} + C_2 y_{n+r-2} + \ldots + C_r y_n = R(n)$$

where $C_0, C_1, C_2, \ldots, C_n$ are constants and $R(n)$ is same function of independent variable n.

A solution of a recurrence relation in any function which satisfies the given equation.

LINEAR HOMOGENEOUS RECURRENCE RELATION WITH CONSTANT COEFFICIENTS

The equation is said to be linear homogeneous difference equation if and only if $R(n) = 0$ and it will be of order n.

The equation is said to be linear non-homogeneous difference equation if $R(n) \neq 0$.

For example : The equation $a_{r+3} + 6a_{r+2} + 12a_{r+1} + 8a_r = 0$ is a linear homogeneous equation of order 3.

For example : The equation $a_{r+2} - 4a_{r+1} + 4a_r = 3r + 2^r$ is a linear non-homogeneous equation of order 2.

A linear homogeneous difference equation with constant coefficients is given by

$$C_0 y_n + C_1 y_{n-1} + C_2 y_{n-2} + \ldots + C_r y_{n-r} = 0 \qquad \ldots(i)$$

where $C_0, C_1, C_2, \ldots, C_r$ are constants.

The solution of the equation (*i*) is of the form $A\alpha_1^K$, where α_1 is the characteristic root and A is constant.

Substitute the values of $A\alpha^K$ for y_n in equation (*i*), we have

$$C_0 A\alpha^K + C_1 A\alpha^{K-1} + C_2 A\alpha^{K-2} + \ldots + C_r A\alpha^{K-r} = 0 \qquad \ldots(ii)$$

After simplifying equation (*ii*), we have

$$C_0 \alpha^r + C_1 \alpha^{r-1} + C_2 \alpha^{r-2} + \ldots + C_r = 0 \qquad \ldots(iii)$$

The equation (*iii*) is called the characteristic equation of the difference equation.

If α_1 is one of the roots of the characteristic equation, then $A\alpha_1^K$ is a homogeneous solution to the difference equation.

To find the solution of the linear homogeneous difference equations, we have the four cases, that are discussed as follows.

Case I. If the characteristic equation has n distinct real roots $\alpha_1, \alpha_2, \alpha_3, \ldots, \alpha_n$.

Thus, $\alpha_1^K, \alpha_2^K, \ldots\ldots, \alpha_n^K$ are all solutions of equation (*i*).

Also, we have, $A_1 \alpha_1^K, A_2 \alpha_1^K, \ldots\ldots, A_n \alpha_n^K$ are all solutions of equation (*i*). The sums of solutions are also solutions.

Hence, the homogeneous solution of the difference equation is

$$y_K = A_1\alpha_1^K + A_2\alpha_2^K + \ldots + A_n \alpha_n^K.$$

Case II. If the characteristic equation has repeated real roots.

If $\alpha_1 = \alpha_2$, then $(A_1 + A_2 K) \alpha_1^K$ is also a solution.

If $\alpha_1 = \alpha_2 = \alpha_3$, then $(A_1 + A_2 K + A_3 K^2) \alpha_1^K$ is also a solution.

Similarly, if root α_1 is repeated n times, then

$$(A_1 + A_2 K + A_3 K^2 + \ldots + A_n K^{n-1}) \alpha_1^K$$

is the solution of the homogeneous equation.

Case III. If the characteristic equation has one imaginary root.

If $\alpha + i\beta$ is the root of the characteristic equation, then $\alpha - i\beta$ is also the root, where α and β are real.

Thus, $(\alpha + i\beta)^K$ and $(\alpha - i\beta)^K$ are solutions of the equations. This implies

$$(\alpha + i\beta)^K A_1 + (\alpha - i\beta)^K A_2$$

is also a solution of the characteristic equation, where A_1 and A_2 are constants which are to be determined.

Case IV. If the characteristic equation has repeated imaginary roots.

When the characteristic equation has repeated imaginary roots,

$$(C_1 + C_2 K)(\alpha + i\beta)^K + (C_3 + C_4 K)(\alpha - i\beta)^K$$

is the solution of the homogeneous equation.

Example 1. *Solve the difference equation* $a_r - 3a_{r-1} + 2a_{r-2} = 0$.

Sol. The characteristic equation is given by

$$s^2 - 3s + 2 = 0 \quad \text{or} \quad (s-1)(s-2) = 0$$

$\Rightarrow \quad s = 1, 2$

Therefore, the homogeneous solution of the equation is

$$a_r = C_1 + C_2 . 2^r.$$

Example 2. *Solve the difference equation* $a_r - 6a_{r-1} + 8a_{r-2} = 0$.

Sol. The characteristic equation is

$$s^2 - 6s + 8 = 0 \quad \text{or} \quad (s-2)(s-4) = 0$$

$\Rightarrow \quad s = 2, 4$

Therefore, the homogeneous solution of the equation is

$$a_r = C_1 . 2^r + C_2 . 4^r.$$

Example 3. *Solve the difference equation* $9y_{K+2} - 6y_{K+1} + y_K = 0$.

Sol. The characteristic equation is

$$9s^2 - 6s + 1 = 0 \quad \text{or} \quad (3s-1)^2 = 0$$

$\Rightarrow \quad s = \frac{1}{3}$ and $\frac{1}{3}$

Therefore, the homogeneous solution is given by

$$y_K = (C_1 + C_2 K) . \left(\frac{1}{3}\right)^K.$$

Example 4. *Solve the difference equation* $a_r - 4a_{r-1} + 4a_{r-2} = 0$.

Sol. The characteristic equation is given by

$$s^2 - 4s + 4 = 0 \quad \text{or} \quad (s-2)^2 = 0$$

$\Rightarrow \quad s = 2$ and 2

Therefore, the homogeneous solution of equation is $a_r = (C_1 + C_2 r) . 2^r$.

Example 5. *Solve the difference equation* $a_r + a_{r-1} + a_{r-2} = 0$.

Sol. The characteristic equation is $s^2 + s + 1 = 0$

The roots of this characteristic equation are imaginary, *i.e.*, $s = \frac{-1+i\sqrt{3}}{2}$ and $\frac{-1-i\sqrt{3}}{2}$.

Therefore, the homogeneous solution of the equation is

$$a_r = \left[\frac{-1+i\sqrt{3}}{2}\right]^r C_1 + \left[\frac{-1-i\sqrt{3}}{2}\right]^r C_2.$$

Example 6. *Solve the difference equation* $y_K - y_{K-1} - y_{K-2} = 0$.

Sol. The characteristic equation is $s^2 - s - 1 = 0$

$$s = \frac{1 \pm \sqrt{1+4}}{2} = \frac{1 \pm \sqrt{5}}{2}$$

Therefore, the homogeneous solution of the equation is

$$y_K = C_1 \left[\frac{1+\sqrt{5}}{2}\right]^K + C_2 \left[\frac{1-\sqrt{5}}{2}\right]^K.$$

Example 7. *Solve the difference equation* $a_{r+4} + 2a_{r+3} + 3a_{r+2} + 2a_{r+1} + a_r = 0$.

Sol. The characteristic equation is $s^4 + 2s^3 + 3s^2 + 2s + 1 = 0$

or $$(s^2 + s + 1)(s^2 + s + 1) = 0$$

or $$s = \frac{-1 \pm i\sqrt{3}}{2}, \frac{-1 \pm i\sqrt{3}}{2}$$

Therefore, the homogeneous solution is given by

$$a_r = (C_1 + C_2 r)\left(\frac{-1+i\sqrt{3}}{2}\right)^r + (C_3 + C_4 r)\left(\frac{-1-i\sqrt{3}}{2}\right)^r.$$

Example 8. *Solve the difference equation* $y_{K+4} + 4y_{K+3} + 8y_{K+2} + 8y_{K+1} + 4y_K = 0$.

Sol. The characteristic equation is $s^4 + 4s^3 + 8s^2 + 8s + 4 = 0$

or $$(s^2 + 2s + 2)(s^2 + 2s + 2) = 0$$

or $$s = -1 \pm i, -1 \pm i$$

Therefore, the homogeneous solution is given by

$$y_K = (C_1 + C_2 K)(-1+i)^K + (C_3 + C_4 K)(-1-i)^K.$$

PARTICULAR SOLUTION

(*a*) Homogeneous Linear Difference Equations

We can find the particular solution of the difference equation, when the equation is of homogeneous linear type by putting the values of the initial conditions in the homogeneous solution.

Example 9. *Solve the difference equation* $2a_r - 5a_{r-1} + 2a_{r-2} = 0$ *and find particular solution such that* $a_0 = 0$ *and* $a_1 = 1$.

Sol. The characteristic equation is $2s^2 - 5s + 2 = 0$

or $$(2s - 1)(s - 2) = 0$$

or $$s = \frac{1}{2} \text{ and } 2.$$

Therefore, the homogeneous solution is given by

$$a_{r(h)} = C_1 \left(\frac{1}{2}\right)^r + C_2 \cdot 2^r \quad ...(i)$$

Putting $r = 0$ and $r = 1$ in equation (*i*), we get

$$a_0 = C_1 + C_2 = 0 \quad ...(a)$$

$$a_1 = \frac{1}{2} C_1 + 2C_2 = 1 \quad ...(b)$$

Solving eq. (*a*) and (*b*), we have

$$C_1 = -\frac{2}{3} \text{ and } C_2 = \frac{2}{3}$$

Hence, the particular solution is

$$a_{r(P)} = -\frac{2}{3} \cdot \left(\frac{1}{2}\right)^r + \frac{2}{3} \cdot (2)^r.$$

Example 10. *Solve the difference equation $a_r - 4a_{r-1} + 4a_{r-2} = 0$ and find the particular solution, given that $a_0 = 1$ and $a_1 = 6$.*

Sol. The characteristic equation is

$$s^2 - 4s + 4 = 0 \text{ or } (s - 2)^2 = 0$$

or $$s = 2, 2$$

Therefore, the homogeneous solution is given by

$$a_{r(h)} = (C_1 + C_2 r) \cdot 2^r \quad ...(i)$$

Putting $r = 0$ and $r = 1$ in equation (*i*), we get

$$a_0 = (C_1 + 0) \cdot 2^0 = 1 \quad \therefore \quad C_1 = 1$$

$$a_1 = (C_1 + C_2) \cdot 2 = 6 \quad \therefore \quad C_1 + C_2 = 3 \Rightarrow C_2 = 2$$

Hence, the particular solution is $a_{r(P)} = (1 + 2r) \cdot 2^r$.

Example 11. *Solve the difference equation $9a_r - 6a_{r-1} + a_{r-2} = 0$ satisfying the conditions $a_0 = 0$ and $a_1 = 2$.*

Sol. The characteristic equation is

$$9s^2 - 6s + 1 = 0 \text{ or } (3s - 1)^2 = 0$$

or $$s = \frac{1}{3}, \frac{1}{3}$$

Therefore, the homogeneous solution is given by

$$a_{r(h)} = (C_1 + C_2 r) \cdot \left(\frac{1}{3}\right)^r \quad ...(i)$$

Putting $r = 0$ and $r = 1$ in equation (*i*), we get

$$a_0 = C_1 = 0 \quad ...(a)$$

$$a_1 = (C_1 + C_2) \cdot \frac{1}{3} = 2. \quad \therefore \quad C_1 + C_2 = 6 \Rightarrow C_2 = 6$$

Hence, the particular solution is

$$a_{r(P)} = 6r \cdot \left(\frac{1}{3}\right)^r.$$

Example 12. *Solve the difference equation $a_r - 7a_{r-1} + 10a_{r-2} = 0$ satisfying the conditions $a_0 = 0$ and $a_1 = 6$.*

Sol. The characteristic equation is

$$s^2 - 7s + 10 = 0 \quad \text{or} \quad (s-5)(s-2) = 0 \quad \text{or } s = 2, 5$$

Therefore, the homogeneous solution is given by

$$a_{r(h)} = C_1 \cdot 2^r + C_2 \cdot 5^r. \qquad ...(i)$$

Putting $r = 0$ and $r = 1$ in equation (*i*), we get

$$a_0 = C_1 + C_2 = 0 \qquad ...(a)$$

$$a_1 = 2C_1 + 5C_2 = 6 \qquad ...(b)$$

From eq. (*a*) $\quad C_1 = -C_2$

From eq. (*b*) $\quad 3C_2 = 6 \quad \therefore \quad C_2 = 2$ and $C_1 = -2$.

Hence, the particular solution is $a_{r(P)} = -2 \cdot 2^r + 2 \cdot 5^r$.

(*b*) **Non-Homogeneous Linear Difference Equations.** There are two methods to find the particular solution of non-homogeneous linear difference equations. These are as follows :

1. Undetermined coefficients method
2. E and Δ operator method.

1. Undetermined Coefficients Method. This method is used to find particular solution of non-homogeneous linear difference equations, whose R.H.S. term R(*n*) consists of terms of special forms.

In this method, firstly we assume the general form of the particular solution according to the form of R(*n*) containing a number of unknown constant coefficients, which have to be determined. Then according to the difference equation, we will determine the exact solution.

The general form of particular solution to be assumed for the special forms of R(*n*), to find the exact solution is shown in the table.

Form of R(*n*)	General form to be assumed
Z, here *z* is a constant	A
Z^r, here Z is a constant	$A \cdot Z^r$
P(*r*), a polynomial of degree *n*	$A_0 r^n + A_1 r^{n-1} + ... + A_n$
Z^r. P(*r*), here P(*r*) is a polynomial of *n*th degree in *r*. Z is a constant.	$[A_0 r^n + A_1 r^{n-1} + ... + A_n] \cdot Z^r$

Example 13. *Find the particular solution of the difference equation*

$$a_{r+2} - 3a_{r+1} + 2a_r = Z^r, \qquad ...(i)$$

where Z is some constant.

Sol. The general form of solution is $= A \cdot Z^r$

Now putting this solution on L.H.S. of equation (*i*), we get

$$= AZ^{r+2} - 3AZ^{r+1} + 2AZ^r = (Z^2 - 3Z + 2)AZ^r \qquad ...(ii)$$

Equating equation (*ii*) with R.H.S. of equation (*i*), we get

$$(Z^2 - 3Z + 2)A = 1$$

or
$$A = \frac{1}{Z^2 - 3Z + 2} = \frac{1}{(Z-1)(Z-2)} \quad (Z \neq 1, Z \neq 2)$$

Therefore, the particular solution is $\frac{Z^r}{(Z-1)(Z-2)}$.

Example 14. *Find the particular solution of the difference equation*

$$a_{r+2} - 5a_{r+1} + 6a_r = 5^r. \quad ...(i)$$

Sol. Let us assume the general form of the solution = A . 5^r.

Now to find the value of A, put this solution on L.H.S. of the equation (*i*), then this becomes

$$= A.\ 5^{r+2} - 5\ .\ A5^{r+1} + 6\ .\ A5^r$$
$$= 25A\ .\ 5^r - 25A\ .\ 5^r + 6A\ .\ 5^r$$
$$= 6A\ .\ 5^r \quad ...(ii)$$

Equating equation (*ii*) to R.H.S. of equation (*i*), we get

$$A = \frac{1}{6}$$

Therefore, the particular solution of the difference equation is = $\frac{1}{6} \cdot 5^r$.

Example 15. *Find the particular solution of the difference equation*

$$a_{r+2} - 4a_r = r^2 + r - 1. \quad ...(i)$$

Sol. The homogeneous solution of the difference equation is given by

$$a_{r(h)} = C_1(2)^r + C_2(-2)^r$$

To find the particular solution, let us assume the general form of the solution is

$$= A_1 r^2 + A_2 r + A_3.$$

Putting this solution in L.H.S. of equation (*i*), we get

$$= A_1(r+2)^2 + A_2(r+2) + A_3 - 4A_1 r^2 - 4A_2 r - 4A_3$$
$$= -3A_1 r^2 + (4A_1 - 3A_2)r + (4A_1 + 2A_2 - 3A_3) \quad ...(ii)$$

Equating equation (*ii*) with R.H.S. of equation (*i*), we get

$$-3A_1 = 1$$
$$4A_1 - 3A_2 = 1$$
$$4A_1 + 2A_2 - 3A_3 = -1$$

After solving these three equations, we get

$$A_1 = -\frac{1}{3},\ A_2 = -\frac{7}{9},\ A_3 = -\frac{17}{27}$$

Therefore, the particular solution is $= -\frac{r^2}{3} - \frac{7}{9}r - \frac{17}{27}$.

Example 16. *Find the particular solution of the difference equation*

$$a_{r+2} - 2a_{r+1} + a_r = 3r + 5. \quad ...(i)$$

Sol. The homogeneous solution of the difference equation is given by

$$a_{r(h)} = C_1 + C_2\, r \quad ...(ii)$$

Corresponding to the term $3r + 5$, we assume the general form of the solution as $A_1 r + A_2$ but due to occurrence of these terms in equation (*ii*), we multiply this by suitable power of r so that none of the term will occur in equation (*ii*). Thus multiply by r^2.

Hence, the general form of the solution becomes

$$= A_1 r^3 + A_2 r^2 \ ......$$

Putting this solution in L.H.S. of equation (*i*), we get

$$= A_1(r+2)^3 + A_2(r+2)^2 - 2A_1(r+1)^3 - 2A_2(r+1)^2 + A_1 r^3 + A_2 r^2$$

$$= A_1(r^3 + 8 + 6r^2 + 12r) + A_2(r^2 + 4 + 4r) - 2A_1(r^3 + 1 + 3r^2 + 3r) - 2A_2(r^2 + 1 + 2r) + A_1 r^3 + A_2 r^2$$

$$= (12A_1 + 4A_2 - 6A_1 - 4A_2)r + (8A_1 + 4A_2 - 2A_1 - 2A_2)$$

$$= (6A_1)r + (6A_1 + 2A_2) \qquad ...(iii)$$

Equating equation (*iii*) with R.H.S. of equation (*i*), we get

$$6A_1 = 3 \quad \therefore \quad A_1 = \frac{1}{2}$$

$$6A_1 + 2A_2 = 5 \quad \therefore \quad A_2 = 1$$

Therefore, the particular solution is $\frac{1}{2} r^3 + r^2$.

Example 17. *Find the particular solution of the difference equation*

$$a_{r+2} + a_{r+1} + a_r = r \,.\, 2^r. \qquad ...(i)$$

Sol. Let us assume the general form of the solution $= (A_0 + A_1 r) \,.\, 2^r$

Now, put this solution in the L.H.S. of equation (*i*), we get

$$= 2^{r+2}[A_0 + A_1(r+2)] + 2^{r+1}[A_0 + A_1(r+1)] + 2^r(A_0 + A_1 r)$$

$$= 4.\, 2^r(A_0 + A_1 r + 2A_1) + 2 \,.\, 2^r(A_0 + A_1 r + A_1) + 2^r(A_0 + A_1 r)$$

$$= r.\, 2^r(7A_1) + 2^r(7A_0 + 10A_1) \qquad ...(ii)$$

Equating equation (*ii*) with R.H.S. of equation (*i*), we get

$$7A_1 = 1 \quad \therefore \quad A_1 = \frac{1}{7}$$

$$7A_0 + 10A_1 = 0 \quad \therefore \quad A_0 = \frac{-10}{49}$$

Therefore, the particular solution is $2^r\left(\frac{-10}{49} + \frac{1}{7} r\right)$.

Example 18. *Find the particular solution of the difference equation*

$$a_r - 4a_{r-1} + 4a_{r-2} = (r+1) \,.\, 2^r. \qquad ...(i)$$

Sol. The homogeneous solution of the difference equation is given by

$$a_{r(h)} = (C_1 + C_2 r) \,.\, 2^r \qquad ...(ii)$$

because it has two real and equal roots *i.e.*, 2 and 2.

To find the particular solution, let us assume the general form of the solution is $= 2^r(A_1 r + A_0)$, but due to occurrence of there terms in equation (*ii*), we multiply this by suitable power of r so that none of the terms will occur in equation (*ii*). Thus multiply by r^2.

Hence, the general form of the solution becomes $= 2^r(A_1 r + A_0) \,.\, r^2$

Putting this solution in L.H.S. of equation (*i*), we get

$= 2^r . (A_1 r + A_0) . r^2 - 4 . 2^{r-1} [A_1(r-2) + A_0] . (r-1)^2 + 4 . 2^{r-2} [A_1 (r-2) + A_0] . (r-2)^2$

$= 2^r .(A_1 r + A_0) . r^2 - 2(r^2 + 1 - 2r) . 2^r (A_1 r - A_1 + A_0) + (r^2 + 4 - 4r) . 2^r . (A_1 r - 2A_1 + A_0)$

$= r . 2^r (6A_1) + 2^r (- 6A_1 + 2A_0)$ ...(*iii*)

Equating equation (*iii*) with R.H.S. of equation (*i*), we get

$$6A_1 = 1 \quad \therefore \quad A_1 = \frac{1}{6}$$

$$- 6A_1 + 2A_0 = 1 \quad \therefore \quad A_0 = 1$$

Therefore, the particular solution is $= r^2 . 2^r \left(\frac{r}{6} + 1\right)$.

2. E and Δ Operator Method. Before discussing this method to find the particular solution of the non-homogeneous linear difference equation, we will discuss the meaning of the operator E and Δ.

Meaning of Operator E. The operation of E on $f(x)$ means that give an increment to the value of x in the function. The operation of E is, put $(x + h)$ in the function wherever there is x. Here h is increment quantity. So

$$Ef(x) = f(x + h)$$

Here, E is operated on $f(x)$, therefore, E is a symbol known as shift operator.

Meaning of Operator Δ. The operator Δ is an operation of two steps.

Firstly, x in the function is incremented by a constant and then former is subtracted from the later *i.e.*,

$$\Delta f(x) = f(x + h) - f(x).$$

Theorem I. *Prove that* $E \cong 1 + \Delta$.

Proof. The operation of Δ on $f(x)$ is of two steps. First, increment the value of x in the function. So, whenever, there is x in $f(x)$ put $x + h$ (here h is constant increment), which means operation of E on $f(x)$ *i.e.*,

$$f(x + h) = Ef(x).$$

Second, subtract the original function from the value obtained in first step, hence

$$\Delta f(x) = Ef(x) - 1 f(x) = (E - 1) f(x)$$

[Where E and I are operators and 1 is an identity operator]

So, the operation of Δ on $f(x)$ is equivalent to the operation of (E – 1) on $f(x)$.

Therefore, we have

$$E \cong 1 + \Delta.$$

Theorem II. *Show that* $E^n f(x) = f(x + nh)$.

Proof. We know that $E f(x) = f(x + h)$

Now

$$\begin{aligned} E^n f(x) &= E . E . E . E. \ldots n \text{ times } f(x) \\ &= E^{n-1} [Ef(x)] = E^{n-1} f(x + h) \\ &= E^{n-2} [E f(x + h)] = E^{n-2} f(x + 2h) \\ &\ldots\ldots\ldots\ldots\ldots\ldots \\ &\ldots\ldots\ldots\ldots\ldots\ldots \\ &= E f [x + (n - 1)h] = f(x + nh). \end{aligned}$$

Theorem III. *Show that* $E\,Cf(x) = CE\,f(x)$.

Proof. We know that $E\,Cf(x) = Cf(x + h) = CE\,f(x + h)$. Hence proved.

There is no effect of operation of E on any constant. Therefore, the operation of E on any constant will be equal to the constant itself.

By E and Δ operator method, we will find the solution of

$$C_0 y_{n+r} + C_1 y_{n+r-1} + C_2 y_{n+r-2} + \ldots\ldots + C_n y_n = R(n) \qquad \ldots(i)$$

Equation (*i*) can be written as

$$C_0 E^r y_n + C_1 E^{r-1} y_n + C_2 E^{r-2} y_n + \ldots + C_n y_n = R(n)$$

i.e.,
$$(C_0 E^r + C_1 E^{r-1} + C_2 E^{r-2} + \ldots\ldots + C_n) y_n = R(n)$$

Putting
$$C_0 E^r + C_1 E^{r-1} + C_2 E^{r-2} + \ldots + C_n = P(E)$$

So
$$P(E)\, y_n = R(n)$$

$$\therefore \quad y_n = \frac{R(n)}{P(E)} \qquad \ldots(ii)$$

To find the particular solution of (*ii*) for different forms of R(*n*), we have the following cases.

Case I. When R(*n*) is some constant A

We know that, the operation of E on any constant will be equal to the constant itself *i.e.*,

$$EA = A$$

Therefore,
$$\begin{aligned} P(E)\,A &= (C_0 E^r + C_1 E^{r-1} + C_2 E^{r-2} + \ldots + C_n)A \\ &= (C_0 + C_1 + C_2 + \ldots + C_n)A \\ &= P(1)A \end{aligned}$$

i.e.,
$$\frac{A}{P(E)} = \frac{1}{P(1)} . A$$

Therefore, using equation (*ii*), the particular solution of (*i*) is

$$y_n = \frac{A}{P(1)}, \quad P(1) \neq 0$$

P(1) is obtained by putting E = 1 in P(E).

Case II. When R(*n*) is of the form $A . Z^n$, where A, and Z are constants

We have,
$$\begin{aligned} P(E)\,(A . Z^n) &= \{C_0 E^r + C_1 E^{r-1} + \ldots + C_n\}\,(A . Z^n) \\ &= A\,\{C_0 Z^{r+n} + C_1 Z^{r+n-1} + \ldots + C_n Z^n\} \\ &= A\,\{C_0 Z^r + C_1 Z^{r-1} + \ldots + C_n\} . Z^n \\ &= AP(Z) . Z^n \end{aligned}$$

To get, P(Z) put E = Z in P(E)

Therefore,
$$\frac{A . Z^n}{P(E)} = \frac{A . Z^n}{P(Z)}, \text{ provided } P(Z) \neq 0$$

Thus,
$$y_n = \frac{A . Z^n}{P(Z)}, \quad P(Z) \neq 0$$

If A = 1, then
$$y_n = \frac{Z^n}{P(Z)}$$

When P(Z) = 0 then for equation

(*i*) $\qquad (E - Z)\, y_n = A . Z^n$

For this, the particular solution becomes $= A . \dfrac{1}{(E - Z)} . Z^n = A . n\, Z^{n-1}$

(*ii*) $(E - Z)^2 y_n = A \cdot Z^n$

For this, the particular solution becomes $= A . \frac{1}{(E-Z)^2} . Z^n = \frac{A . n(n-1)}{2!} . Z^{n-2}$

(*iii*) $(E - Z)^3 y_n = A \cdot Z^n$

For this, the particular solution becomes $= A . \frac{1}{(E-Z)^3} . Z^n = \frac{A . n(n-1)(n-2)}{3!} . Z^{n-3}$ and so on.

Case III. When R(n) be a polynomial of degree m in n

We know that $E \equiv 1 + \Delta$

So, $P(E) = P(1 + \Delta)$

or $$\frac{1}{P(E)} = \frac{1}{P(1+\Delta)},$$

which can be expanded in ascending power of Δ as far as upto Δ^m

$$\Rightarrow \quad \frac{1}{P(E)} = \frac{1}{P(1+\Delta)} = (b_0 + b_1\Delta + b_2\Delta + \ldots + b_m \Delta^m + \ldots)$$

$$\Rightarrow \quad \frac{1}{P(E)} . R(n) = (b_0 + b_1\Delta + b_2\Delta + \ldots + b_m\Delta^m + \ldots) . R(n)$$

$$= b_0 R(n) + b_1\Delta R(n) + \ldots + b_m \Delta^m R(n)$$

All other higher terms will be zero because R(n) is a polynomial of degree m.

Thus, the particular solution of equation (i), in this case will be

$$y_n = b_0 R(n) + b_1 \Delta R(n) + \ldots + b_m \Delta^m R(n).$$

Case IV. When R(n) is of the form R(n) . Z^n, where R(n) is a polynomial of degree m and Z is some constant

We have $E^r [Z^n R(n)] = Z^{r+n} R(n + r) = Z^r. Z^n. E^r. R(n) = Z^n (ZE)^r R(n)$

Similarly, we have

$$\frac{1}{P(E)} [Z^n R(n)] = Z^n \frac{1}{P(ZE)} . (R(n)) = Z^n [P(Z + Z\Delta)]^{-1}. R(n)$$

Thus, the particular solution of equation (i), in this case will be

$$y_n = Z^n [P(Z + Z\Delta)]^{-1} R(n).$$

Example 19. *Find the particular solution of the difference equation*

$$2a_{r+1} - a_r = 12.$$

Sol. The above equation can be written as

$$(2E - 1) a_r = 12$$

The particular solution is given by

$$a_r = \frac{1}{(2E-1)} . 12$$

Put E = 1, in the equation. The particular solution is $a_r = 12$.

Example 20. *Find the particular solution of difference equation* $a_r - 3a_{r-1} + 2a_{r-2} = 2^r$.

Sol. The above equation can be written as

$$(E^2 - 3E + 2) a_r = 2^r$$

Therefore, $P(E) = E^2 - 3E + 2$ or $P(E) = (E - 2)(E - 1)$

Thus, the particular solution is given by

$$a_r = \frac{1}{(E-2)(E-1)} \cdot 2^r$$

$$= r. \frac{1}{(E-1)} \cdot 2^{r-1} = r. \frac{1}{(2-1)} \cdot 2^{r-1}$$

$$a_r = r.\ 2^{r-1}.$$

Example 21. *Find the particular solution of difference equation*

$$a_r - 4a_{r-1} + 4a_{r-2} = 2^r.$$

Sol. The above equation can be written as

$$(E^2 - 4E + 4)\ a_r = 2^r$$

Therefore, $\qquad P(E) = E^2 - 4E + 4 = (E-2)^2$

Thus, the particular solution is given by

$$a_r = \frac{1}{(E-2)^2} \cdot 2^r = \frac{r(r-1)}{2!} \cdot 2^{r-2}$$

$$a_r = r(r-1).\ 2^{r-3}.$$

Example 22. *Find the particular solution of the difference equation*

$$a_r - 2a_{r-1} = 7r^2.$$

Sol. The above equation can be written as $(E-2)\ a_r = 7r^2$

Thus, the particular solution is given by

$$a_r = \frac{1}{(E-2)} \cdot 7r^2 = \frac{1}{(\Delta - 1)} \cdot 7r^2 = -7. \frac{1}{(1-\Delta)} \cdot r^2$$

$$= -7(1-\Delta)^{-1}.\ (r^2) = -7(1 + \Delta + \Delta^2)\ (r^2)$$

By factorial notation, $[r]^n = r(r-1)(r-2) \ldots (r-n+1)$.

Hence $\qquad (r^2) = [r]^2 + [r]$

$$a_r = -7(1 + \Delta + \Delta^2)\ ([r]^2 + [r]) = -7([r]^2 + [r] + 2[r] + 1 + 2 + 0)$$

The operation of Δ on $[r]^n$ is analogous to its differentiation

$$a_r = -7[r]^2 - 21[r] - 21 = -7r(r-1) - 21r - 21$$

$$a_r = -7r^2 - 14r - 21.$$

Example 23. *Find the particular solution of the difference equation*

$$a_r - 2a_{r-1} = 7r.$$

Sol. The above equation can be written as $(E-2)\ a_r = 7r$

Thus, the particular solution is given by

$$a_r = \frac{1}{(E-2)} \cdot 7r = 7. \frac{1}{(\Delta - 1)} \cdot r = -7 \frac{1}{(1-\Delta)} [r]$$

(By factorial notation $r = [r]$)

$$= -7(1-\Delta)^{-1}\ [r] = -7(1 + \Delta + \Delta^2 + \ldots)\ [r]$$

By neglecting higher powers

$$a_r = -7(1 + \Delta)\ [r] = -7([r] + 1)$$

$$a_r = -7(r + 1).$$

Example 24. *Find the particular solution of the difference equation*

$$a_r + a_{r-1} + a_{r-2} = r.\ 2^r.$$

Sol. The above equation can be written as $(E^2 + E + 1)\ a_r = r\ .\ 2^r$

Thus, the particular solution is given by

$$a_r = \frac{1}{(E^2 + E + 1)}\ .\ r.\ 2^r = 2^r \frac{1}{(2E)^2 + 2E + 1}\ .\ r$$

$$= 2^r\ (4(1 + \Delta)^2 + 2(1 + \Delta) + 1)^{-1}.\ r = 2^r\ (7 + 10\Delta + 4\Delta^2)^{-1}.\ r$$

$$= \frac{2^r}{7}\left(1 + \frac{10\Delta + 4\Delta^2}{7}\right)^{-1}.\ [r] = \frac{2^r}{7}\left(1 - \frac{10\Delta + 4\Delta^2}{7}\right).\ [r] = \frac{2^r}{7}\left([r] - \frac{10}{7}\right)$$

$$a_r = \frac{2^r}{7}\left(r - \frac{10}{7}\right).$$

TOTAL SOLUTION

The total solution or the general solution of a non-homogeneous linear difference equation with constant coefficients is the sum of the homogeneous solution and particular solution. If no initial conditions are given, then our work is finished. If n initial conditions are given, obtain n linear equations in n unknowns and solve them, if possible to get a total solution.

As we already know that the homogeneous solution of the recurrence relation is obtained by putting the right hand side of the equation equal to zero and the particular solution is obtained with $R(n)$ on the right hand side of the equation.

If $y_{(h)}$ denotes the homogeneous solution of the recurrence relation and $y_{(p)}$ denotes the particular solution of the recurrence relation then, the total solution or the general solution y of the recurrence relation is given by

$$y = y_{(h)} + y_{(p)}.$$

Example 25. *Solve the difference equation*

$$a_r - 4a_{r-1} + 4a_{r-2} = 3r + 2^r. \qquad ...(i)$$

Sol. The homogeneous solution of this equation is obtained by putting R.H.S. equal to zero *i.e.*,

$$a_r - 4a_{r-1} + 4a_{r-2} = 0$$

The homogeneous solution is $a_{r(h)} = (C_1 + C_2\ r)\ .\ 2^r$ as shown in example 4.

The equation (*i*) can be written as $(E^2 - 4E + 4)\ a_r = 3r + 2^r$

The particular solution is given as

$$a_{r(p)} = \frac{1}{(E^2 - 4E + 4)}\ .\ (3r + 2^r) = \frac{1}{(E-2)^2}\ .\ (3r) + \frac{1}{(E-2)^2}\ .\ 2^r$$

$$= 3\ .\ \frac{1}{(1-\Delta)^2}\ (r) + \frac{r(r-1)}{2!}\ .\ 2^{r-2} = 3(1 - \Delta)^{-2}\ (r) + \frac{r(r-1)}{2!}\ .\ 2^{r-2}$$

$$= 3(1 + 2\Delta)\ .\ [r] + \frac{r(r-1)}{2!}\ .\ 2^{r-2} = 3(r+2) + r(r-1)\ .\ 2^{r-3}$$

$$a_{r(p)} = 3(r+2) + r(r-1)\ .\ 2^{r-3}$$

Therefore, the total solution is $a_r = (C_1 + C_2\ r)\ .\ 2^r + 3(r+2) + r(r-1)\ .\ 2^{r-3}$.

Example 26. *Solve the difference equation*

$$a_r + 4a_{r-1} + 4a_{r-2} = r^2 - 3r + 5. \quad ...(i)$$

Sol. Put the R.H.S. of the equation equal to zero *i.e.*,

$$a_r + 4a_{r-1} + 4a_{r-2} = 0$$

This can be written as

$$(E^2 + 4E + 4)\, a_r = 0$$

or $$(E + 2)^2 = 0 \quad \text{or} \quad E = -2, -2$$

The homogeneous solution is given by

$$a_{r(h)} = (C_1 + C_2 r) . (-2)^r$$

The equation (*i*) can be written as

$$(E^2 + 4E + 4)\, a_r = r^2 - 3r + 5$$

The particular solution is given by

$$a_{r(p)} = \frac{1}{E^2 + 4E + 4} . [(r)^2 - 3(r) + 5] = \frac{1}{(E+2)^2} (r^2 - 3r + 5)$$

$$= \frac{1}{(3+\Delta)^2} . ([r]^2 - 2[r] + 5)$$

{By factorial notation, $[r]^2 = r(r-1)$ and $[r] = r$}

$$= \frac{1}{9} . \frac{1}{\left(1+\frac{\Delta}{3}\right)^2} . ([r]^2 - 2[r] + 5)$$

$$= \frac{1}{9}\left(1+\frac{\Delta}{3}\right)^{-2} . ([r]^2 - 2[r] + 5)$$

$$= \frac{1}{9}\left(1-\frac{2\Delta}{3}+\frac{3\Delta^2}{9}\right) . ([r]^2 - 2[r] + 5)$$

$$= \frac{1}{9}\left([r]^2 - \frac{10}{3}[r] + 7\right)$$

$$a_{r(p)} = \frac{1}{9}\left(r^2 - \frac{13}{3}r + 7\right)$$

Therefore, the total solution is

$$a_r = a_{r(h)} + a_{r(p)}$$

$$a_r = (C_1 + C_2 r) . (-2)^r + \frac{1}{9}\left(r^2 - \frac{13}{3}r + 7\right).$$

Example 27. *Solve the difference equation*

$$a_r + a_{r-1} + a_{r-2} = r . 2^r. \quad ...(i)$$

Sol. Put the R.H.S. of the equation to zero *i.e.*,

$$a_r + a_{r-1} + a_{r-2} = 0$$

The homogeneous solution is

$$a_{r(h)} = \left[\frac{-1+i\sqrt{3}}{2}\right]^r C_1 + \left[\frac{-1-i\sqrt{3}}{2}\right]^r C_2$$

as shown in example 5.

The equation (i) can be written as $(E^2 + E + 1)\, a_r = r.\, 2^r$

The particular solution is given $a_{r(p)} = \frac{2^r}{7}\left(r - \frac{10}{7}\right)$

as shown in example.

Therefore, the total solution is

$$a_r = a_{r(h)} + a_{r(p)}$$

$$a_r = \left(\frac{-1 + i\sqrt{3}}{2}\right)^r C_1 + \left(\frac{-1 - i\sqrt{3}}{2}\right)^r C_2 + \frac{2^r}{7}\left(r - \frac{10}{7}\right).$$

Example 28. *Solve the difference equation*

$$a_r + 6a_{r-1} + 9a_{r-2} = 3 \qquad ...(i)$$

with initial conditions $a_0 = 0$ *and* $a_1 = 1$.

Sol. Put the R.H.S. of the equation (i) equal to zero *i.e.*,

$$a_r + 6a_{r-1} + 9a_{r-2} = 0$$

This can be written as $E^2 + 6E + 9 = 0$

or $(E + 3)(E + 3) = 0$

or $E = -3, -3$

The homogeneous solution is $a_{r(h)} = (C_1 + C_2 r) . (-3)^r$

Now, we have $P(E) = (E + 3)(E + 3)$

Putting E = 1,we get P(1) = 16

Hence, the particular solution is $a_{r(p)} = \frac{3}{16}$

Therefore, the total solution is $a_r = (C_1 + C_2 r)(-3)^r + \frac{3}{16}$

Now, putting $r = 0$, we have $a_0 = (C_1 + 0)(1) + \frac{3}{16} = 0$

$\therefore$ $C_1 = -\frac{3}{16}$

Now, putting $r = 1$, we have $a_1 = (C_1 + C_2)(-3) + \frac{3}{16} = 1$

Substitute the value of $C_1 = -\frac{3}{16}$ in the equation, we have

$$\left(-\frac{3}{16} + C_2\right)(-3) + \frac{3}{16} = 1$$

$$\left(-\frac{3}{16} + C_2\right)(-3) = \frac{13}{16}$$

$$C_2 = -\frac{1}{12}$$

Therefore, the total solution satisfying initial conditions is

$$a_r = \left(-\frac{3}{16} - \frac{r}{12}\right)(-3)^r + \frac{3}{16}.$$

GENERATING FUNCTIONS

Generating functions is a method to solve the recurrence relations.

Let us consider, the sequence $a_0, a_1, a_2, ..., a_r, ...$ of real numbers. For some interval of real numbers containing zero whose value at t is given, the function G(t) is defined by the series

$$G(t) = a_0 + a_1 t + a_2 t^2 + ... + a_r t^r + ... \qquad ...(i)$$

This function G(t) is called the generating function of the sequence a_r.

Now, for the constant sequence 1, 1, 1, 1, ... the generating function is

$$G(t) = \frac{1}{(1-t)}$$

because it can be expressed as

$$G(t) = (1-t)^{-1} = 1 + t + t^2 + t^3 + t^4 + ... \qquad \text{[By Binomial expansion]}$$

Comparing, this with equation (i), we get

$$a_0 = 1, a_1 = 1, a_2 = 1 \text{ and so on.}$$

For, the constant sequence 1, 2, 3, 4, 5, ... the generating function is

$$G(t) = \frac{1}{(1-t)^2} \text{ because it can be expressed as}$$

$$G(t) = (1-t)^{-2} = 1 + 2t + 3t^2 + 4t^3 + ... + (r+1)\, t^r$$

Comparing this with equation (i), we get

$$a_0 = 1, a_1 = 2, a_2, = 3, a_3 = 4 \text{ and so on.}$$

The generating function of Z^r, ($Z \neq 0$ and Z is a constant) is given by

$$G(t) = 1 + Zt + Z^2 t^2 + Z^3 t^3 + ... + Z^r t^r$$

or $$G(t) = \frac{1}{(1-Zt)} \qquad [\text{Assume } |Zt| < 1].$$

So $$G(t) = \frac{1}{(1-Zt)} \text{ generates } Z^r, Z \neq 0.$$

Also, If $a_r^{(1)}$ has the generating function $G_1(t)$ and $a_r^{(2)}$ has the generating function $G_2(t)$, then $\lambda_1 a_r^{(1)} + \lambda_2 a_r^{(2)}$ has the generating function $\lambda_1 G_1(t) + \lambda_2 G_2(t)$. Here λ_1 and λ_2 are constants.

Theorem IV. *If we know the generating function for the sequence a_r, then the generating function of sequences a_{r+1}, a_{r+2} can be find easily.*

Proof. Let us consider, that G(t) be the generating function for the sequence, having the general term a_r, then we have

$$G(t) = a_0 + a_1 t + a_2 t^2 + ... + a_r t^r + ...$$

or $$\frac{G(t) - a_0}{t} = a_1 + a_2 t + a_3 t^2 + ... + a_r t^{r-1} + a_{r+1} t^r + ...$$

Thus $\frac{G(t) - a_0}{t}$ is the generating function for the sequence a_{r+1}.

In the same way,

$$\frac{G(t) - a_0 - a_1 t}{t^2} = a_2 + a_3 t + a_4 t^2 + ...$$

is the generating function for a_{r+2}.

The summary of results, which we have deduced from the above discussion, is shown in Fig. 1.

S. No.	General term of sequence a_r	Generating function G(t)
1.	1	$1/(1-t)$
2.	$r+1$	$1/(1-t)^2$
3.	r	$t/(1-t)^2$
4.	$r(r+1)$	$2t/(1-t)^3$
5.	Z^r, here z is a constant	$1/(1-Zt)$
6.	${}^nC_r A^{n-r}. Z^r$ $(r = 0, 1, 2, 3, ..., n)$ $r > n$	$(A+Zt)^n$
7.	a_r	$G(t)$
8.	a_{r+1}	$\frac{G(t)-a_0}{t}$
9.	a_{r+2}	$\frac{G(t)-a_0-a_1 . t}{t^2}$
10.	a_{r+3}	$\frac{G(t)-a_0-a_1 . t-a_2 t^2}{t^3}$
11.	a_{r+n}	$\frac{G(t)-a_0-a_1 t- ......-a_{n-1} t^{n-1}}{t^n}$
12.	$\lambda_1 a_r^{(1)} + \lambda_2 a_r^{(2)}$	$\lambda_1 G_1(t) + \lambda_2 G_2(t)$

Fig.1.

Example 29. *Solve the recurrence relation*

$$a_{r+2} - 3a_{r+1} + 2a_r = 0 \qquad ...(i)$$

by the method of generating functions with the initial conditions $a_0 = 2$ and $a_1 = 3$.

Sol. Let us assume that

$$G(t) = \sum_0^\infty a_r t^r$$

Multiply equation (*i*) by t^r and summing from $r = 0$ to ∞, we have

$$\sum_0^\infty a_{r+2} t^r - 3\sum_0^\infty a_{r+1} t^r + 2\sum_0^\infty a_r t^r = 0$$

or $$(a_2 + a_3 t + a_4 t^2 + ...) - 3(a_1 + a_2 t + a_3 t^2 + ...) + 2(a_0 + a_1 t + a_2 t^2 + ...) = 0$$

$$[\because \; G(t) = a_0 + a_1 t + a_2 t^2 + ...]$$

$\therefore$ $$\frac{G(t) - a_0 - a_1 t}{t^2} - 3\left(\frac{G(t) - a_0}{t}\right) + 2G(t) = 0 \qquad ...(ii)$$

Now, put $a_0 = 2$ and $a_1 = 3$ in equation (*ii*) and solving, we get

$$G(t) = \frac{2-3t}{1-3t+2t^2} \quad \text{or} \quad G(t) = \frac{2-3t}{(1-t)(1-2t)}$$

Now, Let $$\frac{2-3t}{(1-t)(1-2t)} = \frac{A}{1-t} + \frac{B}{1-2t}$$

i.e., $$2 - 3t = A(1-2t) + B(1-t) \qquad ...(iii)$$

Put $t = 1$ on both sides of equation (*iii*) to find A. Hence

$$-1 = -A \quad \therefore \quad A = 1$$

Put $t = \frac{1}{2}$ on both sides of equation (*iii*) to find B. Hence

$$\frac{1}{2} = \frac{1}{2}B \quad \therefore \quad B = 1$$

Thus $G(t) = \frac{1}{1-t} + \frac{1}{1-2t}$. Hence, $a_r = 1 + 2^r$.

Example 30. *Solve the recurrence relation*

$$a_r - 7a_{r-1} + 10a_{r-2} = 0 \qquad \text{...}(i)$$

by the method of generating functions with the initial conditions $a_0 = 3$ *and* $a_1 = 3$.

Sol. Let us assume that

$$G(t) = \sum_0^\infty a_r t^r$$

Multiply equation (*i*) by t^r and summing from $r = 2$ to ∞, we have

$$\sum_2^\infty a_r t^r - 7\sum_2^\infty a_{r-1} t^r + 10\sum_2^\infty a_{r-2} t^r = 0$$

or
$$(a_2 t^2 + a_3 t^3 + \ldots) - 7(a_1 t^2 + a_2 t^3 + \ldots) + 10(a_0 t^2 + a_1 t^3 + \ldots) = 0$$

$$G(t) - a_0 - a_1 t - 7t[G(t) - a_0] + 10t^2 G(t) = 0 \qquad \text{...}(ii)$$

Now, put $a_0 = 3$ and $a_1 = 3$ in equation (*ii*) and solving, we get

$$G(t) = \frac{3+24t}{10t^2 - 7t + 1} = \frac{3+24t}{(5t-1)(2t-1)}$$

By partial fractions

$$G(t) = \frac{10}{(2t-1)} - \frac{13}{(5t-1)} \text{ or } G(t) = \frac{13}{(1-5t)} - \frac{10}{(1-2t)}$$

Therefore, $\quad a_r = 13(5)^r - 10(2)^r.$

Example 31. *Solve the recurrence relation*

$$a_{r+2} - 2a_{r+1} + a_r = 2^r \qquad \text{...}(i)$$

by the method of generating functions with the initial conditions $a_0 = 2$ *and* $a_1 = 1$.

Sol. By taking generating functions of equation (*i*) both the sides, we have

$$\frac{G(t) - a_0 - a_1 t}{t^2} - 2\left(\frac{G(t) - a_0}{t}\right) + G(t) = \frac{1}{1-2t} \qquad \text{...}(ii)$$

Now, put $a_0 = 2$ and $a_1 = 1$ in equation (*ii*) and solving, we get

$$G(t) - 2 - t - 2t\,G(t) - 2t + t^2 G(t) = \frac{t^2}{1-2t}$$

$$(t^2 - 2t + 1)\,G(t) = 2 + 3t + \frac{t^2}{1-2t}$$

$$(1-t)^2 G(t) = 2 + 3t + \frac{t^2}{1-2t}$$

$$G(t) = \frac{2}{(1-t)^2} + \frac{3t}{(1-t)^2} + \frac{t^2}{(1-2t)(1-t)^2}$$

By partial fraction $\frac{t^2}{(1-2t)(1-t)^2} = \frac{1}{(1-2t)} - \frac{1}{(1-t)^2}$

Hence, $G(t) = \frac{1}{(1-t)^2} + \frac{3t}{(1-t)^2} + \frac{1}{(1-2t)}$

Therefore, $a_r = (r + 1) + 3r + 2^r$

$a_r = 1 + 4r + 2^r$.

Example 32. *Solve the recurrence relation*

$$a_{r+2} - 5a_{r+1} + 6a_r = 2 \quad \text{...}(i)$$

by the method of generating functions satisfying the initial conditions $a_0 = 1$ and $a_1 = 2$.

Sol. Let us assume that

$$G(t) = \sum_{0}^{\infty} a_r\, t^r$$

Now, by taking generating functions of equation (*i*), we have

$$\frac{G(t) - a_0 - a_1 t}{t^2} - 5\left(\frac{G(t) - a_0}{t}\right) + 6G(t) = \frac{2}{1-t}$$

Put $a_0 = 1$ and $a_1 = 3$ in the above equation and solving, we get

$$G(t) = \frac{5t^2 - 4t + 1}{(1-t)(1-2t)(1-3t)}$$

By partial fractions

$$G(t) = \frac{1}{(1-t)} + \frac{-1}{(1-2t)} + \frac{1}{(1-3t)}$$

Therefore, the solution after applying inverse transformations

$$a_r = 1 - 2^r + 3^r.$$

SOLVED PROBLEMS

Problem 1. *Solve the following difference equation*

$$y_{K+3} + y_{K+2} - 8y_{K+1} - 12y_K = 0.$$

Sol. The characteristic equation is given by

$$s^3 + s^2 - 8s - 12 = 0$$

or $(s - 3)(s + 2)(s + 2) = 0$

or $s = 3, -2, -2$

Therefore, the homogeneous solution of the equation is

$$y_K = C_1 . 3^K + (C_2 + C_3 K) . (-2)^K.$$

Problem 2. *Solve the difference equation*

$$y_{K+3} + 6y_{K+2} + 12y_{K+1} + 8y_K = 0.$$

Sol. The characteristic equation is

$$s^3 + 6s^2 + 12s + 8 = 0$$

or $$(s + 2)(s + 2)(s + 2) = 0$$

or $$s = -2, -2, -2.$$

Therefore, the homogeneous solution of the equation is

$$y_K = (C_1 K^2 + C_2 K + C_3) \cdot (-2)^K.$$

Problem 3. *Solve the difference equation* $a_r - 4a_{r-1} + 6a_{r-2} - 4a_{r-3} + a_{r-4} = 0.$

Sol. The characteristic equation is given by

$$s^4 - 4s^3 + 6s^2 - 4s + 1 = 0$$

or $$(s - 1)(s^3 - 3s^2 + 2s - 1) = 0$$

or $$(s - 1)(s - 1)(s^2 - 2s + 1) = 0$$

or $$(s - 1)(s - 1)(s - 1)(s - 1) = 0$$

or $$s = 1, 1, 1, 1$$

Therefore, the homogeneous solution of the equation is

$$y_K = (C_1 r^3 + C_2 r^2 + C_3 r + C_4) \cdot (1)^r.$$

Problem 4. *Solve the difference equation*

$$a_r - 2a_{r-1} + 2a_{r-2} - a_{r-3} = 0.$$

Sol. The characteristic equation is given by

$$s^3 - 2s^2 + 2s - 1 = 0$$

or $$(s - 1)(s^2 - s + 1) = 0$$

or $$s = 1, \frac{1 + i\sqrt{3}}{2}, \frac{1 - i\sqrt{3}}{2}$$

Therefore, the homogeneous solution of the equation is

$$a_r = C_1 + C_2 \left[\frac{1 + i\sqrt{3}}{2}\right]^r + C_3 \left[\frac{1 - i\sqrt{3}}{2}\right]^r.$$

Problem 5. *Solve the difference equation* $a_{r+2} - 2a_{r+1} + a_r = 3r + 5.$ ...(*i*)

Sol. Put R.H.S. of the equation equal to zero *i.e.*,

$$a_{r+2} - 2a_{r+1} + a_r = 0$$

This can be written as $(E^2 - 2E + 1)\, a_r = 0$

or $$(E - 1)^2 = 0 \Rightarrow E = +1, 1$$

Therefore, the homogeneous solution is

$$a_{r(h)} = C_1 + C_2 r$$

The particular solution of equation (*i*) is

$$a_{r(p)} = \frac{1}{2} r^3 + r^2$$

as shown in example.

Therefore, the total solution of the difference equation is

$$a_r = C_1 + C_2 r + \frac{1}{2} r^3 + r^2.$$

Problem 6. *Solve the difference equation*

$$y_K - y_{K-1} - 6y_{K-2} = -30 \quad ...(i)$$

Given that $y_0 = 20$, $y_1 = -5$.

Sol. The equation (*i*) can be written as

$$(s^2 - s - 6)\, y_K = -30$$

Its homogeneous solution is given as

$$(s^2 - s - 6)y_K = 0$$

or $$(s - 3)(s + 2) = 0$$

or $$s = 3, -2$$

Thus, the homogeneous solution is

$$y_{K(h)} = C_1 \,.\, 3^K + C_2 \,.\, (-2)^K$$

For the particular solution of equation (*i*), consider the general form of solution = A

Put in the general form in L.H.S. of equation (*i*), we get

$$A - A - 6A = -30 \quad \Rightarrow \quad A = 5$$

Therefore, the particular solution = 5

The total solution is $y_K = C_1 \,.\, 3^K + C_2(-2)^K + 5$

Now, putting K = 0, we have

$$y_0 = C_1 + C_2 + 5 = 20$$

$\therefore$ $$C_1 + C_2 = 15 \quad ...(ii)$$

Now, putting K = 1, we have

$$y_1 = 3C_1 - 2C_2 + 5$$

$$-5 = 3C_1 - 2C_2 + 5 = 3C_1 - 2C_2 = -10. \quad ...(iii)$$

Solving (*ii*) and (*iii*), we have

$$C_1 = 4,\ C_2 = 11$$

Therefore, the total solution satisfying the initial conditions is

$$y_K = 4 \,.\, 3^K + 11(-2)^K + 5.$$

Problem 7. *Solve the recurrence relation*

$$a_{r+2} - 2a_{r+1} + a_r = r^2 \,.\, 2^r. \quad ...(i)$$

Sol. Put R.H.S. of the equation (*i*) equal to zero *i.e.*,

$$a_{r+2} - 2a_{r+1} + a_r = 0$$

This can be written as $(s^2 - 2s + 1)\, a_r = 0$

or $$(s - 1)^2 = 0$$

or $$s = 1, 1$$

Therefore, the homogeneous solution is

$$a_{r(h)} = C_1 + C_2\, r$$

For, finding particular solution of equation (*i*), consider its general form

$$a_r = 2^r (A_0 + A_1\, r + A_2\, r^2)$$

Put this solution in the L.H.S. of equation (*i*), we get

$$= 2^{r+2} [A_2 (r + 2)^2 + A_1 (r + 2) + A_0] - 2 \,.\, 2^{r+1} [A_2 (r + 1)^2 + A_1(r + 1) + A_0] + 2^r (A_2\, r^2 + A_1\, r + A_0)$$

$$= r^2 \,.\, 2^r (A_2) + r \,.\, 2^r (8A_2 + A_1) + 2^r (12A_2 + 4A_1 + A_0) \quad ...(ii)$$

Equating equation (*ii*) with R.H.S. of equation (*i*), we get

$$A_2 = 1$$
$$8A_2 + A_1 = 0$$
$$12A_2 + 4A_1 + A_0 = 0$$

On solving these equations, we get

$$A_0 = 20,\ A_1 = -8,\ A_2 = 1$$

Therefore, the particular solution is

$$a_{r(p)} = 2^r\,(r^2 - 8r + 20)$$

Hence, the total solution is

$$a_r = C_1 + C_2 r + 2^r\,(r^2 - 8r + 20).$$

Problem 8. *Solve the recurrence relation*

$$a_r + a_{r-1} + a_{r-2} = 0 \qquad ...(i)$$

Satisfying the initial conditions $a_0 = 0$ *and* $a_1 = 2$.

Sol. This equation can be written as $(s^2 + s + 1)\,a_r = 0$

or $$s = \frac{-1 + i\sqrt{3}}{2}, \frac{-1 - i\sqrt{3}}{2}$$

Therefore, the solution is given by

$$a_r = C_1 \left[\frac{-1 + i\sqrt{3}}{2}\right]^r + C_2 \left[\frac{-1 - i\sqrt{3}}{2}\right]^r \qquad ...(ii)$$

Put $r = 0$, in equation (*ii*), we have

$$a_0 = C_1 + C_2 = 0 \qquad ...(iii)$$

Put $r = 1$, in equation (*ii*), we have

$$a_1 = C_1 \left[\frac{-1 + i\sqrt{3}}{2}\right] + C_2 \left[\frac{-1 - i\sqrt{3}}{2}\right] = 2 \qquad ...(iv)$$

On solving equations (*iii*) and (*iv*), we have

$$C_1 = \frac{2}{i\sqrt{3}}$$

$$C_2 = \frac{-2}{i\sqrt{3}}$$

Therefore, the solution satisfying initial conditions is

$$a_r = \frac{2}{i\sqrt{3}}\left[\frac{-1 + i\sqrt{3}}{2}\right]^r - \frac{2}{i\sqrt{3}}\left[\frac{-1 - i\sqrt{3}}{2}\right]^r.$$

Problem 9. *Solve the recurrence relation*

$$a_r - a_{r-1} - a_{r-2} = 0 \qquad ...(i)$$

satisfying the initial conditions $a_0 = 1$ *and* $a_1 = 1$.

Sol. The equation (*i*) can also be written as

$$(s^2 - s - 1)\,a_r = 0 \quad \text{or} \quad s = \frac{1 + \sqrt{5}}{2}, \frac{1 - \sqrt{5}}{2}$$

Therefore, the solution is given by

$$a_r = C_1\left[\frac{1+\sqrt{5}}{2}\right]^r + C_2\left[\frac{1-\sqrt{5}}{2}\right]^r \quad ...(ii)$$

Put $r = 0$, in equation (*ii*), we have

$$a_0 = C_1 + C_2 = 1 \quad ...(iii)$$

Put $r = 1$, in equation (*ii*), we have

$$a_1 = C_1\left[\frac{1+\sqrt{5}}{2}\right] + C_2\left[\frac{1-\sqrt{5}}{2}\right] = 1 \quad ...(iv)$$

On solving equations (*iii*) and (*iv*), we have

$$C_1 = \frac{\sqrt{5}+1}{2\sqrt{5}}$$

$$C_2 = \frac{\sqrt{5}-1}{2\sqrt{5}}$$

Therefore, the solution satisfying initial conditions is

$$a_r = \left(\frac{\sqrt{5}+1}{2\sqrt{5}}\right)\left(\frac{1+\sqrt{5}}{2}\right)^r + \left(\frac{\sqrt{5}-1}{2\sqrt{5}}\right)\left(\frac{1-\sqrt{5}}{2}\right)^r.$$

Problem 10. *Solve the recurrence relation*

$$a_{r+1} - 3a_r = 2^r. \quad ...(i)$$

Sol. Put R.H.S. of the equation (*i*) equal to zero *i.e.*,

$$a_{r+1} - 3a_r = 0$$

This can be written as

$$(s - 3)a_r = 0 \quad \text{or} \quad s = 3.$$

Therefore, the homogeneous solution is

$$a_{r(h)} = C_1 \,.\, 3^r$$

For the particular solution of the equation (*i*), consider the general form of the solution

$$= A \,.\, 2^r$$

Put this solution on L.H.S. of equation (*i*), we have

$$= A \,.\, 2^{r+1} - 3A2^r = (2A - 3A) \,.\, 2^r$$

$$= - A \,.\, 2^r \quad ...(ii)$$

Equating equation (*ii*) with R.H.S. of equation (*i*), we get

$$A = -1$$

Therefore, the particular solution is

$$a_{r(p)} = -(2^r)$$

Hence, the total solution is $a_r = C_1 \,.\, 3^r - (2^r)$.

Problem 11. *Solve the recurrence relation*

$$a_{r+2} - 5a_{r+1} + 6a_r = 5^r. \quad ...(i)$$

Sol. Put the R.H.S. of equation equal to zero *i.e.*,

$$a_{r+2} - 5a_{r+1} + 6a_r = 0$$

This can also be written as

$$(s^2 - 5s + 6)a_r = 0$$

or $$(s - 3)(s - 2) = 0$$

or $$s = 3, 2$$

Therefore, the homogeneous solution is

$$a_{r(h)} = C_1 \,.\, 2^r + C_2 \,.\, 3^r$$

For, finding the particular solution consider the general form of the solution = A . 5^r.

Now, put this solution on L.H.S. of equation (*i*), we get

$$= A \,.\, 5^{r+2} - 5 \,.\, A \,.\, 5^{r+1} + 6 \,.\, A \,.\, 5^r$$

$$= 25A \,.\, 5^r - 25A \,.\, 5^r + 6A \,.\, 5^r$$

$$= 6A \,.\, 5^r \qquad ...(ii)$$

Equating equation (*ii*) to R.H.S. of equation (*i*), we get

$$A = \frac{1}{6}$$

Therefore, the particular solution is

$$a_{r(p)} = \frac{1}{6} \,.\, 5^r$$

Hence, the total solution is

$$a_r = C_1 \,.\, 2^r + C_2 \,.\, 3^r + \frac{1}{6} \,.\, 5^r.$$

Problem 12. *Solve the recurrence relation*

$$a_r - a_{r-1} = 7r. \qquad ...(i)$$

Sol. Put the R.H.S. of equation (*i*) equal to zero *i.e.*,

$$a_r - a_{r-1} = 0$$

This equation can also be written as

$$(s - 1)a_r = 0 \quad \text{or} \quad s = 1$$

Therefore, the homogeneous solution is

$$a_{r(h)} = C_1$$

For finding, the particular solution, consider the general form of the solution = $A_0 r + A_1$. Since one of the terms of the solution occurs in the homogeneous solution, so multiply the solution by r, we get $A_0 r^2 + A_1 \,.\, r$.

Now, put this solution in L.H.S. of equation (*i*), we have

$$= A_0 r^2 + A_1 r - A_0(r - 1)^2 - A_1 (r - 1) = 2A_0 r + A_1 - A_0 \qquad ...(ii)$$

Equating equation (*ii*) with R.H.S. of equation (*i*), we get

$$A_0 = \frac{7}{2}$$

$$A_1 - A_0 = 0 \;\Rightarrow\; A_1 = \frac{7}{2}$$

Therefore, the particular solution is $a_{r(p)} = \frac{7}{2} r^2 + \frac{7}{2} r$

Hence, the total solution is

$$a_r = C_1 + \frac{7}{2} r^2 + \frac{7}{2} r.$$

Problem 13. *Solve the recurrence relation*

$$a_{r+2} - 5a_{r+1} + 6a_r = r^2. \qquad \text{...}(i)$$

Sol. The homogeneous solution of the equation (*i*) is

$$a_{r(h)} = C_1 \,.\, 2^r + C_2 \,.\, 3^r$$

For finding, the particular solution of equation (*i*), consider the general form of the solution

$$= A_1\, r^2 + A_2\, r + A_3$$

Put this solution in L.H.S. of equation (*i*), we have

$$= A_1 (r + 2)^2 + A_2 (r + 2) + A_3 - 5A_1 (r + 1)^2 - 5A_2 (r + 1) - 5A_3 + 6A_1 r^2 + 6A_2 r + 6A_3$$

$$= (2A_1)r^2 + (-6A_1 + 2A_2)r + (-A_1 - 3A_2 + 2A_3) \qquad \text{...}(ii)$$

Equating equation (*ii*) with R.H.S. of equation (*i*), we get

$$2A_1 = 1 \qquad \text{...}(iii)$$

$$-6A_1 + 2A_2 = 0 \qquad \text{...}(iv)$$

$$-A_1 - 3A_2 + 2A_3 = 0 \qquad \text{...}(v)$$

On solving equation (*iii*), (*iv*) and (*v*), we get

$$A_1 = \frac{1}{2},\ A_2 = \frac{3}{2},\ A_3 = \frac{5}{2}$$

Therefore, the particular solution is

$$a_{r(p)} = \frac{1}{2} r^2 + \frac{3}{2} r + \frac{5}{2}$$

Hence, the total solution is

$$a_r = C_1 \,.\, 2^r + C_2 \,.\, 3^r + \frac{1}{2} r^2 + \frac{3}{2} r + \frac{5}{2}.$$

Problem 14. *Solve the recurrence relation*

$$y_{K+1} - 5y_K = 5^K \qquad \text{...}(i)$$

satisfying the initial condition $y_0 = 3$.

Sol. Put the R.H.S. of the equation (*i*) equal to zero *i.e.*,

$$y_{K+1} - 5y_K = 0$$

This can also be written as

$$(s - 5)y_K = 0 \quad \text{or} \quad s = 5$$

Therefore, the homogeneous solution is $y_{K(h)} = C_1 \,.\, 5^K$.

For finding, the particular solution, consider the general form of the solution $= A \,.\, 5^K$.

Since, this term occurs in the homogeneous solution, so multiply the solution by K.

Hence, the general form of the solution becomes $= AK \,.\, 5^K$.

Now, put this solution on L.H.S. of equation (*i*), we have

$$= A(K + 1) \,.\, 5^{K+1} - 5AK \,.\, 5^K = A \,.\, 5^{K+1}$$

$$= 5A \,.\, 5^K \qquad \text{...}(ii)$$

Equating equation (*ii*) with R.H.S. of equation (*i*), we get

$$A = \frac{1}{5}$$

Therefore, the particular solution is

$$y_{K(p)} = \frac{1}{5}K \cdot 5^K$$

Hence, the total solution is

$$y_K = C_1 \cdot 5^K + \frac{1}{5}K \cdot 5^K \quad \text{or} \quad y_K = \left(C_1 + \frac{K}{5}\right) \cdot 5^K \qquad ...(iii)$$

Put K = 0, in equation (*iii*), we have

$$y_0 = C_1 = 3.$$

The total solution satisfying the initial condition is

$$y_K = \left(3 + \frac{K}{5}\right) \cdot 5^K.$$

Problem 15. *Solve the recurrence relation*

$$a_r - 3a_{r-1} + 2a_{r-2} = 0 \qquad ...(i)$$

satisfying the initial conditions $a_0 = 1$ *and* $a_1 = 4$.

Sol. The equation (*i*) can also be written as

$$(s^2 - 3s + 2)a_r = 0 \quad \text{or} \quad (s-2)(s-1) = 0$$

or $$s = 1, 2$$

Therefore, the solution is $a_r = C_1 + C_2 \cdot 2^r$ ...(*ii*)

Put $r = 0$, in equation (*ii*), we have

$$a_0 = C_1 + C_2 = 1 \qquad ...(iii)$$

Put $r = 1$, in equation (*ii*), we have

$$a_1 = C_1 + 2C_2 = 4 \qquad ...(iv)$$

On solving equations (*iii*) and (*iv*), we get

$$C_1 = -2$$
$$C_2 = 3$$

Therefore, the solution satisfying the initial conditions is

$$a_r = -2 + 3 \cdot 2^r.$$

Problem 16. *Solve the recurrence relation*

$$y_K = r \cdot y_{K-1} + Z \qquad ...(i)$$

satisfying the initial conditions $y_0 = 0$. $r, z > 0, r \neq 1$.

Sol. The equation (*i*) can be written as

$$(E - r)\, y_K = Z \qquad ...(ii)$$

Put the R.H.S. of the equation (*ii*) equal to zero *i.e.*,

$$(E - r)y_K = 0$$

or $$(E - r) = 0 \quad \text{or} \quad E = r$$

Therefore, the homogeneous solution is

$$y_{K(h)} = C_1 \cdot r^K$$

The particular solution of equation (*ii*), is given by

$$y_{K(p)} = \frac{1}{(E-r)} \cdot z$$

Put E = 1, because z is a constant. Hence

$$y_{K(p)} = \frac{z}{(1-r)}$$

Hence, the total solution is

$$y_K = C_1 \cdot r^K + \frac{z}{(1-r)} \quad \text{...(iii)}$$

Put K = 0, in equation (*iii*), we have

$$y_0 = C_1 + \frac{z}{(1-r)} = 0$$

$$\therefore \quad C_1 = -\frac{z}{1-r}$$

The total solution satisfying the initial condition is

$$y_K = -\frac{z}{1-r} \cdot r^K + \frac{z}{(1-r)} \quad \text{or} \quad y_K = \frac{z(1-r^K)}{(1-r)}.$$

Problem 17. *Solve the recurrence relation*

$$y_{K+3} + y_{K+2} - 8y_{K+1} - 12y_K = 2K^2 + 5. \quad \text{...(i)}$$

Sol. Put the R.H.S. of equation (*i*) equal to zero *i.e.*,

$$y_{K+3} + y_{K+2} - 8y_{K+1} - 12y_K = 0$$

i.e., $$(E^3 + E^2 - 8E - 12)y_K = 0$$

or $$E = 3, -2, -2$$

Therefore, the homogeneous solution is

$$y_{K(h)} = C_1 \cdot 3^K + (C_2 + C_3 K)(-2)^K$$

The particular solution of equation (*i*) is given as

$$y_{K(p)} = \frac{1}{(E^3 + E^2 - 8E - 12)} \cdot (2K^2 + 5)$$

$$= \frac{1}{(E-3)(E+2)^2} \cdot (2K^2 + 5) \quad \text{[Put } E = 1 + \Delta]$$

$$= \frac{1}{(\Delta - 2)(\Delta + 3)^2} \cdot (2K^2 + 5)$$

$$= (\Delta^3 + 4\Delta^2 - 3\Delta - 18)^{-1} \cdot (2K^2 + 5)$$

$$= -\frac{1}{18}\left(1 + \frac{3\Delta - 4\Delta^2 - \Delta^3}{18}\right)^{-1} \cdot (2K^2 + 5)$$

$$y_{K(p)} = -\frac{1}{18}\left(1 - \frac{3\Delta - 4\Delta^2 - \Delta^3}{18} + \frac{(3\Delta - 4\Delta^2 - \Delta^3)^2}{(18)^2} + \dots\right) \cdot (2K^2 + 5)$$

$$= -\frac{1}{18}\left(1 - \frac{\Delta}{6} + \frac{\Delta^2}{4}\right) \cdot (2K^2 + 5) \quad \text{[By neglecting higher degrees]}$$

By factorial notation $2K^2 + 5 = 2[K]^2 + 2[K] + 5$.

Hence, $$y_{K(p)} = -\frac{1}{18}\left(1 - \frac{\Delta}{6} + \frac{\Delta^2}{4}\right).\ (2[K]^2 + 2[K] + 5)$$

$$= -\frac{1}{18}\left(2[K]^2 + 2[K] + 5 - \frac{1}{6}(4[K] + 2) + \frac{1}{4}(4)\right)$$

$$= -\frac{1}{18}\left(2[K]^2 - \frac{4}{3}[K] + \frac{17}{3}\right)$$

$$= -\frac{1}{18}\left(2K(K-1) - \frac{4}{3}K + \frac{17}{3}\right)$$

$$y_{K(p)} = -\frac{K^2}{9} + \frac{1}{27}K - \frac{17}{54}$$

Hence, the total solution is

$$y_K = C_1 . 3^K + (C_2 + C_3 K)(-2)^K + \left(\frac{-K^2}{9} + \frac{1}{27}K - \frac{17}{54}\right).$$

Problem 18. *Solve the recurrence relation*

$$y_{K+4} + 4y_{K+3} + 8Y_{K+2} + 8y_{K+1} + 4y_K = 10. \qquad \text{...}(i)$$

Sol. Put the R.H.S. of equation (*i*) equal to zero *i.e.,*

$$y_{K+4} + 4y_{K+3} + 8y_{K+2} + 8y_{K+1} + 4y_K = 0$$

i.e., $$(E^4 + 4E^3 + 8E^2 + 8E + 4)\ y_K = 0$$

Its homogeneous solution is

$$y_{K(h)} = (C_1 + C_2 K)(-1 + i)^K + (C_3 + C_4 K)(-1 - i)^K$$

as shown before.

The particular solution of equation (*i*) is given as

$$y_{K(p)} = \frac{1}{(E^4 + 4E^3 + 8E^2 + 8E + 4)}.\ 10$$

By putting E = 1, we have

$$y_{K(p)} = \frac{10}{1+4+8+8+4} = \frac{10}{25} = \frac{2}{5}$$

Hence, the total solution is

$$y_K = (C_1 + C_2 K)(-1 + i)^K + (C_3 + C_4 K)(-1 - i)^K + \frac{2}{5}.$$

12

Algebraic Structures

If there exists a system such that it consists of a non-empty set and one or more operations on that, set, then that system is called an algebraic system. It is generally denoted by (A, op_1, op_2, ..., op_n), where A is a non-empty set and op_1, op_2, ..., op_n are operations on A.

An algebraic system is also called an algebraic structure because the operations on the set A define a structure on the elements of A.

N-ARY OPERATION

A function $f: A \times A \times \ldots A \to A$ is called an n-ary operation.

BINARY OPERATION

Consider a non-empty set A and *a* function *f* such that $f: A \times A \to A$ is called a binary operation on A. If * is a binary operation on A, then it may be written as $a * .b$.

A binary operation can be denoted by any of the symbols +, –, *, ⊕, Δ, □, ∨, ∧ etc.

The value of the binary operation is denoted by placing the operator between the two operands.

e.g., (*i*) The operation of addition is a binary operation on the set of natural numbers.

(*ii*) The operation of subtraction is a binary operation on set of integers. But, the operation of subtraction is not a binary operation on the set of natural numbers because the subtraction of two natural numbers may or may not be a natural number.

(*iii*) The operation of multiplication is a binary operation on the set of natural numbers, set of integers and set of complex numbers.

(*iv*) The operation of set union is a binary operation on the set of subsets of a universal set. Similarly, the operation of set intersection is a binary operation on the set of subsets of a universal set.

TABLES OF OPERATION

Consider a non empty finite set A = {a_1, a_2, a_3, ..., a_n}. A binary operation * on A can be described by means of table as shown in Fig. 1.

*	a_1	a_2	a_3		a_n
a_1	a_1*a_1				
a_2		a_2*a_2			
a_3			a_3*a_3		
a_n					a_n*a_n

Fig. 1.

The empty in the jth row and kth column represent the element a_j*a_k.

Example 1. *Consider the set A = {1, 2, 3} and a binary operation * on the set A defined by*

$$a * b = 2a + 2b.$$

*Represent operation * as a table on A.*

Sol. The table of the operation is shown in Fig. 2.

*	1	2	3
1	4	6	8
2	6	8	10
3	8	10	12

Fig. 2.

PROPERTIES OF BINARY OPERATIONS

There are many properties of the binary operations which are as follows :

1. Closure Property. Consider a non-empty set A and a binary operation * on A. Then A is closed under the operation *, if $a * b \in$ A, where a and b are elements of A.

e.g., The operation of addition on the set of integers is a closed operation.

Example 2. *Consider the set A = {– 1, 0, 1}. Determine whether A is closed under (i) addition (ii) multiplication.*

Sol. (*i*) The sum of the elements is (– 1) + (– 1) = – 2 and 1 + 1 = 2 does not belong to A. Hence A is not closed under addition.

(*ii*) The multiplication of every two elements of the set are

$$-1 * 0 = 0\ ; \quad -1 * 1 = -1\ ; \quad -1 * -1 = 1$$
$$0 * -1 = 0\ ; \quad 0 * 1 = 0\ ; \quad 0 * 0 = 0$$
$$1 * -1 = -1\ ; \quad 1 * 0 = 0\ ; \quad 1 * 1 = 1$$

Since, each multiplication belongs to A hence A is closed under multiplication.

Example 3. *Consider the set A = {1, 3, 5, 7, 9, ...}, the set of odd +ve integers. Determine whether A is closed under (i) addition (ii) multiplication.*

Sol. (*i*) The set A is not closed under addition because the addition of two odd numbers produces an even number which does not belong to A.

(*ii*) The set A is closed under the operation multiplication because the multiplication of two odd numbers produces an odd number. So, for every $a, b \in$ A, we have $a * b \in$ A.

2. Associative Property. Consider a non-empty set A and a binary operation * on A. Then the operation * on A is associative, if for every $a\ b, c, \in$ A, we have $(a * b) * c = a* (b * c)$.

Example 4. *Consider the binary operation * on Q, the set of rational numbers, defined by*

$$a * b = a + b - ab \ \forall\ a, b \in Q.$$

*Determine whether * is associative.*

Sol. Let us assume some elements $a, b, c \in$ Q, then by definition

$$(a * b) * c = (a + b - ab) * c = (a + b - ab) + c - (a + b - ab)c$$
$$= a + b - ab + c - ca - bc + abc = a + b + c - ab - ac - bc + abc.$$

Similarly, we have

$$a * (b * c) = a + b + c - ab - ac - bc + abc$$

Therefore, $(a * b) * c = a * (b * c)$.

Hence * is associative.

3. Commutative Property. Consider a non-empty set A and a binary operation * on A. Then the operation * on A is commutative, if for every $a, b \in$ A, we have $a * b = b * a$.

Example 5. *Consider the binary operation * on Q, the set of rational numbers, defined by*

$$a * b = a^2 + b^2 \ \forall\ a, b \in Q.$$

*Determine whether * is commutative.*

Sol. Let us assume some elements $a, b \in$ Q, then by definition

$$a * b = a^2 + b^2 = b * a$$

Hence * is commutative.

Example 6. *Consider the binary operation * and Q, the set of rational numbers defined by*

$$a * b = \frac{ab}{2} \quad \forall\ a, b \in Q.$$

*Determine whether * is (i) associative (ii) commutative.*

Sol. (*i*) Let $a, b \in$ Q, then we have

$$a * b = \frac{ab}{2} = b * a$$

Hence * is commutative.

(*ii*) Let $a, b, c \in$ Q, then by definition we have

$$(a * b) * c = \left(\frac{ab}{2}\right) * c = \frac{\frac{ab}{2} . c}{2} = \frac{abc}{4}$$

Similarly, $a * (b * c) = a * \left(\frac{bc}{2}\right) = \frac{\frac{abc}{2}}{2} = \frac{abc}{4}$

Therefore, $a * (b * c) = a * (b * c)$

Hence, * is associative.

4. Identity. Consider a non-empty set A and a binary operation * on A. Then the operation * has an identity property if there exists an element, e, in A such that

$$a * e \text{ (right identity)} = e * a \text{(left identity)} = a \ \forall\ a \in A.$$

Theorem I. *Prove that $e_1' = e_1''$ where e_1' is a right identity and e_1'' is a left identity of a binary operation.*

Proof. We know that e_1' is a right identity.

Hence, $\quad e_1'' * e_1' = e_1''$ ...(*i*)

Also, we know that e_1'' is a left identity.

Hence, $\quad e_1'' * e_1' = e_1'$ ...(*ii*)

From (*i*) and (*ii*), we have $e_1' = e_1''$.

Thus, we can say that if e is a right identity of a binary operation then e is also a left identity or there is no left identity.

Example 7. *Consider the binary operation * on I_+, the set of positive integers defined by*

$$a * b = \frac{ab}{2}.$$

*Determine the identity for the binary operation *, if exists.*

Sol. Let us assume that e be a +ve integer number, then

$$e * a, a \in I_+$$

$$\frac{ea}{2} = a, e = 2 \qquad \text{...}(i)$$

Similarly, $\quad a * e = a, a \in I_+$

$$\frac{ae}{2} = a \quad \text{or} \quad e = 2 \qquad \text{...}(ii)$$

Form (*i*) and (*ii*) for $e = 2$, we have $e * a = a * e = a$

Therefore, 2 is the identity element for *.

5. Inverse. Consider a non-empty set A and a binary operation * on A. Then operation * has the inverse property if for each $a \in A$, there exists an element b in A such that

$a * b$ (right inverse) $= b * a$ (left inverse) $= e$, where b is called an inverse of a.

6. Idempotent. Consider a non-empty set A and a binary operation * on A. Then the operation * has the idempotent property, if for each $a \in A$, we have

$$a * a = a \ \forall\ a \in A.$$

7. Distributivity. Consider a non-empty set A and two binary operations * and + on A. Then the operation * distributes over +, if for every $a, b, c \in A$, we have

$$a * (b + c) = (a * b) + (a * c) \qquad \text{[Left distributivity]}$$

and $$(b + c) * a = (b * a) + (c * a) \qquad \text{[Right distributivity]}$$

8. Cancellation. Consider a non-empty set A and a binary operation * on A. Then the operation * has the cancellation property, if for every $a, b, c \in A$, we have

$$a * b = a * c \ \Rightarrow\ b = c \qquad \text{[Left cancellation]}$$

and $$b * a = c * a \ \Rightarrow\ b = c \qquad \text{[Right cancellation]}$$

SEMIGROUP

Let us consider, an algebraic system (A, $*$), where $*$ is a binary operation on A. Then, the system (A, $*$) is said to be a semi-group if it satisfies the following properties :

1. The operation $*$ is a closed operation on set A.
2. The operation $*$ is an associative operation.

Example 8. *Consider an algebraic system (A, $*$), where A = {1, 3, 5, 7, 9, ...}, the set of all positive odd integers and $*$ is a binary operation means multiplication. Determine whether (A, $*$) is a semi-group.*

Sol. Closure property. The operation $*$ is a closed operation because multiplication of two +ve odd integers is a +ve odd number.

Associative property. The operation $*$ is an associative operation on set A. Since for every $a, b, c \in$ A, we have

$$(a * b) * c = a * (b * c)$$

Hence, the algebraic system (A, $*$) is a semigroup.

Example 9. *Consider the algebraic system ({0, 1}, $*$), where $*$ is a multiplication operation. Determine whether ({0, 1}, $*$) is a semigroup.*

Sol. Closure property. The operation $*$ is a closed one on the given set since

$$0 * 0 = 0 \,;\, 0 * 1 = 0 \,;\, 1 * 0 = 0 \,;\, 1 * 1 = 1.$$

Associative property. The operation $*$ is associative since we have

$$(a * b) * c = a * (b * c) \ \forall\, a, b, c$$

Since, the algebraic system is closed and associative. Hence, it is a semi-group.

Example 10. *Let (A, $*$) be semi-group. Show that for a, b, c in A, if $a * c = c * a$ and $b * c = c * b$, then $(a * b) * c = c * (a * b)$.*

Sol. Take L.H.S., we have

$$(a * b) * c \Rightarrow a * (b * c) \qquad [\because * \text{ is associative}]$$
$$\Rightarrow a * (c * b) \qquad [\because b * c = c * b]$$
$$\Rightarrow (a * c) * b \qquad [\because * \text{ is associative}]$$
$$\Rightarrow (c * a) * b \qquad [\because a * c = c * a]$$
$$\Rightarrow c * (a * b) \qquad [\because * \text{ is associative}]$$

Which is equal to R.H.S.

Hence, $(a * b) * c = c * (a * b)$.

SUBSEMIGROUP

Consider a semigroup (A, $*$) and let B $\subseteq$ A. Then the system (B, $*$) is called a subsemigroup, if the set B is closed under the operation $*$.

e.g., Consider a semigroup (N, +), where N is the set of all natural numbers and + is an addition operation. The algebraic system (E, +) is a subsemigroup of (N, +), where E is a set of all +ve even integers.

FREE SEMIGROUP

Consider a nonempty set $A = \{a_1, a_2, \ldots\ldots, a_n\}$.

Now A* is the set of all finite sequences of elements of A, *i.e.*, A* consists of all words that can be formed from the alphabet of A.

If α, β and γ are any elements of A*, then $\alpha \,.\, (\beta \,.\, \gamma) = (\alpha \,.\, \beta) \,.\, \gamma$.

Here• is a concatenation operation, which is an associative operation as shown above.

Thus (A*, •) is a semigroup. This semigroup (A*, •) is called the free semigroup generated by set A.

PRODUCT OF SEMIGROUP

Theorem. *If* $(S_1, *)$ *and* $(S_2, *)$ *are semigroups, then* $(S_1 \times S_2, *)$ *is a semigroup, where * is defined by* $(s_1', s_2') * (s_1'', s_2'') = (s_1' * s_1'', s_2' * s_2'')$.

Proof. The semigroup $S_1 \times S_2$ is closed under the operation*.

Associativity of *. Let $a, b, c \in S_1 \times S_2$

So,
$$\begin{aligned} a * (b * c) &= (a_1, a_2) * ((b_1, b_2) * (c_1, c_2)) \\ &= (a_1, a_2) * (b_1 *_1 c_1, b_2 *_2 c_2) \\ &= (a_1 *_1 (b_1 *_1 c_1), a_2 *_2 (b_2 *_2 c_2)) \\ &= ((a_1 *_1 b_1) *_1 c_1, (a_2 *_2 b_2) *_2 c_2) \\ &= (a_1 *_1 b_1, a_2 *_2 b_2) * (c_1, c_2) \\ &= ((a_1, a_2) * (b_1, b_2)) * (c_1, c_2) \\ &= (a * b) * c. \end{aligned}$$

Since * is closed and associative. Hence $S_1 \times S_2$ is a semigroup.

CONGRUENCE RELATION

An equivalence relation R on the semigroup (S, *) is called a congruence relation if

$$aRa' \quad \text{and} \quad bRb'$$
$$\Rightarrow \qquad (a * b)\, R\, (a' * b')$$

Example 11. *Let (I, +) be a semigroup and R is an equivalence relation on I defined by* aRb *iff* $a \equiv b$ *(MOD 3).*

Sol. If a and b yield the same remainder when divided by 3, then we have 3 divides $a - b$ *i.e.*, $\frac{3}{(a-b)}$.

Now, if $a \equiv b$ (MOD 3) and $c \equiv d$ (MOD 3) then 3 divides $a - b$ and 3 divides $c - d$.

Thus, we can write $a - b = 3m$ ...(*i*)

and $c - d = 3n$ ...(*ii*)

Here m and n are same integer of I.

Adding eqns. (*i*) and (*ii*), we have

$$(a - b) + (c - d) = 3m + 3n \quad \text{or} \quad (a + c) - (b + d) = 3(m + n)$$

or
$$a + c = b + d \text{ (MOD 3)}$$

Thus, the relation is a congruence relation.

Example 12. *Consider the set A = {a, b}. Let (A*, •) is the semigroup generated by A, also let R is a relation on A defined by αRβ iff α and β have the same number of a's.*

Show whether the relation R is a congruence on (A, •).*

Sol. First of all we will show that R is an equivalence relation. So, for that we will check reflexive, symmetric and transitive properties of the relation R.

Reflexive. αRα for any α ∈ A* since α has same number of *a*'s as itself. Thus, R is reflexive.

Symmetric. If α and β have same number of *a*'s, then αRβ or we can say βRα. Thus, R is symmetric.

Transitive. If αRβ, it means α and β have same number of *a*'s. If βRγ, it means β and γ have same number of *a*'s. It implies α and γ have same number of *a*'s *i.e.*, αRγ. Thus, R is transitive.

Hence, R is an equivalence relation.

To show that R is a congruence relation, let us assume that $\alpha R \alpha_1$ and $\beta R \beta_1$. It means α and α_1 have same number of *a*'s and β and β_1 have same number of *a*'s. We know that the number of *a*'s in α . β is the sum of number of *a*'s in α and the number of *a*'s in β.

From the above discussion, we can say that the number of *a*'s in α .β is same as in $\alpha_1 . \beta_1$.

Hence $(\alpha . \beta)\ R\ (\alpha_1 . \beta_1)$

which shows that R is a congruence relation.

Example 13. *Consider the semigroup (I, +), where + is an addition operation. Let* $f(x) = x^2 - 2x - 3$ *and also let R is a relation on I defined by aRb iff f(a) = f(b).*

Show whether R is a congruence relation.

Sol. It can be easily shown that the relation R is an equivalence relation on the set I.

To check whether R is congruence relation or not, we will try to find two pair of numbers *a*R*b* and *c*R*d* but $(a + b) \not R (c + d)$ if possible. Then we will say R is not a congruence relation.

Thus, we have

$$2\ R\ 0 \quad i.e., \quad f(2) = f(0) = -3$$

and

$$-2\ R\ 4 \quad i.e., \quad f(-2) = f(4) = 5$$

But $(2 + (-2)) \not R (0 + 4)$ *i.e.*, $0 \not R 4$

As $f(0) = -3$ and $f(4) = 5$

Hence, R is not a congruence relation.

MONOID

Let us consider an algebraic system (A, *o*), where *o* is a binary operation on A. Then the system (A, *o*) is said to be a monoid if it satisfies the following properties.

(*i*) The operation *o* is a closed operation on set A.

(*ii*) The operation *o* is an associative operation.

(*iii*) There exists an identity element w.r.t. the operation *o*.

Example 14. *Consider an algebraic system (N, +), where the set N = {0, 1, 2, 3, 4, ...} the set of natural numbers and + is an addition operation. Determine whether (N, +) is a monoid.*

Sol. Closure property. The operation + is closed since sum of two natural numbers is a natural number.

Associative property. The operation + is an associative property since we have

$$(a + b) + c = a + (b + c) \ \forall\ a, b, c \in N.$$

Identity. There exists an identity element in set N w.r.t. the operation +. The element 0 is an identity element w.r.t. the operation +. Since, the operation + is a closed, associative and there exists an identity. Hence, the algebraic system (N, +) is a monoid.

SUBMONOID

Let us consider a monoid (M, *o*), also let S ⊆ M. Then (S, *o*) is called a submonoid of (M, *o*), if and only if it satisfies the following properties.

(*i*) S is closed under the operation *o*.

(*ii*) There exists an identity element $e \in$ T.

e.g., Let us consider, a monoid (M, *),where * is a binary operation and M is a set of all integers. Then (M_1, *) is a submonoid of (M, *), where M_1 is defined as

$$M_1 = \{a^i \mid i \text{ is from 0 to } n, \text{ a positive integer and } a \in M\}.$$

GROUP

Let us consider an algebraic system (G, *), where * is a binary operation on G. Then the system (G, *) is said to be a group if it satisfies following properties.

(*i*) The operation * is a closed operation.

(*ii*) The operation * is an associative operation.

(*iii*) There exists an identity element w.r.t. the operation *.

(*iv*) For every $a \in$ G, there exists an element $a^{-1} \in$ G such that

$$a^{-1} * a = a * a^{-1} = e$$

e.g., The algebraic system (I, +), where I is the set of all integers and + is an addition operation, is a group. The element 0 is the identity element w.r.t. the operation +. The inverse of every element $a \in$ I is $-\,a \in$ I.

Example 15. *Determine whether the algebraic system (Q, +) is a group where Q is the set of all rational numbers and + is an addition operation.*

Sol. Closure property. The set Q is closed under operation +, since the addition of two rational numbers is a rational number.

Associative Property. The operation + is associative, since $(a + b) + c = a + (b + c) \ \forall\ a, b, c \in$ Q.

Identity. The element 0 is the identity element. Hence $a + 0 = 0 + a = a \ \forall\ a \in$ Q.

Inverse. The inverse of every element $a \in$ Q is $-\ a \in$ Q. Hence the inverse of every element exists.

Since, the algebraic system (a, +) satisfy all the properties of a group, hence (a, +) is a group.

Example 16. *Consider an algebraic system (Q, *),where Q is the set of rational numbers and * is a binary operation defined by*

$$a * b = a + b - ab \ \forall\ a, b \in Q.$$

Determine whether (Q, +) is a group.

Sol. Closure property. Since the element $a, b \in Q$ for every $a, b \in Q$, hence, the set Q is closed under the operation *.

Associative property. Let us assume $a, b, c \in Q$, then we have

$$\begin{aligned}(a * b) * c &= (a + b - ab) * c \\ &= (a + b - ab) + c - (a + b - ab)c \\ &= a + b - ab + c - ac - bc + abc \\ &= a + b + c - ab - ac - bc + abc\end{aligned}$$

Similarly, $a * (b * c) = a + b + c - ab - ac - bc + abc.$

Therefore, $(a * b) * c = a * (b * c)$

$\therefore$ * is associative.

Identity. Let e is an identity element. Then we have $a * e = a \; \forall \; a \in Q$

$\therefore$ $a + e - ae = a$ or $e - ae = 0$

or $e(1 - a) = 0$ or $e = 0$

Similarly, $e * a = a \; \forall \; a \in Q$

Therefore, for $e = 0$, we have $a * e = e * a = a$

Thus 0 is the identity element.

Inverse. Let us assume an element $a \in Q$. Let a^{-1} is an inverse of a, where $a^{-1} \in Q$. Then we have

$a * a^{-1} = 0$ [Identity]

$\therefore$ $a + a^{-1} - aa^{-1} = 0$

or $a^{-1}(1 - a) = -a$ or $a^{-1} = \dfrac{a}{a-1}$

Now, $\dfrac{a}{a-1} \in Q$, if $a \neq \dfrac{1}{a}$

Therefore, every element has inverse such that $a \neq 1$.

Since, the algebraic system (Q, *) satisfy all the properties of a group. Hence, (Q,*) is a group.

Theorem II. *Show that the identity element in a group is unique.*

Proof. Let us assume that there exists two identity elements of G *i.e.*, e and e'.

Since, $e \in G$ and e' is an identity.

We have $e'e = ee' = e$

Also, $e' \in G$ and e is an identity. We have $e'e = ee' = e'$

$\therefore$ $e = e'$

Hence, identity in a group is unique.

Theorem III. *Show that inverse of an element a in the group is unique.*

Proof. Let us assume that $a \in G$ be an element. Also, assume that a_1^{-1} and a_2^{-1} be two inverse elements of a.

Then we have

$$a_1^{-1} a = aa_1^{-1} = e \quad \text{and} \quad a_2^{-1} a = aa_2^{-1} = e$$

Now, $a_1^{-1} = a_1^{-1} e = a_1^{-1}(aa_2^{-1}) = (a_1^{-1} a)a_2^{-1} = ea_2^{-1} = a_2^{-1}$

Thus, the inverse of an element is unique.

Theorem IV. *Show that $(a^{-1})^{-1} = a$ for all $a \in G$, where G is a group and a^{-1} is an inverse of a.*

Proof. Given that a^{-1} is an inverse of a. Then, we have

$$aa^{-1} = a^{-1}a = e$$

This implies that a is also an inverse of a^{-1}. Therefore $(a^{-1})^{-1} = a$.

Theorem V. *Show that $(ab)^{-1} = b^{-1}a^{-1}$ for all $a, b \in G$.*

Proof. We have to prove that ab is inverse of $b^{-1}a^{-1}$.

So, $\quad (ab)(b^{-1}a^{-1}) = (b^{-1}a^{-1})(ab) = e$

Now, take L.H.S.

$$(ab)(b^{-1}a^{-1}) = [(ab)b^{-1}]a^{-1} = [a(bb^{-1})]a^{-1}$$
$$= (ae)a^{-1} = aa^{-1} = e$$

Similarly, the R.H.S. *i.e.,* $(b^{-1}a^{-1})(ab) = e$

Hence proved.

Theorem VI. *Prove the left cancellation law i.e., $ab = ac \Rightarrow b = c \ \forall\ a, b, c \in G$.*

Proof. Consider that $ab = ac$.

Then, we have $\quad b = eb = (a^{-1}a)b = a^{-1}(ab) = a^{-1}(ac) \qquad [\because ab = ac]$

$$= (a^{-1}a)c = ec = c$$

Hence, $\quad ab = ac \Rightarrow b = c.$

Theorem VII. *Prove the right cancellation law i.e., $ba = ca \Rightarrow b = c \ \forall\ a, b, c \in G$.*

Proof. Consider that $ba = ca$.

Then, we have $\quad b = be = b(aa^{-1}) = (ba)a^{-1} = (ca)a^{-1} \qquad [\because ba = ca]$

$$= c(aa^{-1}) = ce = c$$

Hence, $\quad ba = ca \Rightarrow b = c.$

FINITE AND INFINITE GROUP

A group (G, *) is called a finite group if G is a finite set.

A group (G, *) is called an infinite group if G is an infinite set.

For example : The group (I, +) is an infinite group as the set I of integers is an infinite set.

For example : The group G = {1, 2, 3, 4, 5, 6, 7} under multiplication modulo 8 is a finite group as the set G is a finite set.

ORDER OF GROUP

The order of the group G is the number of elements in the group G. It is denoted by |G|. A group of order 1 has only the identity element *i.e.,* ({e} *).

A group of order 2 has two elements *i.e.,* one identity element and one some other element.

For example : *Let ({e, x}, *) be a group of order 2. The table of operation is shown in (Fig. 3).*

*	*e*	*x*
e	*e*	*x*
x	*x*	*e*

Fig. 3.

The group of order 3 has three elements *i.e.*, one identity element and two other elements.

For example : *Let ({e, x, y}, *) be a group of order 3. The table of operation is shown in (Fig. 4).*

*	*e*	*x*	*y*
e	*e*	*x*	*y*
x	*x*	*y*	*e*
y	*y*	*e*	*x*

Fig. 4.

For example : *Consider an algebraic system ({0, 1}, +) where the operation + is defined as shown in (Fig. 5).*

+	0	1
0	0	1
1	1	0

Fig. 5.

The system ({0, 1}, +) is a group. In this 0 is identity element and every element is its own inverse.

SUBGROUP

Let us consider a group (G, $*$). Also, let $S \subseteq G$; then (S, $*$) is called a subgroup if it satisfies following conditions :

(*i*) The operation * is closed operation on S.

(*ii*) The operation * is an associative operation.

(*iii*) As *e* is an identity element belonged to G. It must belong to the set S *i.e.*, The identity element of (G, $*$) must belong to (S, $*$).

(*iv*) For every element $a \in S$, a^{-1} also belongs to S.

e.g., Let (G, +) be a group, where G is a set of all integers and (+) is an addition operation. Then (H, +) is a subgroup of the group G, where H = [{2K | K ∈ G}, the set of all even integer]

e.g., Let G be a group. Then the two subgroups of G are G and $G_1 = \{e\}$ is the indentity element.

Example 17. *Let (I, +) be a group, where I is the set of all integers and (+) is an addition operation. Determine whether the following subsets of G are subgroups of G.*

(*a*) *The set G_1 of all odd integers.* (*b*) *The set G_2 of all positive integers.*

Sol. (*a*) The set G_1 of all odd integers is not a subgroup of G. It does not satisfy the closure property, since addition of two odd integers is always even.

(*b*) **Closure property.** The set G_2 is closed under the operation +, since addition of two even integers is always even.

Associative property. The operation + is associative since $(a + b) + c = a + (b + c)$ for every $a, b, c \in G_2$.

Identity. The element 0 is the identity element. Hence, $0 \in G_2$.

Inverse. The inverse of every element $a \in G_2$ is $-a \notin G_2$. Hence, the inverse of every element does not exists.

Since the system $(G_2, +)$ does not satisfy all the conditions of a subgroup. Hence, $(G_2, +)$ is not a subgroup of (I, +).

ABELIAN GROUP

Let us consider, an algebraic system (G, $*$), where $*$ is a binary operation on G. Then the system (G, $*$)is said to be an abelian group if it satisfies all the properties of the group plus an additional following property :

(*i*) The operation $*$ is commutative *i.e.*,

$$a * b = b * a \ \forall \ a, b \in G$$

e.g., Consider an algebraic system (I, +), where I is the set of all integers and + is an addition operation. The system (I, +) is an abelian group because it satisfies all the properties of a group also the operation + is commutative for every $a, b \in I$.

Example 18. *Consider an algebraic system (G, *), where G is the set of all non-zero real numbers and * is a binary operation defined by*

$$a * b = \frac{ab}{4}.$$

*Show that (G, *) is an abelian group.*

Sol. Closure property. The set G is closed under the operation $*$. Since, $a * b = \frac{ab}{2}$ is a real number. Hence, belongs to G.

Associative property. The operation $*$ is associative. Let $a, b, c \in G$, then we have

$$(a * b) * c = \left(\frac{ab}{4}\right) * c = \frac{(ab)c}{16} = \frac{abc}{16}.$$

Similarly, $$a * (b * c) = a * \left(\frac{bc}{4}\right) = \frac{a(bc)}{16} = \frac{abc}{16}.$$

Identity. To find the identity element, let us assume that e is a +ve real number. Then $e * a = a$, where $a \in G$.

$$\frac{ea}{4} = a \quad \text{or} \quad e = 4$$

Similarly, $$a * e = a$$

$$\frac{ae}{4} = a \quad \text{or} \quad e = 4.$$

Thus, the identity element in G is 4.

Inverse. Let us assume that $a \in G$. If $a^{-1} \in Q$ is an inverse of a, then $a * a^{-1} = 4$

Therefore, $$\frac{aa^{-1}}{4} = 4 \quad \text{or} \quad a^{-1} = \frac{16}{a}$$

Similarly, $a^{-1} * a = 4$

Therefore, $\frac{a^{-1} a}{4} = 4$ or $a^{-1} = \frac{16}{a}$.

Thus, the inverse of element a in G is $\frac{16}{a}$.

Commutative. The operation * on G is commutative.

Since, $a * b = \frac{ab}{4} = b * a$.

Thus, the algebraic system (G, $*$) is closed, associative, identity element, inverse and commutative. Hence, the system (G, $*$) is an abelian group.

PRODUCT OF GROUPS

Theorem. *Prove that if* $(G_1, *_1)$ *and* $(G_2, *_2)$ *are groups, then* $G = G_1 \times G_2$ *i.e.,* $(G, *)$ *is a group with operation defined by* $(a_1, b_1) * (a_2, b_2) = (a_1 *_1 a_2, b_1 *_2 b_2)$.

Proof. To prove that $G_1 \times G_2$ is a group, we have to show that $G_1 \times G_2$ has the associative property, has an identity and also exists inverse of every element.

Associativity. Let $a, b, c \in G_1 \times G_2$, then

$$\begin{aligned} a * (b * c) &= (a_1, a_2) * ((b_1, b_2) * (c_1, c_2)) \\ &= (a_1, a_2) * (b_1 *_1 c_1, b_2 *_2 c_2) \\ &= (a_1 *_1 (b_1 *_1 c_1), a_2 *_2 (b_2 *_2 c_2)) \\ &= ((a_1 *_1 b_1) *_1 c_1), ((a_2 *_2 b_2) *_2 c_2) \\ &= (a_1 *_1 b_1, a_2 *_2 b_2) * (c_1, c_2) \\ &= ((a_1, a_2) * (b_1, b_2)) * (c_1, c_2) = (a * b) * c. \end{aligned}$$

Identity. Let e_1 and e_2 are identities for G_1 and G_2 respectively. Then, the identity for $G_1 \times G_2$ is $e = (e_1, e_2)$. Assume same $a \in G_1 \times G_2$.

Then
$$\begin{aligned} a * e &= (a_1, a_2) * (e_1, e_2) \\ &= (a_1 *_1 e_1, a_2 *_2 e_2) = (a_1, a_2) = a \end{aligned}$$

Similarly, we have $e * a = a$.

Inverse. To determine the inverse of an element in $G_1 \times G_2$, we will determine it componentwise *i.e.,*

$$a^{-1} = (a_1, a_2)^{-1} = (a_1^{-1}, a_2^{-1}).$$

Now to verify that this is the exact inverse, we will compute $a * a^{-1}$ and $a^{-1} * a$.

Now
$$\begin{aligned} a * a^{-1} &= (a_1, a_2) * (a_1^{-1}, a_2^{-1}) \\ &= (a_1 *_1 a_1^{-1}, a_2 *_2 a_2^{-1}) = (e_1, e_2) = e \end{aligned}$$

Similarly, we have $a^{-1} * a = e$.

Thus, $(G_1 \times G_2, *)$ is a group.

In general, if $G_1, G_2, ..., G_n$ are groups, then $G = G_1 \times G_2 \times .....\times G_n$ is also a group.

COSETS

Consider an algebraic system (G, $*$), where $*$ is a binary operation. Now, if (G, $*$) is a group and let a be an element of G and $H \subseteq G$, then the LEFT COSET $a * H$ of H is the set of elements such that

$$a * H = \{a * h / h \in H\}.$$

The RIGHT COSET $H * a$ of H is the set of elements such that

$$H * a = \{h * a / h \in H\}.$$

The subset H is itself a left and right coset since $e * H = H * e = H$.

Example 19. *Let us consider a group (G, *), where G is a set having elements {0, 1} and * is a binary operation. Also, let H = {1} is a subgroup of G. Determine all the left cosets of H in G.*

Sol. There is only 2 left cosets *i.e.*,

$$1 * H = H = \{1\}$$

$$0 * H = \{0\}.$$

Example 20. *Let (I, +) is a group, where I is the set of all integers and + is an addition operation and let H = {..., – 4, – 2, 0, 2, 4, 6, 8, ...} be the subgroup consisting of multiples of 2. Determine all the left cosets of H in I.*

Sol. There are two distinct left cosets of H in I.

$$0 + H = \{ ..., -6, -4, -2, 0, 2, 4, 6, ...\} = H$$

$$1 + H = \{... -5, -3, -1, 1, 3, 5, 7, ...\}$$

There is no other distinct left coset because any other left coset coincides with the cosets given above.

NORMAL SUBGROUP

Consider a group (G, *) and subgroup (H, *) of the group, then the (H, *) is called a normal sub-group if for any $a \in G$, we have

$$aH = Ha.$$

The meaning of the above definition is that if H is a normal sub-group, then both the left and right cosets of H in G are equal.

Theorem VIII. *Show that every sub-group H of an abelian group G is normal.*

Proof. Let us assume any element $a \in H$

Also, assume any element $b \in G$

Then we have $\quad b^{-1}ab = ab^{-1}b = ae \qquad \because \quad b^{-1}b = e$

We know that $\quad a \in H = a \qquad \because \quad ae = a$

Hence, H is a normal sub-group.

HOMOMORPHISMS

Let $(G_1, *)$ and $(G_2, 0)$ be two algebraic systems, where $*$ and 0 both are binary operations. Then, the mapping $f : G_1 \to G_2$ is said to be homomorphism from $(G_1, *)$ to $(G_2, 0)$ such that for every $a, b \in G$, we have

$$f(a * b) = f(a) \,.\, f(b).$$

ISOMORPHISM

Let (G_1, $*$) and (G_2, 0) be two algebraic system, where $*$ and 0 both are binary operations. The systems (G_1, $*$) and (G_2, 0) are said to be isomorphic if there exists an isomorphic mapping $f: G_1 \to G_2$.

When two algebraic systems are isomorphic, the system are structurally equivalent and one can be obtained from another by simply remaining the elements and the operation.

Example 21. *Let (A_1, $*$) and (A_2, $\square$) be the two algebraic systems as shown in (Fig. 6). Determine whether the two algebraic systems are isomorphic.*

$*$	a	b	c
a	a	b	c
b	b	c	a
c	c	a	b

$\square$	1	w	w^2
1	1	w	w^2
w	w	w^2	1
w^2	w^2	1	w

Fig. 6.

Sol. The two algebraic systems (A_1, $*$) and (A_2, $\square$) are isomorphic and (A_2, $\square$) is an isomorphic image of A_1, such that

$$f(a) = 1$$
$$f(b) = w$$
$$f(c) = w^2.$$

AUTOMORPHISM

Let (G_1, $*$) and (G_2, 0) be two algebraic systems, where $*$ and 0 both are binary operations on G_1 and G_2 respectively. Then an isomorphism from (G_1, $*$) to (G_2, 0) is called an automorphism if $G_1 = G_2$.

RINGS

An algebraic system (R, + ,.) where R is a set with two arbitrary binary operations + and ., is called a ring if it satisfies following conditions

(*i*) (R, +) is an abelian group.

(*ii*) (R, $\cdot$) is a semigroup.

(*iii*) The multiplication operation,., is distributive over the addition operation + *i.e.*,

$$a(b + c) = ab + ac \text{ and } (b + c)a = ba + ca \text{ for all } a, b, c \in R.$$

For example : Consider M be the set of all matrices of the type $\begin{bmatrix} c & d \\ 0 & 0 \end{bmatrix}$ over integers under matrix addition and matrix multiplication. Thus M form a ring.

For example : The set $Z_9 = \{0, 1, 2, 3, 4, 5, 6, 7, 8\}$ under the operation addition and multiplication modulo 9 forms a ring.

TYPES OF RINGS

1. Commutative Rings. A ring (R, + ,.) is called a commutative ring if it holds the commutative law under the operation of multiplication *i.e.*,

$$a \,.\, b = b \,.\, a, \text{ for every } a, b \in R$$

For example : Consider a set E of all even integers under the operation of addition and multiplication. The set E forms a commutative ring.

2. Ring with Unity. A ring (R, + ,.) is called a ring with unity, if it has a multiplicative identity *i.e.,*

$$a \,.\, e = e \,.\, a = a \text{ for every } a \in R$$

For example : Consider a set M of all 2 × 2 matrices over integers under matrix multiplication and matrix addition. The set M forms a ring with unity $\begin{bmatrix} 1 & 0 \\ 0 & 1 \end{bmatrix}$.

3. Ring with Zero Divisions. If $a \,.\, b = 0$, where a and b are any two non-zero elements of R in the ring (R, + ,.), then a and b are called divisions of zero and the ring (R, + ,.) is called ring with zero division.

4. Ring without Zero Division. An algebraic system (R, + ,.) where R is a set with two arbitrary binary operation + and., is called a ring without divisors of zero if for every $a, b \in R$, we have

$$a \,.\, b \neq 0 \quad \Rightarrow \quad a \neq 0 \quad \text{and} \quad b \neq 0.$$

SUBRINGS

A subset A of a ring (R, + ,.) is called a subring of R, if it satisfies following conditions.

(*i*) (A, +) is a subgroup of the group (R, +)

(*ii*) A is closed under the multiplication operation *i.e.,* $a \,.\, b \in A$, for every $a, b \in A$.

For example : The ring (I, + ,.) of integers is a subring of ring (R, + ,.) of real numbers.

Points to be Noted

(*i*) If R is any ring then {0} and R are subrings of R.

(*ii*) Sum of two subrings may not be a subring.

(*iii*) Intersection of subrings is a subring.

For example. Consider the ring M of 2 × 2 matrices of integers under matrix addition and matrix multiplication. Now, S, the set of all matrices of the type $\begin{bmatrix} x & 0 \\ y & 0 \end{bmatrix}$ where x, y are integers is a subring of M.

Theorem IX. *Show that* $(-a)(-b) = ab$ *is a ring R.*

Proof. We have $(-a)(-b) = -[a(-b)] = -(-ab) = ab$

Hence proved.

Theorem X. *Show that* $a(b-c) = ab - ac$ *is a ring R.*

Proof. We have $a(b-c) = -a[b+(-c)] = ab + a(-c) = ab - ac$

Hence proved.

Theorem XI. *Show that* $a \,.\, 0 = 0 \,.\, a = 0$ *for all* $a \in R$, *where R is a ring.*

Proof. We have $\quad a.\,0 = a \,.\, (0 + 0)$

$\Rightarrow \quad a \,.\, 0 = a \,.\, 0 + a \,.\, 0$

$\Rightarrow \quad a\,.\,0 + 0 = a\,.\,0 + a\,.\,0 \quad \Rightarrow \quad 0 = a\,.\,0$ [Using left cancellation law]

Hence proved.

Theorem XII. *Show that* $a(-b) = (-a)b = -ab$ *for all* $a, b \in R$*, where* R *is a ring.*

Proof. We know that $a.\,0 = 0 \quad \Rightarrow \quad a(-b + b) = 0$

$\Rightarrow \quad a(-b) + ab = 0 \quad \Rightarrow \quad a(-b) = -(ab)$

Similarly, we have $(-a)b = -ab$. Hence proved.

INTEGRAL DOMAIN

Let us consider, an algebraic system (D, +, $*$) where + and $*$ are two binary operations. The algebraic system (D, +, $*$) is called an integral domain if it satisfies the following conditions.

1. (D, +) is an abelian group.
2. The operation $*$ is commutative. Also, if we have $z \neq 0$ and $z * x = z * y$; then we have $x = y$. Hence 0 is the additive identity.
3. The operation $*$ is distributive over the operation +.

Example 22. *Let us consider an algebraic system (I, +, $*$)where I is the set of all integers and + and $*$ are the operations of addition and multiplication integers respectively. Determine whether (I, +, $*$) is an integral domain.*

Sol. The algebraic system (I, +) is an abelian group. Here, the identity element is 0 and inverse of every element a is $-a$.

The group $*$ is commutative.

Also, for every non-zero integer z, we have

$$z * x = z * y \quad \Rightarrow \quad x = y$$

Also, the operation $*$ is distributive over +. Since, the system (I, +, $*$) satisfies all the conditions of integral domain. Hence, (I, + ,$*$) is an integral domain.

FIELD

An algebraic system (F, + ,$\cdot$), where F is a set with two arbitrary binary operations + and $\cdot$, is called a field if it satisfies the following conditions

(*i*) (F , +) is an abelian group.

(*ii*) Every non-zero element has a multiplicative inverse.

(*iii*) The operation $\cdot$ is distributive over the operation +.

e.g., The algebraic system (R, +, $\cdot$) is a field, where R is the set of all real numbers.

Points to Remember

* Every field is a ring.
* Every field is an integral domain but every integral domain is not a field.
* Every finite integral domain is a field.

e.g., The set of rational numbers with addition and multiplication is not a field because the multiplicative inverse of elements do not exists.

SOLVED PROBLEMS

Problem 1. *Let* $(S,*)$ *be a commutative semigroup. Show that if* $x * x = x$ *and* $y * y = y$, *then*

$$(x * y) * (x * y) = x * y.$$

Sol. Take L.H.S. $(x * y) * (x * y)$

$\Rightarrow \quad (x * y) * (y * x)$ [$\because$ (S, $*$) is a commutative semigroup]

$\Rightarrow \quad x * y * y * x \Rightarrow x * y * x$ [$\because$ $y * y = y$]

$\Rightarrow \quad x * x * y$ [$\because$ Commutative semigroup]

$\Rightarrow \quad x * y$ [$\because$ $x * x = x$]

Hence, $\quad (x * y) * (x * y) = x * y.$

Problem 2. *Let* $(\{x, y\}, .)$ *be a semigroup where* $x . x = y$ *show that*

(*i*) $x . y = y . x$ (*ii*) $y . y = y$.

Sol. (*i*) To show that $\quad x . y = y . x$

We know that $\quad x . x . x = x . x . x \Rightarrow x . y = y . x$ [$\because$ $x . x = y$]

Hence proved.

(*ii*) To show that $\quad y . y = y$

We know that the set (x, y) is closed under the operation. Therefore, we have two options

$$x. y = x, \quad x . y = y$$

(I) Let $\quad x . y = x$

Now assume $\quad y . y = y . (x . x) = (y . x) . x$ [$\because$. is associative]

$= (x . y) . x$ [$\because$ $x . y = y . x$]

$= x . x$ [$\because$ $x . y = x$]

$= y$

(II) Let $\quad x . y = y$

Now assume $\quad y . y = (x . x) . y$ [$\because$ $x . x = y$]

$= x . (x . y)$ [$\because$. is associative]

$= x . y$ [$\because$ $x . y = y$]

$= y$

Hence proved.

Problem 3. *Let* $(A, *)$ *be a semigroup. Further more, for every* a *and* b *in* A, *if* $a \neq b$ *then* $a * b \neq b * a$.

(*a*) *Show that for every* a *in* A

$$a * a = a$$

(*b*) *Show that for every* a, b *in* A

$$a * b * a = a$$

(*c*) *Show that for every* a, b, c *in* A

$$a * b * c = a * c.$$

Sol. (*a*) We know that A is a semigroup.

$\therefore \quad (a * b) * c = a * (b * c)$

Now putting $b = a$ and $c = a$, we have

$$(a * a) * a = a * (a * a)$$

Since A is not commutative semigroup.

Hence $\quad a * a = a \quad$ ...(*i*)

(*b*) Let us assume that $b \in A$, then we have

$$b * b = b$$

Multiplying both sides by a, we get

$a * b * b = a * b$ or $(a * b) * b = a * b \quad$ [$\because$ * is associative]

Hence, $\quad a * b = a \quad$ ...(*ii*)

So, $\quad a * b * a = (a * b) * a \quad$ [$\because$ * is associative]

$= a * a \quad$ [$\because$ $a * b = a$ from (*ii*)]

$= a \quad$ [$\because$ $a * a = a$ from (*i*)]

(*c*) We know that $\quad a * b * c = (a * b) * c \quad$ [$\because$ * is associative]

$= a * c \quad$ [$\because$ $a * b = a$ from (*ii*)]

Problem 4. *Let* (**Z**, $*$) *be an algebraic structure, where z is the set of integers and the operation * is defined by* $n * m$ = *maximum* (n, m). *Determine whether* (**Z**, $*$) *is a monoid or a group or an abelian group.*

Sol. Closure Property

We know that $n * m = \max(n, m) \in \mathbf{Z}$,

$$\forall\, n, m \in \mathbf{Z}$$

Hence * is closed.

Associative property. Let us assume $a, b, c \in \mathbf{Z}$.

Then, we have $a * (b * c) = a * \max.(b, c) = \max.(a, \max.(b, c))$

$= \max.(a, b, c)$

Similarly, $\quad (a * b) * c = \max.(a, b, c)$

Hence * is associative.

Identity. Let e be the identity element. Then $\max.(a, e) = a$

Hence, the minimum element is the identity element.

Inverse. The inverse of any element does not exist. Since, the inverse does not exist, hence (**Z**, $*$) is not a group or abelian group but a monoid as it satisfies the properties of closure, associative and identity.

Problem 5. *Let* $S = \{0, 1, 2, 3, 4, 5, 6, 7\}$ *and multiplication modulo 8, that is*

$$x \otimes y = (xy) \text{ Mod } 8$$

(*i*) *Prove that* $(\{0, 1\}, \otimes)$ *is not a group.*

(*ii*) *Write three distinct groups* $(G, \otimes)$ *where* $G \subset S$ *and G has 2 elements.*

Sol. (*i*) (*a*) **Closure property.** The set $\{0, 1\}$ is closed under the operation $\otimes$, as shown in table of operation (Fig. 7).

$\otimes$	0	1
0	0	0
1	0	1

Fig. 7.

(*b*) **Associative property.** The operation $\otimes$ is associative. Let $a, b, c \in G$, then we have

$$(a \otimes b) \otimes c = a \otimes (b \otimes c) \quad \text{e.g.,} \quad (0 \otimes 1) \otimes 1 = (0) \otimes 1 = 0$$

Similarly, $\quad 0 \otimes (1 \otimes 1) = 0 \otimes (1) = 0.$

(*c*) **Identity.** The element 1 is the identity element as for every

$$a \in \{0, 1\}$$
$$e \otimes a = a.$$

(*d*) **Inverse.** There must exist an inverse of every element $a \in \{0, 1\}$, such that

$$a \otimes a^{-1} = e$$

But the inverse of element 0 does not exist.

Therefore, since the inverse of every element $a \in \{0, 1\}$ does not exist. Hence $(\{0, 1\}, \otimes)$ is not a group.

(*ii*) The three distinct groups $(G, \otimes)$, where $G \subset S$ and G has 2 elements is as follows

(*a*) $(\{1, 3\}, \otimes)$ (*b*) $(\{1, 5\}, \otimes)$ (*c*) $(\{1, 7\}, \otimes)$.

Problem 6. *Determine whether a semigroup with more than one idempotent element can be group.*

Sol. Let $(A, *)$ be a semigroup with two idempotent elements a and b. Then we have

$$a * a = a \qquad \text{...}(a)$$
$$b * b = b \qquad \text{...}(b) \quad a \neq b.$$

Now assume that A is a group with identity element e.

Then, $\quad a * e = a$ and $b * e = b$

From (a) and (b), we have $a * a = a * e$ and $b * b = b * e$

By the law of left cancellation, we get $a = e = b$

which is a contradiction to $a \neq b$.

Hence $(A, *)$ can not be group.

Problem 7. *Let G_1 and G_2 be subgroups of a group G*

(*i*) *Show that $G_1 \cap G_2$ is also a subgroup of G.*

(*ii*) *Is $G_1 \cup G_2$ always a subgroup of G ?*

Sol. (*i*) Let G_1 and G_2 be two subgroups of G. Then we have

$$G_1 \cap G_2 \neq \phi \qquad [\because \text{ Identity element} \cap \text{is common to both } G_1 \text{ and } G_2]$$

To show that $G_1 \cap G_2$ is a subgroup, we shall have to prove that

$$a \in G_1 \cap G_2 \text{ and } b \in G_1 \cap G_2 \Rightarrow a, b \in G_1 \cap G_2.$$

Let us assume

$$a \in G_1 \cap G_2 \quad \Rightarrow \quad a \in G_1 \text{ and } a \in G_2$$

and $\quad b \in G_1 \cap G_2 \quad \Rightarrow \quad b \in G_1 \text{ and } b \in G_2.$

But we know that G_1 and G_2 are subgroups.

Therefore, $\quad a \in G_1$ and $b \in G_1 \Rightarrow ab^{-1} \in G_1$ ...(*i*)

and $\quad a \in G_2$ and $b \in G_2 \Rightarrow ab^{-1} \in G_2$ ...(*ii*)

From (*i*) and (*ii*), we have $ab^{-1} \in G_1 \cap G_2$

Thus, we have

$$a \in G_1 \cap G_2 \text{ and } b \in G_1 \cap G_2 \Rightarrow ab^{-1} \in G_1 \cap G_2.$$

Hence $G_1 \cap G_2$ is a subgroup of G.

(*ii*) It is not always necessary that $G_1 \cup G_2$ is a subgroup of G.

Problem 8. *Let (G, o) be a group. Show that (G, o) is an Abelian group if and only if $(a \circ b)^2 = a^2 \circ b^2$ for all a and b in G.*

Sol. We know that

$$(a \circ b)^2 = (a \circ b) \circ (a \circ b) = a \circ (b \circ a) \circ b \qquad [\because \ o \text{ is associative}]$$

Now let us assume G is an Abelian group

$$= a \circ (a \circ b) \circ b = (a \circ a) \circ (b \circ b)$$

Hence, $(a \circ b)^2 = a^2 \circ b^2 \quad \forall\, a, b \in G$

Thus, the group G is Abelian if and only if

$$(a \circ b)^2 = a^2 \circ b^2 \quad \forall\, a, b \in G.$$

Problem 9. *Let G = (I, +) be a group, where I is the set of integers and + is an addition operation, also let G_1 = {......– 14, – 7, 0, 7, 14, 21, ...} be a subgroup consisting of the multiples of 7. Determine the cosets of G_1 in I.*

Sol. The set I has 7 different cosets (left or right) of G_1, which are as shown below.

0 + H = {...... – 14, – 7, 0, 7, 14, 21, ...}

1 + H = {...... – 13, – 6, 1, 8, 15, 22, ...}

2 + H = {...... – 12, – 5, 2, 9, 16, 23, ...}

3 + H = {...... – 11, – 4, 3, 10, 17, 24, ...}

4 + H = {...... – 10, – 3, 4, 11, 18, 25, ...}

5 + H = {...... – 9, – 2, 5, 12, 19, 26, ...}

6 + H = {...... – 8, – 1, 6, 13, 20, 27, ...}

All other cosets coincides with any one of the cosets shown above, hence we will not count them.

Problem 10. *Consider G = {1, 5, 7, 11, 13, 17} under multiplication modulo 18.*

(*i*) *Construct the multiplication table of G.*

(*ii*) *Find 5^{-1}, 7^{-1} and 17^{-1}.*

(*iii*) *Find the order and group generated by*

(*a*) 5 (*b*) 13.

(*iv*) *Is G cyclic ?*

Sol. (*i*) We can find $x * y$ in G by finding the remainder when the product xy is divided by 18. The table of multiplication modulo 18 is shown in (Fig. 8).

*	1	5	7	11	13	17
1	1	5	7	11	13	17
5	5	7	17	1	11	13
7	7	17	13	5	1	11
11	11	1	5	13	17	7
13	13	11	1	17	7	5
17	17	13	11	7	5	1

Fig. 8.

(*ii*) We know that a^{-1} is the inverse of a in G. The element 1 is the identity element of G. Also, we know that

$$aa^{-1} = a^{-1}a = e$$

i.e., $aa^{-1} = 1$

Hence, (*a*) $5^{-1} = 11$ $[\because 5 * 11 = 11 * 5 = 1]$

(*b*) $7^{-1} = 13$ $[\because 7 * 13 = 13 * 7 = 1]$

(*c*) $17^{-1} = 17$ $[\because 17 * 17 = 17 * 17 = 1]$

(*iii*) (*a*) We have $5^1 = 5, 5^2 = 7, 5^3 = 17$;

Thus, the order of the group $|5| = 3$ and the group generated by 5 = {5, 7, 17}.

(*b*) Again we have $13^1 = 13$; $13^2 = 7$; $13^3 = 1$;

Thus, the order of the group $|13| = 3$ and group generated by 13 = {1, 7, 13}.

(*iv*) G is not cyclic since G ≠ any of the subgroups generated by 5 and 13.

Problem 11. *Let G be a finite group and H be a subgroup of G. For $a \in G$, define $aH = \{ah / h \in H\}$*

(*a*) *Show that $|aH| = |H|$.*

(*b*) *Show that for every pair of elements, a, b $\in$ G either aH = bH or aH and bH are disjoint.*

(*c*) *Use the above to argue that the order of H must divide the order of G.*

Sol. (*a*) To show that $|a\text{H}| = |\text{H}|$

Let H = $\{h_1, h_2, h_3, ..., h_n\}$ be the n elements of H.

Then $a\text{H} = \{ah_1, ah_2, ah_3, ..., ah_n\}$

But we know that $ah_i = ah_j$ or $h_i = h_j$

Hence, the n elements in aH are disjoint.

Therefore, $|a\text{H}| = |\text{H}|$.

(*b*) Let us assume that $(a * \text{H}) \cap (b * \text{H})$ is non empty

Also, let $c \in (a * \text{H}) \cap (b * \text{H})$

Then, we have $c \in a * \text{H} \Rightarrow a * \text{H} = c * \text{H}$

Also, $c \in b * \text{H} \Rightarrow b * \text{H} = c * \text{H}$

Therefore, $a * \text{H} = b * \text{H}$.

Since the cosets form a partition of G. Hence aH = bH or aH and bH are disjoint.

(*c*) Since we know that H is a subgroup of G. Let us assume, that the subgroup H has K elements and let there are b distinct left cosets. From the (*b*) above, we have seen that the cosets partition the group G and from the (*a*) above, we have seen that each coset has K elements.

Therefore, the group G must contain K_p elements. Thus, it implies that the order of H must divide the order of G.

This is also called the Lagrange's theorem.

Problem 12. *Show that if $G = G_1 \times G_2 \times ...... \times G_n$ is a group and $(a_1, a_2, ..., a_n) \in$ G,* then

(*i*) $(a_1, a_2, ..., a_n)^{-1} = (a_1^{-1}, a_2^{-1}, ..., a_n^{-1})$

(*ii*) $(a_1, a_2, ..., a_n)^m = (a_1^m, a_2^m, ..., a_n^m)$.

Sol. (*i*) Let $a \in G$ and also $a^{-1} \in G$. Also assume that

$$a = (a_1, a_2, a_3, \ldots\ldots, a_n) \quad \text{and} \quad a^{-1} = (a_1^{-1}, a_2^{-1}, a_3^{-1}, \ldots\ldots a_n^{-1})$$

Now
$$a * a^{-1} = (a_1, a_2, a_3, \ldots\ldots, a_n) * (a_1^{-1}, a_2^{-1}, a_3^{-1}, \ldots\ldots, a_n^{-1})$$
$$= (a_1 * a_1^{-1}, a_2 * a_2^{-1}, \ldots\ldots, a_n * a_n^{-1}) = (e_1, e_2, e_3, \ldots\ldots, e_n)$$

Here e_i is the identity of group G_i.

Since, a^{-1} is the inverse of a. So we can write $(a_1, a_2, a_3, \ldots\ldots, a_n)^{-1} = (a_1^{-1}, a_2^{-1}, \ldots\ldots, a_n^{-1})$.

(*ii*) We can prove this by the method of induction. Let us assume that $n = 2$ *i.e.*,

$$(a_1, a_2, a_3, \ldots\ldots, a_n)^2 = (a_1, a_2, a_3, \ldots\ldots, a_n)(a_1, a_2, a_3, \ldots\ldots, a_n)$$
$$= (a_1a_1, a_2a_2, a_3a_3, \ldots\ldots, a_na_n) = (a_1^2, a_2^2, a_3^2, \ldots\ldots, a_n^2)$$

Thus, the result is true for $n = 2$.

Let us assume that it is true for $n = m - 1$ *i.e.*,

$$(a_1, a_2, a_3, \ldots\ldots, a_n)^{m-1} = (a_1^{m-1}, a_2^{m-1}, a_3^{m-1}, \ldots\ldots, a_n^{m-1})$$

Now $(a_1, a_2, a_3, \ldots\ldots, a_n)^m = (a_1^{m-1}, a_2^{m-1}, a_3^{m-1}, \ldots\ldots, a_n^{m-1})(a_1, a_2, a_3, \ldots\ldots, a_n)$
$$= (a_1^m, a_2^m, a_3^m, \ldots\ldots, a_n^m)$$

Thus, it is true for $n = m$.

Problem 13. *Show that if $G = G_1 \times G_2 \times G_3 \ldots\ldots \times G_n$ is a group and $(a_1, a_2, a_3, \ldots\ldots, a_n) \in G$, then*

(*i*) *The identity of G is $(e_1, e_2, \ldots\ldots, e_n)$, where e_i is the identity of corresponding G_i.*

(*ii*) *The group G is an abelian group iff each of groups $G_1, G_2, \ldots\ldots, G_n$ is abelian.*

Sol. (*i*) Let $e = (e_1, e_2, e_3, \ldots\ldots, e_n)$ is the identity for $G_1 \times G_2 \times G_3 \times \ldots\ldots, G_n$, where e_i is the identity for G_i.

Now let us assume that $a \in G_1 \times G_2 \times G_3 \times \ldots\ldots \times G_n$

Then we have
$$a * e = (a_1, a_2, a_3, \ldots\ldots, a_n) * (e_1, e_2, \ldots\ldots, e_n)$$
$$= (a_1 * e_1, a_2 * e_2, a_3 * e_3, \ldots\ldots, a_n * e_n)$$
$$= (a_1, a_2, \ldots\ldots, a_n) \qquad [\because a_i * e_i = a_i]$$

Similarly, we can show that $e * a = a$.

Hence, e is the identity of G.

(*ii*) Let $a, b \in G$. Then, we have

$a = (a_1, a_2, a_3, \ldots\ldots, a_n)$ and $b = (b_1, b_2, b_3, \ldots\ldots, b_n)$ belongs to $G_1 \times G_2 \times G_3 \times \ldots\ldots \times G_n$, where a_i, b_i belongs to the corresponding G_i.

Now
$$a * b = (a_1, a_2, a_3, \ldots\ldots, a_n) * (b_1, b_2, b_3, \ldots\ldots, b_n)$$
$$= (a_1 * b_1, a_2 * b_2, a_3 * b_3, \ldots\ldots, a_n * b_n)$$

Also
$$b * a = (b_1, b_2, b_3, \ldots\ldots, b_n) * (a_1, a_2, a_3, \ldots\ldots, a_n)$$
$$= (b_1 * a_i, b_2 * a_2, b_3 * a_3, \ldots\ldots, b_n * a_n)$$

A group G is an abelian group iff $a * b = b * a$.

Thus, iff $a_i * b_i = b_i * a_i$ for each G_i *i.e.*, the group G is an abelian group if each of the groups $G_1, G_2, G_3, \ldots\ldots, G_n$ is abelian.

Problem 14. *Consider a ring $(R, +, *)$ defined by $a * a = a$.*
Determine whether the ring is commutative or not.

Sol. We have given the ring (R, +, $*$) satisfying the following properties

(*i*) (R, +) is an abelian group.

(*ii*) R, $*$) is a semigroup.

(*iii*) The operation $*$ distributes over +.

Also we have, $a * a = a \quad \forall\ a$ in (R, +,$*$).

To prove that the ring (R,+,$*$) is commutative. We have to prove that there exists as identity element and (R, $*$) is a commutative monoid.

Let us assume $a, b \in$ R

Also, let $\qquad c = a + b$

Since, + is a closed operation, hence $c \in$ R.

Then, we have $\qquad c * c = c * (a + b)$

We know that, the operation $*$ distributes over +.

So, $\qquad c * c = (a + b) * (a + b) = a * a + a * b + b * a + b * b$

So, $\qquad c = a + b + a * b + b * a \qquad [\because\ a * a = a]$

or $\qquad c = c + (a * b + b * a)$

Therefore, $a * b + b * a = e$ is an identity for operation + also $a * b$ is inverse of $b * a$

Again, let us assume $a + a = b$

or $\qquad a * (a + a) = a * b$

or $\qquad a * a + a * a = a * b$

or $\qquad a + a = a * b \qquad$ ...(*i*)

Similarly, $\qquad (a + a) * a = b * a$

or $\qquad a * a + a * a = b * a$

or $\qquad a + a = b * a \qquad$ ...(*ii*)

From (*i*) and (*ii*), we have

$$a * b = b * a$$

Thus, the operation $*$ is commutative. Since, (R, $*$) is a commutative monoid. Therefore, (R, +, $*$)is a commutative ring.

13

Posets, Lattices and Boolean Algebra

PARTIALLY ORDERED SET

Definition. Consider a relation R on a set S satisfying the following properties :

1. R is reflexive *i.e.*, $x\text{R}x$ for every $x \in \text{S}$.
2. R is antisymmetric *i.e.*, if $x\text{R}y$ and $y\text{R}x$, then $x = y$.
3. R is transitive *i.e.*, if $x\text{R}y$ and $y\text{R}z$, then $x\text{R}z$.

Then R is called a partially order relation and the set S together with partial order is called a partially order set or POSET and is denoted by (S, ≤).

For example :

1. The set N of natural numbers form a poset under the relation '≤' because firstly $x \le x$, secondly, if $x \le y$ and $y \le x$, then we have $x = y$ and lastly if $x \le y$ and $y \le z$, it implies $x \le z$ for all $x, y, z \in \text{N}$.

2. The set N of natural numbers under divisibility *i.e.*, 'x divides y' forms a poset because x/x for every $x \in \text{N}$. Also if x/y and y/x, we have $x = y$. Again if x/y, y/z we have x/z, for every $x,y,z \in \text{N}$.

3. Consider a set S = {1, 2} and power set of S is P(S). The relation of set inclusion ⊆ is a partial order. Since, for any sets A, B, C in P(S), firstly we have A ⊆ A, secondly, if A ⊆ B and B ⊆ A, then we have A = B. Lastly, if A ⊆ B and B ⊆ C, then A ⊆ C. Hence, (P(S), ⊆) is a poset.

Example 1. *Consider a set S = {a, b, c}. Is the relation of set inclusion '⊆' is a partial order on P(S) where P(S) is a power set of S ?*

Sol. The power set of S is

$$\text{P(S)} = \{\{a\}, \{b\}, \{c\}, \{a, b\}, \{a, c\}, \{b, c\}, \{a, b, c\}, \phi\}$$

Now consider any sets A, B and C in P(S).

1. Since every A ⊆ A, hence it is reflexive.
2. If A ⊆ B and B ⊆ A, we have A = B. Hence it is antisymmetric.
3. If A ⊆ B, B ⊆ C, we have A ⊆ C. Hence it is transitive.

∴ (P(S), ⊆) is a poset.

Example 2. *Consider a set A = {4, 9, 16, 36}. Is the relation 'divides' a partial order.*

Sol. The relation 'divides' is a partial order if it satisfies the property of reflexivity, antisymmetry and transitivity.

1. Since for every $a \in \text{A}$, we have a/a. Hence, 'divides' is reflexive.
2. If a/b and b/a, we have $a = b$ for any $a, b \in \text{A}$. Hence, 'divides' is antisymmetric.
3. If a/b and b/c, we have a/c for any $a, b, c \in \text{A}$. Hence, the relation 'divides' is a partial order and (A, /) is a poset.

COMPARABLE ELEMENTS

Consider an ordered set A. Two elements a and b of set A are called comparable if

$$a \underset{R}{\leq} b \quad \text{or} \quad b \underset{R}{\leq} a.$$

NON-COMPARABLE ELEMENTS

Consider an ordered set A. Two elements a and b of set A are called non-comparable if neither $a \leq b$ nor $b \leq a$.

Example 3. *Consider A = {1, 2, 3, 5, 6, 10, 15, 30} is ordered by divisibility. Determine all the comparable and non-comparable pairs of elements of A.*

Sol. The comparable pairs of elements of A are :

{1, 2}, {1, 3}, {1, 5}, {1, 6}, {1, 10}, {1, 15}, {1, 30}
{2, 6}, {2, 10}, {2, 30}
{3, 6}, {3, 15}, {3, 30}
{5, 10}, {5, 15}, {5, 30}
{6, 30}, {10, 30}, {15, 30}.

The non-comparable pair of elements of A are :

{2, 3}, {2, 5}, {2, 15}
{3, 5}, {3, 10}, {5, 6}, {6, 10}, {6, 15}, {10, 15}.

LINEARLY ORDERED SET

Consider an ordered set A. The set A is called linearly ordered set or totally ordered set, if every pair of elements in A are comparable.

For Example : The set of positive integers I_+ with the usual order ≤ is a linearly ordered set.

Example 4. *Consider the set I = {1, 2, 3,......} is ordered by divisibility. Determine whether each of the following subsets of I are linearly ordered or not.*

(*i*) {2, 4, 8} (*ii*) {3, 6, 9, 11}
(*iii*) {1} (*iv*) {2, 4, 6, 8, 10, ...}.

Sol. (*i*) The subset is linearly ordered, since every pair of elements is comparable *i.e.*, $2 \mid 4 \mid 8$.

(*ii*) The subset is not linearly ordered, since the pair (3, 11) is not comparable.

(*iii*) The subset is linearly ordered, since the set containing one element is always linearly ordered.

(*iv*) The subset is not linearly ordered since every pair of elements is not comparable *i.e.*, neither 4/6 nor 6/4.

HASSE DIAGRAMS

It is a useful tool, which completely describes the associated partial order. Therefore, it is also called an ordering diagram. It is very easy to convert a directed graph of a relation on a set A to an equivalent Hasse diagram. Therefore, while drawing a Hasse diagram following points must be remembered.

1. The vertices in Hasse diagram are denoted by points rather than by circles.

2. Since a partial order is reflexive, hence each vertex of A must be related to itself, so the edges from a vertex to itself are deleted in Hasse diagram.

3. Since a partial order is transitive, hence whenever aRb, bRc, we have aRc. Eliminate all edges that are implied by the transitive property in Hasse diagram *i.e.*, Delete edge from a to c but retain the other two edges.

4. If a vertex 'a' is connected to vertex 'b' by an edge *i.e.*, aRb, then vertex 'b' appears above vertex 'a'. Therefore, the arrows may be ommitted from the edges in Hasse diagram.

The Hasse diagram is much simpler than the directed graph of the partial order.

Example 5. *Consider the set A = {4, 5, 6, 7}. Let R be the relation ≤ on A. Draw the directed graph and the Hasse diagram of R.*

Sol. The relation ≤ on the set A is given by

R = {{4, 5}, {4, 6}, {4, 7}, {5, 6}, {5, 7}, {6, 7}, {4, 4}, {5, 5}, {6, 6}, {7, 7}}

The directed graph of the relation R is as shown in Fig. 1.

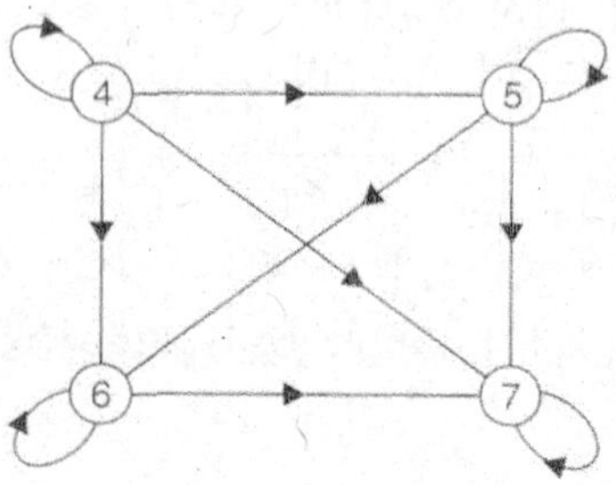

Fig. 1.

To draw the Hasse diagram of partial order, apply the following points :

1. Delete all edges implied by reflexive property *i.e.*,

(4, 4), (5, 5), (6, 6), (7, 7).

2. Delete all edges implied by transitive property *i.e.*,

(4, 7), (5, 7) and (4, 6).

3. Replace the circles representing the vertices by dots.

4. Omit the arrows.

The Hasse diagram is as shown in Fig. 2.

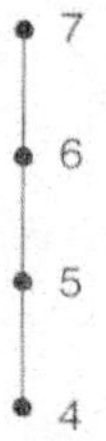

Fig. 2.

Example 6. *Draw the directed graph of relation determined by the Hasse diagram on the set A = {1, 4, 6, 8} as shown in Fig. 3.*

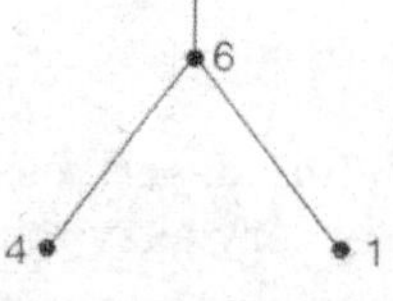

Fig. 3.

Sol. The directed graph is shown in Fig. 4.

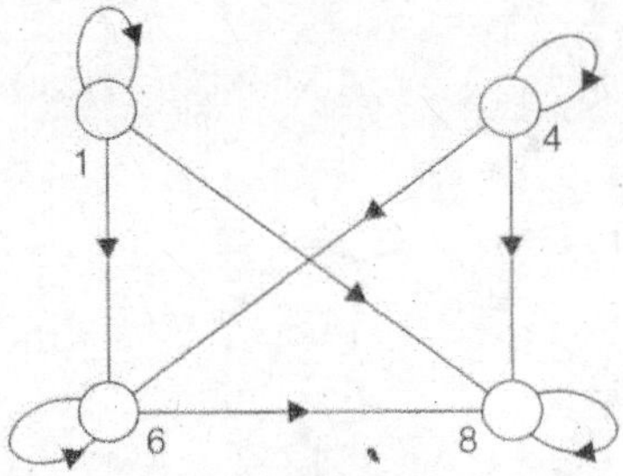

Fig. 4.

Example 7. *Determine the Hasse diagram of the partial order having the directed graph as shown in Fig. 5.*

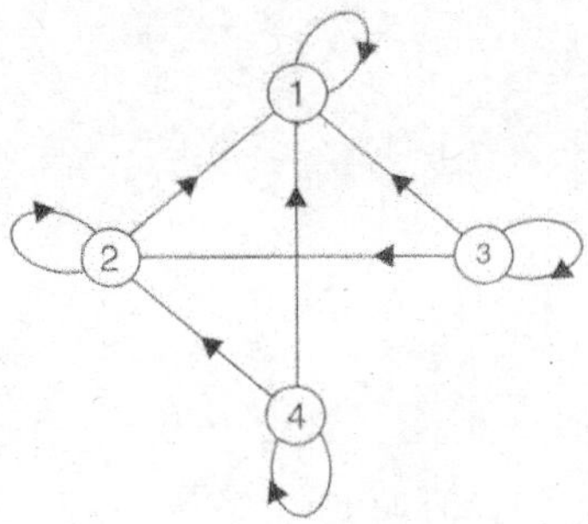

Fig. 5.

Sol. The Hasse diagram of the given partial order determine by the directed graph is as shown in Fig. 6.

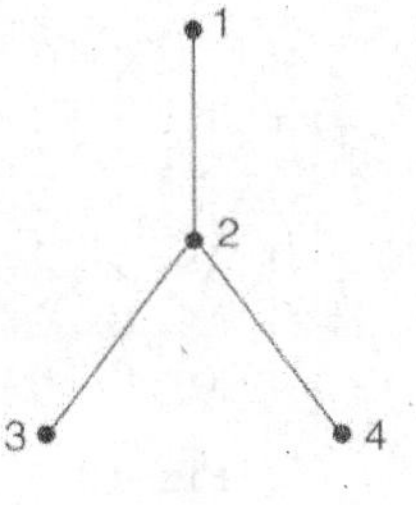

Fig. 6.

Example 8. *Consider the set A = {k, l, m, n, p} and the corresponding relation R = {(k, k), (l, l), (m, m), (n, n), (p, p), (k, m), (k, l), (k, n), (k, p), (m, n), (m, p), (n, p), (l, p)}.*

Construct the directed graph and the corresponding Hasse diagram of this partial order.

Sol. The directed graph of the partial order is as shown in Fig. 7.

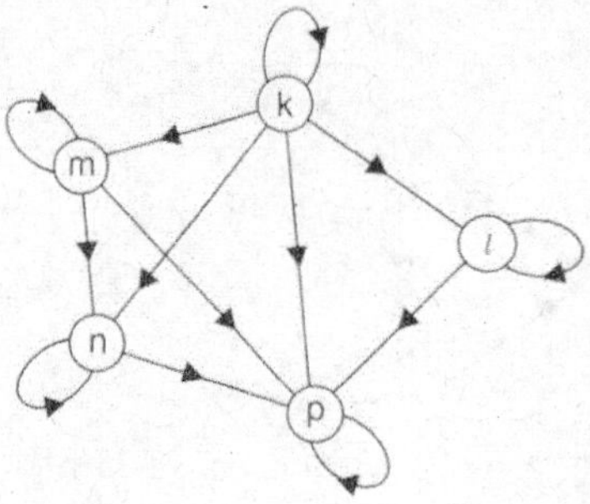

Fig. 7.

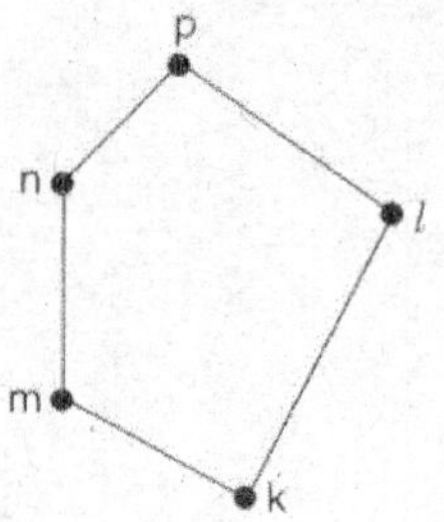

Fig. 8.

The Hasse diagram of the partial order is as shown in Fig. 8.

Example 9. *Consider the Hasse diagram as shown in Fig. 9. Determine the value of set A and also determine the set R.*

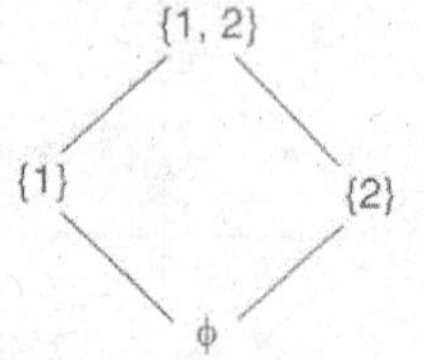

Fig. 9.

Sol. The set A = {{1}, {2}, {1, 2}, ϕ} and

R = {({1}, {1}), ({2}, {2}), ({1, 2}, {1, 2}), {ϕ, ϕ}, ({1}, {1, 2}), ({2}, {1, 2}), (ϕ, {1}), (ϕ, {2}), (ϕ, {1, 2})}.

Here the relation R is (set inclusion) $\subseteq$.

Example 10. *Determine the Hasse diagram of the relation on A = {a, b, c, d, e} whose matrix is shown in Fig. 10.*

$$\begin{array}{c} \\ a \\ b \\ c \\ d \\ e \end{array} \begin{array}{c} \begin{array}{ccccc} a & b & c & d & e \end{array} \\ \begin{bmatrix} 1 & 0 & 1 & 1 & 1 \\ 0 & 1 & 1 & 1 & 1 \\ 0 & 0 & 1 & 1 & 1 \\ 0 & 0 & 0 & 1 & 0 \\ 0 & 0 & 0 & 0 & 1 \end{bmatrix} \end{array}$$

Fig. 10.

Sol. The Hasse diagram of the relation on A = {*a*, *b*, *c*, *d*, *e*} is shown in Fig. 11.

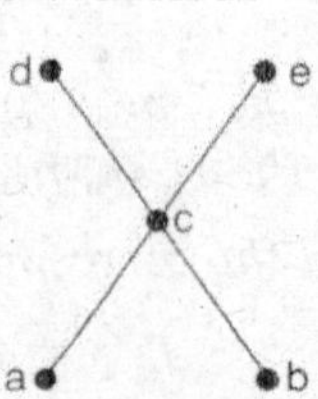

Fig. 11.

Example 11. *Let A = {1, 2, 3, 4, 5} be ordered by Hasse diagram.*

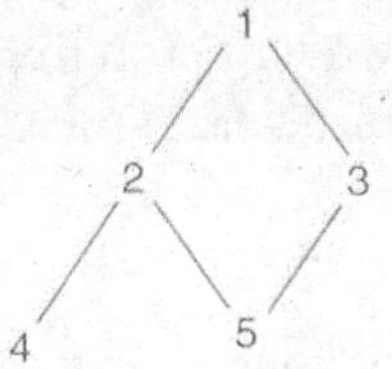

Fig. 12.

Insert correct symbol <, >, or ∥ (not comparable) between :

(i) 1 – 5 *(ii) 2 – 3* *(iii) 4 – 1* *(iv) 3 – 4.*

Sol. *(i)* As there is a path from 1 to 5, hence 1 > 5.

(ii) As there is no path from 2 to 3 or vice-versa, hence 2 ∥ 3.

(iii) As there is a path from 4 to 1, hence 4 < 1.

(iv) As there is no path from 3 to 4, hence 3 ∥ 4.

ELEMENTS OF POSET

Maximal element. An element $a \in A$ is called a maximal element of A if there is no element c in A such that $a \le c$.

Minimal element. An element $b \in A$ is called a minimal element of A if there is no element c in A such that $c \le b$.

Note. There can be more than one maximal or more than one minimal element.

Example 12. *Determine all the maximal and minimal elements of the poset whose Hasse diagram is shown in Fig. 13.*

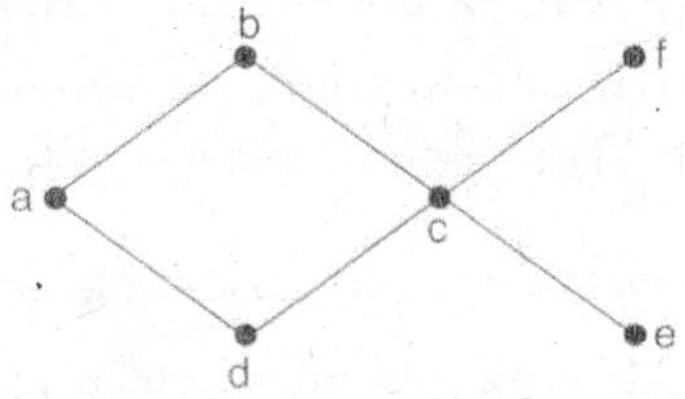

Fig. 13.

Sol. The maximal elements are b and f.

The minimal elements are d and e.

Example 13. *Let A = {2, 3, 4, 6, 8, 24, 48} with partial ordering of divisibility. Determine all the maximal and minimal elements of A.*

Sol. The maximal element is 48.

The minimal elements are 2 and 3.

Theorem I. *Prove that the finite non empty poset (A, ≤) has at least one maximal and one minimal element in A.*

Proof. Consider an element $a \in$ A. Now if a is not maximal element, we will find another element $b \in$ A such that $a < b$. Again if b is not a maximal element, we will find another element $c \in$ A such that $b < c$. At last, we will stop finding the element, as we cannot continue this process indefinitely since A is a finite set. Hence, we obtain a finite sequence

$$a < b < c < d \ldots\ldots < n$$

which cannot be extended further.

Therefore, we cannot have $n < x$ for any $x \in$ A. So, we conclude that n is a maximal element of (A, ≤). By giving similar arguments we can prove that the finite non empty poset (A, ≤) has atleast one minimal element in A.

Greatest element. An element $x \in$ A is called a greatest element of A if for all $a \in$ A, $a \leq x$.

Least element. An element $y \in$ A is called a least element of A if, for all $a \in$ A, $y \leq a$.

* The greatest element of a poset is denoted by 1 and is called the unit element and the least element of a poset is denoted by 0 and is called the zero element.

Example 14. *Determine the greatest and least elements of the poset whose Hasse diagrams are shown in Fig. 14, if they exist.*

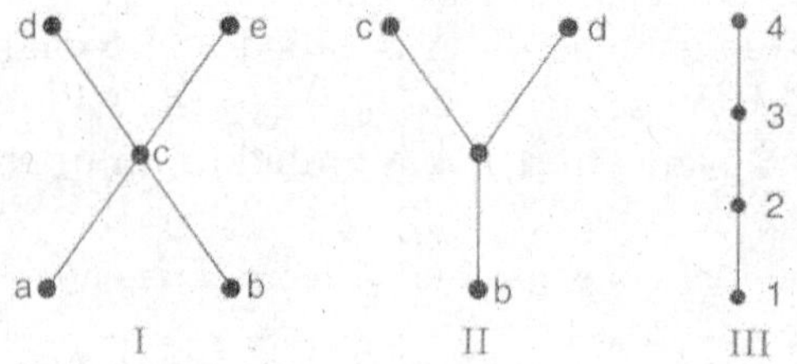

Fig. 14.

Sol. The poset shown in Fig. 14(I) has neither greatest nor least element.

The poset shown in Fig. 14(II) has no greatest element but b is the least element.

The poset shown in Fig. 14(III) has 4 as greatest element and 1 as least element.

Theorem II. *Prove that a poset has at the most one greatest element and one least element.*

Proof. Let x and y are two greatest elements of a poset A. Now, since y is a greatest element, so we have $x \leq y$. Also, as x is a greatest element, so we have $y \leq x$ which implies that $x = y$ (*antisymmetric property*). Hence, we can say that if the poset has a greatest element, there exists only one such element.

By similar arguments, we can prove that if the poset has a least element, there exists only one such element.

UPPER BOUND

Consider B be a subset of a partially ordered set A. An element $x \in$ A is called an upper bound of B if $y \leq x$ for every $y \in$ B.

LOWER BOUND

Again, consider B be a subset of a partially order set A. An element $z \in$ A is called a lower bound of B if $z \leq x$ for every $x \in$ B.

Example 15. *Consider the poset A = {a, b, c, d, e, f, g} be ordered as shown in Fig. 15. Also let B = {c, d, e}. Determine the upper and lower bounds of B.*

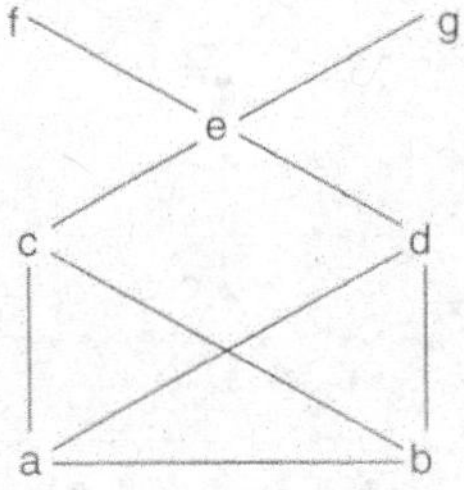

Fig. 15.

Sol. The upper bounds of B are *e*, *f* and *g* because every element of B is '≤' *e*, *f* and *g*.

The lower bounds of B are *a* and *b* because *a* and *b* are '≤' every element of B.

Example 16. *Consider the poset A = {1, 2, 3, 4, 5, 6} be ordered as shown in Fig. 16. Also let B = {3, 4}. Determine the upper and lower bounds of B.*

Fig. 16.

Sol. The upper bound of B is 4, 5 and 6 because every element of B is '≤' 4, 5 and 6.

The lower bound of B is 3, 2 and 1 because 3, 2 and 1 are '≤' every element of B.

Example 17. *Consider the poset A = {1, 2, 3, 4, 5, 6, 7, 8} be ordered as shown in Fig. 17. Also let B = {3, 4, 5}. Determine the upper and lower bounds of B.*

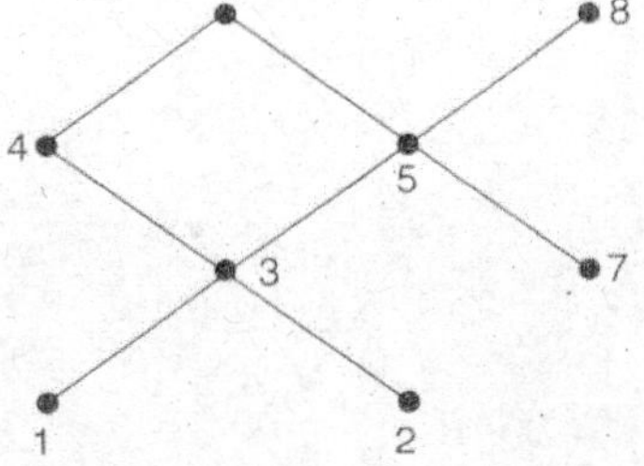

Fig. 17.

Sol. The upper bound of B is 6 because every element of B is '$\leq$' 6. The 8 is not an upper bound of B since $4 \not\leq 8$.

The lower bound of B is 1, 2 and 3 because 1 and 2 are '$\leq$' every element of B. The 7 is not a lower bound of B since $7 \not\leq 3$ and also $7 \not\leq 4$.

Example 18. *Consider the poset A = {a, b, c, d} as shown in Fig. 18 and let B = {c}. Determine the upper and lower bounds of B.*

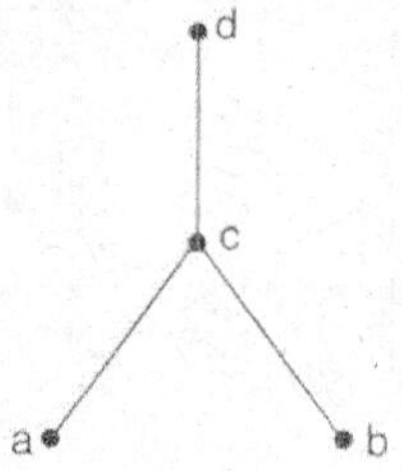

Fig. 18.

Sol. The upper bound of B is *c*, *d*.

The lower bound of B is *a*, *b*, *c*.

Note 1. A subset B of a poset A may or may not contain upper bounds or lower bounds in poset A.

2. An upper bound or lower bound of B may or may not belong to the subset B itself.

LEAST UPPER BOUND (SUPREMUM)

Consider B be a subset of a partially ordered set A. An element $x \in$ A is called a least upper bound of B or supremum written as L U B (B) or S U P(B) if x is an upper bound of B and we have $x \leq x'$ for every x' which is an upper bound of B.

GREATEST LOWER BOUND (INFIMUM)

Consider B be a subset of a partially ordered set A. An element $y \in$ A is called a greatest lower bound or infimum of B, written as GLB(B) or INF(B) if y is a lower bound of B and we have $y' \leq y$, for every y' which is a lower bound of B.

Example 19. *Determine the least upper bound and greatest lower bound of B = {a, b, c} if they exist, of the poset whose Hasse diagram is shown in Fig. 19.*

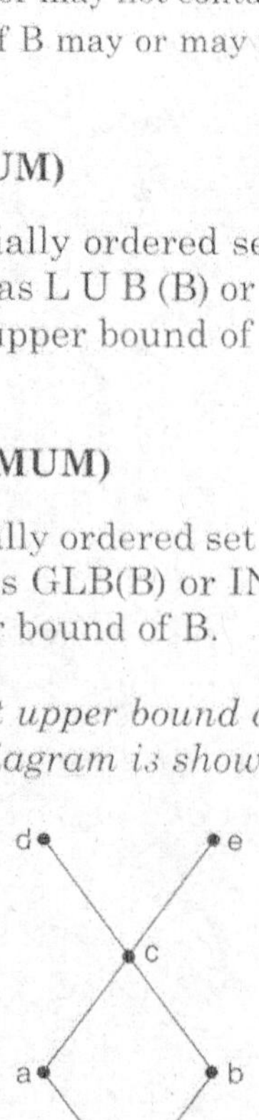

Fig. 19.

Sol. The least upper bound is c.

The greatest lower bound is k.

Example 20. *Let* $D_{100} = \{1, 2, 4, 5, 10, 20, 25, 50, 100\}$ *and let the relation* $\leq$ *be the relation / (divides) be a partial ordering on* D_{100}.

(*a*) *Determine the GLB of B, where* $B = \{10, 20\}$.

(*b*) *Determine the LUB of B, where* $B = \{10, 20\}$.

(*c*) *Determine the GLB of B, where* $B = \{5, 10, 20, 25\}$.

(*d*) *Determine the LUB of B, where* $B = \{5, 10, 20, 25\}$.

Sol. The Hasse diagram of the poset D_{100} is as shown in Fig. 20.

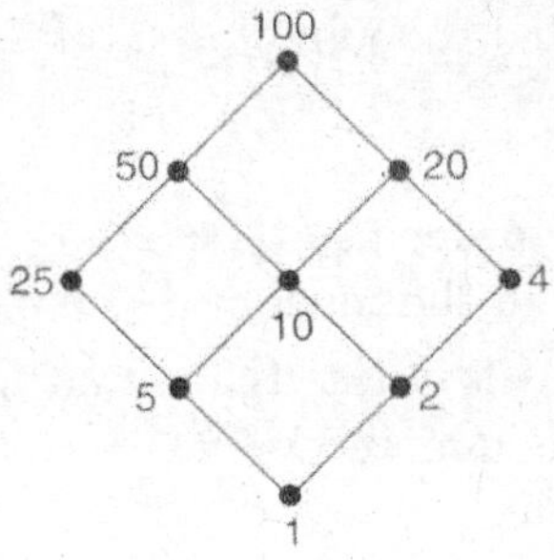

Fig. 20.

(*a*) The GLB (B) is 10. (*b*) The LUB (B) is 20.

(*c*) The GLB (B) is 5. (*d*) The LUB (B) is 100.

Note. D_n denote the set of all positive integers which are divisors of n, where n is a + ve integer *e.g.*,

$$D_{10} = \{1, 2, 5, 10\} \text{ as all are divisors of 10.}$$

Example 21. *Determine the least upper bound and greatest lower bound of each B if they exist, of the poset whose Hasse diagram is shown in Fig. 21.*

Fig. 21.

(*i*) $B = \{2, 3\}$ (*ii*) $B = \{2, 3, 4\}$

(*iii*) $B = \{1, 2, 3, 4\}$ (*iv*) $B = \{1, 2, 3, 4, 5\}$.

Sol. (*i*) The least upper bound of B is 3. The greatest lower bound of B is 2.

(*ii*) The least upper bound of B is 4. The greatest lower bound of B is 2.

(*iii*) The least upper bound of B is 4. The greatest lower bound of B is 1.

(*iv*) The least upper bound is 5. The greatest lower bound is 1.

Note. Let $B = \{b_1, b_2, ..., b_r\}$. If $a = LUB(B)$, then a is the first vertex that can be reached from b_1, b_2,......, b_r by upward paths. Similarly, if $a = GLB(B)$, then a is the first vertex that can be reached from b_1, b_2,......, b_r by downward paths.

LATTICES

Lattice

A lattice L is a poset in which every pair of elements has a least upper bound (LUB) or supremum and a greatest lower bound (GLB) or infimum.

Join

Consider a poset L under the ordering $\leq$. Let $a, b \in L$. Then LUB (a, b) or SUP (a, b) is denoted by $a \vee b$ or $a \cup b$ and is called the join of a and b *i.e.*, $a \vee b = SUP(a, b)$.

Meet

Consider a poset L under the ordering $\leq$. Let $a, b \in L$. Then GLB (a, b) or inf (a, b) is denoted by $a \wedge b$ or $a \cap b$ and is called the meet of a and b *i.e.*, $a \wedge b = \inf(a, b)$.

From the above, it follows that a lattice L is a mathematical structure with two binary operations $\vee$(Join) and $\wedge$ (meet). It is denoted by $\{L, \vee, \wedge\}$. The lattice L for any elements a, b and c satisfies the following properties :

(*a*) **Commutative Property**

(*i*) $a \wedge b = b \wedge a$

(*ii*) $a \vee b = b \vee a$.

(*b*) **Associative Property**

(*i*) $(a \wedge b) \wedge c = a \wedge (b \wedge c)$

(*ii*) $(a \vee b) \vee c = a \vee (b \vee c)$.

(*c*) **Absorption Property**

(*i*) $a \wedge (a \vee b) = a$

(*ii*) $a \vee (a \wedge b) = a$.

Theorem III. *Prove that if L be a lattice then* $a \wedge b = a$ *if and only if* $a \vee b = b$.

Proof. Let us first assume that $a \wedge b = a$.

Using absorption property, we have

$$b = b \vee (b \wedge a) = b \vee (a \wedge b) = b \vee a = a \vee b \qquad ...(i)$$

Conversely, let us assume $a \vee b = b$.

Again using the absorption property, we have

$$a = a \wedge (a \vee b) = a \vee b \qquad ...(ii)$$

From eqn. (*i*) and (*ii*), we have

$$a \wedge b = a \text{ if and only if } a \vee b = b.$$

Theorem IV. *Prove that for elements of lattice*

(*i*) $a \wedge a = a$ (*ii*) $a \vee a = a$ Idempotent property

Proof. (*i*) $a \wedge a = a \wedge (a \vee (a \wedge b))$ {Using absorption property $c(ii)$}

$= a$ {Using absorption property $c(i)$}

(*ii*) $a \vee a = a \vee (a \wedge (a \vee b))$ {Using absorption property $c(i)$}

$= a$ {Using absorption property $c(ii)$}

Theorem V. *Consider a lattice L. Prove that the relation $a \le b$ defined by either $a \wedge b = a$ or $a \vee b = b$ is a partial ordering on lattice L.*

Proof. For any element $a \in$ L, we have $a \wedge a = a$

i.e., $a \le a$. Therefore, the relation $\le$ is reflexive. Now assume $a \le b$ and $b \le a$. Then we have

$$a \wedge b = a \quad \text{and} \quad b \wedge a = b$$

Thus, $a = a \wedge b = b \wedge a = b$, therefore the relation $\le$ is antisymmetric.

At last, we assume $a \le b$ and $b \le c$. So, we have

$$a \wedge b = a \text{ and } b \wedge c = b$$

i.e.,
$$a \wedge c = (a \wedge b) \wedge c = a \wedge (b \wedge c)$$
$$= a \wedge b = a$$

Therefore, $a \le c$. So, the relation $\le$ is transitive. From above, we can say, $\le$ is a partial order on L.

Example 22. *Let P(S) be the power set of the set S = {1, 2, 3}. Construct the Hasse diagram of the partial order induced on P(S) by the lattice (P(S), $\wedge$, $\vee$).*

Sol. The Hasse diagram obtained by a lattice is same as obtained under the partial ordering of set inclusion. In the lattice, $a \le b$ whenever $a \wedge b = a$. Thus, in the above case $a \le b$, whenever $a \wedge b = a$

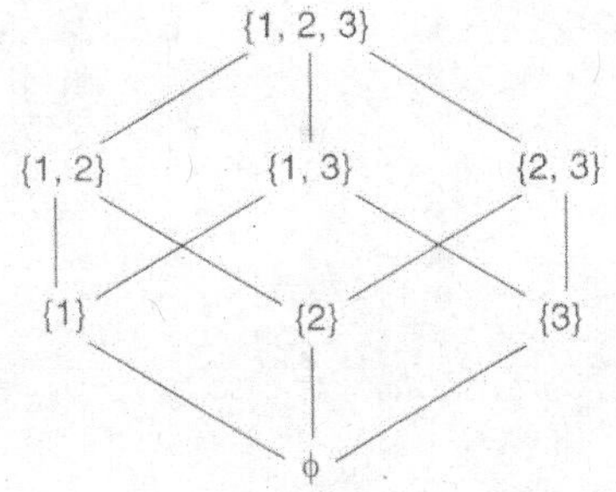

Fig. 22.

Example 23. *Determine which of the posets shown in Fig. 23 are lattices.*

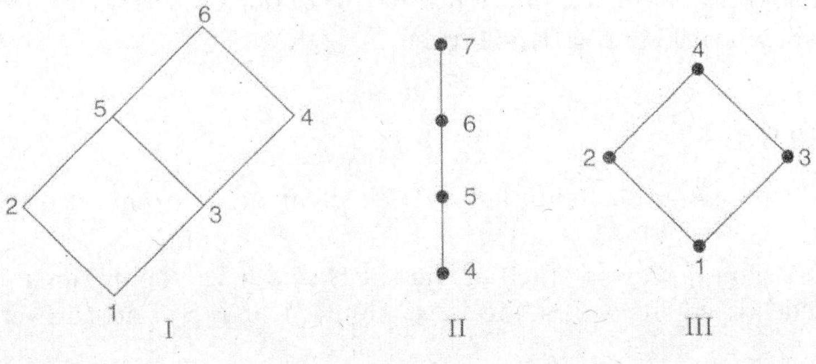

Fig. 23.

Sol. All the posets shown in Fig. 23 are lattices.

Example 24. *Determine whether the posets shown in Fig. 24 are lattices or not.*

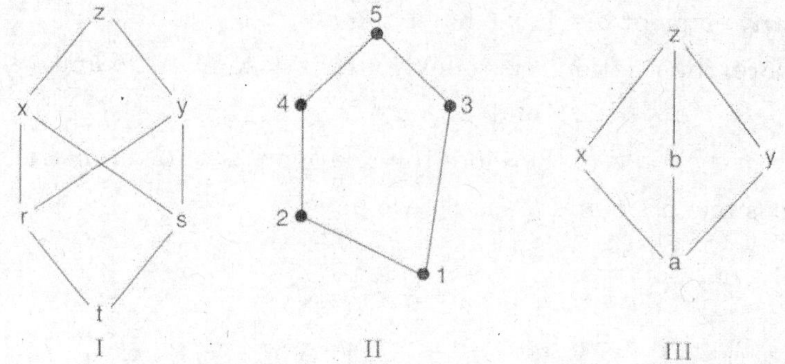

Fig. 24.

Sol. The posets shown in Figs. 24(II) and 24(III) are lattices. The posets shown in Fig. 24(I) is not lattice as (x, y) has three lower bounds *i.e.*, r, s and t but the inf (x, y) does not exist. Also, sup (r, s) does not exist.

Example 25. *Determine whether the posets shown in Fig. 25 are lattices or not.*

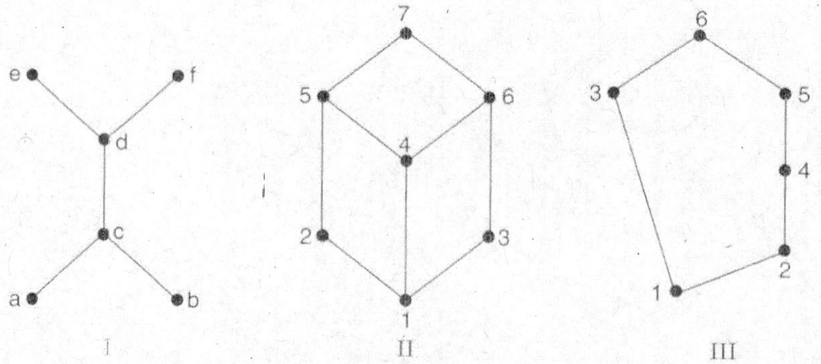

Fig. 25.

Sol. The poset shown in Fig. 25(II) is a lattice. The poset shown in Fig. 25(I) is not lattice since the elements e and f have no upper bound, hence sup (e, f) does not exist. Similarly, the elements a and b have no lower bound, hence inf (a, b) does not exist.

The poset shown in Fig. 25(III) is lattice.

BOUNDED LATTICES

A lattice L is called a bounded lattice if it has a greatest element 1 and a least element 0.

Example (*i*). The power set P(S) of the set S under the operations of intersection and union is a bounded lattice since ϕ is the least element of P(S) and the set S is the greatest element of P(S).

(*ii*) The set of +ve integers I_+ under the usual order of $\leq$ is not a bounded lattice since it has a least element 1 but the greatest element does not exist.

Properties of Bounded Lattices

It L is a bounded lattice, then for any element $a \in$ L, we have the following identities :

(*i*) $a \vee 1 = 1$ (*ii*) $a \wedge 1 = a$ (*iii*) $a \vee 0 = a$ (*iv*) $a \wedge 0 = 0$.

Theorem VI. *Prove that every finite lattice* $L = \{a_1, a_2, a_3, \ldots\ldots, a_n\}$ is bounded.

Proof. We have given the finite lattice :

$$L = \{a_1, a_2, a_3, \ldots\ldots, a_n\}$$

Thus, the greatest element of lattice L is $a_1 \vee a_2 \vee a_3 \vee \ldots\ldots \vee a_n$.

Also, the least element of lattice L is $a_1 \wedge a_2 \wedge a_3 \wedge \ldots\ldots \wedge a_n$.

Since, the greatest and least elements exist for every finite lattice. Hence L is bounded.

SUBLATTICES

Consider a non empty subset L_1 of a lattice L. Then L_1 is called a sublattice of L if L_1 itself is a lattice w.r.t. the operations of L *i.e.*, if $a \vee b \in L_1$ and $a \wedge b \in L_1$ whenever $a \in L_1$ and $b \in L_1$.

Example 26. *Consider the lattice of all +ve integers* I_+ *under the operation of divisibility. The lattice* D_n *of all divisors of* $n > 1$ *is a sublattice of* I_+.

Determine all the sublattices of D_{30} *that contain at least four elements,* $D_{30} = \{1, 2, 3, 5, 6, 10, 15, 30\}$.

Sol. The sublattices of D_{30} that contain at least four elements are as follows :

(*i*) {1, 2, 6, 30} (*ii*) {1, 2, 3, 30}

(*iii*) {1, 5, 15, 30} (*iv*) {1, 3, 6, 30}

(*v*) {1, 5, 10, 30} (*vi*) {1, 3, 15, 30}

(*vii*) {2, 6, 10, 30}.

Example 27. *Consider the lattice L = {1, 2, 3, 4, 5} as shown in Fig. 26. Determine all the sublattices with three or more elements.*

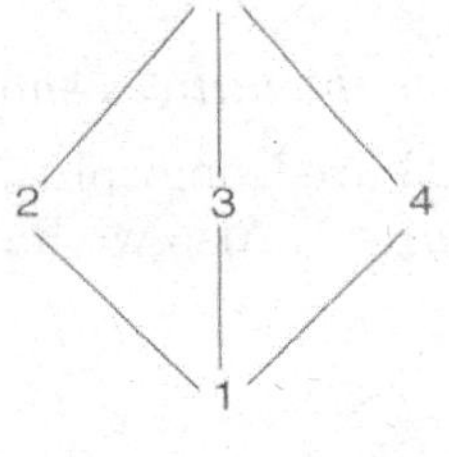

Fig. 26.

Sol. All the sublattices with three or more elements are those whose supremum and infimum exists for every pair of elements which are as follows :

(*i*) {1, 2, 5} (*ii*) {1, 3, 5}

(*iii*) {1, 4, 5} (*iv*) {1, 2, 3, 5}

(*v*) {1, 3, 4, 5} (*vi*) {1, 2, 3, 4, 5}

(*vii*) {1, 2, 4, 5}.

Example 28. *Consider the lattice L as shown in Fig. 27. Determine whether or not each of the following is a sublattice of L.*

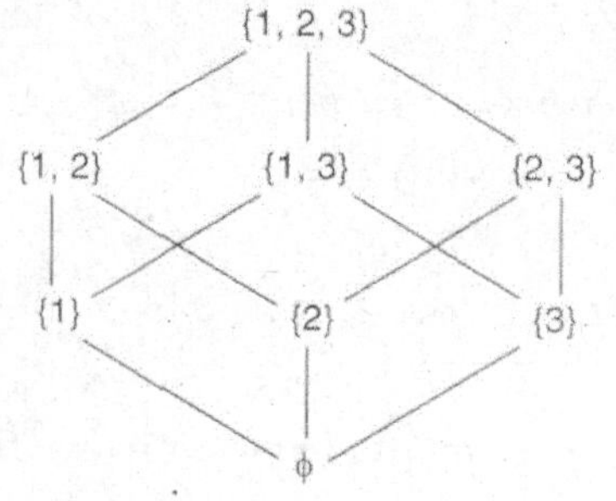

Fig. 27.

$A = \{\phi, \{1, 2\}, \{2, 3\}, \{1, 2, 3\}\}$
$B = \{\phi, \{1\}, \{1, 2\}, \{1, 2, 3\}\}$
$C = \{\phi, \{3\}, \{1, 3\}, \{1, 2, 3\}\}$
$D = \{\{1\}, \{3\}, \{1, 3\}, \{1, 2, 3\}\}$
$E = \{\phi, \{3\}, \{1, 2\}, \{1, 2, 3\}\}$.

Sol. A is not a sublattice since $\{1, 2\} \wedge \{2, 3\} = \{2\}$ which does not exist in A.
B is a sublattice since sup and inf of every pair of elements exist.
C is a sublattice since sup and inf of every pair of elements exist.
D is not a sublattice since $\{1\} \wedge \{3\} = \phi$ which does not exist in D.
E is a sublattice since sup and inf of every pair of elements exist.

ISOMORPHIC LATTICES

Two lattices L_1 and L_2 are called isomorphic lattices if there is a bijection from L_1 to L_2 *i.e.*, $f : L_1 \to L_2$, such that

$$f(a \wedge b) = f(a) \wedge f(b) \text{ and } f(a \vee b) = f(a) \vee f(b)$$

for every element a, b belongs to L_1.

Example 29. *Determine whether the lattices shown in Fig. 28 are isomorphic.*

Sol. The lattices shown in Fig. 28 are isomorphic. Consider the mapping $f = \{(a, 1), (b, 2), (c, 3), (d, 4)\}$. For example $f(b \wedge c) = f(a) = 1$. Also we have $f(b) \wedge f(c) = 2 \wedge 3 = 1$.

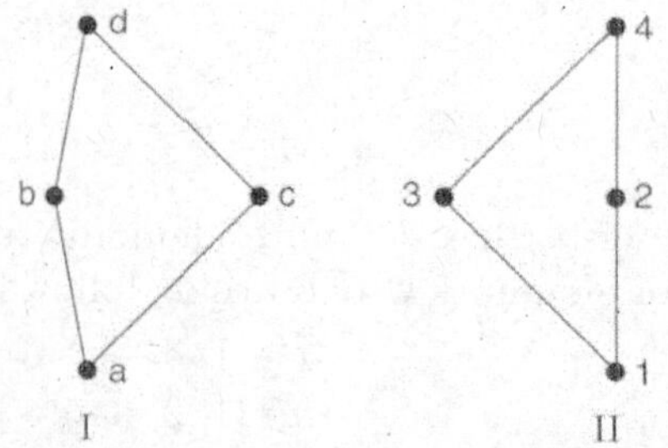

Fig. 28.

Example 30. *Determine whether the lattices shown in Fig. 29 are isomorphic.*

Sol. The lattices shown in Fig. 29 are not isomorphic since bijection is not possible as the element of the two lattices are not same.

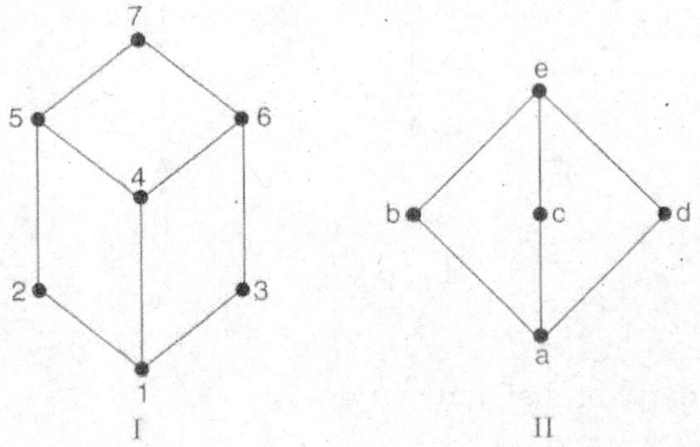

Fig. 29.

DISTRIBUTIVE LATTICE

A lattice L is called distributive lattice if for any elements a, b and c of L, it satisfies following distributive properties :

(*i*) $a \wedge (b \vee c) = (a \wedge b) \vee (a \wedge c)$

(*ii*) $a \vee (b \wedge c) = (a \vee b) \wedge (a \vee c)$.

If the lattice L does not satisfies the above properties, it is called a non-distributive lattice.

For example :

1. The power set P(S) of the set S under the operations of intersection and union is a distributive function. Since,

$$a \cap (b \cup c) = (a \cap b) \cup (a \cap c)$$

and also $a \cup (b \cap c) = (a \cup b) \cap (a \cup c)$ for any sets a, b and c of P(S).

2. The lattice shown in Fig. 30 is distributive. Since it satisfies the distributive properties for all ordered triples which are taken from 1, 2, 3 and 4.

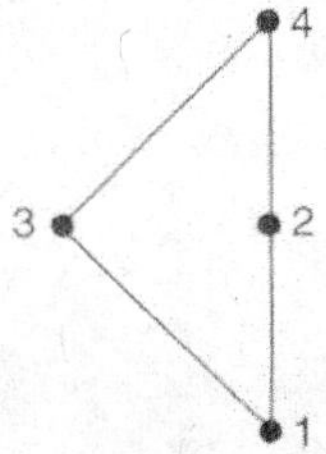

Fig. 30.

Example 31. *Show that the lattices shown in Fig. 31 are non-distributive.*

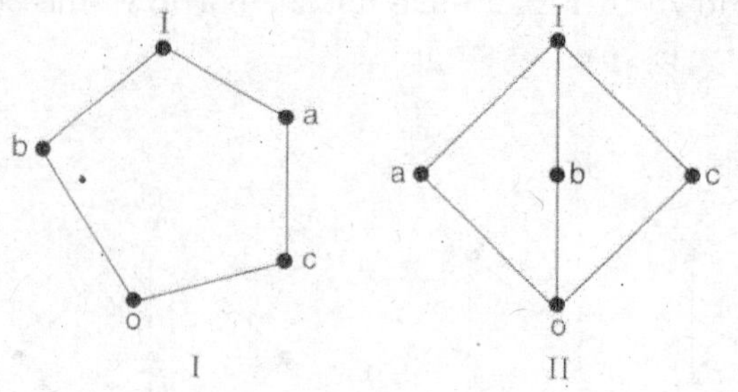

Fig. 31.

Sol.

I. From the (*i*) property of distributivity, we have

$$a \wedge (b \vee c) = a \wedge I = a$$

But, $(a \wedge b) \vee (a \wedge c) = 0 \vee c = c$

Since $a \wedge (b \vee c) \neq (a \wedge b) \vee (a \wedge c)$

Hence, the lattice is not distributive.

II. Again, from the (*i*) property of distributivity, we have

$$a \wedge (b \vee c) = a \wedge I = a$$

But $(a \wedge b) \vee (a \wedge c) = 0 \vee 0 = 0$

Since $a \wedge (b \vee c) \neq (a \wedge b) \vee (a \wedge c)$

Hence, the lattice is not distributive.

Note. A lattice L is non-distributive if and only if it contains a sublattice that is isomorphic to one of the two lattices shown in Figs. 31(I) and (II).

Theorem VII. *Prove that in a distributive lattice* $(L, \wedge, \vee)$, $(a \wedge b) \vee (b \wedge c) \vee (c \wedge a) = (a \vee b) \wedge (b \vee c) \wedge (c \vee a)$ *holds for all* $a, b, c \in L$.

Proof. We have given that L is distributive, so using distributive property, we have

$$\begin{aligned}
(a \wedge b) \vee (b \wedge c) \vee (c \wedge a) &= \{[(a \vee b) \vee b] \wedge [(a \wedge b) \vee c)]\} \vee (c \wedge a)\\
&= [b \wedge \{(a \vee c) \wedge (b \vee c)\}] \vee (c \wedge a)\\
&= [(a \vee c) \wedge \{b \wedge (b \vee c)\}] \vee (c \wedge a)\\
&= [(a \vee c) \wedge b] \vee (c \wedge a)\\
&= [(a \vee c) \vee (c \wedge a)] \wedge [b \vee (c \wedge a)]\\
&= [\{(a \vee c) \vee c\} \wedge \{(a \vee c) \vee a\}] \wedge [(b \vee c) \wedge (b \vee a)]\\
&= (a \vee c) \wedge (a \vee c) \wedge (b \vee c) \wedge (a \vee b)\\
&= (a \vee c) \wedge (b \vee c) \wedge (a \vee b) = \text{R.H.S.}
\end{aligned}$$

JOIN-IRREDUCIBLE

Consider a lattice $(L, \wedge, \vee)$. An element $a \in L$ is called join-irreducible if it can be expressed as the join of two distinct elements of L *i.e.*, $a \in L$ is join-irreducible if

$$a = x \vee y \Rightarrow a = x \text{ or } a = y, \text{ where } x, y \in L.$$

Example 32. *Determine the join-irreducible elements of the lattices as shown in Fig. 32.*

Sol. 1. The join-irreducible elements of Fig. 32(I) are a, b and d.

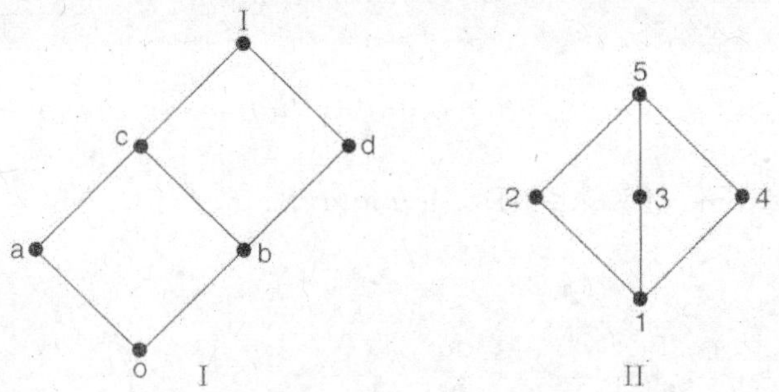

Fig. 32.

2. The join-irreducible elements of Fig. 32(II) are 2, 3 and 4.

MEET-IRREDUCIBLE

Consider a lattice (L, $\wedge$, $\vee$). An element $a \in$ L is called meet-irreducible if it can be expressed as the meet of two distinct elements of L *i.e.*, $a \in$ L is meet-irreducible if

$$a = x \wedge y \quad \Rightarrow \quad a = x \text{ or } a = y,$$

where $x, y \in$ L.

Example 33. *Determine the meet-irreducible elements of the lattices as shown in Fig. 33.*

Sol. 1. The meet-irreducible elements of Fig. 33(I) are a, b and c.

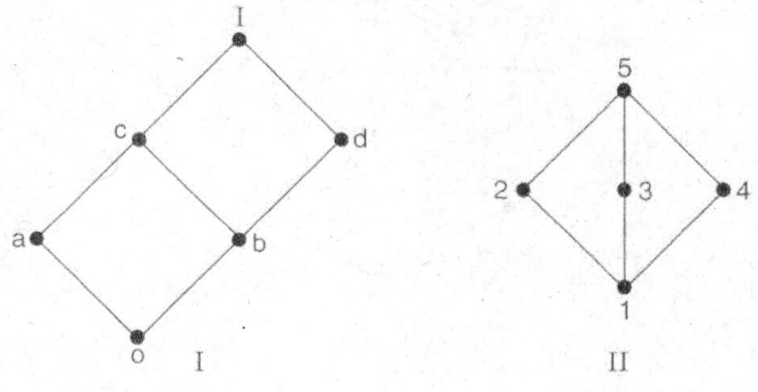

Fig. 33.

2. The meet-irreducible elements of Fig. 33(II) are 2, 3 and 4.

ATOM

Consider a lattice (L, $\wedge$, $\vee$) with a lower bound (1). An element a is called an atom if it is an immediate successor of 0 *i.e.*, $a \neq 0$ is an atom if $0 \leq b \leq a \Rightarrow b = 0$ or $b = a$.

ANTIATOM

Consider a lattice (L, ∧, ∨) with an upper bound 1. An element a is called an antiatom if it is an immediate predecessor of 1 *i.e.*, $a \neq 1$ is an antiatom if

$$a \leq b \leq 1 \;\Rightarrow\; b = a \;\text{ or }\; b = 1.$$

Example 34. *Determine the atoms and antiatoms of the following lattices shown in Fig. 34.*

Sol. Atoms. 1. Atoms of Fig. 34(I) are a and b.

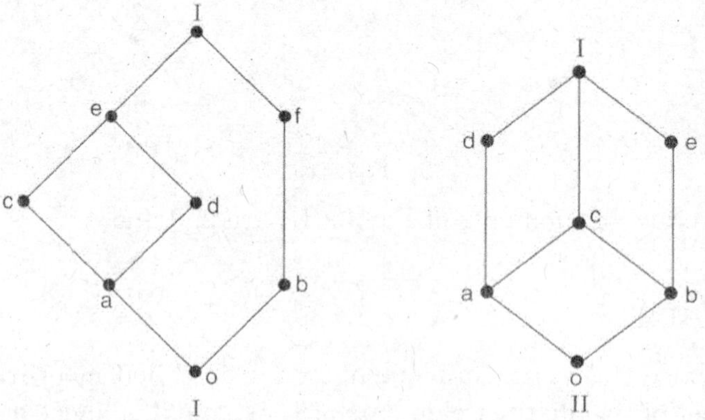

Fig. 34.

2. Atoms of Fig. 34(II) are a and b.

Antiatoms. 1. Antiatoms of Fig. 34(I) are e and f.

2. Antiatoms of Fig. 34(II) are d and e.

Example 35. *Construct the meet and join table of the lattice (L, ∨, ∧) as shown in Fig. 35.*

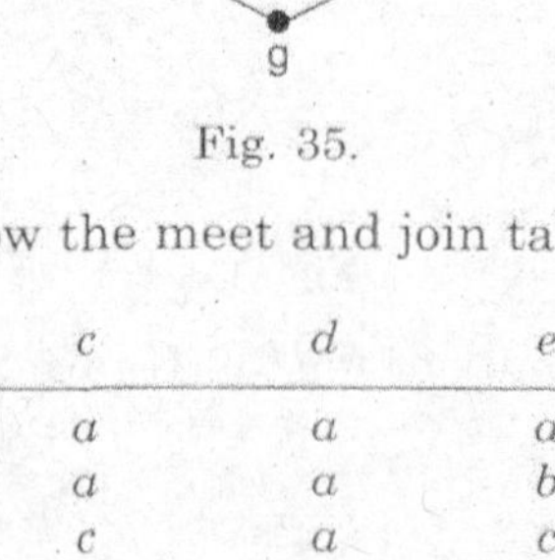

Fig. 35.

Sol. The following tables show the meet and join table of the lattice (L, ∧, ∨).

∨	a	b	c	d	e	f	g
a	a	a	a	a	a	a	a
b	a	b	a	a	b	a	b
c	a	a	c	a	c	c	c
d	a	a	a	d	a	d	d
e	a	b	c	a	e	c	e
f	a	a	c	d	c	f	f
g	a	b	c	d	e	f	g

$\wedge$	a	b	c	d	e	f	g
a	a	b	c	d	e	f	g
b	b	b	e	g	e	g	g
c	c	e	c	f	e	f	g
d	d	g	f	d	g	f	g
e	e	e	e	g	e	g	g
f	f	g	f	f	g	f	g
g	g	g	g	g	g	g	g

COMPLEMENTED LATTICES

Consider a bounded lattice L with greatest element I and the least element 0. An element $x \in$ L is called a complement of x if $x \vee x' = 1$ and $x \wedge x' = 0$.

From the definition of complement, if x' is a complement of x, then x is a complement of x'. It is not necessary that an element x has a complement. Also the complements need not be unique *i.e.*, an element have more than one complement.

Note. That $1' = 0$ and $0' = 1$.

Definition. A lattice L is called a complemented lattice if L is bounded and every element in L has a complement.

Example 36. *Determine the complement of a and c in Fig. 36.*

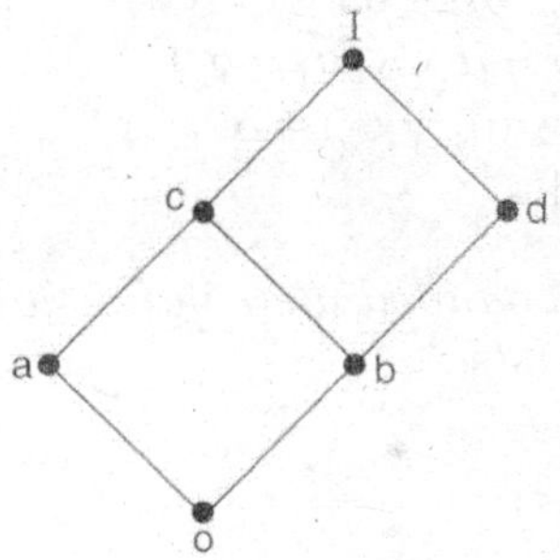

Fig. 36.

Sol. The complement of a is d. Since, $a \vee d = 1$ and $a \wedge d = 0$.

The complement of c does not exist. Since, there does not exist any element c such that $c \vee c' = 1$ and $c \wedge c' = 0$.

Theorem VIII. *Prove that 0 and 1 are complement of each other.*

Proof. To show that 1 is the only complement of 0, consider that $c \neq 1$ is a complement of 0 and $c \in$ L.

Then, $\quad 0 \wedge c = 0 \quad$ and $\quad 0 \vee c = 1$

But $\quad 0 \vee c = c \quad$ {Property of bounded lattice}

and $c \neq 1$ leads to a contradiction.

Similarly, we can show that 0 is the only complement of 1.

Example. The power set P(S) of the set S under the operations of intersection and union is a complemented lattice L since each element of L has a unique complement.

Example. The lattices shown below are complemented lattices Fig. 37.

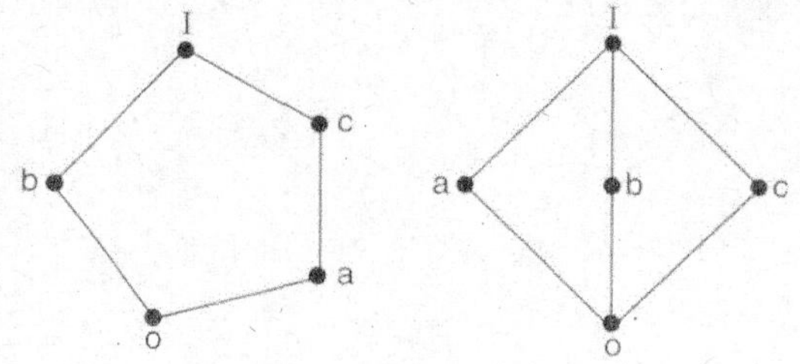

Fig. 37.

But the complements of some of the elements are not unique *e.g.*, b has two complements a and c in both the cases.

Theorem IX. *Prove that for a bounded distributive lattice L, the complements are unique if they exist.*

Proof. Consider a_1 and a_2 be complements of some elements $a \in$ L. Then, we have

$$a \vee a_1 = \mathrm{I} \quad \text{and} \quad a \vee a_2 = \mathrm{I}$$

Also $$a \wedge a_1 = 0 \quad \text{and} \quad a \wedge a_2 = 0$$

Now using the distributive property, we have

$$a_1 = a_1 \vee 0 = a_1 \vee (a \wedge a_2)$$
$$= (a_1 \vee a) \wedge (a_1 \vee a_2) = (a \vee a_1) \wedge (a_1 \vee a_2) = \mathrm{I} \wedge (a_1 \vee a_2) = a_1 \vee a_2$$

Similarly, $$a_2 = a_2 \vee 0 = a_2 \vee (a \wedge a_1) = (a_2 \vee a) \wedge (a_2 \vee a_1)$$
$$= (a \vee a_2) \wedge (a_1 \vee a_2)$$
$$= \mathrm{I} \wedge (a_1 \vee a_2) = a_1 \vee a_2$$

Thus, $a_1 = a_2$. Hence proved.

Example 37. *Consider the bounded distributive lattice as shown in Fig. 38. Show that every complement is unique if it exists.*

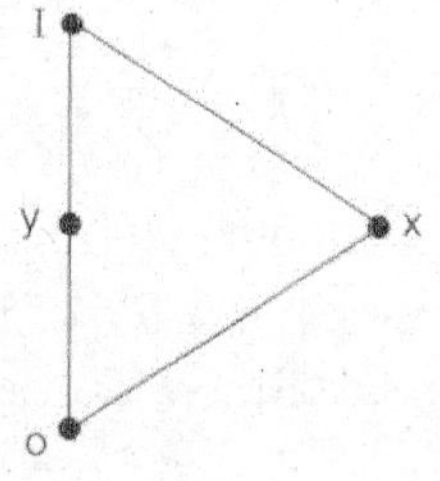

Fig. 38.

Sol. The complement of x is y and vice-versa. Similarly, the complement of 0 is I and vice-versa. Hence, all the complements are unique.

MODULAR LATTICE

A lattice (L, $\wedge$, $\vee$) is called a modular lattice if $a \vee (b \wedge c) = (a \vee b) \wedge c$ whenever $a \leq c$.

Theorem X. *Show that* $a \vee (b \wedge c) = (a \vee b) \wedge c$ *whenever* $a \leq c$.

Proof. If $a \le c$, then $a \vee c = c$

If L is distributive, then, we have

$$a \vee (b \wedge c) = (a \vee b) \wedge (a \vee c) = (a \vee b) \wedge c \qquad | \because a \vee c = c$$

Note. (*i*) Every distributive lattice is modular.

(*ii*) Every modular lattice is not distributive.

Example. The lattice shown in Fig. 39 is a non-distributive modular lattice.

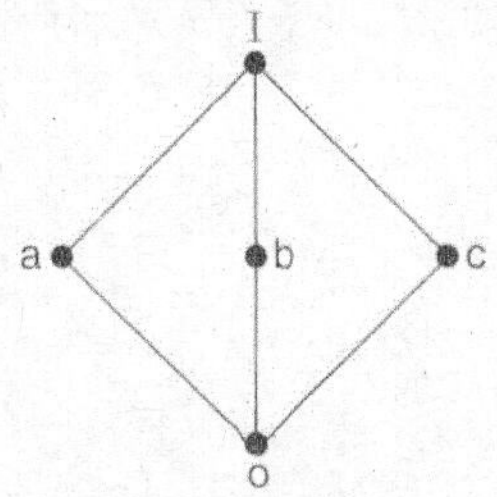

Fig. 39.

We can check that the lattice shown in Fig. 39 is not distributive.

$$a \wedge (b \vee c) = a \wedge 1 = a$$

but $$(a \wedge b) \vee (a \wedge c) = 0 \vee 0 = 0$$

To, check the modularity holds for $a = c$.

DIRECT PRODUCT OF LATTICES

Let $(L_1, \vee_1, \wedge_1)$ and $(L_2, \vee_2, \wedge_2)$ be two lattices. Then $(L, \vee, \wedge)$ is the direct product of lattices, where $L = L_1 \times L_2$ in which the binary operations $\vee$(join) and $\wedge$(meet) on L are such that for any (a_1, b_1) and (a_2, b_2) in L.

$$(a_1, b_1) \vee (a_2, b_2) = (a_1 \vee_1 a_2, b_1 \vee_2 b_2)$$

and $$(a_1, b_1) \wedge (a_2, b_2) = (a_1 \wedge_1 a_2, b_1 \wedge_2 b_2).$$

Example 38. *Consider a lattice $(L, \le)$ as shown in Fig. 40, where $L = \{1, 2\}$. Determine the lattice $(L^2, \le)$, where $L^2 = L \times L$.*

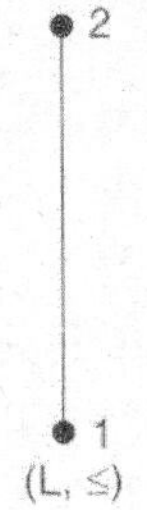

Fig. 40.

Sol. The lattice $(L^2, \le)$ is as shown in Fig. 41.

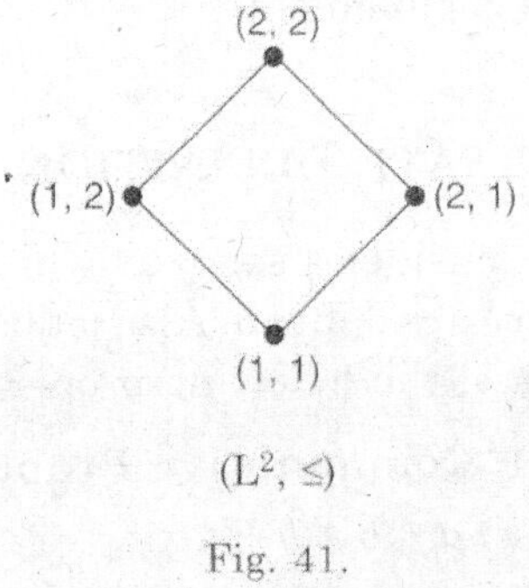

Fig. 41.

Example 39. *Let $(L_1, \leq)$ and $(L_2, \leq)$ be two lattices as shown in Figs. 42 and 43. Determine the lattice $(L, \leq)$, where $L = L_1 \times L_2$.*

Fig. 42.

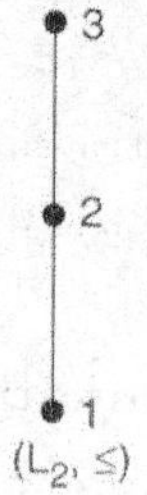

Fig. 43.

Sol. The lattice $(L, \leq)$ is a direct product of the lattices $(L_1, \leq)$ and $(L_2, \leq)$ as shown in Fig. 44.

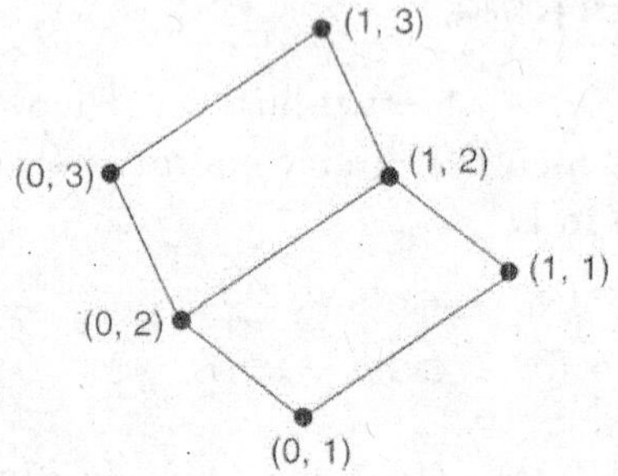

Fig. 44.

BOOLEAN ALGEBRA

Definition

A complemented distributive lattice is called a Boolean algebra. It is denoted by $(B, \wedge, \vee, ', 0, 1)$, where B is a set on which two binary operations $\wedge$ (*) and $\vee$ (+) and a unary operation $'$ (complement) are defined. Here 0 and 1 are two distinct elements of B.

Since $(B, \wedge, \vee)$ is a complemented distributive lattice, therefore each element of B has a unique complement.

ALTERNATE DEFINITION

Consider a set B on which two binary operations * and + and a uniary operation$'$ (complement) are defined. Also let 0 and 1 are two distinct elements of B. Then it is called a Boolean algebra, if the following properties are satisfied for any elements a, b and c of the set B by it.

1. Commutative Properties

(*i*) $a + b = b + a$

(*ii*) $a * b = b * a$

2. Distributive Properties

(*i*) $a + (b * c) = (a + b) * (a + c)$

(*ii*) $a * (b + c) = (a * b) + (a * c)$

3. Identity Properties

(*i*) $a + 0 = a$

(*ii*) $a * 1 = a$

4. Complement Laws

(*i*) $a + a' = 1$

(*ii*) $a * a' = 0$

The Boolean algebra is denoted by (B, +, *, ′, 0, 1).

Example. Consider a set B = {0, 1}. Also consider the operations *, + and ′ on the set B as shown in Fig. 45. Since the algebra (B, *, +, ′, 0, 1) satisfies all the properties. Hence, it is a Boolean algebra.

*	0	1
0	0	0
1	0	1

+	0	1
0	0	1
1	1	1

x	x'
0	1
1	0

Fig. 45.

Example. Consider a non-empty set S and power set of S is P(S). Since P(S) is a collection of set S closed under the operations ∪ (union), ∩ (intersection) and ~ (complement). Hence P(S) is a Boolean algebra, with the empty set ϕ as the zero element and unit element is U (universal set). The complement of any subset $A \subseteq U$ is $\sim A = U - A$, called the relative complement of the set A.

Note 1. If a finite lattice L does not contain 2^n element for some positive integer n, then L cannot be a Boolean algebra.

2. If $|L| = 2^n$, then L may or may not be a Boolean algebra.

Example 40. *Determine whether the poset given in Fig. 46 is a Boolean algebra or not. Give reason.*

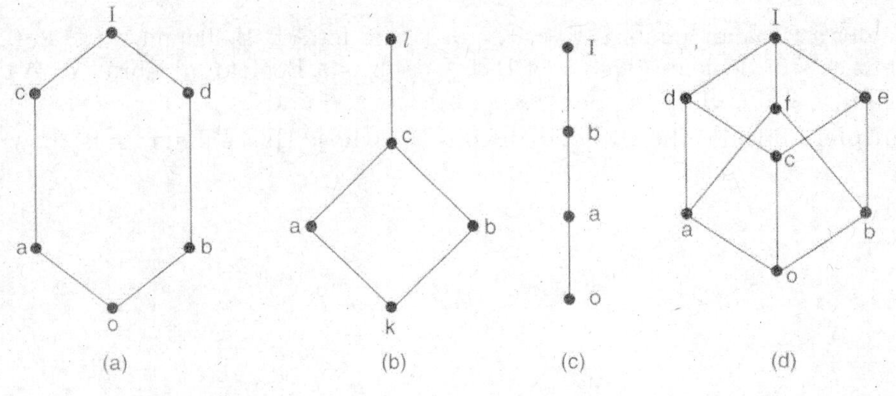

Fig. 46.

Sol. 1. The poset shown in Fig. 46(*a*) is not a Boolean algebra since it contains 6 elements and $6 \neq 2^n$ for any integer $n \geq 0$.

2. The poset shown in Fig. 46(*b*) is not a Boolean algebra since it contains 5 elements and $5 \neq 2^n$ for any integer $n \geq 0$.

3. The poset shown in Fig. 46(*c*) is not a Boolean algebra since it does not satisfy the properties of Boolean algebra *i.e.*, complement of a and b does not exist.

4. The poset shown in Fig. 46(*d*) is a Boolean algebra since it is isomorphic to (P(S), ⊆) which is a Boolean algebra. Here P(S) is a power set of S having three elements.

Theorem XI. *Prove that the complement of every element on a Boolean algebra B is unique.*

Proof. Consider an element $a \in$ B, let a has two complements a_1' an a_2'' in B.

Therefore, $$\left.\begin{array}{l} a + a_1' = 1 = a_1' + a \\ a * a_1' = 0 = a_1' * a \end{array}\right\} \quad \text{...(1)}$$

and also $$\left.\begin{array}{l} a + a_2' = 1 = a_2' + a \\ a * a_2' = 0 = a_2' * a \end{array}\right\} \quad \text{...(2)}$$

We know that
$$\begin{aligned}
a_1' &= a_1' * 1 && \{\because\ a_1' * 1 = a_1'\} \\
&= a_1' * (a + a_2') && \{\because\ a + a_2' = 1\} \\
&= (a_1' * a) + (a_1' * a_2') && \{\because\ \text{B is distributive}\} \\
&= 0 + (a_1' * a_2') && \{\because\ a_1' * a = 0\} \\
&= a * a_2' + a_1' * a_2' && \{\because\ a * a_2' = 0\} \\
&= (a + a_1') * a_2' && \{\because\ \text{B is distributive}\} \\
&= 1 * a_2' && \{\because\ a + a_1' = 1\} \\
&= a_2' && \{\because\ 1 * a_2' = a_2'\}
\end{aligned}$$

Hence, the complement is unique.

SUB-ALGEBRA

Consider a Boolean algebra (B, ∗, +, ′, 0, 1) and let A ⊆ B. Then (A, ∗, +, ′, 0, 1) is called a sub-algebra or sub-Boolean algebra of B if A itself is a Boolean algebra *i.e.*, A contains the elements 0 and 1 and is closed under the operations ∗, + and ′.

Example. Consider the Boolean algebra D_{70} whose Hasse diagram is shown in Fig. 47.

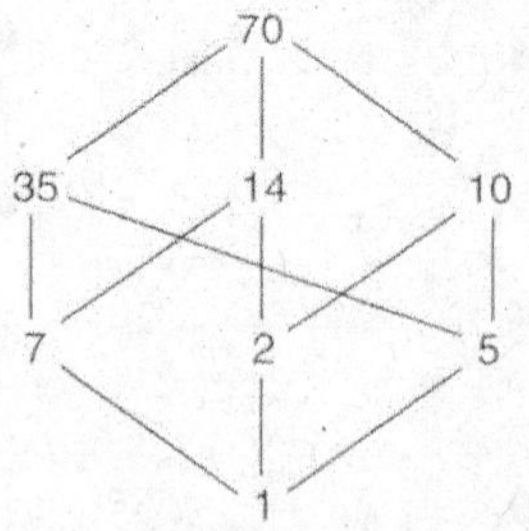

Fig. 47.

Clearly A = {1, 7, 10, 70} and B = {1, 2, 35, 70} is sub-algebra of D_{70}. Since, both A and B are closed under operation ∧, ∨ and ′.

Note. A subset of a Boolean algebra can be a Boolean algebra but it may or may not be a sub-algebra as it may not closed w.r.t. the operations in B.

Example 41. *Consider the Boolean algebra (B, +, *, ′, 0, 1) as shown in Fig. 48. Let the subsets of S be as follows :*

(i) $S_1 = \{c, b, 0, 1\}$ *(ii)* $S_2 = \{a, b, 0, 1\}$

(iii) $S_3 = \{a, b, 0, f\}$ *(iv)* $S_4 = \{a, c, 0, 1\}$

(v) $S_5 = \{a, c, f, 1\}$.

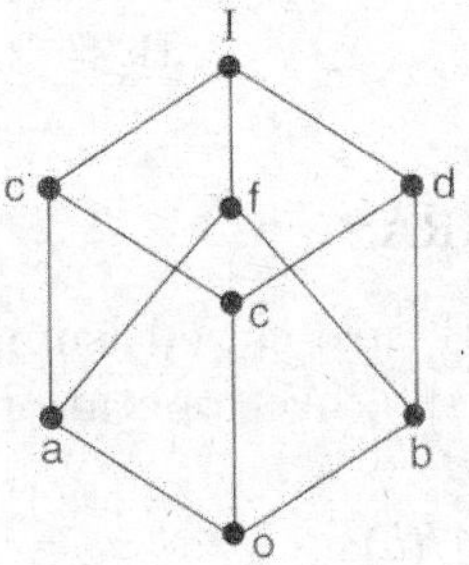

Fig. 48.

Determine whether the S_1, S_2, S_3, S_4, S_5 *are Boolean algebra, sub-algebra, either both or none.*

Sol. *(i)* The subset S_1 is a sub-algebra, since it is closed under the operations *, +, ′. So, it is also a Boolean algebra.

(ii) The subset S_2 is neither a sub-algebra nor Boolean algebra.

(iii) The subset S_3 is a Boolean algebra, but not a sub-algebra of B.

(iv) The subset S_4 is neither a Boolean algebra nor a sub-algebra of B.

(v) The subset S_5 is not a sub-algebra, but it is a Boolean algebra.

Note 1. Every Boolean sub-algebra of a Boolean algebra B is a Boolean algebra. But there may be Boolean algebras which are not Boolean sub-algebras of B.

2. The Boolean algebra B and {0, 1} are always sub-algebras of B.

Example 42. *Consider the Boolean algebra* D_{30}. *Determine the following :*

(a) All the Boolean sub-algebra of D_{30}.

(b) All Boolean algebras which are not Boolean sub-algebras of D_{30} *having at least four elements.*

Sol. The Hasse diagram of D_{30} is shown in Fig. 49.

(a) All Boolean sub-algebras of D_{30} are

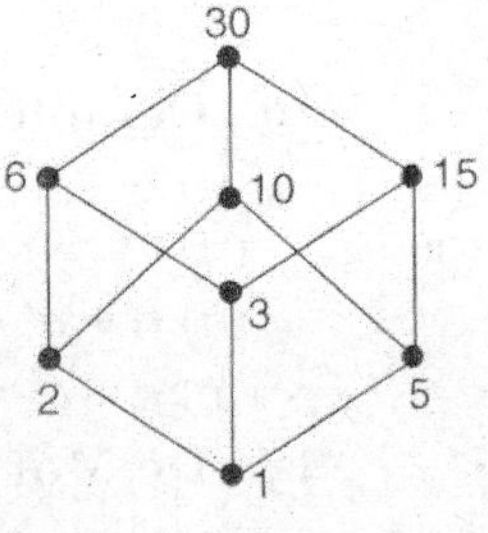

Fig. 49.

$B_1 = \{1, 2, 15, 30\}$ $B_4 = \{1, 30\}$

$B_2 = \{1, 5, 6, 30\}$ $B_5 = \{1, 2, 3, 5, 6, 10, 15, 30\}$.

$B_3 = \{1, 3, 10, 30\}$

(*b*) All Boolean algebras which are not Boolean sub-algebras of D_{30} are :

$B_1 = \{1, 2, 3, 6\}$ $B_2 = \{1, 3, 5, 15\}$

$B_3 = \{3, 6, 5, 30\}$ $B_4 = \{1, 2, 5, 10\}$

$B_5 = \{5, 10, 15, 30\}$ $B_6 = \{2, 6, 10, 30\}$.

ISOMORPHIC BOOLEAN ALGEBRAS

Two Boolean algebras B and B_1 are called isomorphic if there is a one-to-one correspondence $f : B \to B_1$ which preserves the three operations +, $*$ and $'$ for any elements a, b in B *i.e.*,

$$f(a + b) = f(a) + f(b)$$

$$f(a * b) = f(a) * f(b) \quad \text{and} \quad f(a') = f(a)'.$$

Example. The following are two distinct Boolean algebras with two elements, which are isomorphic.

1. The first one is a Boolean algebra that is derived from a power set P(S) under $\subseteq$ (set inclusion) *i.e.*, let $S = \{a\}$, then $B = \{P(S), \cup, \cap, '\}$ is a boolean algebra with two elements $P(S) = \{\phi, \{a\}\}$.

2. The second one is a Boolean algebra $\{B, \vee, \wedge, '\}$ with two elements 1 and p {here p is a prime number} under operation divides *i.e.*, let $B = \{1, p\}$. So, we have $1 \wedge p = 1$ and $1 \vee p = p$ also $1' = p$ and $p' = 1$.

The table shows all the basic properties of a Boolean algebra (B, $*$, +, $'$, 0, 1) for any elements a, b, c belongs to B. The greatest and least elements of B are denoted by 1 and 0 respectively.

Table I

1. $a \le b$ iff $a + b = b$

2. $a \le b$ iff $a * b = a$

3. Idempotent Laws

(*i*) $a + b = a$

(*ii*) $a * a = a$

4. Commutative Property

(*i*) $a + b = b + a$

(*ii*) $a * b = b * a$

5. Associative Property

(*i*) $a + (b + c) = (a + b) + c$

(*ii*) $a * (b * c) = (a * b) * c$

6. Absorption Laws

(*i*) $a + (a * b) = a$

(*ii*) $a * (a + b) = a$

7. Identity Laws

(*i*) $a + 0 = a$

(*ii*) $a * 1 = a$

8. Null Laws

(*i*) $a * 0 = 0$

(*ii*) $a + 1 = 1$

9. Distributive Laws

(*i*) $a * (b + c) = (a * b) + (a * c)$

(*ii*) $a + (b * c) = (a + b) * (a + c)$

10. Complement Laws

(*i*) $0' = 1$

(*ii*) $1' = 0$

(*iii*) $a + a' = 1$

(*iv*) $a * a' = 0$

11. Involution Law

$(a')' = a$

12. De Morgan's Laws

(*i*) $(a * b)' = (a' + b')$

(*ii*) $(a + b)' = (a' * b')$.

Note. (*i*) $0 \le a \le 1$ for every $a \in B$.

(*ii*) Every element b has a unique complement b'.

Theorem XII. *Prove the idempotent laws i.e.,*

(*i*) $a * a = a$ (*ii*) $a + a = a$.

Proof. (*i*)

$$a = a * 1 = a * (a + a')$$
$$= (a * a) + (a * a') \quad \text{...(Distributive Law)}$$
$$= (a * a) + 0 = a * a.$$

(*ii*) It follows from the duality of (*i*).

Theorem XIII. *Prove the Boundedness (NULL) laws i.e.,*

(*i*) $a * 0 = 0$ (*ii*) $a + 1 = 1$.

Proof. (*i*)

$$a * 0 = (a * 0) + 0$$
$$= (a * 0) + (a * a') \quad \text{...\{Complement law\}}$$
$$= a * (0 + a') \quad \text{...\{Distributive law\}}$$
$$= a * (a' + 0)$$
$$= a * a'$$
$$= 0.$$

(*ii*) It is directly followed from the duality of proof (*i*).

Theorem XIV. *Prove the absorption laws i.e.,*

(*i*) $a + (a * b) = a$ (*ii*) $a * (a + b) = a$.

Proof. (*ii*)

$$a * (a + b) = (a + 0) * (a + b)$$
$$= a + (0 * b) \quad \text{...[Distributive law]}$$
$$= a + (b * 0)$$
$$= a + 0 \quad \text{...[Null law]}$$
$$= a$$

(*i*) It is directly followed from duality of proof (*ii*).

Theorem XV. *Prove De Morgan's laws i.e.,*

(*i*) $(a + b)' = a' * b'$ (*ii*) $(a * b)' = a' + b'$.

Proof. (*i*) We have to show that

$$(a + b) + (a' * b') = 1 \quad \text{and} \quad (a + b) * (a' * b') = 0$$

Then, we have $a' * b' = (a + b)'$

Now,

$$(a + b) + (a' * b') = b + a + (a' * b')$$
$$= b + (a + a') * (a + b') \quad \text{...[Distributive law]}$$
$$= b + [1 * (a + b')] \Rightarrow b + (1 * a) + (1 * b')$$
$$= b + a + b' = b + b' + a$$
$$= 1 + a = 1$$

Also,

$$(a + b) * (a' * b') = ((a + b) * a') * b'$$
$$= ((a * a') + (b * a')) * b'$$
$$= (0 + (b * a')) * b'$$
$$= (b * a') * b'$$
$$= (b * b') * a'$$

$= 0 + a'$

$= 0$

Thus, $a' * b' = (a + b)'$.

(*ii*) It follows directly from duality of proof (*i*).

BOOLEAN FUNCTIONS

Consider the Boolean algebra (B, ∨, ∧, ′, 0, 1). A function from A^n to A is called a Boolean function if it can be specified by a Boolean expression of n variables.

For the two valued Boolean algebra any function from $\{0, 1\}^n$ to $\{0, 1\}$ is a Boolean function.

e.g., 1. The table shows a function f from $\{0, 1\}^3$ to $\{0, 1\}$

Table II

(x, y, z)	f
(0, 0, 0)	0
(0, 0, 1)	0
(0, 1, 0)	1
(0, 1, 1)	0
(1, 0, 0)	1
(1, 0, 1)	1
(1, 1, 0)	0
(1, 1, 1)	1

2. The table shows a function f from $\{0, 1, 2, 3\}^2$ to $\{0, 1, 2, 3\}$.

Table III

(x, y)	f
(0, 0)	1
(0, 1)	0
(0, 2)	0
(0, 3)	3
(1, 0)	1
(1, 1)	1
(1, 2)	0
(1, 3)	3
(2, 0)	2
(2, 1)	0
(2, 2)	1
(2, 3)	1
(3, 0)	3
(3, 1)	0
(3, 2)	0
(3, 3)	2

Note. A function can always be described in tabular form. An alternative way of describing the functions is specifying the function by an expression.

BOOLEAN EXPRESSION

Consider a Boolean algebra (B, $\vee$, $\wedge$, $'$, 0, 1). A Boolean expression over Boolean algebra B is defined as

(*i*) Every element of B is a Boolean expression

(*ii*) Every variable name is a Boolean expression

(*iii*) If a_1 and a_2 are Boolean expressions, then a_1', $a_1 \vee a_2$ and $a_1 \wedge a_2$ are Boolean expressions.

e.g., Consider the Boolean algebra ({0, 1, 2, 3}, $\vee$, $\wedge$, $'$, 0, 1)

(*i*) $0 \vee x$ (*ii*) $\overline{(2 \wedge 3)}$ (*iii*) $(x_1 \vee \overline{x_2}) \wedge \overline{(x_1 \wedge x_3)}$

are Boolean expressions over the Boolean algebra.

A Boolean expression that contains n distinct variables is usually referred to as a Boolean expression of n variables.

EVALUATION OF BOOLEAN EXPRESSION

Let $E(x_1, x_2, \ldots\ldots, x_n)$ be a Boolean expression of n variables over a Boolean algebra B. By an assignment of values to the variables $x_1, x_2, \ldots\ldots, x_n$, means an assignment of elements of A to be the values of the variables.

We can evaluate the expression $E(x_1, x_2, \ldots\ldots, x_n)$ by substituting the variables in the expression by their values.

Example. Consider the Boolean expression

$$E(x_1, x_2, x_3) = (x_1 \vee x_2) \wedge (\overline{x_1} \vee \overline{x_2}) \wedge \overline{(x_2 \vee x_3)}$$

over the Boolean algebra ({0, 1}, $\vee$, $\wedge$, $'$)

By assigning the values $x_1 = 0$, $x_2 = 1$, $x_3 = 0$ yields

$$E(0, 1, 0) = (0 \vee 1) \wedge (\bar{0} \vee \bar{1}) \wedge (\overline{1 \vee 0}) = 1 \wedge 1 \wedge 0 = 0.$$

EQUIVALENT BOOLEAN EXPRESSIONS

Two Boolean expressions of n variables are said to be equivalent if they assume the same value for every assignment of values to the n variables.

e.g., The following two Boolean algebras

$$(x_1 \wedge x_2) \vee (x_1 \wedge \overline{x_3}) \quad \text{and} \quad x_1 \wedge (x_2 \vee \overline{x_3})$$

are equivalent.

We may write $E_1(x_1, x_2, \ldots\ldots, x_n) = E_2(x_1, x_2, \ldots\ldots, x_n)$ to mean the two expressions $E_1(x_1, x_2, \ldots\ldots, x_n)$ and $E_2(x_1, x_2, \ldots\ldots, x_n)$ are equivalent.

e.g., The Boolean expression

$$(\overline{x_1} \wedge x_2 \wedge \overline{x_3}) \vee (x_1 \wedge \overline{x_2}) \vee (x_1 \wedge x_3)$$

over the Boolean algebra ({0, 1}, $\vee$, $\wedge$, $'$) defines the function f in figure.

Minterm. A Boolean expression of n variables $x_1, x_2, \ldots\ldots, x_n$ is said to be a minterm if it is of the form

$$\tilde{x}_1 \wedge \tilde{x}_2 \wedge \tilde{x}_3 \wedge \ldots\ldots \wedge \tilde{x}_n$$

where $\tilde{x}_i$ is used to denote x_i or x_i'.

DISJUNCTIVE NORMAL FORM OR SUM OF PRODUCTS OR SOP

A Boolean expression over ({0, 1}, ∨, ∧, ′) is said to be in disjunctive normal form if it is a join of minterms *e.g.*,

$$(x_1' \wedge x_2' \wedge x_3') \vee (x_1' \wedge x_2 \wedge x_3') \vee (x_1 \wedge x_2 \wedge x_3)$$

is a Boolean expression in disjunctive normal form.

Since there are three minterms $x_1' \wedge x_2' \wedge x_3'$, $x_1' \wedge x_2 \wedge x_3$ and $x_1 \wedge x_2 \wedge x_3$.

Maxterm. A Boolean expression of n variables $x_1, x_2, \ldots, x_n$ is said to be a maxterm if it is of the form $\tilde{x}_1 \vee \tilde{x}_2 \vee \ldots\ldots \vee \tilde{x}_n$, where $\tilde{x}_i$ is used to denote x_i or x_i'.

CONJUNCTIVE NORMAL FORM OR PRODUCTS OF SUMS OR POS

A Boolean expression over ({0, 1}, ∨, ∧, ′) is said to be in conjunctive normal form if it is a meet of maxterms *e.g.*,

$$(x_1 \vee x_2 \vee x_3) \wedge (x_1 \vee x_2 \vee x_3) \wedge (x_1 \vee x_2 \vee x_3) \wedge (x_1' \vee x_2 \vee x_3') \wedge (x_1' \wedge x_2' \wedge x_3)$$

is a Boolean expression in conjunctive normal form consisting of five maxterms.

OBTAINING A DISJUNCTIVE NORMAL FORM

Consider a function from $\{0, 1\}^n$ to {0, 1}. A Boolean expression can be obtained in disjunctive normal form corresponding to this function by having a minterm corresponding to each ordered n-tuple of 0's and 1's for which the value of function is 1.

OBTAINING A CONJUNCTIVE NORMAL FORM

Consider a function from $\{0, 1\}^n$ to {0, 1}. A Boolean expression can be obtained in conjunctive normal form corresponding to this function by having a maxterm corresponding to each ordered n-tuple of 0's and 1's at which the value of function is 0.

Example 43. *Express the following function in*
(i) Disjunctive normal form *(ii) Conjunctive normal form*

	f		*f*
(0, 0, 0)	*1*	*(1, 0, 0)*	*0*
(0, 0, 1)	*0*	*(1, 0, 1)*	*1*
(0, 1, 0)	*1*	*(1, 1, 0)*	*0*
(0, 1, 1)	*0*	*(1, 1, 1)*	*1*

Sol. (*i*) $(x_1' \wedge x_2' \wedge x_3') \vee (x_1' \wedge x_2 \wedge x_3') \vee (x_1 \wedge x_2' \wedge x_3) \vee (x_1 \wedge x_2 \wedge x_3)$
(*ii*) $(x_1' \vee x_2' \vee x_3') \wedge (x_1' \vee x_2 \vee x_3) \wedge (x_1 \vee x_2' \vee x_3') \wedge (x_1 \vee x_2 \vee x_3')$.

Example 44. *If $f(x, y, z) = (x \vee y) \wedge (x \vee y') \wedge (x' \vee z)$ be a given Boolean function. Determine its DN form.*

Sol. First determine all values of $f(x, y, z)$ when x, y, z take values 0, 1 and then from this table, we will find the DN form

x	y	z	$f(x, y, z)$	$(x \vee y) \wedge (x \vee y') \wedge (x' \vee z)$
0	0	0	0	$= (0 \vee 0) \wedge (0 \vee 1) \wedge (1 \vee 0)$
0	0	1	0	$= (0 \vee 0) \wedge (0 \vee 1) \wedge (1 \vee 1)$
0	1	0	0	$= (0 \vee 1) \wedge (0 \vee 0) \wedge (1 \vee 0)$
0	1	1	0	$= (0 \vee 1) \wedge (0 \vee 0) \wedge (1 \vee 1)$
1	0	0	0	$= (1 \vee 0) \wedge (1 \vee 1) \wedge (0 \vee 0)$
1	0	1	1	$= (1 \vee 0) \wedge (1 \vee 1) \wedge (0 \vee 1)$
1	1	0	0	$= (1 \vee 1) \wedge (1 \vee 0) \wedge (0 \vee 0)$
1	1	1	1	$= (1 \vee 1) \wedge (1 \vee 0) \wedge (0 \vee 1)$

The disjunctive normal form of the function is

$$f(x, y, z) = (x \wedge y' \wedge z) \vee (x \wedge y \wedge z).$$

Example 45. *Determine the disjunctive normal form of the following Boolean expression :*

$$x \wedge (y \vee z).$$

Sol. First determine all values of $f(x, y, z)$ when x, y, z take values 0, 1 and then from this table we will write disjunctive form. Thus,

x	y	z	$f(x, y, z)$	$x \wedge (y \vee z)$
0	0	0	0	$= 0 \wedge (0 \vee 0)$
0	0	1	0	$= 0 \wedge (0 \vee 1)$
0	1	0	0	$= 0 \wedge (1 \vee 0)$
0	1	1	0	$= 0 \wedge (1 \vee 1)$
1	0	0	0	$= 1 \wedge (0 \vee 0)$
1	0	1	1	$= 1 \wedge (0 \vee 1)$
1	1	0	1	$= 1 \wedge (1 \vee 0)$
1	1	1	1	$= 1 \wedge (1 \vee 1)$

Thus, the DNF of the function is

$$f(x, y, z) = (x \wedge y \wedge z) \vee (x \wedge y' \wedge z) \vee (x \wedge y \wedge z').$$

Example 46. *Find the Boolean expression in conjunctive normal form that defines the function f given by*

x	y	z	f
0	0	0	1
0	0	1	0
0	1	0	1
0	1	1	0
1	0	0	0
1	0	1	0
1	1	0	0
1	1	1	1

Sol. To obtain the expression in conjunctive normal form, take the values of $f(x, y, z)$ when it is 0 *i.e.*, take the max. terms of the function. Thus, the expression in CN form is

$$(x \vee y \vee z') \wedge (x \vee y' \vee z) \wedge (x \vee y' \vee z') \wedge (x' \vee y \vee z) \wedge (x' \vee y' \vee z).$$

PRINCIPLE OF DUALITY

The dual of any expression E is obtained by interchanging the operations + and * and also interchanging the corresponding identity elements 0 and 1, in original expression E.

Example 47. *Write the dual of following Boolean expressions :*

(*i*) $(x_1 * x_2) + (x_1 * \overline{x_3})$ (*ii*) $(1 + x_2) * (x_1 + 1)$

(*iii*) $(a \wedge (b \wedge c))$.

Sol. (*i*) $(x_1 + x_2) * (x_1 + \overline{x_3})$ (*ii*) $(0 * x_2) + (x_1 * 0)$

(*iii*) $(a \vee (b \wedge c))$.

Note. The dual of any theorem in a Boolean algebra is also a theorem.

DESIGN OF DIGITAL CIRCUITS

Schematic or Graphical Representation of Boolean Expressions

The Boolean expressions can be graphically represented by using logic circuits. These logic circuits can be constructed using solid state devices called gates, which are capable of switching voltage levels. If x and y are variables, then the basic expressions $x \wedge y$ (AND), $x \vee y$ (OR) and $\overline{x'}$ (NOT) are shown graphically as follows :

AND Gate. An AND gate receives inputs x and y and produces output denoted $x \wedge y$, as shown in logic table.

x	y	$x \wedge y$
0	0	0
0	1	0
1	0	0
1	1	1

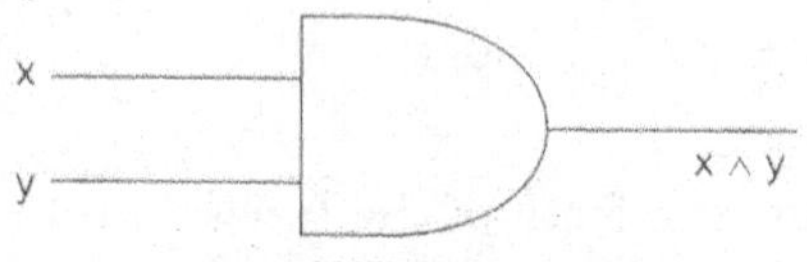

AND Gate

Fig. 50.

OR Gate. An OR gate receives inputs x and y and produces output denoted $x \vee y$ as shown in logic table.

x	y	$x \vee y$
0	0	0
0	1	1
1	0	1
1	1	1

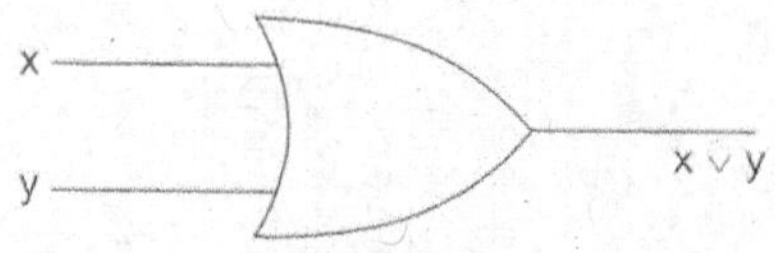

OR Gate

Fig. 51.

NOT Gate. A NOT gate receives input x and produces output y denoted $\bar{x}$, as shown in logic table.

x	$\bar{x}$
0	1
1	0

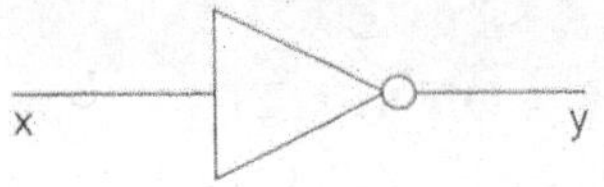

Fig. 52.

We can interconnect these devices to form an electronic circuit that realizes any given Boolean expression.

Example 48. *Design a combinatorial circuit for the following Boolean expression :*

(*i*) $(x \wedge \bar{y}) \vee (\bar{x} \wedge y)$ (*ii*) $(x_1 \vee x_2) \wedge x_1$.

Sol. (*i*) The combinatorial circuit of the Boolean expression is shown in Fig. 53.

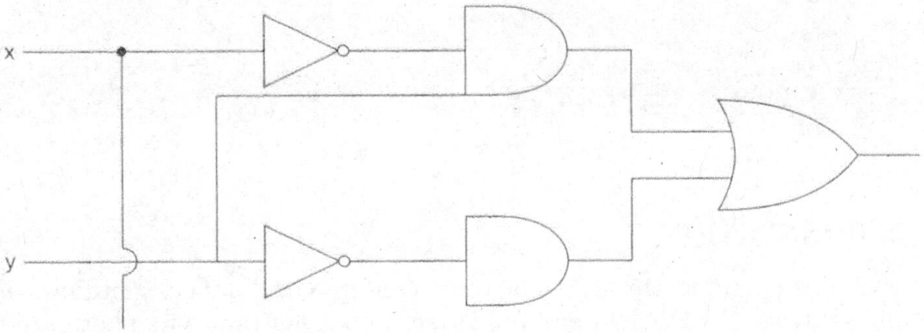

Fig. 53.

(*ii*) The combinatorial circuit of the Boolean expression is shown in Fig. 54.

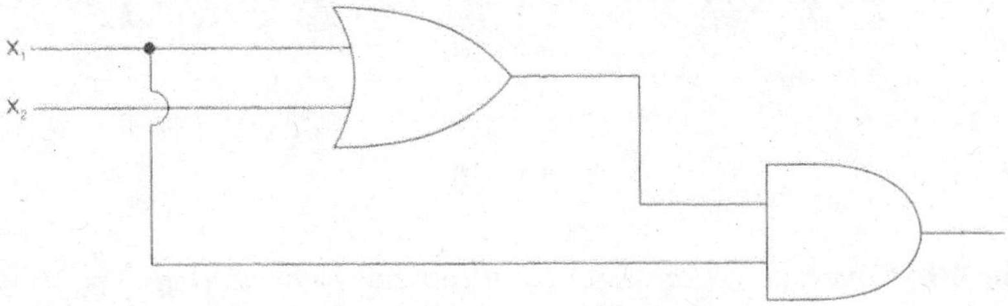

Fig. 54.

APPLICATIONS OF BOOLEAN ALGEBRA IN SWITCHING THEORY

The simplest switching device is NO-OFF switch. A switch is a device in an electric circuit which lets or does not let the current to flow through the circuit.

The switch has two states NO or OFF (closed or open)

— If the switch is closed, current will pass through it.

— If the switch is open, current will not pass through it.

— ON is denoted by 1.

OFF is denoted by 0.

The electrical circuits containing switches by Boolean expressions, if ON is denoted by true or 1 and OFF is denoted by False or 0.

SERIES AND PARALLEL CONNECTION

There are 2 ways in which switches are connected with each other.

(*i*) Series connection (*ii*) Parallel connection.

SERIES CONNECTION

Two switches x_1 and x_2 are said to be connected in series if current can pass only when both the switches are ON (closed) and the current does not flow if both the switches or any one are OFF (open). Symbolically, series connection of two switches x_1 and x_2 is denoted by $x_1 \wedge x_2$ or $x_1 * x_2$. It is represented diagrammatically as follows :

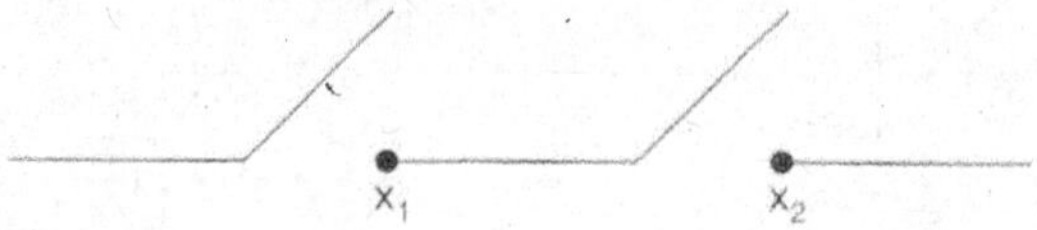

PARALLEL CONNECTION

Two switches x_1 and x_2 are said to be connected in parallel if current flows when both or any one of the switches are ON (closed) and current does not flow when both are OFF (open). Symbolically, parallel connection of two switches x_1 and x_2 is denoted by $x_1 \vee x_2$ or $x_1 + x_2$. It is represented diagrammatically as follows :

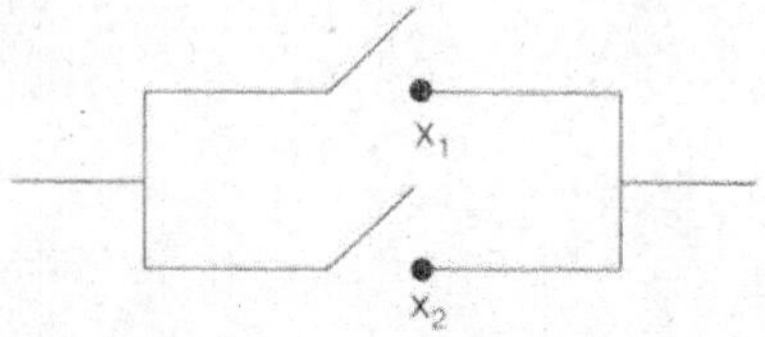

Example 49. *Construct the circuit diagram represented by Boolean expression* $a \wedge (b \vee c)$.

Sol. For the given Boolean expression, we will see that the current would flow when a and b or a and c are ON (closed). So, it is represented as follows :

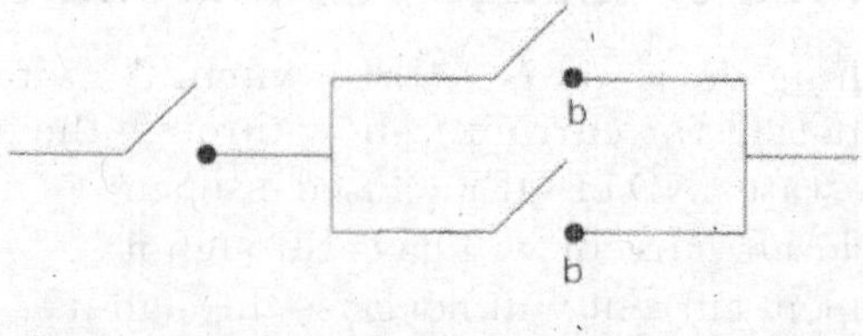

Example 50. *Draw the circuit diagram represented by following Boolean expressions :*

(*i*) $(a \wedge b \wedge c) \vee (a' \wedge b' \wedge c')$ (*ii*) $(x_1 \wedge x_2) \vee (x_1' \wedge x_2')$.

Sol. (*i*) For the given expression, we will see that the current would flow when a, b, c or a', b', c' are ON (closed). It is represented as follows :

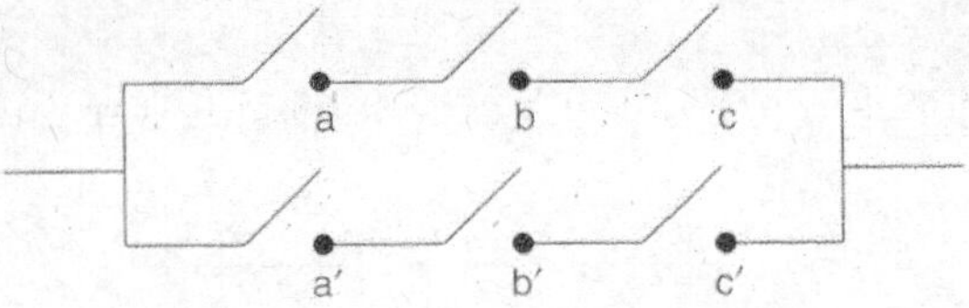

(*ii*) The circuit diagram for the expression is represented as follows :

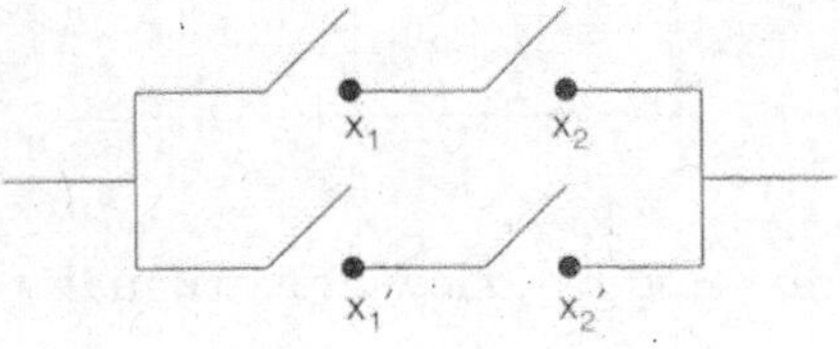

KARNAUGH MAPS

A karnaugh map is a planar area subdivided into 2^n equal cells each representing a point B_2^n for functions of n variables. Each variable x is used to split the area into two equal halves in a different ways *i.e.*, one for x and other for x'. The cells corresponding to the arguments for which the function has the value 1 contains 1.

Example 1. When the number of variables $n = 1$, the karnaugh map is like as shown in Fig. 55.

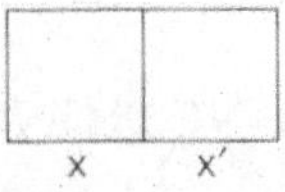

Fig. 55.

2. When the number of variables $n = 2$, the karnaugh map is like as shown in Fig. 56.

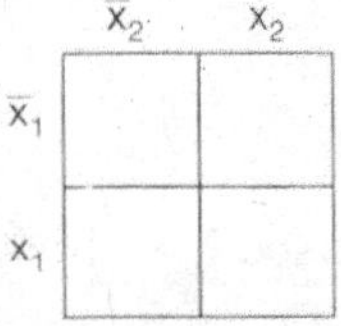

Fig. 56.

3. When the number of variables $n = 3$, the karnaugh map is like as shown in Fig. 57.

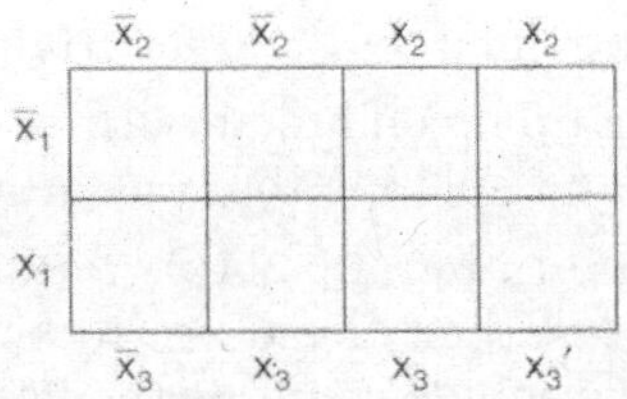

Fig. 57.

Example 51. *Represent the following functions by using karnaugh maps :*

(*a*) $f(x_1, x_2, x_3) = x_1 x_2 x_3 + x_1 x_2' x_3 + x_1' x_2' x_3$

(*b*) $f(x_1, x_2) = x_1 x_2' + x_1 x_2$.

Sol. (*a*) The function is represented by the following karnaugh map (Fig. 58).

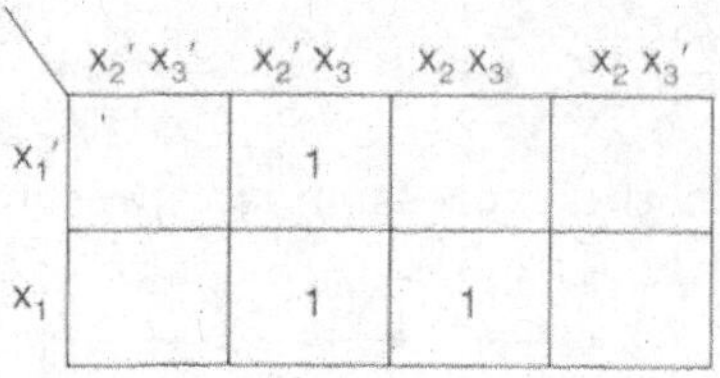

Fig. 58.

(*b*) The function is represented by following karnaugh map shown in Fig. 59.

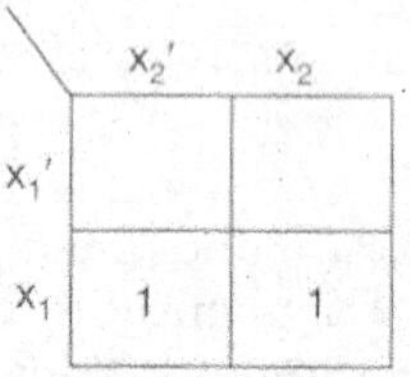

Fig. 59.

Note. It is widely used for representation and simplification of Boolean functions. It is most efficient method for simplification of functions having upto 5 variables.

SIMPLIFICATION OF BOOLEAN FUNCTIONS USING K-MAP

Boolean functions can be simplified with *k*-map. It is based on the principle of combining terms in adjacent cells. Two cells are said to be adjacent if they differ in only one variable. In adjacent cells one of the variable is same, whereas the other variable appears in uncomplemented form in one and in the complemented form in the other cell.

MINIMISATION OF SOP FORM

The following algorithm can be used by which minimized expression can be obtained :

1. Identify the ones which cannot be combined with any other ones encircle them.

2. Identify the ones that can be combined in groups of two in only one way and encircle them as groups.

3. Identify the ones that can be combined with three other ones, to make a group of four adjacent ones, in only one way and encircle them as groups.

4. Identify the ones that can be combined with seven other ones, to make a group of eight adjacent ones, in only way and encircle them as groups.

5. After identifying the essential groups of 2, 4 and 8 ones, if there still remains some ones which have not been encircled then these are to be combined with each other or with other already encircled ones *i.e.*, we should combine the left over ones in largest possible groups and in as few groupings as possible.

Example 52. *Minimize the following Boolean expression using k-map :*

$$f(A, B) = \bar{A}B + BA.$$

Sol. First of all draw the 2-variable k-map and insert 1's in the corresponding cells as shown in Fig. 60.

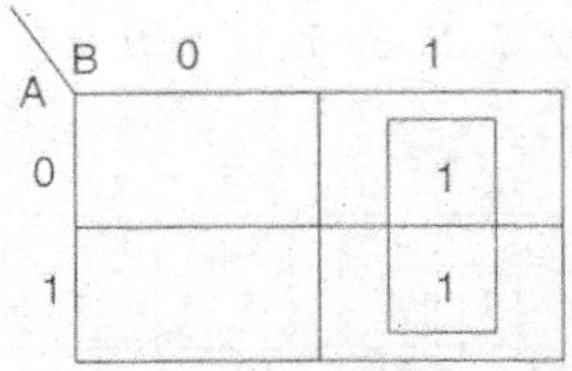

Fig. 60.

Now, group the 1's as described earlier. Then the required minimized Boolean expression is

$$f = \text{B}.$$

Example 53. *Minimize the following Boolean expression using k-map :*

$$AB + \bar{A}B + B\bar{A}.$$

Sol. Draw the two variable k-map and insert 1's in the corresponding cells as shown in Fig. 61.

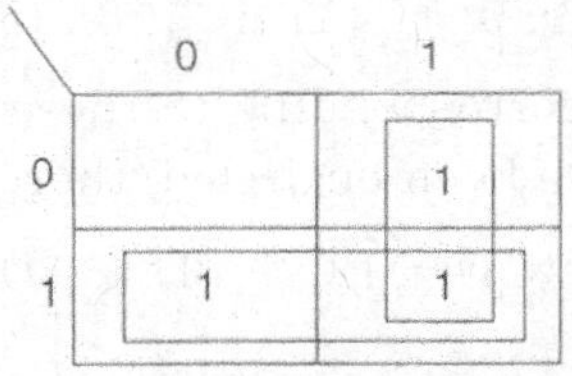

Fig. 61.

Now, group the 1's as described earlier. Then the required minimized Boolean expression is

$$f = \text{A} + \text{B}.$$

Example 54. *Minimize following Boolean expression using k-map :*

$$f(A, B, C) = A\bar{B}C + \bar{A}BC + AB + \bar{A}\bar{B}C.$$

Sol. Draw the 3-variable k-map and insert 1's in the corresponding cells as shown in Fig. 62.

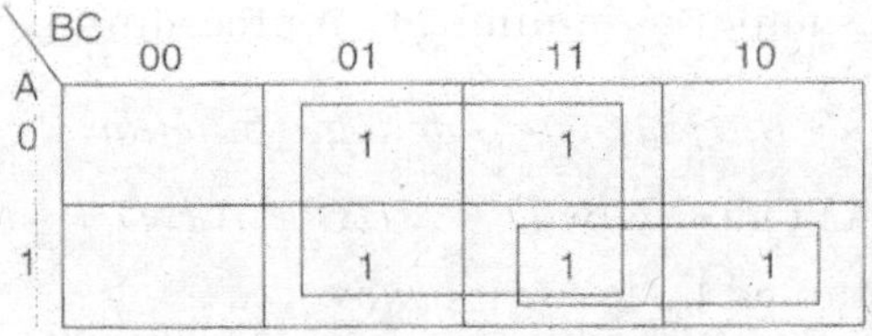

Fig. 62.

Now, group the 1's as described earlier. Then the required minimized Boolean expression is

$$f = AB + \bar{C}.$$

Example 55. *Minimize the four-variable logic function using k-map :*

$$f(A, B, C, D) = \Sigma\,(0, 1, 2, 3, 5, 7, 8, 9, 11, 14).$$

Sol.

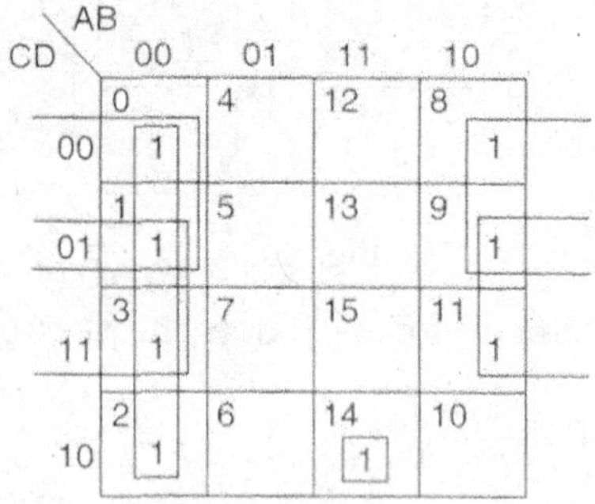

Fig. 63.

(*a*) Encircle 1 in cell 14 which cannot be combined with any other 1. The term for this is $ABC\bar{D}$.

(*b*) There is only one possible group of four adjacent ones involving each of the cells 8, 11, 5 or 7 and 2, and these are (8, 9, 0, 1), (11, 9, 1, 3), (5, 7, 3, 1) and (2, 3, 10) respectively. Encircle these groups. The terms corresponding to these groups are $\bar{B}\bar{C}$, $\bar{B}D$, $\bar{A}D$ and $\bar{A}\,\bar{B}$ respectively. Since all the ones have been encircled, therefore, the minimized equation is

$$f(A, B, C, D) = ABC\bar{D} + \bar{B}\bar{C} + \bar{B}D + \bar{A}D + \bar{A}\,\bar{B}.$$

MINIMISATION OF BOOLEAN FUNCTIONS NOT IN MINTERMS/MAXTERMS

One way to minimize such functions is to convert them into standard forms *i.e.*,SOP or POS, then make the *k*-map and obtain the minimized function.

Another way is to directly prepare the *k*-map using the following algorithm :

1. Enter ones for minterms and zeros for maxterms.
2. Enter a pair of ones/zeros for each of the terms with one variable less than the total number of variables.
3. Enter a four adjacent ones/zeros for terms with two variables less than the total number of variables.
4. Repeat for other terms in similar way.

After preparing the *k*-map, the minimization procedure is same.

Example 56. *Minimize the four variable logic function*

$$f(A, B, C, D) = AB\bar{C}D + \bar{A}BCD + \bar{A}\,B\,\bar{C} + \bar{A}\,B\,\bar{D} + A\bar{C} + A\bar{B}C + B.$$

Sol. The *k*-map is obtained by following way

(*a*) Enter 1 in the cell with A = 1, B = 1, C = 0, D = 1 corresponding to the minterm $AB\bar{C}D$.

(*b*) Enter 1 in the cell with A = 0, B = 1, C = 1, D = 1 corresponding to the minterm $\overline{A}BCD$.

(*c*) Enter 1's in the two cells with A = 0, B = 0, C = 0 corresponding to the term $\overline{A}\,\overline{B}\,\overline{C}$.

(*d*) Enter 1's in the two cells with A = 0, B = 0, D = 0 corresponding to the term $\overline{A}\,\overline{B}\,\overline{D}$.

(*e*) Enter 1's in the two cells with A = 1, B = 0, C = 1 corresponding to the term ABC.

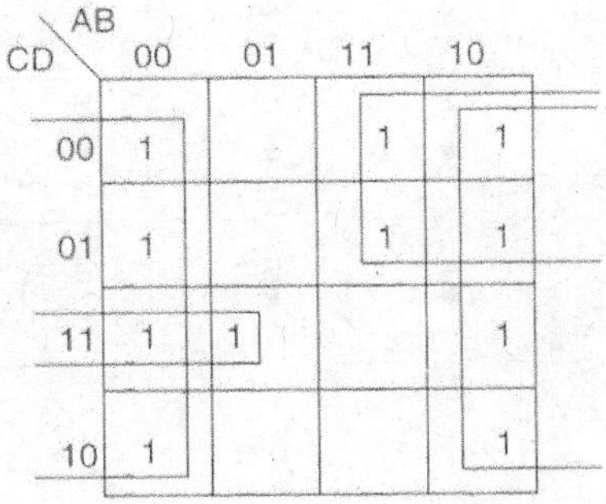

Fig. 64.

(*f*) Enter 1's in the four cells with A = 1, C = 0 corresponding to the term $A\overline{C}$.

(*g*) Enter 1's in the eight cells with B = 0 corresponding to the term $\overline{B}$.

The minimized expression is $\overline{B} + A\overline{C} + \overline{A}CD$.

Example 57. *Given the Boolean expression* $f = ABC + B\overline{C}D + \overline{A}BC$.

(*i*) *Make a truth table* (*ii*) *Simplify using k-map*

(*iii*) *Make the switching circuit of the expression.*

Sol. (*i*) The truth table of the Boolean expression is shown in Fig. 65.

A	B	C	D	$ABC + B\overline{C}D + \overline{A}CD$
0	0	0	0	0
0	0	0	1	0
0	0	1	0	0
0	0	1	1	0
0	1	0	0	0
0	1	0	1	1
0	1	1	0	1
0	1	1	1	1
1	0	0	0	0
1	0	0	1	0
1	0	1	0	0
1	0	1	1	0
1	1	0	0	0
1	1	0	1	1
1	1	1	0	1
1	1	1	1	1

Fig. 65.

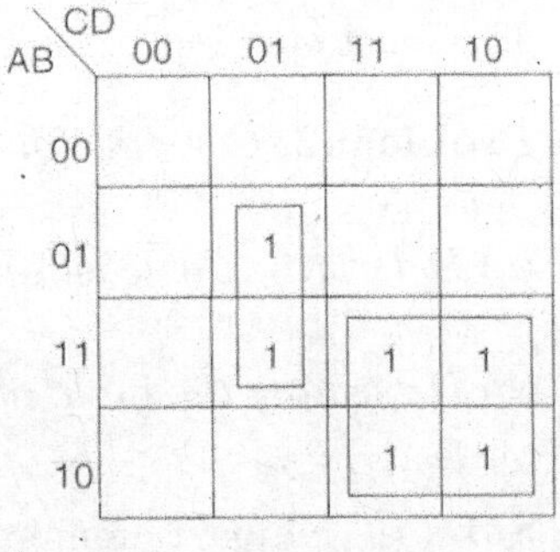

Fig. 66.

(*ii*) Draw the 4-variable k-map and insert 1's in the corresponding cells as shown in Fig. 66. Now, group the 1's. The required simplified Boolean expression is as follows :

$$f = AC + BC\overline{D}.$$

(*iii*) The switching circuit of the Boolean expression is shown in Fig. 67.

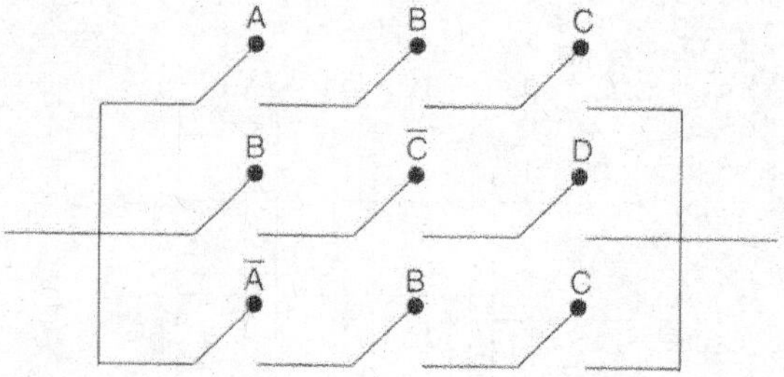

Fig. 67.

SOLVED PROBLEMS

Problem 1. *Consider the set* D_{50} = *{1, 2, 5, 10, 25, 50} and the relation divides (/) be a partial ordering relation on* D_{50}.

(*a*) *Draw the Hasse diagram of* D_{50} *with relation divides.*

(*b*) *Determine all upper bounds of 5 and 10.*

(*c*) *Determine all lower bounds of 5 and 10.*

(*d*) *Determine g.l.b. of 5 and 10.*

(*e*) *Determine l.u.b. of 5 and 10.*

(*f*) *Determine the greatest and least element of* D_{50}.

Sol. (*a*) The Hasse diagram of D_{50} is shown in Fig. 68.

(*b*) The upper bounds of 5 and 10 are 10 and 50.

(*c*) The lower bounds of 5 and 10 are 5 and 1.

(*d*) The greatest lower bound of 5 and 10 is 5.

(*e*) The least upper bound of 5 and 10 is 10.

(*f*) The greatest element of D_{50} is 50.

The least element of D_{50} is 1.

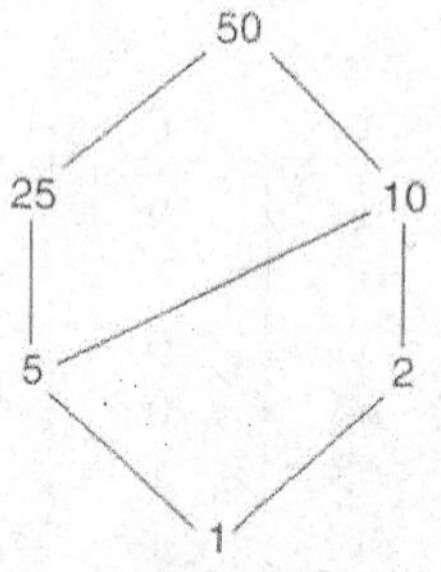

Fig. 68.

Problem 2. *Consider the Hasse diagram of the poset as shown in Fig. 69.*

(*a*) *Determine the least and greatest element of the poset, if they exist.*

(*b*) *Determine the l.u.b. of all pairs of elements.*

(*c*) *Determine the g.l.b. of all pairs of elements.*

Sol. (*a*) The least element of the poset is a_1.

The greatest element of the poset does not exist.

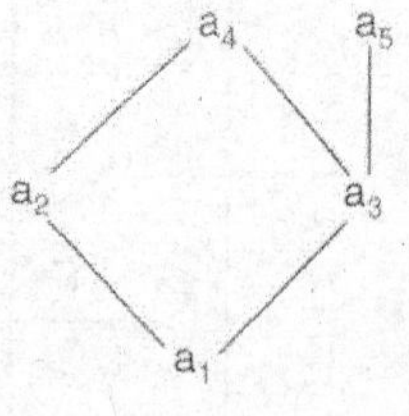

Fig. 69.

(*b*) The least upper bound (l.u.b.) of all pairs of elements of the poset is shown in the table as follows :

l.u.b.	a_1	a_2	a_3	a_4	a_5
a_1	a_1	a_2	a_3	a_4	a_5
a_2	a_2	a_2	a_4	a_4	—
a_3	a_3	a_4	a_3	a_4	a_5
a_4	a_4	a_4	a_4	a_4	—
a_5	a_5	—	a_5	—	a_5

Note. The (–) indicates that for these pairs of elements l.u.b. does not exist.

(*c*) The greatest lower bound (g.l.b.) of all pairs of elements of the poset is shown in the table as follows :

g.l.b.	a_1	a_2	a_3	a_4	a_5
a_1	a_1	a_1	a_1	a_1	a_1
a_2	a_1	a_2	a_1	a_2	a_1
a_3	a_1	a_1	a_3	a_3	a_3
a_4	a_1	a_2	a_3	a_4	a_3
a_5	a_1	a_1	a_3	a_3	a_5

Problem 3. *Determine whether the posets shown in Fig. 70 are lattices or not ?*

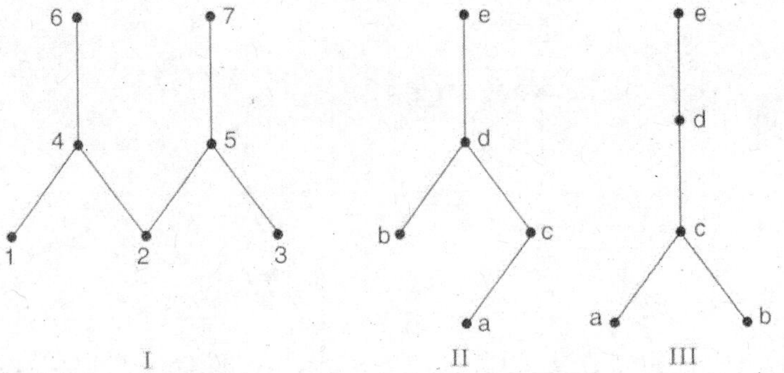

Fig. 70.

Sol. (*i*) The poset shown in Fig. 70(I) is not a lattice since the elements 6 and 7 have no upper bound. Similarly, the elements 1 and 2, 1 and 3, 2 and 3 have no lower bound.

(*ii*) The poset shown in Fig. 70(II) is not a lattice since the elements *b* and *c* have no lower bound *i.e.*, inf (*b*, *c*) does not exist.

(*iii*) The poset shown in Fig. 70(III) is not a lattice since the elements *a* and *b* have no lower bound *i.e.*, inf (*a*, *b*) does not exist.

Problem 4. *Consider the lattice as shown in Fig. 71.*

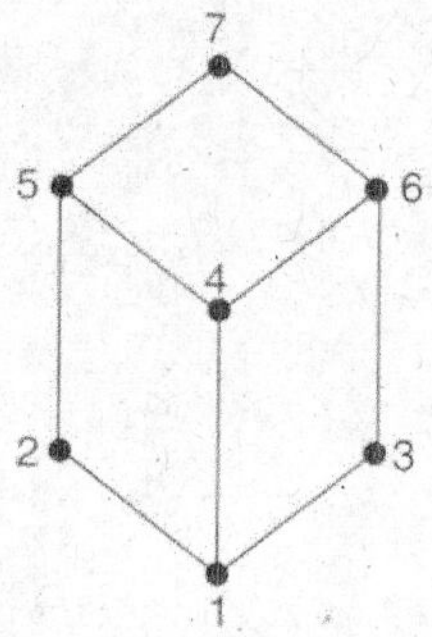

Fig. 71.

(*i*) *Determine the join-irreducible elements of the lattice.*

(*ii*) *Determine the meet-irreducible elements of the lattice.*

(*iii*) *Determine the atoms of the lattice.*

(*iv*) *Determine the antiatoms of the lattice.*

Sol. (*i*) The join-irreducible elements of the lattice are 2, 3 and 4.

(*ii*) The meet-irreducible elements of the lattice are 5, 6, 2 and 3.

(*iii*) The atoms of the lattice are 2, 3 and 4.

(*iv*) The antiatoms of the lattice are 5 and 6.

Problem 5. *Draw the diagram of a lattice which is the direct product of the following lattices shown in Fig. 72.*

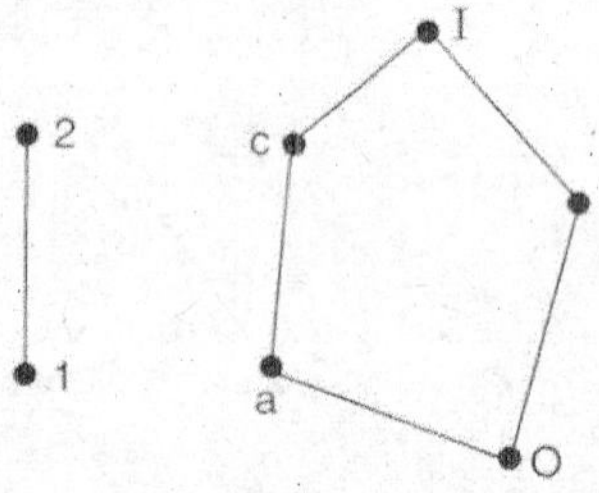

Fig. 72.

Sol. The lattice which is the direct product of the lattices shown in Fig. 72 is as shown in Fig. 73.

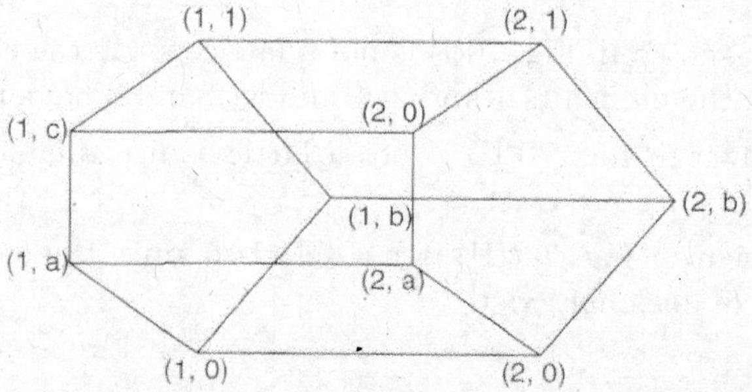

Fig. 73.

Problem 6. *Determine whether or not each of the following sets is a lattice with respect to divisibility :*

(*i*) $A_1 = \{1, 2, 4, 5, 20\}$

(*ii*) $A_2 = \{2, 4, 8, 10, 20, 40\}$

(*iii*) $A_3 = \{1, 2, 3, 4, 5, 6, 10, 15, 20, 30, 60\}$.

Sol. (*i*) A_1 is not a lattice since l.c.m. of (2, 5) = 10, which does not belong to set A_1.

(*ii*) A_2 is a lattice since for every pair (a, b) of elements $\inf(a, b)$ and $\sup(a, b)$ exists.

(*iii*) A_3 is a lattice.

Problem 7. *Show that the following Boolean expressions are equivalent :*

(*i*) $x \wedge (y \vee (y' \wedge (y \vee y')))$; x

(*ii*) $(z' \vee x) \wedge ((x \wedge y) \vee z) \wedge (z' \vee y)$; $x \wedge y$.

Sol. (*i*) Take L.H.S. of the Boolean expression

$$= x \wedge (y \vee (y' \wedge (y \vee y')))$$
$$= x \wedge (y \vee (y' \wedge 1)) \qquad \{\because\ y \vee y' = 1\}$$
$$= x \wedge (y \vee y') \qquad \{\because\ y' \wedge 1 = y'\}$$
$$= x \wedge 1 \qquad \{\because\ y \vee y' = 1\}$$
$$= x \qquad \{\because\ x \wedge 1 = x\}$$

Hence, the Boolean expressions are equivalent.

(*ii*) Take the L.H.S. of the Boolean expression

$$= (z' \vee x) \wedge ((x \wedge y) \vee z) \wedge (z' \vee y)$$
$$= (z' \vee (x \wedge y)) \wedge ((x \wedge y) \vee z) \qquad \{\because\ x \wedge x = x,\ z \wedge z' = 0\}$$
$$= (x \wedge y) \vee (z' \wedge z) = (x \wedge y) \vee 0 \qquad \{\because\ z' \wedge z = 0\}$$
$$= x \wedge y.$$

Hence, the Boolean expressions are equivalent.

Problem 8. *Show that if $M_1 \vee M_2 \vee \ldots\ldots \vee M_k$ is the disjunctive normal form of $f(x_1, x_2, \ldots\ldots, x_n)$, then $\overline{M_1} \wedge \overline{M_2} \wedge \ldots\ldots \wedge \overline{M_k}$ is the conjunctive normal form of $\overline{f(x_1, x_2, \ldots\ldots, x_n)}$.*

Sol. We have

$$f(x_1, x_2, \ldots\ldots, x_n) = M_1 \vee M_2 \vee \ldots\ldots \vee M_k \qquad \text{...Disjunctive normal form}$$
$$\overline{f(x_1, x_2, \ldots\ldots, x_n)} = \overline{M_1 \vee M_2 \vee \ldots\ldots \vee M_k}$$
$$= \overline{M_1} \wedge \overline{M_2} \wedge \overline{M_3} \wedge \ldots\ldots \wedge \overline{M_k}$$
$$\{\text{By De Morgan's law } \overline{A \vee B} = \overline{A} \wedge \overline{B}\}$$

which is the conjunctive normal form of $\overline{f(x_1, x_2, \ldots\ldots, x_n)}$

Problem 9. *Show that the following switching circuits are equivalent (Fig. 74).*

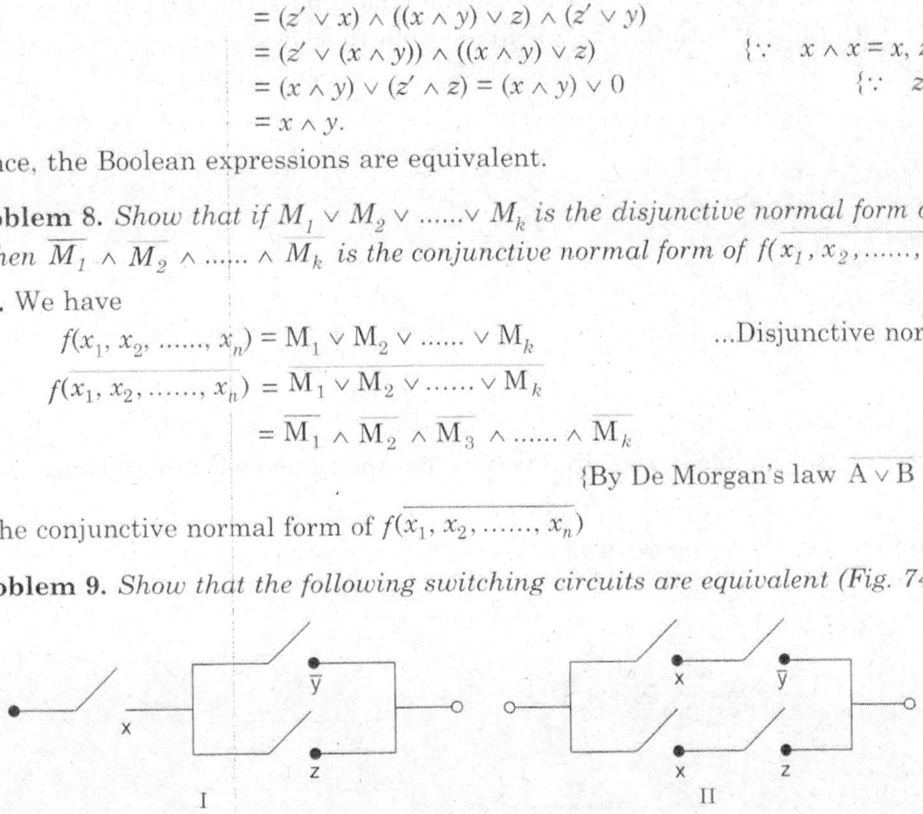

Fig. 74.

Sol. Two switching circuits are equivalent if the Boolean expressions that represent them are equal.

The Boolean expression of both the switching circuits are as follows :

$$x \wedge (\bar{y} \vee z) \quad ...(i)$$

and
$$(x \wedge \bar{y}) \vee (x \wedge z) \quad ...(ii)$$

Now take the Boolean expression (*ii*), we have

$$= (x \wedge \bar{y}) \vee (x \wedge z) = x \wedge (\bar{y} \vee z) \quad \{\text{Distributive law}\}$$

which is equal to the Boolean expression (*i*). Hence both the switching circuits are equivalent.

Problem 10. *Show whether the following circuits are equivalent or not (Fig. 75).*

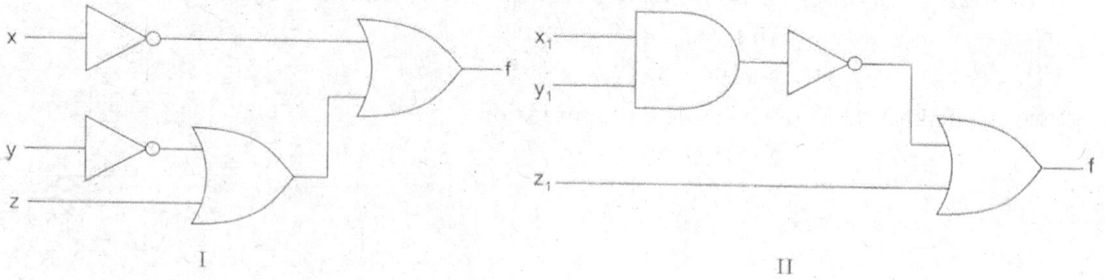

Fig. 75.

Sol. Two combinatorial circuits, both having inputs $x_1, x_2, \ldots\ldots, x_n$ and a single output are equivalent if, whenever the circuits receive the same inputs, they produce the same outputs.

The truth table of both the circuits are as shown below :

x	y	z	f
0	0	0	1
0	0	1	1
0	1	0	0
0	1	1	1
1	0	0	1
1	0	1	1
1	1	0	1
1	1	1	1

x_1	y_1	z_1	f
0	0	0	1
0	0	1	1
0	1	0	1
0	1	1	1
1	0	0	1
1	0	1	1
1	1	0	0
1	1	1	1

Since the two combinatorial circuits have not the same logic tables, hence they are not equivalent.

Problem 11. *Write the Boolean expression that represents the following combinatorial circuit, write the logic table and also write the output of each gate symbolically on the figure.*

(*i*)

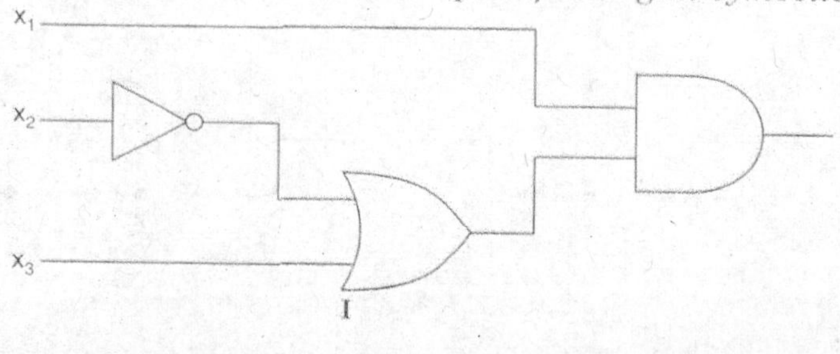

(*ii*)

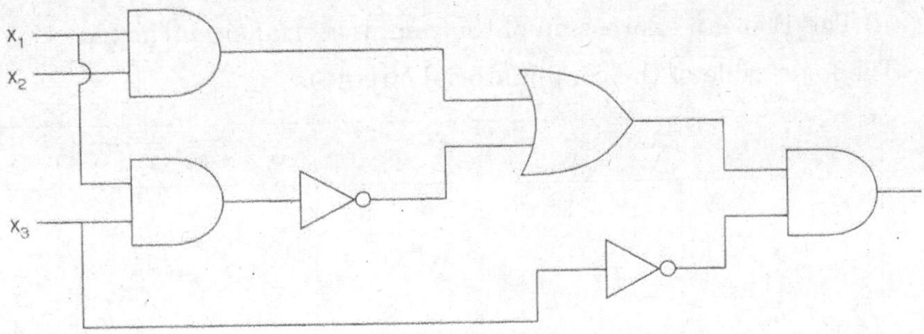

II

(*iii*)

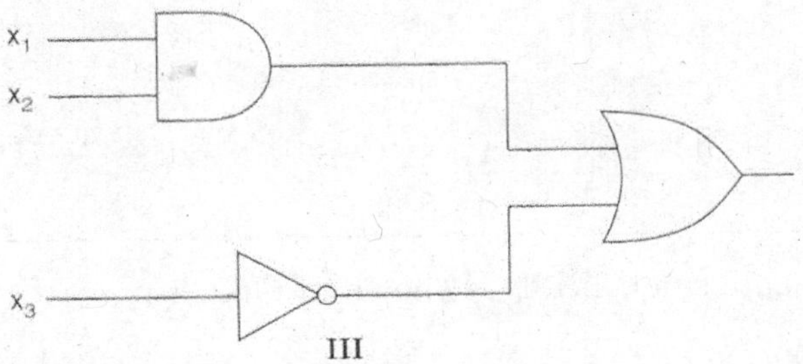

III

Fig. 76.

Sol. (*i*) The Boolean expression of the combinatorial circuit is

$$x_1 \wedge (\bar{x}_2 \vee x_3).$$

The logic table of the combinatorial circuit is

x_1	x_2	x_3	$x_1 \wedge (\bar{x}_2 \vee x_3)$
1	1	1	1
1	1	0	0
1	0	1	1
1	0	0	1
0	1	1	0
0	1	0	0
0	0	1	0
0	0	0	0

The output of each gate symbolically is shown in the Fig. 77.

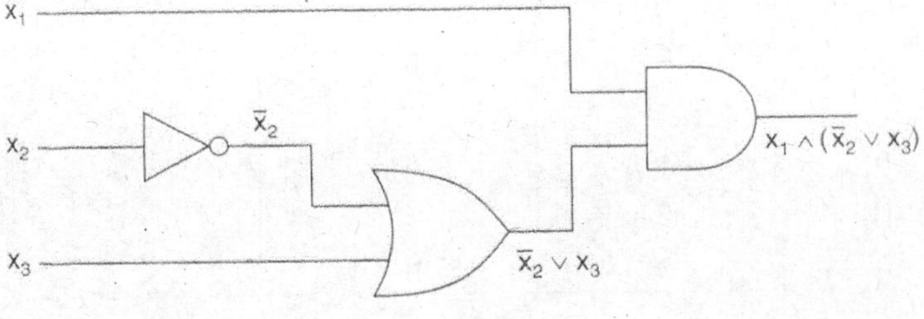

Fig. 77.

(*ii*) The Boolean expression of the combinatorial circuit is $((x_1 \wedge x_2) \vee \overline{(x_1 \wedge x_3)}) \wedge \bar{x}_3$.

The logic table of the combinatorial circuit is

x_1	x_2	x_3	$((x_1 \wedge x_2) \vee \overline{(x_1 \wedge x_3)}) \wedge \bar{x}_3$
1	1	1	0
1	1	0	1
1	0	1	0
1	0	0	1
0	1	1	0
0	1	0	1
0	0	1	0
0	0	0	1

The output of each gate symbolically is shown in Fig. 78.

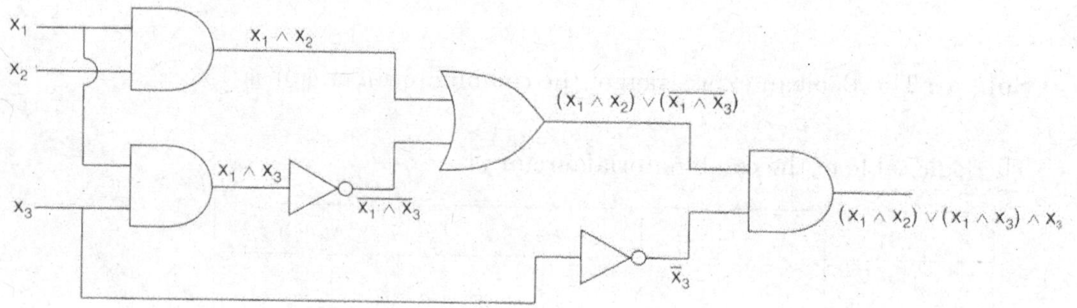

Fig. 78.

(*iii*) The Boolean expression of the combinatorial circuit is

$$(x_1 \wedge x_2) \vee \bar{x}_3.$$

The logic table of the combinatorial circuit is

x_1	x_2	x_3	$(x_1 \wedge x_2) \vee \bar{x}_3$
1	1	1	1
1	1	0	1
1	0	1	0
1	0	0	1
0	1	1	0
0	1	0	1
0	0	1	0
0	0	0	1

The output of each gate is shown symbolically in Fig. 79.

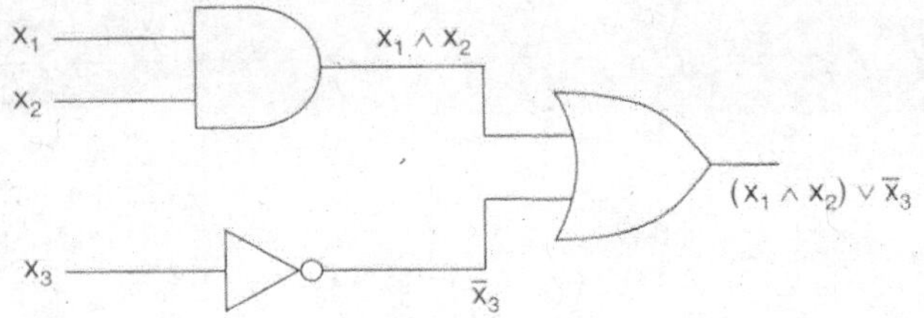

Fig. 79.

Problem 12. *Represent the following circuits symbolically and also give its switching table.*

(*i*)

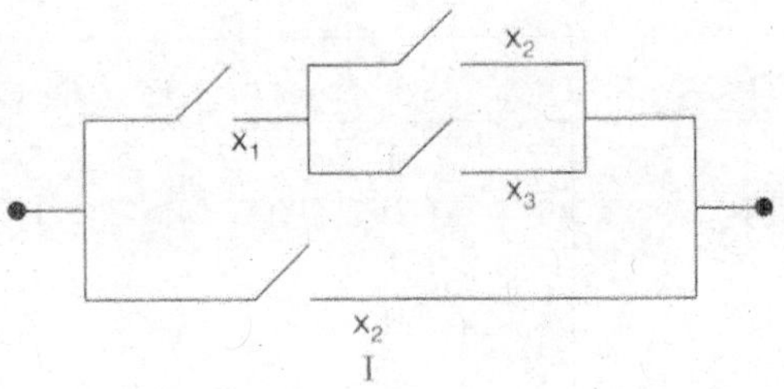

(*ii*)

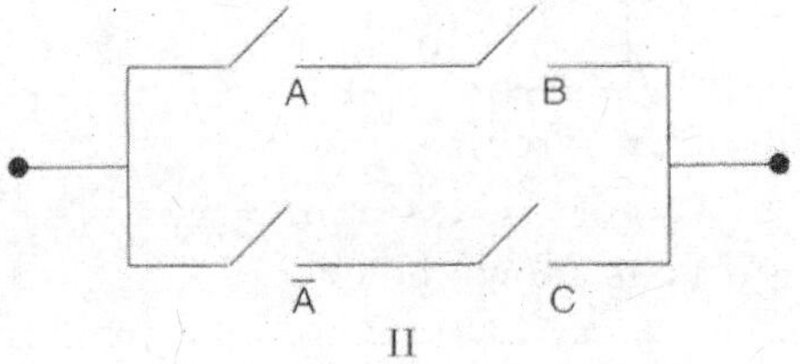

(*iii*)

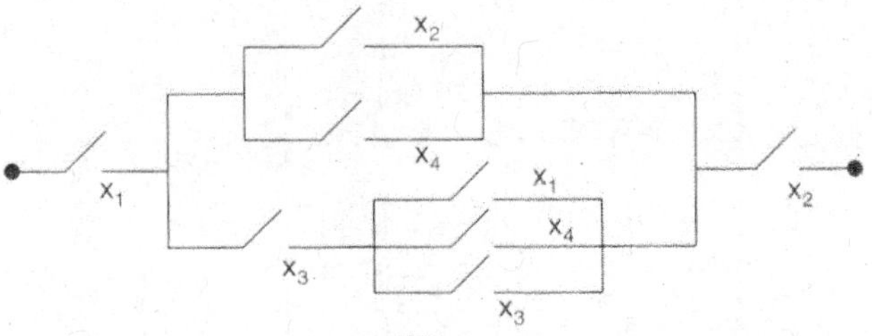

Fig. 80.

Sol. (*i*) The symbolic representation of the circuit is $(x_1 \wedge (x_2 \vee x_3)) \vee \bar{x}_2$.

The switching table of the circuit is as shown below :

x_1	x_2	x_3	$(x_1 \wedge (x_2 \vee x_3)) \vee \bar{x}_2$
1	1	1	1
1	1	0	1
1	0	1	1
1	0	0	1
0	1	1	0
0	1	0	0
0	0	1	1
0	0	0	1

(*ii*) The symbolic representation of the circuit is $(A \wedge B) \vee (\bar{A} \wedge C)$.
The switching table of the circuit is as shown below :

A	B	C	$(A \wedge B) \vee (\bar{A} \wedge C)$
1	1	1	1
1	1	0	1
1	0	1	0
1	0	0	0
0	1	1	1
0	1	0	0
0	0	1	1
0	0	0	0

(*iii*) The symbolic representation of the circuit is

$$x_1 \wedge ((x_2 \vee x_4) \vee (x_3 \wedge (x_1 \vee x_4 \vee x_3))) \wedge x_2.$$

The switching table of the circuit is as shown below :

x_1	x_2	x_3	x_4	$x_1 \wedge ((x_2 \vee x_4) \vee (x_3 \wedge (x_1 \vee x_4 \vee x_3))) \wedge x_2$
0	0	0	0	0
0	0	0	1	0
0	0	1	0	0
0	0	1	1	0
0	1	0	0	0
0	1	0	1	0
0	1	1	0	0
0	1	1	1	0
1	0	0	0	0
1	0	0	1	0
1	0	1	0	0
1	0	1	1	0
1	1	0	0	1
1	1	0	1	1
1	1	1	0	1
1	1	1	1	1

Problem 13. *Find the combinatorial circuit corresponding to each of the following Boolean expressions :*

(*i*) $x_1 \wedge (x_2 \vee x_3)$ (*ii*) $(x_1 \wedge \bar{x}_2) \vee (x_1 \wedge \bar{x}_3)$

(*iii*) $\overline{(A \wedge B)}$ (*iv*) $(x_1 \wedge ((x_2 \wedge \bar{x}_3) \vee (\bar{x}_2 \wedge x_3))) \vee (\bar{x}_1 \wedge x_2 \wedge x_3)$.

Sol. (*i*) The combinatorial circuit of the Boolean expression is shown in Fig. 81.

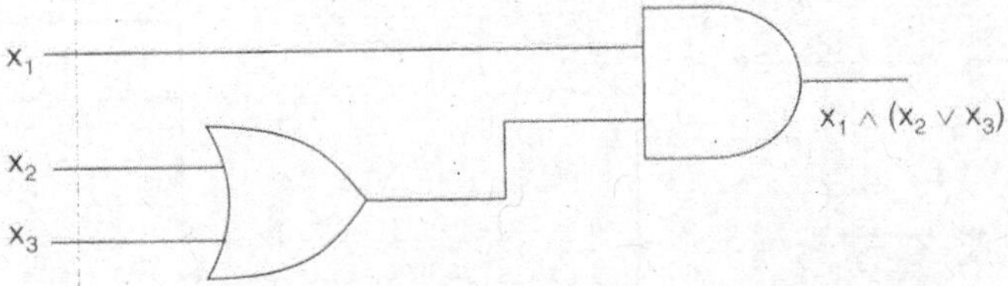

Fig. 81.

(*ii*) The combinatorial circuit of the Boolean expression is shown in Fig. 82.

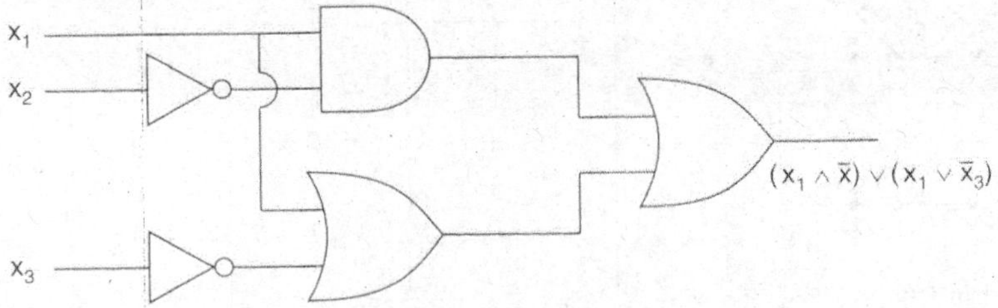

Fig. 82.

(*iii*) The combinatorial circuit of the Boolean expression is shown in Fig. 83.

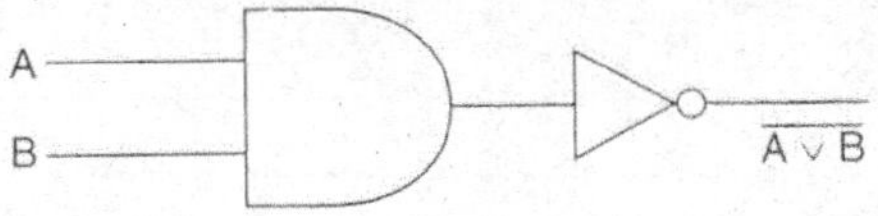

Fig. 83.

(*iv*) The combinatorial circuit of the Boolean expression is shown in Fig. 84.

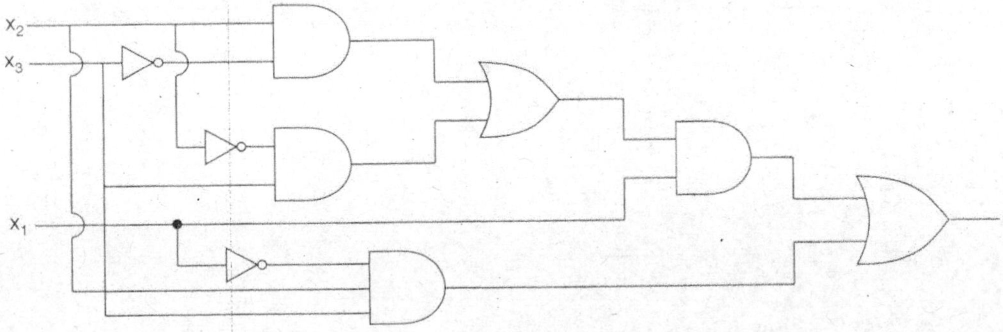

Fig. 84.

Problem 14. *Represent the following expressions as switching circuits and also write the switching tables :*

(*i*) $(x_1 \wedge \bar{x}_2) \vee x_1$ (*ii*) $(x_1 \wedge x_3) \vee (\bar{x}_1 \wedge x_2)$

(*iii*) $(x_1 \wedge (x_2 \vee (x_1 \wedge \bar{x}_2))) \vee ((x_1 \wedge \bar{x}_2) \vee (x_1 \wedge \bar{x}_3))$.

Sol. (*i*) The switching circuit and the corresponding switching table of the Boolean expression is shown in Fig. 85.

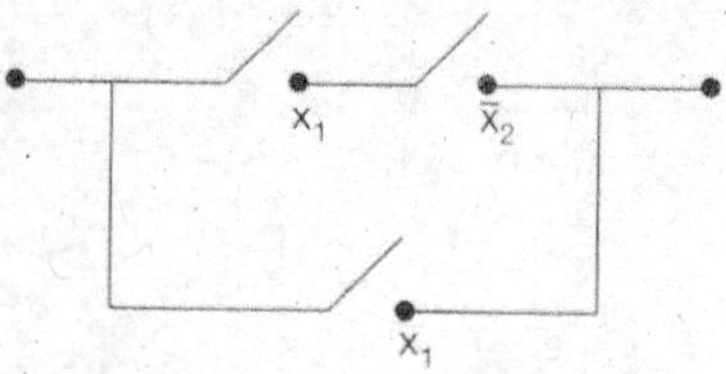

Fig. 85.

x_1	x_2	$(x_1 \wedge \bar{x}_2) \vee x_1$
1	1	1
1	0	1
0	1	0
0	0	0

(*ii*) The switching circuit and the corresponding switching table of the Boolean expression is shown in Fig. 86.

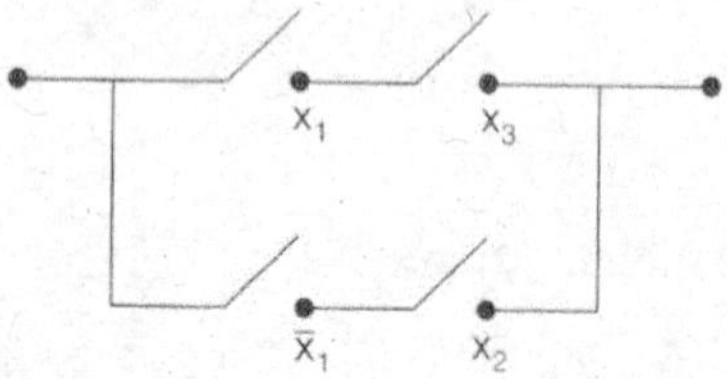

Fig. 86.

x_1	x_2	x_3	$(x_1 \wedge x_3) \vee (\bar{x}_1 \wedge x_2)$
1	1	1	1
1	1	0	0
1	0	1	1
1	0	0	0
0	1	1	1
0	1	0	1
0	0	1	0
0	0	0	0

(*iii*) The switching circuit and the corresponding switching table of the Boolean expression is shown in Fig. 87.

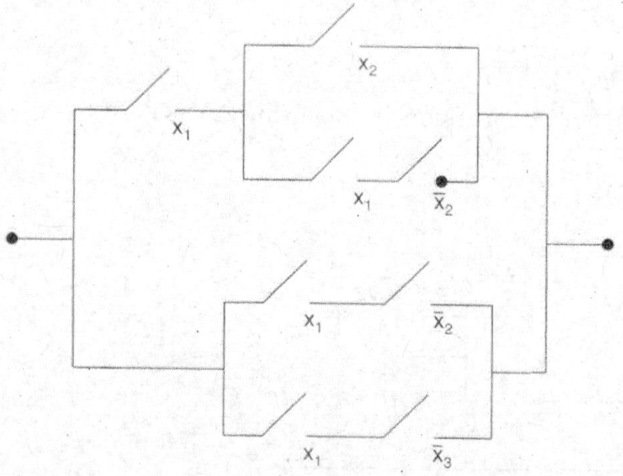

Fig. 87.

x_1	x_2	x_3	$(x_1 \wedge (x_2 \vee (x_1 \wedge \bar{x}_2)))$ $\vee ((x_1 \wedge \bar{x}_2) \vee (x_1 \wedge \bar{x}_3))$
1	1	1	1
1	1	0	1
1	0	1	1
1	0	0	1
0	1	1	0
0	1	0	0
0	0	1	0
0	0	0	0

Problem 15. *Find the disjunctive normal form and also the corresponding combinatorial circuit of the following Boolean functions :*

(i)

A	B	C	f(A, B, C)
0	0	0	1
0	0	1	1
0	1	0	1
0	1	1	0
1	0	0	0
1	0	1	1
1	1	0	1
1	1	1	1

(ii)

A	B	C	f(A, B, C)
0	0	0	0
0	0	1	1
0	1	0	0
0	1	1	1
1	0	0	1
1	0	1	0
1	1	0	1
1	1	1	0

Sol. (*i*) The disjunctive normal form of the Boolean function is

$$f = \bar{A}\,\bar{B}\,\bar{C} + \bar{A}\,\bar{B}\,C + \bar{A}\,B\,\bar{C} + A\,\bar{B}\,C + A\,B\,\bar{C} + A\,B\,C$$

and the corresponding combinatorial circuit is shown in Fig. 88.

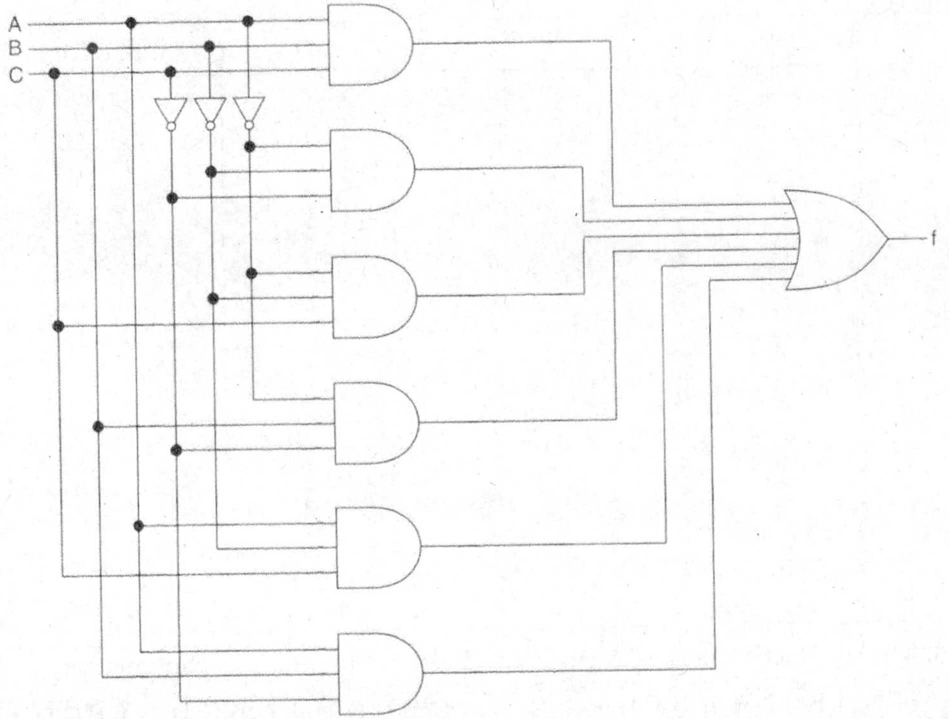

Fig. 88.

(*ii*) The disjunctive normal form of the Boolean function is

$$f = \bar{A}\,\bar{B}\,\bar{C} + \bar{A}\,B\,C + A\,\bar{B}\,\bar{C} + A\,B\,\bar{C}$$

and the corresponding combinatorial circuit is shown in Fig. 89.

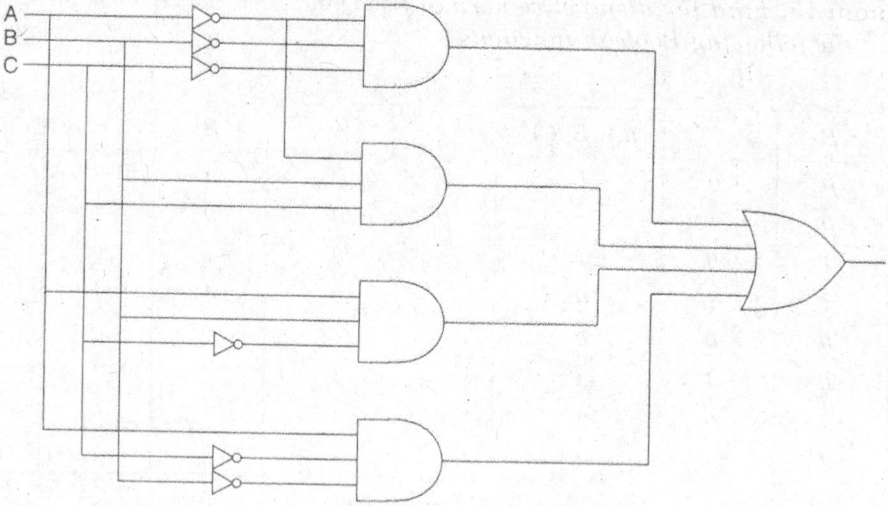

Fig. 89.

Problem 16. *Find the disjunctive and conjunctive normal form of the following Boolean function :*

A	*B*	*C*	*D*	*F(A, B, C, D)*
0	0	0	0	1
0	0	0	1	0
0	0	1	0	1
0	0	1	1	0
0	1	0	0	1
0	1	0	1	1
0	1	1	0	1
0	1	1	1	0
1	0	0	0	1
1	0	0	1	0
1	0	1	0	1
1	0	1	1	1
1	1	0	0	1
1	1	0	1	1
1	1	1	0	0
1	1	1	1	0

Sol. The disjunctive normal form of the Boolean function is given by

$$F = \bar{A}\,\bar{B}\,\bar{C}\,\bar{D} + \bar{A}\,\bar{B}\,C\,\bar{D} + \bar{A}\,B\,\bar{C}\,\bar{D} + \bar{A}\,B\,\bar{C}\,D + \bar{A}\,B\,C\,\bar{D} + A\,\bar{B}\,\bar{C}\,\bar{D} + A\,\bar{B}\,C\,D$$
$$+ A\,\bar{B}\,C\,D + A\,B\,\bar{C}\,\bar{D} + A\,B\,\bar{C}\,\bar{D} + A\,B\,\bar{C}\,D.$$

The conjunctive normal form of the Boolean function is given by

$$F = (A + B + C + \bar{D}) * (A + B + \bar{C} + \bar{D}) * (A + \bar{B} + \bar{C} + \bar{D}) * (\bar{A} + B + C + D) *$$
$$(\bar{A} + \bar{B} + \bar{C}' + D) * (\bar{A} + \bar{B} + \bar{C} + \bar{D}).$$

Problem 17. *Design a circuit that accepts a 3-bit number and gives an output 0 if input represents even decimal number and gives an output 1 if input represents an odd decimal number.*

Sol. The block diagram of the circuit and the corresponding logic table is shown in Fig. 90.

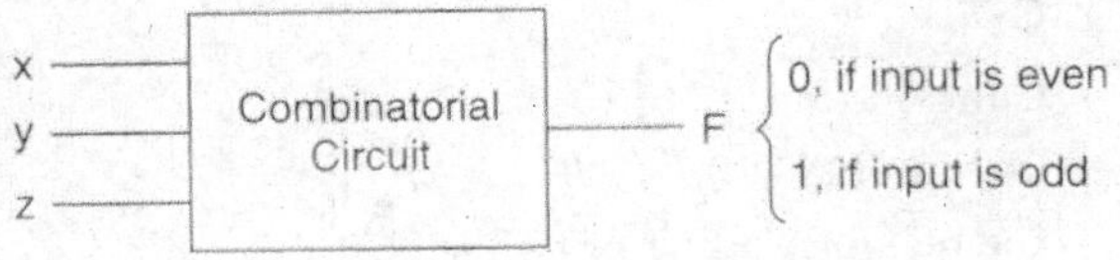

Fig. 90.

Decimal Number	x	y	z	F
0	0	0	0	0
1	0	0	1	1
2	0	1	0	0
3	0	1	1	1
4	1	0	0	0
5	1	0	1	1
6	1	1	0	0
7	1	1	1	1.

From the logic table we obtained the following Boolean expression :

$$F = \bar{x}\,\bar{y}z + \bar{x}\,yz + x\,\bar{y}\,z + xyz$$

The logic circuit corresponding to the above Boolean expression is shown in Fig. 91.

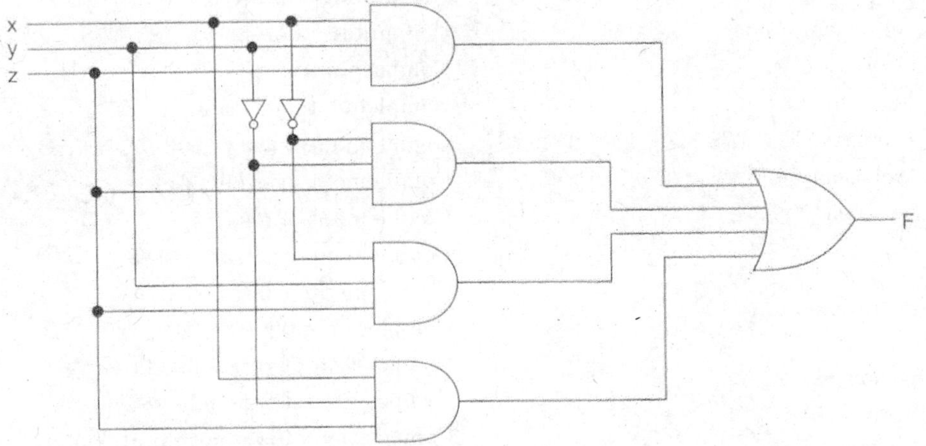

Fig. 91.

Index